D0719323

Plant Growth Substances
Volume 1

PLANT SCIENCE MONOGRAPHS

Published so far

THE BIOLOGY OF MYCORRHIZA (2nd edn)	J. L. Harley
SALT MARSHES AND SALT DESERTS OF THE WORLD*	V. J. Chapman
THE MICROBIOLOGY OF THE ATMOSPHERE (2nd edn in press)	P. H. Gregory
SEED PRESERVATION AND LONGEVITY	L. V. Barton
WEEDS OF THE WORLD: BIOLOGY AND CONTROL*	L. J. King
CROP PRODUCTION IN DRY REGIONS, Volumes 1 and 2	I. Arnon
GRASSLAND IMPROVEMENT	A. T. Semple

Other volumes are in preparation

*Out of print

A PLANT SCIENCE MONOGRAPH
Series Editor: Professor Nicholas Polunin

PLANT GROWTH SUBSTANCES

Volume 1: Chemistry and Physiology

L. J. AUDUS M.A., Ph.D.(Cantab.), F.L.S.

Hildred Carlile Professor of Botany at Bedford College, University of London

LEONARD HILL
LONDON

An Intertext Publisher

Published by Leonard Hill Books
a division of
International Textbook Company Limited
158 Buckingham Palace Road,
London, S.W.1 W9TR, and
24 Market Square, Aylesbury, Bucks.

First published 1953
Second edition 1959
This edition 1972

ISBN 0 249 44085 7

Photoset in Malta by St Paul's Press Ltd
Printed in Great Britain by J. W. Arrowsmith Ltd., Bristol.

DEDICATED

to

Dr. H. Hamshaw Thomas

from whom I first heard of auxins

Preface to the Third Edition

In the six-year interval between the publication of the first and second editions of this book, information on the principles and practice of chemical plant growth regulation had grown so much as to necessitate a 50 per cent increase in size. Since then the relative growth rate of the literature on plant growth substances has progressively steepened over the years; to maintain the comprehensive cover and detail of the first two editions a two- to three-fold increase in size over the second edition is required. Qualitatively, too, the whole field has changed. In 1958 the gibberellins were relative newcomers; now they attract as much, if not more attention than the auxins. Then kinetin was little more than an academic curiosity; now the role of cytokinins in the control of plant growth is perhaps better understood than that of the auxins. Natural inhibitors, then only dimly suspected as playing a part in growth regulation, have now come into their own, particularly in the shape of abscisic acid and its structural relatives. Furthermore the application of many new physical, chemical and biochemical techniques to the study of the phytohormones has revolutionised our concepts of role and mechanism and has revealed phytohormone participation in many previously unsuspected areas of plant biology.

Such considerations, and many others of a purely practical nature, have called for a complete restructuring in the third edition. The vastly increased bulk of material demands two volumes instead of one; in order to conserve the maximum of independent usefulness for each volume, the restructuring has distinguished as far as possible between the various aspects of the subject and has grouped them into two logical units. The present volume deals with the chemistry, metabolism, general physiology and mode of action of plant growth substances, both natural and synthetic. Some chapters remain as updated versions from the second edition (viz. those on the natural and synthetic auxins, on natural plant growth inhibitors, on growth substances in the soil and on reproductive hormones). Others have been completely redrafted, e.g. that on hormone mechanisms. There are five completely new chapters, i.e. on the gibberellins, the cell-division hormones, synthetic plant growth inhibitors and retardants, ethylene and the control of hormone levels at the sites of their actions. The second volume will deal with the role of plant growth substances in the control of various aspects of growth and development in the plant and its organs and particularly with their practical uses in agriculture and horticulture, i.e. the remaining facets of the subject covered in the first two editions.

The first edition unashamedly tried to cater for 'all classes of reader'. The second edition struggled against increasing technical difficulties to maintain that aim and was thereby mildly criticised by some reviewers for being unrealistic on that score. The intervening years have turned difficulties into impossibilities and the present edition has therefore been written for the reader with a broad understanding of the basic principles of chemistry, biochemistry and physics, i.e. the advanced student who, in one scientific discipline or another, needs to inform himself of the nature and functioning of plant growth regulating substances. However in writing this new edition I have still tried to cater for those physical scientists who may feel the need for help in understanding the more technical aspects of plant biology. As far as possible, to be consistent with conciseness, biological phenomena have been described in simple non-technical terms and because this was sometimes impossible to maintain a glossary of some technical terms has been included at the end of the text.

Many new illustrations have been incorporated both as photographs and as graphs and diagrams. I wish to acknowledge with gratitude the generosity of the many fellow scientists who have allowed me to use their photographic records and their research observations in the construction of the figures in this volume. Their names are thankfully recorded at the appropriate points in the text.

L. J. Audus,
Stanmore, Middlesex.
March 1972

Contents

List of Plates

List of Figures

List of Tables

Chapter I

The Nature of Plant Growth and its Control

Most living organisms are built up of complex aggregates of small microscopic units of structure known as cells, much as a house is built of bricks. In any one organism, the form and detailed constitution of these units will vary from place to place in the body of which they form a part, but all living cells have certain properties and attributes which are fundamental, irrespective of their variations in form or position. In the first place, a major component of each cell is a viscous and extremely heterogeneous fluid known as protoplasm,* which is the living stuff of the organism. It is in this protoplasm that occur the marvellously complex sequences of chemical and physical events which integrate into what we know as life. The most important of these processes are those by which the protoplasm increases itself. It is this capacity to build itself up from simple food materials that distinguishes living from dead protoplasm; it is this which is the basis of all growth.

Thus, in the cell which is actively growing in the presence of a plentiful food supply, the protoplasm is rapidly increasing in quantity by these building-up processes and this is accompanied by a steady increase in the size of the cell. Such a steady increase of the cell volume does not continue indefinitely during growth. Eventually a time comes when the cell reaches a maximum size and may divide into two daughter cells, each half the size of the parent. This process of cell multiplication can best be illustrated by taking an example from the simplest of known organisms, the bacterium. In the bacteria a complete organism consists of one simple microscopic cell. In the

*A glossary of technical terms commences on page 497.

presence of a plentiful food supply, the bacterial cell will increase in size at a steady rate until a certain critical level is reached; it then splits into two offspring of equal size which themselves continue to enlarge at the same rate as their parent did. After a similar interval of time, these, too, will have reached the critical size and will divide to form four granddaughters. After further successive and often equal time intervals, 8, 16, 32, etc., cells will be formed. Although at the start the food supply may be plentiful, yet the progressive doubling of the size of this bacterial population cannot go on indefinitely, as the dwindling supply of food material must eventually retard the building-up processes of the cell protoplasm and slow down all growth and the accompanying cell multiplication. When this food is exhausted, growth will cease.

The same fundamental principles underlie the growth of the more highly organized multicellular plant with its wide variety of different types of cell involving a great range of function. Although, as we shall see later, only certain specific cells of each complex organism retain the power to grow after reaching a certain stage of development, yet those cells which are capable of growth do so first by building up their protoplasm and other associated structural materials, and then, at a certain cell size, by splitting into two daughter cells, which then repeat the process. In contrast, however, to the growth of the simple bacterial cell, all such daughter cells in the organized plant do not continue indefinitely to grow and multiply in this way. The majority of daughter cells soon undergo considerable changes in internal structure and organization and stop growing. These 'differentiated' cells then take on other functions consistent with their modified structure and their positions in the plant in which they find themselves.

It would be as well at this point to consider the detailed structure of an actively growing cell of a highly organized green plant, as the reactions of these cells to growth substances are to be the primary concern of this book. An examination of such a cell under the light microscope shows the protoplasm to be optically heterogeneous, containing a multitude of granules of various sizes and densities. The largest, densest, and most important of these is the nucleus, a spherical or oval structure characterized by its affinity for certain basic dyes. This nucleus, which is the most complex of any known cell structure, is the sub-cellular unit (now termed an 'organelle') controlling most of the activities of the cell. Inside it we find the chromosomes, thread-like structures which contain the cell's hereditary information, encoded in the constituent long-chain molecules of deoxyribonucleic acid (DNA). We now know that these acids act as molecular 'jigs' on which are formed the great variety of chemical messengers (ribonucleic acids – RNA) which, after passing from the nucleus into the cytoplasm which lies outside, order and control the multifarious chemical processes which go on there and thereby govern the growth, form, and structure of the organism (see a more detailed account of these processes in Chapter XII). If this order and control is to be maintained, then the constitution of all nuclei in any one species of organism must be the same. During cell division therefore the nucleus of a parent cell divides into two exactly equal parts, one going to each of the two daughter cells. A special and complicated process, into which we need not go, is necessary to ensure the exact duplication of all these factors in the two daughter nuclei.

The plant cell differs from the animal cell in that it is encased in a solid box, called the cell-wall; this wall is composed mostly of a substance called cellulose (a carbohydrate) which we meet in an almost pure form in cotton wool. In the young growing plant cell, this cell-wall contains protein (a major constituent of protoplasm) and enzymes (the protein molecules which control biochemical processes) and can therefore be included in the 'living' part of the cell. Later on, when the cell matures, this living proteinaceous content disappears, except perhaps for a thin layer on its inner surface. Each cell has its own wall, and the cells are cemented together by extremely thin layers of a material (the middle lamella) of which the major component is a substance called calcium pectate.

We are now in a position to consider the various stages in the growth-cycle of a cell of a green plant. Firstly, there is a stage of rapid manufacture of protoplasm culminating, at a certain size limit, in the division of the nucleus into the two identical daughter nuclei. The division of the cell itself is completed by the laying down by the cytoplasm of a new cellulose wall between the two nuclei, thus cutting the cytoplasm into two portions and forming two complete daughter cells. As in the case of bacteria, this cell multiplication in all embryonic plant parts may continue for some time, these parts consequently increasing in size as cell numbers increase. Sooner or later, however, some daughter cells stop dividing and enter instead into the second stage of the growth process. In this phase, as a result of subtle biochemical changes, there is a rapid and considerable increase in cell size, accompanied by the appearancc in the cytoplasm of large vesicles filled with a dilute watery solution of various substances. This is the cell-sap. When this second phase is complete and the cell has reached its maximum size, which is usually many times that of the original embryonic cell from which it came, the appearance of the cell will have greatly altered. The watery vesicles have merged to form one large central vesicle or vacuole, the cytoplasm now being confined to a thin layer in contact with the cell-wall and containing the nucleus embedded in it. This phase of the growth of the cell is known as 'extension-growth' and accounts for most of the volume increase in plants. Following this comes a third phase in which the structure of the cell becomes modified in relation to the function it will ultimately perform, these modifications usually taking place with little change in cell, and therefore tissue, volume. For example, cells which are to function in maintaining the strength and rigidity of plant organs develop very considerable thickening of the cell-wall, which becomes at the same time chemically modified and thereby toughened by impregnation with a substance called *lignin*. These particular cells die soon after this modification is complete, and the protoplasm disintegrates, leaving only the dead shell of the cell-wall. Other cells become modified for special vital functions and thus remain alive and active throughout the life of the organ containing them. The cells of the green spongy tissue in the interior of a leaf belong to this category. The small green bodies (chloroplasts) which develop in these cells are responsible for the manufacture of carbohydrates which takes place therein in the light. The line of demarcation between these two latter growth-phases is not always clear-cut, and in some tissues cell differentiation starts before cell extension is completed. Successive stages in the growth and development of a water-conducting cell of a green flowering plant are illustrated in Fig. 1.

In the very young embryonic plant emerging during germination from the seed coat, nearly all of its constituent cells are in the first stage of growth and therefore actively dividing. Very soon, however, as distinct organs begin to take shape, the necessary cell modifications (differentiations) follow and, as the plant grows, maturing tissues progressively lose their power to divide. Certain parts of the plant, however, retain this power indefinitely and maintain, during active plant growth, a steady production of daughter cells which develop and differentiate into the constituent units of the various plant organs as they form and mature. These regions of permanently embryonic cells, that are either active or potential dividers, depending on conditions, are known as *meristems* and are to be found in certain localized regions in mature plants; thus they are characteristic of stem and root tips. It is in the region just behind the tips that the most active growth is visible, due to the extension-growth of the products of these meristems. They are also found as a thin delicate layer between bark and wood of many plants, and this meristematic layer, called the *cambium*, is responsible for the growth in girth of these plants. Buds are small islands of meristem tissue left behind in mature tissue as the plant grows and are protected by mature non-meristem structures, the bud scales. These buds subsequently grow into new branches and

Fig. 1. Semi-diagrammatic representation of the successive stages in the growth and development of a water-conducting cell of a flowering plant. The arrows mark the progress of the cell through these stages. 1 to 3 show the first stage of growth, which is predominantly by the manufacture of living protoplasm and is characterized by active cell multiplication; 4 illustrates the second stage (stretching growth)

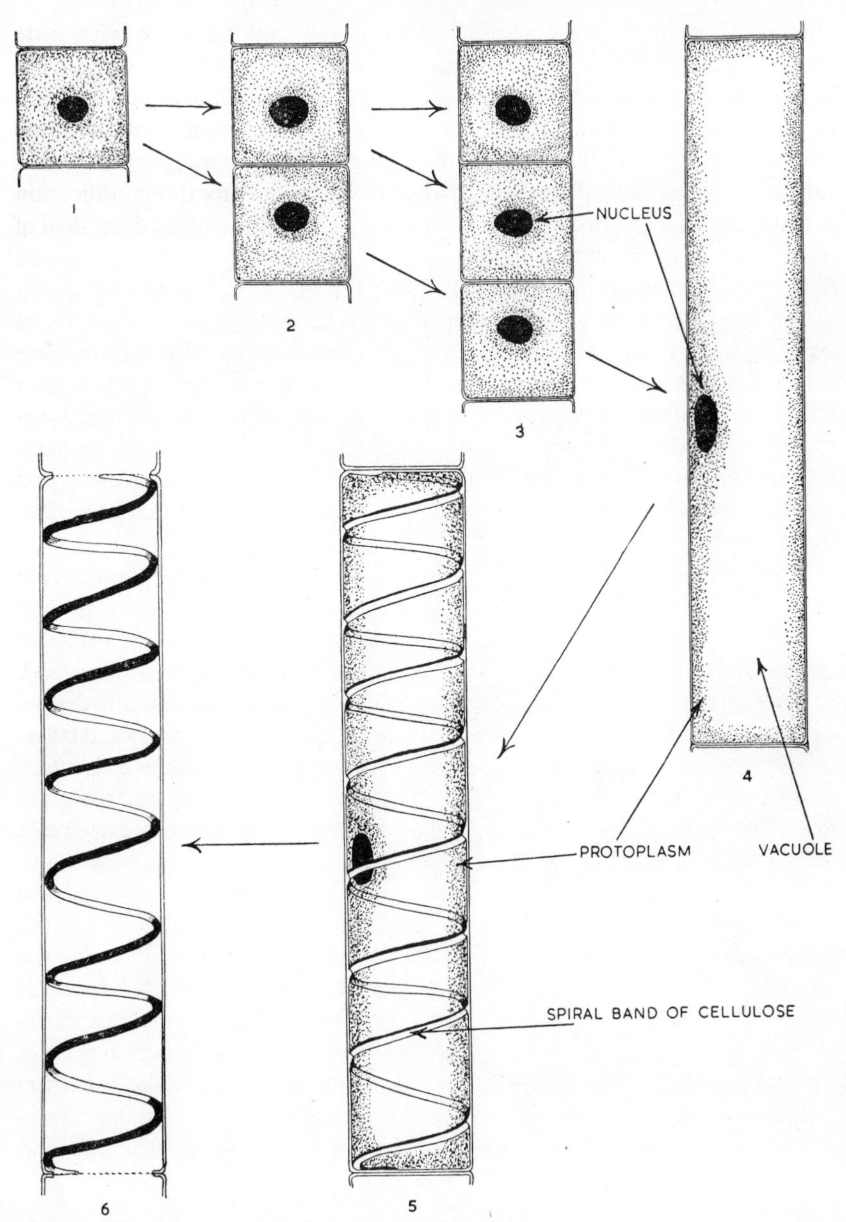

marked by a great elongation of the cell and the appearance of a large central water-containing vesicle, the vacuole. In the third stage, 5, a special spiral strengthening band of cellulose is deposited on the interior of the cell-wall and shortly afterwards the cell dies, the end walls break down and the living cytoplasm disappears (stage 6). In this last stage the cell conducts water. (All cells × 500.)

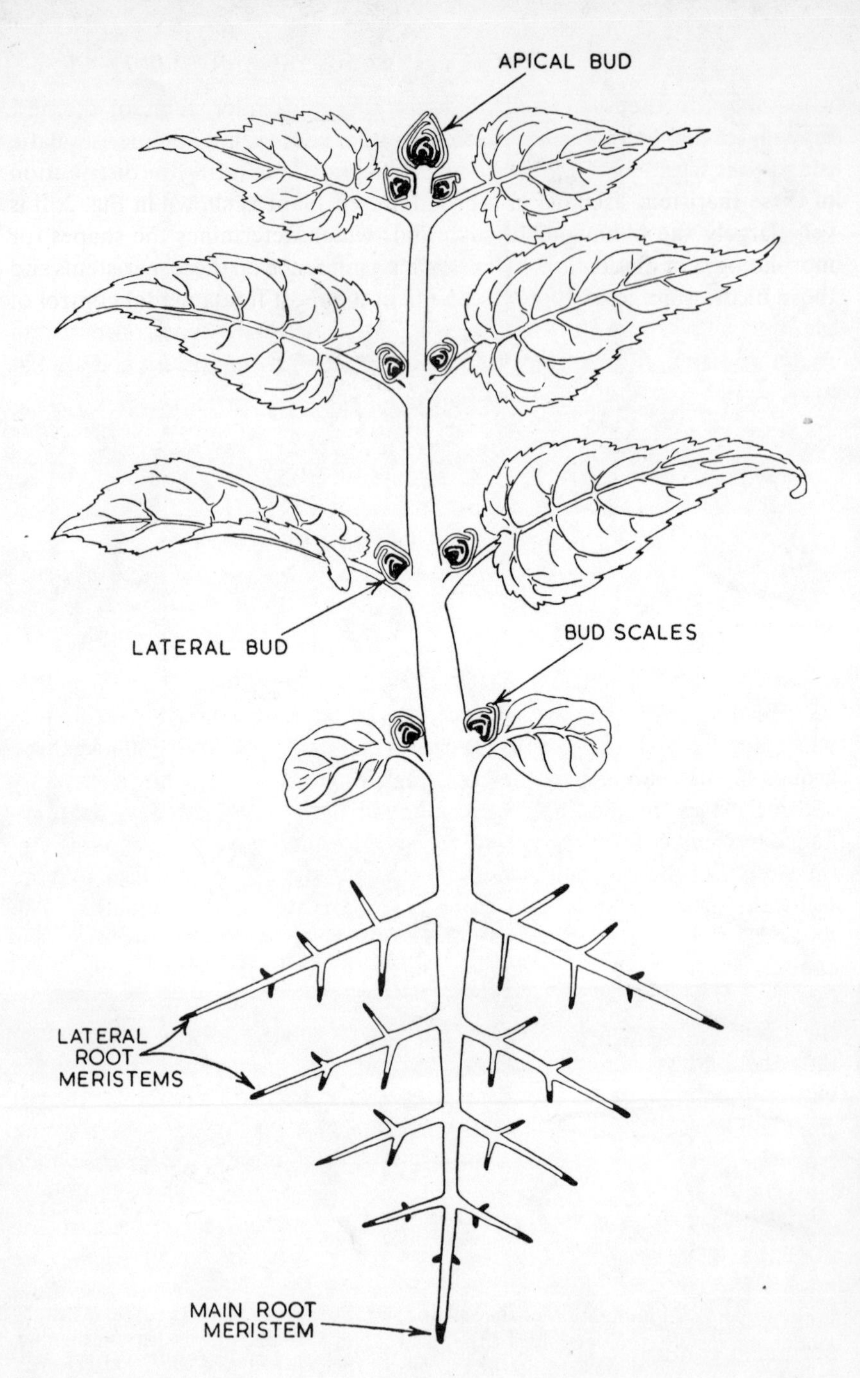

Fig. 2. Diagram showing the distribution of meristems in a broad-leaved flowering plant. The meristems are shown in black.

when they do their meristems become the apical meristems of the new branch, leaving behind in their turn, in regular succession, new meristematic islands, the lateral buds of the branch. A diagram illustrating the distribution of these meristem 'islands' in a broad-leaved plant is shown in Fig. 2. It is very largely the behaviour of such buds which determines the shapes (or morphology) of plants. It is by exerting an influence on these meristems and those of the stem apex that we can effect the most fundamental control of plant growth. These do not by any means exhaust the types of meristem found in plants, but they are undoubtedly the most important and all that need be mentioned at the moment.

The Factors Controlling Plant Growth

INTRODUCTION

As growth is such a complex affair, it is not surprising to find that the factors which contribute to its control are no less complex. We can distinguish three groups of such factors: the first and most obvious of these may be broadly categorized as 'food'. Under such a heading can be included all those influences which contribute directly to the *accumulation* of 'plant capital', e.g. the supply of the raw materials for the manufacture of protoplasm and the cell-wall; these are most conveniently designated *nutritional* factors. The second group is that concerned, not with the gross augmentation of this capital, but with its *distribution* between the various plant organs. Such factors show a very close resemblance, in their general mode of action, to the animal hormones, those 'chemical messengers' which are secreted into the bloodstream by various internal glands and which control the nature and intensity of growth in the several organs of the body; these I shall call *hormonal* factors. Thirdly, there are those influences which spring from the plant's own constitution. These are the hereditary factors which have come down to it from its parents and which the biologist speaks of as *genetic*. It should not be necessary to point out that there can be no sharp line of demarcation between these three categories; there is a very considerable overlapping and dovetailing. Thus it is quite possible for any one individual factor, e.g. light, to have deep-seated influences on both nutritional and hormonal control of growth. Again, the plant's genetic constitution will determine the nature and extent of its behaviour in relation to both nutritional and hormonal factors.

NUTRITIONAL ASPECTS

Nutritional factors in growth are those which contribute directly to the manufacture of the various organic materials of which the plant body is composed. These complex organic compounds are derived ultimately, by an intricate series of chemical events, from very simple inorganic materials such as the gas carbon dioxide absorbed from the atmosphere, water, and simple salts containing the elements nitrogen, phosphorus, sulphur, etc. All the elaborate organic compounds of the plant body are characterized by possessing in their molecules a greater store of chemical energy than the simple substances from which they are constructed. That this is true is quite apparent from the fact that the breaking up of these organic compounds into their constituent units, e.g. in the burning of sugar to form carbon dioxide and water, releases that extra energy in the form of heat. So, in order to make its own body matter, the organism must obtain a supply of these simple components as raw material and also sufficient energy, of chemical or some other nature, to allow the synthesis to take place: indeed this energy is the driving force of the manufacturing processes. All animals and such plants as moulds and toadstools obtain both raw materials and chemical energy from other complex organic substances, which they must first break up into smaller units. During this splitting-up process chemical energy as well as simpler molecular units for the new synthesis are made available. But large amounts of the energy are unavoidably lost as heat, together with many simple products of the breakdown. Although these simple molecular units are always available for subsequent use as 'building bricks' in future growth, because they are retained in the environment of the organism, yet the energy once lost as heat can never be recaptured. It is obvious, therefore, that for life to be maintained on our planet there must be available an independent source of energy for these syntheses: this energy is supplied as light from the sun.

The conversion of this light energy into chemical energy in the living world can be carried out only by the green colouring-matter of plants, chlorophyll. The chemical energy generated in the green cell in sunlight is used in the manufacture, from the carbon dioxide and water absorbed from its surroundings, of simple molecules of high chemical energy content. These form the hub from which radiates all the spokes of the great wheel of plant growth syntheses.* There is no need to stress the fact that the gross accumulation of plant capital as dry matter in green plants, i.e. its overall growth, will be very largely determined by the speed and efficiency of this sunlight-using process, known as photosynthesis. It may be controlled by many factors of which the duration and intensity of the light, the concentration

*An excellent sketch, for the non-specialist, of the present state of our knowledge of the chemistry of this vital process has been written by Fogg (1968).

and availability of carbon dioxide, and temperature, will be the most important.

However, there are many elements other than carbon, oxygen and hydrogen which go to make up the plant body, which consists of two main components, the protoplasm and the cell-wall. The latter, composed almost entirely of complex compounds of the elements carbon, hydrogen and oxygen, can be manufactured completely from the raw materials carbon dioxide and water. Other compounds of the same elements occur in cells and act as food reserves, e.g. starch, oils, etc. Protoplasm, if we disregard the high water content, is formed predominantly of proteins, and therefore contains the additional element nitrogen in high proportion, and the elements phosphorus and sulphur in smaller, but still appreciable, amounts. For synthesis of proteins to take place in the green plant, an adequate supply of raw materials must be available in the form of simple salts of these elements in dilute solution in the soil water surrounding the roots. In addition to the supply of these elements forming the 'fabric' of the protoplasm and therefore exercising a direct effect on the rate of protoplasm synthesis, certain others must be available if plant growth is to be normal. Their presence in the right proportions in the cell is in some way, at present not fully understood, essential for the maintenance of cell function. Finally, there is a long list of other elements which, though essential, are necessary only in very small amounts, and have therefore been termed micro-nutrients. The more important of these are iron, manganese, molybdenum, magnesium, copper, zinc, boron and silicon. Magnesium is an essential constituent of the green molecule chlorophyll. Iron and copper enter into the make-up of some enzymes whose function we shall be describing later. Lack or low availability of these latter elements will cause deficiency diseases characterized by marked derangements of the growth and developmental processes, with correlated distortions and abnormalities of the form and structure of the plant. Botanists are still trying to discover the exact role of many of these elements. The reader interested in this subject should consult the reviews by McElroy & Nason (1954), Broyer & Stout (1959), Nicholas (1961) and Steward (1963). It is possible that the supply of some of these micronutrients may be exerting an influence on growth by affecting the production of hormones, which we are considering in the next section. This is suspected to be one of the functions of zinc in the plant. (See p. 313.)

HORMONAL ASPECTS

So far I have considered, in outline only, that aspect of the control of plant growth which determines primarily the rate of increase of 'plant capital'. The second aspect, which I have called hormonal, must now be similarly outlined.

In the very simple plants such as yeasts and bacteria there is a correspondingly simple differentiation of structure. As far as one can tell, growth is mainly a matter of the replication of existing cell structures and multiplication is a necessary, but still somewhat puzzling, outcome of reaching a critical size. When, however, we come to consider multicellular plants, with distinct organs (e.g. roots, stems, leaves, flowers) each performing its own special function, then the need for control of the direction of development of these different plant parts becomes vital if a well balanced individual is to result.

Thus the development, both in space and time, of the several parts of plants following as it does a regular and, within the limits of variation of the species, an unchanging pattern, demands the existence of some precise mechanism for the co-ordination of the growth intensities of these organs. In other words, we must have a system to deal with the proper apportionment or *distribution* of 'plant capital'. An organization of the appropriate type has long been known in the higher animals: this is the endocrine system. In the animal body, in organs known as the endocrine glands, special organic chemical substances are produced. These substances, which are called *hormones*, are released into the bloodstream in small quantities and pass to various parts of the animal body where they control the activity, and hence growth and development, of the organ affected. These hormones are therefore 'chemical messengers', enabling a central co-ordinating gland to exert a control on other organs of the body. This 'action at a distance' is the essential feature of hormone action. In the normal individual a delicate balance in the secretion of these several hormones is maintained by their parent glands. Any disturbance of this equilibrium, resulting from over- or under-activity of one or more of these glands, gives rise to unbalanced growth and malformations. Giants and dwarfs are examples of what may result from such disturbances.

In the green plant we do not find such a rigidly constant form as in the higher animal. Although, for example, the flower of a primrose or a tiger lily has always the same structure, yet the number of flowers produced on any one individual plant may show wide variations. Even so, the plant also must have some similar co-ordinating system for maintaining the proper balance of growth rates between the several organs. Actually, the existence of such a system has been known since the early days of plant physiological study in the late seventeenth century, and the presence in some plants of a plant hormone was suggested as early as 1909 (Fitting). It was not until 1930, however, that a definite chemical substance was isolated and identified. The isolation of this hormone, and more particularly its synthesis in relatively large quantities in the laboratory, gave such a fillip to research on the subject, that within the space of a few years a number of far-reaching discoveries had been made as to its functions in the plant. It is now known that this hormone, together with others which have been subsequently isolated, and also synthetic chemicals of closely related constitution and properties, will exert far-reaching effects on both cell extension-growth and cell division, the precise action depending on the relative concentrations of the substances present and the reactivity of the organ concerned. Their main role, therefore, is the control of the *growth rate* of the various plant organs. Equal attention has been given to the practical implications of these chemical regulators, which have been given a variety of names to be discussed in greater detail later. For the moment we shall refer to them as *plant growth hormones*.

Other types of plant hormone have also been postulated, for example, a specific 'flowering hormone', the particular role of which is to control flower and hence fruit production. Although this has still to be convincingly isolated and chemically identified, the phenomena associated with it are of sufficient interest to warrant discussion later in the book.

In quite recent years, another group of naturally-occurring chemical compounds has come to be accepted by botanists as playing a possible part in growth regulation. They can be called the growth inhibitors, as their action on growth, in contrast to that of the hormones, is very seldom a stimulation but almost always an inhibition. As we shall see later, these probably co-operate with the growth hormones in the control of growth and development.

Before we leave the subject of hormone control we must return for a moment to consider another aspect of nutrition in animals and in many of the colourless plants (moulds, etc.) which has not yet been mentioned. In these organisms, in addition to the supply of raw materials for growth and for the chemical energy necessary to bring growth about, the food must also contain small quantities of certain complex organic compounds which are called *vitamins*. With insufficient supplies of these vitamins growth will be seriously deranged and the organism will die. Diseases such as beriberi and pellagra, which are so rife among the undernourished nations of the Eastern world, are due to deficiencies of particular vitamins in their diet. This requirement comes about because the organisms concerned lack the ability to make these essential growth factors for themselves. For example, many moulds will grow only when a supply of certain vitamins (e.g. thiamin = vitamin B_1) is supplied to them. Many organisms which cause plant and animal diseases grow in their hosts' tissues only because of the presence there of such essential growth factors. Most of the known vitamins are synthesized in green plants, which are, as far as we can see, the only primary source of these compounds in any quantity. This is yet another way in which the animal is ultimately dependent on the green plant for its existence. In spite of this, the role of vitamins in plants remained a mystery until fairly recently; now it has become apparent that their function is essentially the same as in the animals. Thus, until the discovery and isolation of the vitamins in pure form, no one had succeeded in growing *isolated* plant parts by supplying them with simple raw materials for growth. Then it was discovered that isolated roots of plants could be grown indefinitely in the dark in solutions containing sugar and inorganic salts of the necessary elements, provided traces of certain vitamins (notably thiamin) were added. It was therefore clear that these vitamins were as essential for root growth as they were for animal growth, and that they are normally built up in the green parts of the plant, from which an adequate supply for growth is obtained by the attached root. This is obviously 'action at a distance' and should therefore place these vitamins in the category of plant hormones. Nevertheless, there seems little doubt that the vitamins perform the same vital biochemical functions in plant cells as they do in animal cells, namely as co-factors essential for the catalytic functions of enzyme systems. A real problem of nomenclature arises, of course, when we consider what we are to call the same vitamin when it is functioning in the cell in which it has been synthesized. We shall return to this question later in this chapter.

GENETIC ASPECTS

The effects of hereditary make-up on the development of size and form of living organisms is so well known as to need very little comment. The hereditary factors are carried mainly in the cell nucleus, which is therefore in ultimate control of character expression in growth and development. The exact mechanisms whereby the nucleus exerts this fine control are rapidly being disclosed by recent biochemical studies on the nucleic acids (DNA and RNA) which contain the encoded information of inheritance. As we have already seen, it is now established that hereditary information from the 'jigs' of DNA in the chromosomes is transferred to RNA which is synthesized there and which then passes into the cytoplasm where it determines what kinds of enzyme shall be made. The growth and direction of development of any individual cell, be it plant or animal, are determined by the nature, the relative quantities, and the internal distributions of the constituent enzymes, the catalysts which regulate with great precision each and every activity of the complex 'factory' which is the living cell. During growth and development this complex pattern of enzymes alters, and these changes will be different for cells which develop in different ways. The great puzzle which has always faced the scientist interested in this phenomenon of 'differentiation' is how the total of hereditary information present in the nucleus, all of which is necessary to allow the production of every kind of cell the organism possesses, could be sifted and selected to direct the development of individual cells along particular and precisely prescribed pathways. Hormones, as we shall see later, clearly direct cell development and it is not surprising therefore that biologists have come to regard them as intermediaries in the nuclear control of cell activities. In some way it is supposed they may selectively modulate the flow of hereditary information (in the form of RNA) from the nucleus, thereby determining the fate of the cell concerned. For a stimulating account of current research and ideas in this field the reader is referred to the book by Bonner (1965); the mechanisms of hormone regulation of these processes will be discussed in the last chapter of this volume.

The Plant Growth Substances

Speculations on the nature of the factors underlying the development of plants have occupied the thoughts of many of the early natural historians, and even as far back as the end of the seventeenth century there were emerging tentative suggestions of the existence of a moving 'sap', responsible for correlating the growth activities of the separate plant organs. It was not, however, until the later half of the nineteenth century, when the modern science of plant physiology was really born, that the famous German botanist Sachs, the father of that science, started a detailed study of the phenomena. As a result of his investigations he proposed a generalized theory involving 'organ-forming substances', whose specific actions and localized distributions, which could be affected by external forces (e.g. gravity), were made to explain simply the facts of plant growth and organ development. Sachs's views found little favour with his contemporaries and it was the best part of half a century before the existence of such growth-controlling substances was finally and convincingly proved. During this considerable interval of time, botanists were very concerned with the gross nutritional aspects of plant growth, e.g. the supply of its raw food materials, and, as a result, these aspects of physiology tended to dominate theories of growth correlations. Such ideas have lingered on well into the present century – for example, in the carbon-nitrogen ratio theory of flower induction (see Chapter X). According to this theory, put forward in 1918 in America, the ratio of the level of the carbon to the nitrogen nutrition of the plant was supposed to control its reproductive activity, i.e. the prolixity or otherwise of its flower, and therefore fruit, production. As we shall see later, this has now been superseded by a hormonal concept of flowering control.

The original observations that led directly to the first isolation of a plant hormone were published nearly eighty years ago in a book called *The Power of Movement in Plants*, by the British naturalist Charles Darwin, so famous for his theories of organic evolution. The way in which these ideas of Darwin set in train a series of purely academic enquiries which culminated in the development of techniques now revolutionizing agriculture, is another proof, if such is needed, of the vital importance of fundamental research. This is the story in outline.

All of us who have had any dealings with growing plants know that they are sensitive to external stimuli such as light and the force of gravity. The young parts of plants brought indoors will bend over towards the light coming from a window, and the leaves will arrange themselves so that their blades are perpendicular to the light falling on them. Similarly, young plants that have been laid flat by rain or wind will, if the stems have not been broken, eventually bend upwards at the growing tips so that the stems become vertical once more. These movements of the plant in response to external stimuli – unilateral light in the first instance, the force of gravity in the second – are called *tropisms*, and are brought about by opposite sides of the organ or organs concerned growing at different rates in response to the stimuli. It was the elucidation of the mechanism of such responses that was Darwin's main concern. His experimental plant was the seedling of the ornamental canary grass (*Phalaris canariensis*). Such seedlings, in common with all members of the same family, produce in the early stages a tubular first leaf, which remains short in the light, but in the dark grows very much longer. This first leaf, through which a typical flattened leaf-blade eventually bursts, is called a *coleoptile* and is particularly sensitive to light stimuli. Darwin observed that when these organs were illuminated from one side they bent over towards the source of the illumination. The region in which the bending took place was quite a long way below the coleoptile tip. If, however, the extreme tip of the coleoptile were removed or covered by a small cap of light-proof material, such as tinfoil, no bending took place in the lower region on exposure to light. This and similar experiments led Darwin to postulate that the light was perceived by the tip. These phenomena are very clearly demonstrated in Plate 1, which shows the results of such experiments on the coleoptiles of oat seedlings. As the bending region was situated lower down the organ, some 'influence' must have been transmitted from the stimulated tip to the reacting zone. Experiments on roots and other organs indicated strongly that transmission of similar influences might underlie reactions to gravitational pull.

Research at the turn of the century was more leisurely than it is today and over a period of thirty years following the publication of Darwin's book, only an occasional report verified and amplified his main findings. One of the

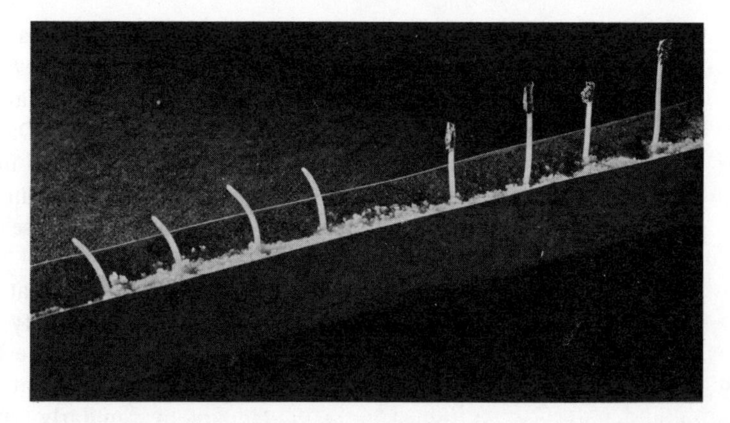

Plate 1. Experiments with oat seedlings showing the effects, on the subsequent reaction to lighting from one side, of covering the tips of the coleoptiles with aluminium foil caps. The four uncapped seedlings on the left show normal bending towards the light while the four capped seedlings on the right have remained upright. (Photographs by kind permission of Dr E. J. S. Hatcher and East Malling Research Station.)

most important of these subsequent observations was that the transmission of the 'influence' down the coleoptile was not affected by a transverse nick between the tip and the reacting zone, provided the cut surfaces of the nick were in contact. Such evidence showed that the continuity of living protoplasm was not essential for the transmission of the 'influence', and led the Danish botanist, Boysen-Jensen, who had chosen to work on the light reactions of the oat coleoptile, to suggest, just before the beginning of the First World War, that a purley chemical mechanism underlay this transfer of stimulus (Boysen-Jensen, 1911). This was proved beyond all doubt when it was shown that the 'influence' would pass through certain non-living materials such as gelatine which will allow the passage of dissolved substances, for if the tip were removed and fastened back with such material on to the stump, then the response of the coleoptile to unilateral light was unimpaired. Thus the influence was obviously chemical but its exact mode of operation remained obscure until it was shown that the tip of the coleoptile could influence the growth of the stump quite independently of any light stimulus. Thus in 1919 it was demonstrated (Paál, 1919) that if the coleoptile tip were removed and replaced asymmetrically on the stump, then the stump bent over in the dark in such a way that the longest side was under the tip. This demonstrated the existence in the tip of a chemical substance which passed into and down the side of the stump in contact with the tip, stimulating the extension growth of the zones below and giving rise to the curvatures noted. The reaction of the coleoptile to unilateral light was then presumably caused by an appropriate redistribution of this substance in the tip, greater amounts going into the darkened side and therefore making the coleoptile grow longer on that side. It was not, however, until 1926 that this chemical messenger was finally isolated by the now classical work of the Dutchman, F. W. Went. It was this isolation and the technique that sprang from it that mark the beginning of the modern era of plant growth substances.

Went, knowing from the work of Paál that the substance would move out of the isolated coleoptile tip into the stump, hit upon the idea of allowing it to diffuse instead into a suitable non-living material, thus isolating it from the living cells in which it acted. The material he chose was agar-agar (agar), a carbohydrate prepared from seaweed; this dissolves in water to form an inert transparent jelly. Small rectangular blocks of this jelly were prepared under sterile conditions and on these blocks were placed a number of severed coleoptile tips with their cut surfaces in contact with the agar. These were left to stand in a humid atmosphere to allow the chemical to diffuse outwards into the jelly. Its presence in the agar was demonstrated when the larger block was cut up into smaller cubes and one of these cubes was placed on one side of a stump of a coleoptile kept in the dark. After a short time-lag, the coleoptile was found to bend away from the side on which the block had been placed, demonstrating conclusively the presence in the block of a growth-stimulating substance. All this, however, was nothing more than a combination and refinement of the techniques of other workers and final confirmation of their findings. Went's original and greatest contribution to these studies was made when he undertook *quantitative* measurements of the curvatures produced by the small agar cubes containing the hormone. By allowing the hormone in one small block to diffuse into larger blocks, known dilutions could be obtained, and, by using small cubes from these latter blocks to produce curvatures, it was shown that, within the limits of experimental error, the curvature resulting was directly proportional to the concentration of the hormone in the agar. The importance of these observations will be appreciated when it is realized that this hormone is active in extremely low concentrations, far, far below the sensitivity of any chemical tests. No known chemical test could possibly have revealed the presence of the hormone in the agar, and yet here was a reasonably accurate method of measuring – in arbitrary units, of course – the quantity of hormone coming from any tip. It was the development and refinement of this biological assay technique, now known among plant physiologists as the Went *Avena* curvature test,* that were chiefly responsible for the final isolation of the hormones concerned in a pure crystalline form and their subsequent chemical identification.

**Avena sativa* is the botanical name for the oat.

The chemists who first made this isolation, Kögl & Haagen Smit (1931) christened them *auxins* and for many years these remained the only plant hormones whose chemical identities were known (see Chap. II); they soon dominated the study of plant growth. In the interests of clarity in the sections which follow in this Chapter it is worthwhile recording at this point that the principle natural auxin is indol-3yl-acetic acid (abbreviated to IAA) and has the formula I shown in Fig. 5, p. 50.

But these investigations into curvature-inducing hormones were not the only incursions into this field by the early pioneers. Apart from the theoretical speculations of Sachs (see p. 13) direct attempts were made to isolate other hormones, such as for example the unconfirmed isolation by Haberlandt (1913–14) of a *wound hormone* which had been postulated by Weisner at the beginning of the century to be responsible for the production of scar tissue, well known to botanists and gardeners as *wound callus*. However, there existed at this time evidence, largely unknown in the western world, of other growth-controlling substances, which in recent years have threatened, justifiably, to oust the auxins from their dominating position. The story is as fascinating as that of the auxins.

In the lands of the Far East, where rice is grown as a staple food, that cereal is often attacked by a fungus, *Fusarium moniliforme*, causing the plant to become pale and spindly in its growth (see Plate 2). Beacuse of the weak lankiness of the infected plants the Japanese have called the affliction *bakanae* or 'foolish seedling disease'. Although the disease had been known for centuries, the causal agent was not recognised until the beginning of this century and in 1912 a Japanese botanist, Sawada, suggested that the symptoms could be the result of the action of a chemical substance secreted by the fungal parasite. Fourteen years later this theory was proved correct by his pupil Kurosawa (1926) who was able to reproduce the disease symptoms and to induce the remarkable stimulation of growth by the application of a sterile extract of the fungal culture medium to healthy rice plants. In the same manner, as we have already seen above for the auxin story, the isolation of this substance in pure crystalline form had to await the development of a refined biological assay with rice seedlings and took place in the University of Tokyo just before the last world war (Yabuta & Sumiki, 1938). The new substances isolated were called *gibberellins* after the name given to the sexually-reproducing (perfect) form of the fungus, i.e. *Gibberella fujikuroi*. These discoveries went largely unnoticed in the west and it was not until after the war that plant scientists, in Great Britain and the United States as well as Japan, finally elucidated the chemical structure of the gibberellins (see Chap. IV). The reader may be asking himself what can be the relevance of a fungal 'poison' to the subject of plant hormones. Plant physiologists also asked the same question and, being convinced that the striking effects of the

Plate 2. The response of rice seedlings to gibberellic acid, producing the symptoms of the 'bakanae' disease.
Left. Control, untreated; *Right.* Plants sprayed with 10 ppm gibberellic acid solution two and four days before the photographs were taken and showing a greatly increased growth. Note also the paler colour of the treated plants, a characteristic symptom of gibberellin treatment due to the smaller content of chlorophyll per unit leaf area.

gibberellins on plant growth could not be fortuitous, have sought the presence of these compounds in healthy plants. They were soon found in immature seeds of bean (*Phaseolus* spp.) (MacMillan & Suter, 1958; West & Phinney, 1959), are now known to be widely distributed and are firmly accepted as plant hormones.

The basic effects of the auxins and the gibberellins are seen in phenomena associated with the extension growth of cells, although certain differences in the detailed patterns of their actions (see later Chapters) make it clear that they belong to two clearly distinct groups of hormones. However, many other growth phenomena seem to be affected, if not controlled by them, including cell division, and in the last decade a third group of plant growth substances, mainly involved in this phase of cell growth, has been discovered. These are the cytokinins with which I shall deal in Chapter V. All these three types of plant growth substance can, either alone or in combination, evoke such a great variety of growth and developmental responses that the old concepts of one hormone for one job, as embodied in the 'flower-forming' and 'root-forming' substances of Sachs, have been slowly going out of favour. Fortunately the fact that this trio cannot explain all phenomena of control has kept alive the search for more specific substances; the evidence that these do exist stubbornly persists.

Nomenclature

It is convenient at this point to turn to a consideration of nomenclature, as a wide range of terms have been used to characterize the plant growth regulating substances, and perusal of the very extensive literature will reveal considerable overlapping and ambiguities in terminology. A widely accepted definition of a plant hormone was given by Thimann (Pincus & Thimann, 1948) as: '*An organic substance produced naturally in higher plants, controlling growth or other physiological functions at a site remote from its place of production, and active in minute amounts*.' Thimann favoured the term 'phytohormone' (Greek: *phyton* = a plant) for these substances to distinguish them from the animal hormones. This definition will therefore include the three groups of substances already described, together with those which induce flowering, wound healing, etc., and also some vitamins, whose action as growth factors are quite well established. Up to the time when the gibberellins came on the scene the auxins were regarded as the only hormones responsible for the control of the extension growth of cells and the definition adopted for them was based firmly on this assumption. Thus the most widely accepted definition of an auxin reads (Pincus & Thimann, 1948):
'*An organic substance which promotes growth* (i.e. *irreversible increase in volume*) *along the longitudinal axis, when applied in low concentrations to shoots of plants freed as far as practicable from their own inherent growth promoting substances. Auxins may, and generally do, have other properties, but this one is critical.*'

However, difficulties arise immediately we bring in the gibberellins since they also promote extension growth in this way, sometimes in the very tests thought to be specific for the auxins (see Chap. II, pp. 31–32). Yet owing to their markedly different actions in many other growth phenomena, no-one has yet suggested that the gibberellins are auxins. Obviously we need a new definition of an auxin, based preferably on a highly specific response; the role of the natural auxins as mediators of the curvature responses to directional stimuli (tropisms) immediately suggests itself as a possible basis since gibberellins so far have not been implicated in these phenomena. Since, as we shall see later, this action of auxins as curvature-inducers seems to depend on a characteristic polar movement, i.e. along the main axis of the organ in one direction, we might use this as a key property in a new definition of these extension growth hormones. Unfortunately we still know so little about the movement-behaviour of the many hundreds of compounds (see Chap. III) that share most if not all of the remaining physiological properties of the natural auxin isolated by Went, that such a definition would be premature. Fortunately we have no such difficulty in defining a gibberellin, of which over two dozen are now known, since, unlike the auxins, they are all very closely related chemically and have a quite specific physiological action, in that they promote in a spectacular manner the growth of a number of dwarf varieties of cultivated plants (see Chapter IV).

However, the big bug-bear to any would-be definition-writer is the overlap and interactions that seem to characterize the activities of the known phytohormones in various growth phenomena. Went's auxin and the gibberellins both control extension growth and may mutually modify each other's activities but no-one yet knows whether they operate in the same or in completely different ways. Lasting definitions must ultimately relate to the mode of action at the molecular level.

Another problem concerns the names we should apply to the already vast and still rapidly expanding list of purely synthetic* compounds whose actions, as far as we can see, are identical with those of the corresponding groups of naturally occurring hormones. In the past a rigid distinction has been made between such 'synthetic growth-regulating substances' and 'natural phytohormones'. For example, Nicol, in an ealier book on this subject, wrote: 'Unless it can be proved that the "synthetic" growth substances occur naturally in plants and regulate growth within them, it would today be as absurd to give the name "hormone" to the exterior substances which affect plant growth as it would be to confuse vitamins and hormones in speaking of animal physiology.' The two cases are not, however, strictly comparable, as the boundary between vitamin and hormone in the plant is by no means as clear-cut as it is in the animal. In the latter, hormones are produced in the organism itself and are then carried to the site of their action.

*The real meaning of synthesis is 'placing together' (Greek: *syn* = together, and *thesis* = a placing) and refers to a building up of a chemical compound, whether performed in the living organism or in the laboratory or factory. Popular usage of the adjective *synthetic* restricts the term to those compounds, made artificially, which are often substitutes for the naturally synthesized materials. Here in the text the term synthetic is, of course, used in its restricted sense of 'made artificially'.

Vitamins are manufactured by other organisms, mainly green plants, and must be taken by the animal in its food as it cannot manufacture them itself. In the green plant both hormones and vitamins are manufactured, and, as they are both regulators of vital processes, the distinction between them now becomes a matter of the distance between the centres of production and those of action or utilization. That this distinction is highly arbitrary has been pointed out by van Overbeek (1950), who asks what the limit of distance should be; between organs, tissues, cells or even molecules? If molecules, then all naturally occurring organic regulators of plant physiological processes could be called phytohormones, as the distance is always finite, however small it may be. A case more nearly approaching that of the animal is seen in some colourless plants such as moulds. Moulds all require for their growth a range of accessory growth substances but different species differ in their ability to build them from simple inorganic food materials. Thus the common mould *Aspergillus niger* will grow normally on a medium containing mineral salts and simple sources of carbon and nitrogen: it can synthesize all its own growth substances. Others, such as *Phycomyces blakesleeanus*, require an external supply of a whole complex of such growth substances. One of these is thiamin (vitamin B_1), which functions in those organisms in precisely the same way as it does in the animal. If we adhere to the nomenclature of the animal physiologist, these obligate food factors are vitamins so far as the mould is concerned. What, therefore, must we call them when they are manufactured by the organism, whether it be fungus or higher green plant? Are they still vitamins or have they become hormones, or must we coin yet another name for them?

Obviously, the ideal system of nomenclature must eventually be based on biochemical and physiological function. Such an ideal nomenclature must therefore embrace not only naturally occurring compounds, but also all artificially made substances having the same action in the cell. Indeed one treatise (van Overbeek, 1950) has given a very broad definition of plant hormones as 'organic compounds which regulate plant physiological processes . . . regardless of whether these compounds are naturally occurring and/or synthetic, stimulating and/or inhibitory, local activators or substances which act at a distance from the place where they are formed'. Similarly, the term auxin is made to include all synthetic compounds having the actions characteristic of Went's *auxin*. There is no *a priori* reason why the term hormone should not be used in this very broad sense, as there is no suggestion, in its derivation from the Greek word *hormaein* (to step up), of an action necessarily at a distance from the centre of production. But the strict limitation of the term in animal physiology to naturally produced substances, ever since its definition by Bayliss & Starling in 1940, makes it essential for plant physiologists similarly to restrict its use, if confusion is to be avoided. A term which could be used to meet this difficulty of nomenclature is *ergon* (van Euler, 1946). An ergon can be defined as any organic compound acting, in low concentrations, as an activator or regulator of a physiological process. An ergon in the plant would then be virtually synonymous with van Overbeek's definition of plant hormone.

In 1951, K. V. Thimann, wishing no doubt to stem the perturbing spread of nomenclatorial confusion in this branch of botany, suggested the appointment of a committee to consider and propose a uniform nomenclature on growth substances (Tukey *et al.*, 1954). The fact that this committee could reach no unanimous agreement illustrates the difficult, and still largely subjective, nature of the problem. A majority report (Tukey *et al.*, 1954) proposed the blanket term '*plant regulator*' to cover all 'organic compounds, other than nutrients, which in small amounts promote, inhibit or otherwise modify any physiological process in plants'. These were then subdivided into growth regulators, flowering regulators, and so on, according to the particular process controlled. Hormones were defined as regulators *produced by the plant* and were similarly subdivided. A minority report by Larsen (1955*a*) objected to the word 'regulator'. He rightly pointed out that to regulate means 'to adjust so as to work accurately and regularly', and this many of the synthetic growth substances do not do. Actually, the practical use of these substances often depends on the opposite action, i.e. a violent upsetting of the 'accurate and regular working' of the growth processes. Larsen made counter proposals with which the present writer is in close agreement and on which the terms to be used in this book are based.

Thus plant growth substances will include 'all organic compounds which at low concentrations promote, inhibit, or qualitatively modify growth. Their effects do not depend on their caloric value or their content of essential elements.' Plant hormones are substances which *regulate* (see above definition) some aspect of plant growth, are produced by the organism itself, and usually act at a point remote from the site of production. They may be growth hormones, flowering hormones, and so forth. Auxins are those plant growth substances which are characterized by their property of stimulating extension-growth of shoot cells in the standard biological tests to be described in the next chapter. It should be noted that there is here no restriction of this term to naturally-produced substances and so we shall speak of 'natural auxins' for those substances produced by the plant itself and 'synthetic auxins' for those which have, as far as we can tell, precisely the same action on the cell as the natural auxins but which have not been shown to occur naturally in plants. Obviously this arrangement is not entirely without ambiguities. Indol-3yl-acetic acid (Went's auxin) is itself now manufactured in fairly large quantities for experimental purposes but is, of course, a 'natural auxin'. We shall refer to these compounds as auxins (natural or synthetic), even though we may be discussing aspects of their action other than the stimulation of extension growth. The gibberellins are regarded as hormones; except to confirm the chemical structures of these natural compounds, no synthetic gibberellins have as yet been made. However, many compounds with physiological activities identical with those of the known gibberellins have been isolated from plants and still remain to be characterized chemically. They may or may not prove to have the basic molecular structure of the gibberellins and so the non-committal connotation of 'gibberellin-like' has been accorded them. The cytokinins, which also seem to enjoy a certain uniformity of chemical structure (see Chap. V) have both natural and synthetic representatives. Similarly, other growth substances which have not the advantage of definitive and easy categorization will be catalogued under the names by which they are best known.

Chapter II

The Natural Auxins

Introduction

We saw in the last chapter that curiosity on the part of Darwin concerning the bending of plants in response to unilateral light gave rise to a series of researches culminating, in 1928, in the isolation of a chemical substance which was active, in very small concentrations, in causing a great acceleration in the rate of extension-growth of oat coleoptiles. It now seems likely that the young shoots of all higher plants are similarly affected. After this isolation it was soon shown that the effect of light in causing curvatures of these organs was indeed transmitted by this active substance. Light from one side caused hormone to move from the tip down the dark side in greater quantities than down the illuminated side; cells on the dark side then extended faster than those on the illuminated side and the coleoptile bent towards the light. Thus, after about fifty years, Darwin's academic questions were answered, but this was only the beginning of the subject of plant hormones as we understand it today. The really great advances that led to their development for use in practical agriculture and horticulture began when they were identified chemically by the classic work of the chemist Kögl and his colleagues in Utrecht. Milestone as this is in the history of the subject, it would not have been possible but for the botanist Went, who provided, in the *Avena* test, a quantitative assay* on which their whole purification technique depended.

Because such quantitative assay has played a fundamental part in both the isolation and identification of natural plant growth hormones and, as we

*Assay is a term originally employed for the determination of the proportion of a metal in an ore or an alloy. In recent years it has been applied by biologists to the determination of specific activity of such biological products as hormones, vitamins, antibiotics, etc.

shall see later, in the preparation of the majority of similarly active synthetic compounds, it would be as well, before we go on to describe the work of Kögl, to deal in rather more detail with the various assay methods available for these auxins. We should, however, first pin-point the three main functions that such assays have fulfilled in advancing our knowledge of the auxins. Firstly, they enabled chemists, in their efforts to obtain pure crystalline hormones in fair yield from natural sources, to test the activity of their various samples during the separate steps of purification. As the methods used in these researches were purely empirical, an assay such as the *Avena* test was an indispensable tool in determining their efficiencies. Secondly, when the constitution and activity of the hormones became known, the assay became important as a method of determining the distribution and relative concentrations of hormones in various parts of the plant, for, as we have already noted, direct chemical methods were then too insensitive for the extremely small quantities involved. Thirdly, when it became apparent that a wide range of synthetic chemical compounds had properties similar to those of the natural growth hormones, it became a point of great practical importance to have accurate methods of assessing their relative activities. Assays of this type play a vital part in the manufacture and standardization of antibiotic drugs such as penicillin and streptomycin, which are the natural products of micro-organisms.

The main assay methods which have been employed in auxin determinations will now be briefly considered. Only the principles of the methods will be given, the details being of interest chiefly to workers actively engaged in their use. Assays for other hormones and growth regulators will be considered in the relevant chapters.

Assay Methods

(A) COLEOPTILE CURVATURE TESTS

The test which has been most used by plant physiologists, and which still continues to be of the greatest importance, is Went's *Avena* test (Went, 1928). It has undergone slight modifications in detail from time to time but still remains in its essentials the same as that originally described. Oat grains are dehusked, soaked in water for 2 to 3 hours, and then sown in a suitable moist medium, usually damp absorbent paper (the chemists' 'filter paper'), in shallow dishes in the dark at a temperature of about 25° C. (77° F.). These environmental conditions, which are held constant during the whole period of the test, have been found to give maximum sensitivity of the coleoptile to the applied hormone. After about thirty-six hours the seedlings are transferred to containers where they can grow with their roots in a dilute solution of mineral salts of carefully controlled constitution, with their coleoptiles in a strictly vertical position. When the coleoptiles have attained a length of about 2.5 cm the seedlings are ready for the test. The extreme tip of the coleoptile is then cut off (see Fig. 3), thus removing the main source of its natural auxin and thereby stopping its growth. After a period of some hours the top few millimetres of the stump are similarly removed (see Fig. 3), as it has been found that this apical tissue may gradually acquire the power to produce auxin during the subsequent test – an undesirable state of affairs. The tip of the young foliage leaf is thus disclosed; this is then pulled with a pair of cork-tipped forceps so that it breaks at the bottom of the coleoptile tube where it is attached. It is then withdrawn so that it projects slightly above the cut end of the coleoptile and acts as a support for the agar block containing the auxin when it is finally applied (see Plate 3). The principle behind this elaborate preparation is to produce a plant organ with an auxin content reduced to vanishingly small proportions, so that the reaction to added auxin is clear-cut and disturbances caused by wide variations in internal (endogenous) hormone content are at a minimum. One refinement of Went's original method is to remove the 'seed' from each seedling some hours before carrying out the test. This simple procedure removes the store of material (precursor) from which the growth hormone is made in the seedling, thereby further increasing the sensitivity of the coleoptile by preventing any possible production of auxin in the tip of the coleoptile stump.

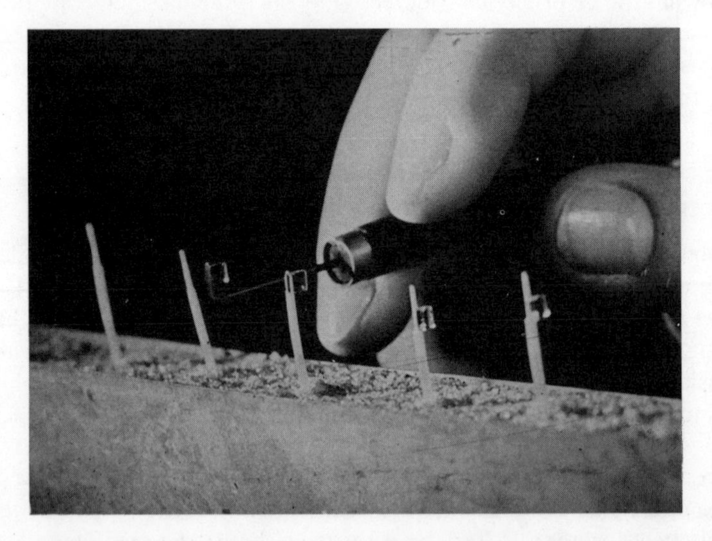

Plate 3. Application of agar blocks containing auxin to prepared *Avena* coleoptile stumps (Photograph by kind permission of Dr. E. J. S. Hatcher and East Malling Research Station.)

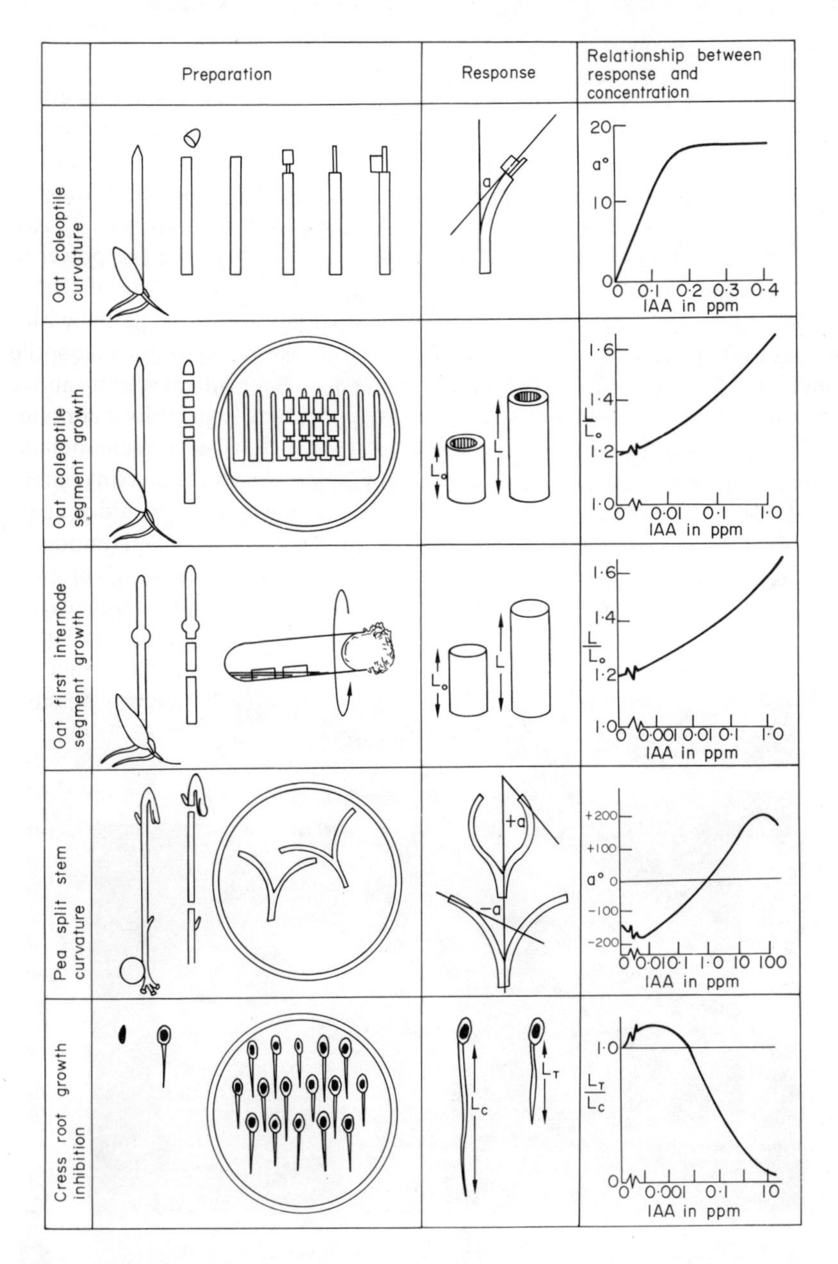

Fig. 3. Diagram illustrating the salient features of five of the main auxin assay techniques (Went's *Avena* curvature test; *Avena* coleoptile segment straight growth test; *Avena* first internode (mesocotyl) segment straight growth test; split pea epicotyl curvature test; cress root growth test). In the first column are illustrated the major steps in the preparation of the material for the assay. The third column shows the quantitative relationships between the response and the concentration of the applied auxin (IAA in all these graphs).

As in these tests one is dealing with plants which are extremely sensitive to external influences, very special precautions have to be taken to ensure optimum response and minimum interference from environmental factors. In addition, a collection of plants, even of the same variety, unlike crystals of a pure chemical substance, is not completely uniform, and this necessitates the use of pure strains of selected plants to minimize this intrinsic variability and increase the accuracy of the assay.

In addition to preparing the oat seedling, the small agar blocks for carrying the auxin have also to be made; these are cut in cubes of 5 to 10 cubic millimeters volume from thin sheets of 3 percent agar (see Plate 4). The hormone can be added to them either when the agar is in a molten condition before the gel is allowed to set, and therefore before the blocks are cut, or the blocks can be soaked in a solution of the hormone for an hour or so before the test. The blocks, which must be uniform in size and therefore need a special cutter for their preparation (see Plate 4), are then placed unilaterally one on each coleoptile and supported in the angle made by the cut surface and the slightly projecting first leaf. (See Plate 3.) All these operations so far

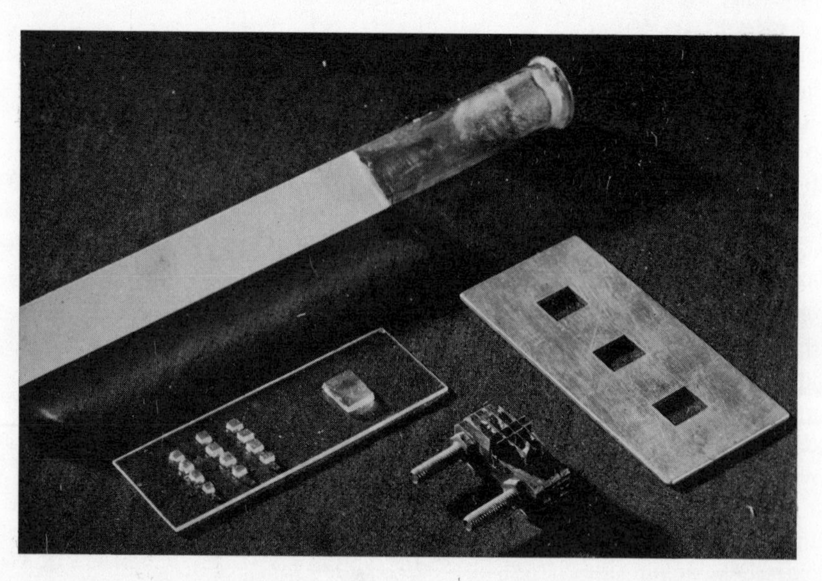

Plate 4. Preparation of agar blocks as carriers of auxin in the Went's *Avena* coleoptile curvature test. The agar, seen in the tube at the top, is melted and a small block cast by pouring it into the brass mould on the right. When set, this block (seen on the right-hand side of the glass slide) is cut into twelve uniform blocks, each of 10 mm^3 (seen on left-hand end of slide) by means of the small cutter shown below. (Photograph by kind permission of Dr. E. J. S. Hatcher and East Malling Research Station).

described must be carried out in weak red light; white light cannot be used as it reduces sensitivity and, more important still, can produce growth curvatures which completely invalidate the test. The plant then remains in darkness for 1½ to 2 hours, which is the normal period optimal for maximum response before any complicating effects set in from the possible renewal of internal auxin supply. One important test condition is that the humidity must remain constant at a value somewhat below 90 percent; if it is allowed to rise above this point, water may be exuded in drops at the cut surface and may even wash off the agar block. When the period for maximum response has elapsed, the curvatures of all the seedlings in the batch are recorded photographically by projecting their shadows on to a long strip of bromide paper (shadowgraph technique). The response is then determined from the print either by measuring the total angle of curvature between tip and base by means of a protractor, or by estimating from the radius of curvature the difference in length between the two sides of the curved coleoptile.

The responses obtained from plants in any one batch receiving identical treatments will vary considerably from plant to plant, a feature which we have already noted as characteristic of biological assay. Such variations of response arise mainly from slight differences in age and length of individual coleoptiles. Each assay consequently requires that a considerable number of seedlings be used, so that this variability can be estimated by statistical methods and used to assess the confidence we can place in the overall response as given by the average curvature in the batch as a whole. It is then found that the average response is directly proportional to the concentration of applied auxin up to a certain maximum. This maximum is equivalent to about 1 part of the naturally occurring auxin indol-3yl-acetic acid in 5 million parts of water, when the corresponding angle of curvature varies from 15° to 35°, depending on such unexpected factors as the time of day. This day-to-day variation in response between batches of apparently identical plants necessitates the 'calibration' of every batch of seedlings with known quantities of hormone (see Fig. 3) and is a complication of the technique that bedevilled workers on auxins from the time of Went's pioneer work onwards. The cause remained a mystery until a short time ago when F. W. Went and his colleagues, working in Pasadena, California, discovered that the sensitivity of the *Avena* coleoptile to auxins was much depressed by extremely small traces of 'smog', the active constituents being partially oxidized petroleum gases (Hull *et al.*, 1954). The daily variations of concentration of these materials in the air in the neighbourhood of the city caused inversely proportioned changes in coleoptile sensitivity which could be eliminated only by growing the seedlings in air filtered through activated carbon. This auxin sensitivity change can, in fact, be used as a very sensitive test for 'smog'.

The advantages of the great sensitivity of Went's *Avena* curvature test are offset by the complications and tedium of the preparative procedures involved. It is not surprising therefore that attempts have been made to 'streamline' such tests without loss of sensitivity. Much saving of time has been achieved by using excised coleoptiles (Kuraishi & Yamaki, 1964) with their basal ends embedded in a block of agar instead of the carefully prepared seedlings of the classical method. A similar test with excised coleoptiles plus first internode not only has the same time-saving advantage but appears also to be more sensitive by a factor of seven than Went's test (Wilczynska, 1959). The first internode is a short solid stem which grows up under the coleoptile when germination takes place in complete darkness.

These assays with oat coleoptiles involve surgical operations which may impair the subsequent growth potential of the organ and hence its response to applied auxin. An assay method employing intact coleoptiles has been used satisfactorily by Linser (1938). In this 'Pastentest', auxins dissolved in lanolin paste are applied in small drops near the tip and to one side of dark-grown coleoptiles. The growth response is twofold: a curvature away from the side of application and an increase in the overall growth rate of the organ; both are measured by a shadowgraph technique. Linser claims that these two responses do not always run parallel, some auxins giving a greatly accelerated growth rate but a small curvature. This may be attributed to a rapid lateral transport of those particular substances, so that both sides of the coleoptile are similarly stimulated. In this way lateral mobility of the auxin, a property which is of some importance, could be determined. An additional advantage is that the technique can be used to assay *growth-inhibiting substances* (see Chap. VI) which give curvatures towards the drop of paste and also retarded growth rates. Went's *Avena* test is quite unsuitable for the assay of such substances.

(B) OTHER CURVATURE TESTS WITH CYLINDRICAL ORGANS

Seedlings of a number of other species have, from time to time, been used for auxin assay with techniques similar to the Went *Avena* test. One of the most promising was that due to Söding (1937) in which the cotyledons were removed by an oblique cut from the tops of hypocotyls of *Cephalaria* sp. and *Raphanus sativus*. Applications of auxins in small agar blocks to the lower half of this cut surface caused curvature away from the block. Söding claimed that these organs were more sensitive to auxin than *Avena* coleoptiles.

A very sensitive new method involving *Avena* coleoptile segments involves their response to gravity. Intact coleoptiles, placed horizontally, will curve upwards because the lower half will have been induced to grow faster than the upper half by the action of gravity. This is an auxin-mediated response and therefore isolated, horizontally disposed segments will not curve because they have been largely robbed of their auxin supply by the removal of the tip. If such segments are supplied with auxin in agar placed in contact with the lower halves of the apical cut surface they will curve. This response has been made the basis of a very sensitive auxin assay which can detect quantities of indol-3yl-acetic acid as low as 1-2 ng* (Kaldewey, Wakhloo, *et al.*, 1969).

(C) SEGMENT STRAIGHT-GROWTH TESTS

It should be obvious, even from these simplified descriptions of *Avena* curvature tests that they require a delicate and skilled technique and very exacting growth conditions. Such tests can therefore be carried out, as a rule, only in research institutions with specially designed growth-rooms. It is therefore not surprising that a number of simpler assay methods, less susceptible to environmental influences, have been devised from time to time. The one much favoured today by plant physiologists also uses the coleoptile of the oat, but instead of measuring curvatures obtained by unilateral application of the auxin, it measures the total increase in growth-rate (stretching growth) of small cylinders cut from the coleoptile and immersed in aerated solutions of the auxins under investigation. This is known as the 'straight growth' test to distinguish it from the previous 'curvature' test. It must be pointed out that such cylinders when immersed in water will show growth owing to their own hormone content, and that therefore two observations are necessary for each test: firstly, of the average growth rate of a batch of cylinders growing in water, and secondly, of the growth rate of a similar set growing in the test solution. The segments, which are usually 3 to 5 mm long, are cut from identical regions of similar uniform coleoptiles, and can be measured to the nearest 0·02 mm. The measurements can be made directly with a travelling microscope, or, more conveniently, from photographic records either as shadowgraphs on film using a special camera (Kiermayer, 1956) or as magnified prints on bromide paper in a photographic enlarger. One worker (Bentley, 1950*a*) recommends longer (10 mm) segments which allow more rapid measurements against a scale accurate to 0·5 mm. Growth must take place in darkness and at a constant temperature. Cane sugar is usually added to all solutions, as lack of such a source of carbon for growth might limit the response to the hormone, although some workers question whether sugar is necessary (Bentley, 1950*a*). In addition, adequate aeration (supply of oxygen) must be ensured, and this is often done by using shallow dishes (Petri dishes) for the tests, and having the cylinders just breaking the

*ng = nanogramme = 10^{-9} gramme = 10^{-6} mg.

surface of the solution. For ease of handling and measurement, the cylinders are sometimes slipped on to the teeth of very fine plastic combs or on to small glass capillary tubes, before being immersed in (or floated on) the solution. Gentle continuous agitation with some mechanical rocking device is often used to advantage. In one modification, sections are placed in only $\frac{1}{2}$ ml of test solution, in small 2-in $\times$ $\frac{3}{4}$-in glass tubes, which are rotated horizontally about their main axis (Hancock & Barlow, 1953). This prevents section curvatures caused by gravitational stimulation, increases growth, and, therefore, with the smaller volume of test solution necessary, greatly increases the sensitivity of the test. The growth response of these cylinders (increase in growth above those in water) is found to be directly proportional, not to the concentration, but to the logarithm of the concentration of the applied hormone. This means that it can be used for estimations over a very wide range of concentrations but that its accuracy over a restricted range is much less than that of the *Avena* test of Went (see Fig. 3).

Even with small volumes of liquid the smallest quantity of auxin (indol-3yl-acetic acid) that the test can detect is of the order of 2 to $5 \times 10^{-3}\ \mu g$*, which compares unfavourably with the sensitivity of Went's curvature test ($10^{-3}\ \mu g$). Modifications of the test, aimed at reducing the endogenous auxin content and hence increasing sensitivity, have included starvation by soaking in water for several hours before the test. Particularly effective pre-treatment seems to be a dilute solution (1 ppm) of manganese sulphate (Nitsch & Nitsch, 1956), which may promote the destruction of the native auxin (see Chap. XI). It is interesting that ferrous sulphate (5×10^{-3} M) in the agar about doubles the sensitivity of Went's curvature test (Shibaoka & Yamaki, 1959).

Some workers prefer wheat (*Triticum*) to oat (*Avena*) coleoptiles as unlike those of *Avena*, they not require exposure to red light at an early stage of germination, to inhibit the growth of the first internode and promote that of the coleoptile.

On the other hand the first internode (mesocotyl) of the oat seedling is much more sensitive to auxins than are coleoptile segments (Nitsch & Nitsch, 1956). If germination is carried out in complete darkness the first internode grows while the coleoptile remains short. The material is harvested when the first internode (mesocotyl) is about 2.5 cm long (coleoptile about 0.5 cm) and 4 mm segments are cut from just below the junction with the coleoptile. Thence assay proceeds as for coleoptile segments (see Fig. 3). A measurable response of mesocotyl segments is obtained with concentrations as low as 2.0 $\mu g/1$, which corresponds to a total quantity of $10^{-3}\ \mu g$ of indol-3yl-acetic acid in 0.5 ml of test solution. This is as sensitive as the *Avena* curvature test.

*μg is the symbol for a microgramme, i.e. 10^{-6} gramme.

It has been widely used in paper partition chromatography assays (see later, pp. 41–45) and is now one of the popular auxin assays. One slight disadvantage of this test is that all operations have to be performed in very dim green light.

(D) CURVATURE TESTS WITH SPLIT ORGANS

Another test, much favoured for estimations of the activity of synthetic compounds, is also due to Went (1934): it is the so-called 'split pea curvature test'. The material used consists of portions of the young stem of the garden pea (*Pisum sativum*) which have been growing in the dark for about a week. The peas should be grown in a room where there is no risk of contamination from coal gas, as the compound ethylene (see Chap. VIII), present in small quantities in that gas, can cause considerable contortions of the young stem. Stems grown in this way are long and colourless, have small leaves, and are said to be *etiolated*. The apical 5 mm of such a stem are removed and the stem split lengthwise for a distance of 3 cm by one strictly median cut made with a razor blade. This split portion is then removed from the main stem by severing it a few millimetres below the slit and it is then washed for a period of an hour or so in glass-distilled water.* After this treatment the split stems have the appearance seen in Fig. 3, the split portions bending outwards in response to a release of tensions present in normal plant tissues. If a segment is now placed in a dilute solution of an auxin, the cut portions of the stem will respond by bending inwards at its tips (see Fig. 3 and Plate 5), owing to the differential stimulation of the growth of the inside and outside of the stem, the cut inside surface growing slower than the intact outside.

This difference in response is partly inherent in the two tissues and partly owing to an inhibition of the growth response of the inner tissue by wounding. Maximum response is attained in about six hours. This curvature differs from that in the *Avena* test in being due to a different response of two tissues to the same hormone concentration, not of the same tissue to different concentrations as in the coleoptile curvature test. The simplest test procedure is to set up in shallow dishes a range of dilutions of the growth substances to be tested, and to place a number of identical stem segments in each solution. From the greatest dilution which gives an observable response, an estimate of activity can be made with an accuracy of about 50 percent. With careful

*Water distilled in an all-glass still must be used throughout the test as it has been shown that this material is very sensitive to traces of heavy metals present in ordinary distilled water and may completely fail to respond in this latter medium, which always contains traces of copper. Traces of cobalt however may enhance the response (Thimann, 1956).

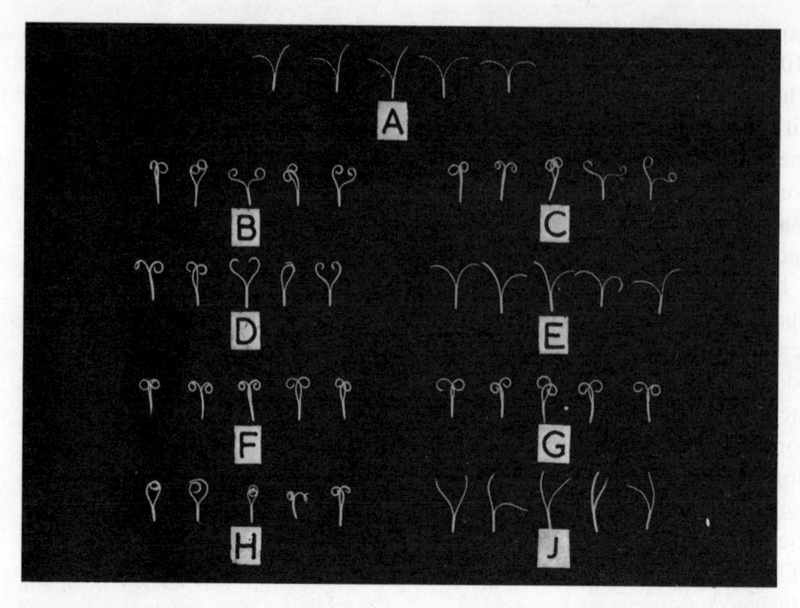

Plate 5. The Went's split pea stem curvature test in the comparison of auxin activities of synthetic chemicals. The substances tested here were all at the same concentration:

A. Control in water
B. 4-chlorophenoxyacetic acid (active)
C. α-(4-chlorophenoxy)-propionic acid (active)
D. γ-(4-chlorophenoxy)-butyric acid (active)
E. α-(4-chlorophenoxy)-isobutyric acid (inactive)
F. 2,4-dichlorophenoxyacetic acid (active)
G. α-(2,4-dichlorophenoxy)-propionic acid (active)
H. γ-(2,4-dichlorophenoxy)-*n*-butyric acid (active)
J. α-(2,4-dichlorophenoxy)-isobutyric acid (inactive)

(Photograph by kind permission of Prof. R. L. Wain, Wye College, University of London).

handling, however, the method is capable of somewhat more accurate results for assay of material of known specific activity, and as a measure of the absolute response, the angle of inward curvature can be determined from shadowgraphs. This is usually measured, as illustrated in Fig. 3, between the tangents at the extreme tips of the split halves and the main axis of the segment. Calibration experiments show that within certain concentration limits the curvature is directly proportional to the logarithm of the concentration. The sensitivity of this stem material to applied auxins can be greatly increased (to about ten times the curvature for the same applied concentration) by exposing the seedling for a short period (4 hours) to very weak red light (100 lx) about 1½ days before harvesting for an assay (Kent & Gortner, 1951). This illustrates the extreme care necessary to ensure uniform growth conditions if reproducible assay material is to be obtained. In quantitative assays, therefore, calibration curves of response against concentration must first be constructed using the pure substances. Variation in response is much greater than in the methods previously described and causes a much smaller overall accuracy. The method involves, like the straight growth test with coleoptiles, a large bulk of solution and therefore is not suitable for work with small quantities of extracts of natural hormones. The fact, however, that it is insensitive to temperature and diffuse light over the ranges normally encountered in a laboratory has made it a popular test where rapid routine qualitative methods are sufficient. A modification of this method, convenient when comparing the activities of different auxins, is to determine the concentrations which will reduce the normal outward curvature of the split stems by a convenient angle (usually 100°). The molecular activities of the auxins are then inversely proportional to these standard activity concentrations (Thimann, 1952). Other plant material such as split dandelion (*Taraxacum officinale*) flower stalks (Jost & Reiss, 1936) and split *Avena* coleoptile segments (Thimann & Schneider, 1938*b*), have been used with some success.

(E) ROOT GROWTH TESTS

Another assay method makes use of the effect of auxins on elongation growth in roots. Concentrations of auxins which stimulate the growth of coleoptiles greatly inhibit that of roots, but at very low concentrations (about 1 part in 10^{11}) a slight stimulation of growth is sometimes detectable. A number of small-seeded species provide seedling roots suitable for this assay, e.g. cress (*Lepidium sativum*), flax (*Linum usitatissimum*), lentil (*Lens culinaris*), wheat (*Triticum vulgare*), *Artemisia absinthium*, etc. By growing selected seedlings for short periods in a range of dilutions of indol-3yl-acetic acid solutions, a curve can be constructed relating growth to the concentration of the hormone. This curve should cover a very wide range of concentrations from optimum growth at 1 part in 10^{11} to complete inhibition in about 1 part in 10^{6}. (See Fig. 3.) In the assay itself, similar dilutions of the solution being tested are set up and root growth is followed in them. A curve of growth-rate against dilution is then constructed, and by comparing this with the standard curve for indol-3yl-acetic acid the auxin content of the test solution (in terms of indol-3yl-acetic acid equivalents) can be calculated. The exact technique employed varies with the quantities of materials available and the accuracy required (Moewus, 1948; Audus, 1949*a*, 1951). Tests involving root growth responses have two advantages over the other tests described. Firstly, stimulations of growth can be obtained at concentrations as low as one part in 10^{11} or 10^{12}; furthermore, by using small segments cut from just behind the tip, very small quantities of solution can be used (Audus & Thresh, 1953; Leopold & Guernsey, 1953*c*) and so the test can be made extremely sensitive, it being possible to detect quantities of indol-3yl-acetic acid of the order of 10^{-5} μg. At this level, however, accuracy is very low. The second advantage is that only the temperature needs to be maintained constant, the test being unaffected by the low light intensities of the laboratory. It has one great disadvantage: it is by no means a specific test for auxins, since compounds which have no effects on coleoptile or stem growth may either stimulate or inhibit growth of roots (see also p. 183–4). This of course is of no consequence when one is dealing with a known substance.

(F) EPINASTY TESTS

A convenient and popular test for auxin activity depends on the movement of leaves of certain test plants (e.g. tomato) in response to application to the surface of the stem or leaf blade. The suspected auxin is usually dissolved in lanolin and the paste is then applied in small drops to the leaf blade. The compound, if active, will be absorbed into the plant, will pass to the leaf stalks, and will induce a growth-response causing the leaves to bend downwards. This phenomenon is known as *epinasty* (see Plate 6). The test is of little value from a quantitative point of view. Apart from this test and the 'pastentest' of Linser (see p. 30) intact plants have been little used for auxin assays as they are relatively insensitive—presumably because their internal supplies of auxin are optimal for growth.

(G) GENERAL CONSIDERATIONS

A weakness inherent in all these biological tests, particularly when they are used to compare the activities of synthetic compounds, is that the relative activity of any one substance may vary from test to test, e.g. some synthetic compounds are extremely effective in the Went pea test but are almost inactive in the *Avena* coleoptile curvature test. The reasons for this will be discussed later (Chapter III). In spite of the vastly increased labour involved, more critical investigators use two or more different biological tests in their researches.

Tests so far described have all been concerned with the determination of auxin activity in controlling the longitudinal growth of a plant organ, i.e.

Plate 6. Epinastic responses of tomato plants to 2,4-D and ethrel (see p. 212).
Top. 2,4-D effects. Left: Control, untreated plant. Right: Plants sprayed to run-off with 5 ppm 2,4-D solution containing a wetting agent.
Bottom. Ethrel effects. Left: Control, untreated plant. Right: Plants sprayed to run-off with 0.05 percent ethrel solution.
(All photographs taken three days after treatment.)

in the phase of stretching growth of a cell-wall. However auxins exert other and varied effects on growth processes, e.g. the initiation of roots, the development of fruit, etc. It has therefore been found necessary in investigations of these phenomena to devise special tests relating to these properties, but these will not be described in this volume. Since the gibberellins are virtually inactive in the above auxin tests, special assay methods have had to be devised for them. These are described later on pages 108–113.

Already in this chapter the extreme sensitivity of these tests has been pointed out. The range of sensitivities in the main tests are brought together for comparison in Table I.

TABLE I

SENSITIVITIES OF AUXIN ASSAYS TO INDOL-3YL-ACETIC ACID
(Adapted from Larsen 1961)

Test	Lowest detectable concentration μg/l	Lowest detectable amount μg	Concentration range for quantitative determination μg/l
Avena curvature	4–7	10^{-3}	5–100
Avena coleoptile segment	3–5	3×10^{-3}	3–1000
Avena first internode segment	1	5×10^{-4}	1–1000
Split pea epicotyl curvature	30–50	0.5	30–5000
Root extension (intact)	10^{-3}	5×10^{-6}	10^{-3}–1000
Root extension (segments)	10^{-3}	10^{-6}	10^{-3}–1000

The *Avena* curvature test can detect quantities of auxins of the order of 3/1000 μg, and the root test can detect quantities of 1/1,000,000 μg. The most sensitive *chemical* tests for indol-3yl-acetic acid will detect quantities of the order of 0.5 μg (Mavrodineanu *et al.*, 1955). The *Avena* test is thus about 200 times more sensitive, and the root test half a million times more sensitive, than this most delicate chemical test. There are, however, occasions when quantitative chemical tests of this sort are to be preferred to biological assay methods on account of their relative simplicity and greater accuracy; for example, when high concentrations of pure substances are being studied. Details of such tests will be given in a later section.

THE ISOLATION AND CHEMICAL IDENTIFICATION OF THE AUXINS

(A) EARLY HISTORY

The first attempt to isolate the growth hormones, in sufficient quantity to allow a chemical analysis of their structure, was not, as might have been expected, carried out on oat coleoptiles or even plant tissue, where they occur in much too small quantities. Fortunately, biochemists found other materials which were far richer in substances active in the *Avena* test, and attention was therefore concentrated on these. Firstly, in 1930, Nielsen, working in Boysen-Jensen's laboratory in Copenhagen, isolated a highly active syrup from the culture medium in which certain moulds (*Rhizopus suinus* was the most important) had been growing. No further attempts were made to purify this syrup, but the active principle was shown later to be a weak acid of approximately the same strength as acetic acid (Dolk & Thimann, 1932). At about the same time, in the laboratories of the University of Utrecht, Kögl & Haagen Smit (1931) discovered that human urine was a particularly active source and therefore chose it as their experimental material. Starting with 150 litres (about 33 gallons) of urine, they concentrated it and removed inactive materials by a series of empirical steps, testing each time for activity by the *Avena* test. The final stages of purification consisted of distillation in high vacuum, giving a final yield of 40 mg (equal to about 1/700th ounce) of crystalline product. From analysis of crystals of this kind, a chemical structure for the active compound was proposed and the substance was given the name of *auxin A*.

Understandably, after this success, these workers turned their attentions to plant sources and found that certain cereal products, in particular germ oil (from maize) and malt (from barley), showed very high auxin activities. A similar series of purifications of these products yielded auxin A and a new compound of similar structure and almost identical activity which was called *auxin B*. The formulae proposed for these substances were completely new to chemical science. It was fortunate however for the future of the auxins that these isolations did not end the study of urine. Curiosity as to the nature of the persistant auxin activity in urine residues after auxin A extraction led ultimately to the isolation of a third active substance which was christened *heteroauxin* ('other auxin') (Kögl *et al.*, 1934). This substance was shown to be the compound indol-3yl-acetic acid (Fig. 5 I), which had long been known to the chemists but had never before been suspected of having any biological functions. It had first been synthesized in Germany in 1904 (Ellinger, 1905). The Dutch workers had shown that this compound could also be obtained from yeast, and in America it was demonstrated conclusively (Thimann, 1935*a*) that it was also identical with the active substance that Nielsen had obtained from the fungus *Rhizopus suinus*. For convenience in this book this auxin will be called by the widely-accepted abbreviation of its chemical name, viz. IAA.

For the following decade or so, before the advent of modern elegant techniques for the separation of small amounts of closely related chemical compounds, plant physiologists spent much time and energy trying to decide which of these three compounds was the natural primary effector of plant growth. The methods employed and the final outcome will be described later in this chapter but we should now turn our attention to the methods employed today for the extraction and chemical study of the natural auxins.

(B) TECHNIQUES FOR EXTRACTION, PURIFICATION, AND IDENTIFICATION

(i) Extraction

The complete chemical characterization of unknown organic compounds, even with modern micro-methods, requires milligramme quantities of pure crystalline material. Even the approximate identification by chromatography coupled with colorimetric techniques (see later) requires a microgramme or so. It is not surprising therefore that the extremely small amounts of auxin that can be obtained by the classical techniques of diffusion into agar from the cut surfaces of plant parts have encouraged the use of easier, more productive solvent extraction methods. However, many physiologists hold that only the freely diffusible auxin is significant physiologically in the control of growth, and this has prompted a return to this method of collection with some success, although a considerable number of individual diffusions have to be employed (Kuraishi & Muir, 1964*b*; Scott & Jacobs, 1964).

Because it is supposed to remove only the 'free' active auxin, water has been used as an extractant in the past. It suffers however from the disadvantage that it allows enzymatic processes to continue and by destruction or production from a precursor, these may alter the auxin content of the tissue preparation. This can be greatly reduced by lowering the temperature to 0–4°C (Terpstra, 1953). But it has the additional disadvantage in extracting unwanted materials and consequently several highly volatile organic solvents are currently regarded as more suitable. The most popular are methanol, ethanol, diethylether, chloroform and ethyl acetate. For details of their respective virtues the reader is referred to articles by Bentley (1961*a*), Larsen (1955*b*) and Pilet (1961).

Tissues may be pulped and extracted fresh but even if this is done at low temperatures enzymatic processes may still go on and affect the free auxin levels. These may be prevented by first plunging the tissue for a short time into boiling water to destroy these enzymes although there is then the risk of destroying labile auxins also. Freeze-drying of tissues before extraction has the double advantage that it reduces enzyme action and in addition allows long pre-extraction storage of plant material where necessary without deterioration.

(ii) Fractionation

In his pioneering studies Kögl purified his crude urine extractions by partition between several immiscible solvents, e.g. ether/water, benzene/water etc., the impurities in one or other solvent being discarded at each step. One of these procedures, the extraction of acid auxins from an ether solution by shaking with sodium bicarbonate solution, has since become established as a standard initial step in the fractionation of auxin extracts. In this way the acids (e.g. IAA) are separated from neutral or basic compounds and can be separately studied. This procedure can result in losses of active material which are undesirable when quantitative estimations are involved so some attention has been given to alternative methods. Promise has been shown by ion-exchange techniques (Burnett *et al.*, 1965) in which the extract is allowed to percolate down a uniformly packed column of a positively charged (basic) powder (e.g. a derivative of cellulose) which retains the negatively charged ions (anions) of the acid auxins, allowing basic and neutral compounds to pass through. This separation is effected with little loss and the acid auxins can be subsequently removed from the column by washing with a dilute salt solution.

But this is only a first-stage process and it is usually necessary to subject the two fractions thereby obtained to further fractionation. The most valuable technique so far employed was one introduced to plant hormone studies by Bennet-Clark *et al.*, in 1952; this is chromatography. This technique hinges on the capacity of certain porous materials to take up substances to be separated from solution in a fluid passing over them. These substances are concentrated on the surface of, or in solution in, the material used, and the tenacity of retention varies from compound to compound for any given material and any given solvent fluid. This property of differential retention can be used to separate the compounds concerned. Thus if a mixture is washed slowly down a uniformly packed column of powder (e.g. aluminium oxide, icing sugar etc.) with a suitable solvent mixture, the compounds will be held back in the column to an extent depending on the relative attraction of the powder particles for them, those attracted most strongly being held back at the top of the column and those weakly attracted passing down to lower levels. With continuing passage of a suitable solvent mixture, which may be of constant or continuously changing constitution, the components of the mixture may be collected in sequence in successive fractions of the solvent emerging from the lower end of the column. The ion exchange method for the separation of acid auxins employs the same principles. In this way it is possible to separate closely similar chemical compounds that are separable with extreme difficulty by other methods.

Chromatography can be of various kinds. The pioneering work of Bennet-Clark involved paper partition chromatography in which a strip of filter paper is used as the particulate solid phase. The principle of operation of paper partition chromatography is that the substances to be separated partition themselves in solution between the running solvent and the water held in the cellulose fibres of the filter paper according to their relative solubilities in the two solvents. A very small drop of concentrated extract is placed at one end of a narrow strip of a suitable paper and then an appropriate solvent mixture is allowed to flow down the paper from that end for a distance of 20 cm or so. It is then removed from the container used for the process, dried in air and, if the solvents used have left no toxic residues, the paper strip can be cut into uniform transverse segments and the auxins present on each segment can be determined by suitable assay. It is usual to extract the segments with a solvent, which may be water, although this step may be eliminated by allowing the plant material to grow in contact with the water-moistened segment. The rate of movement of auxins down the paper can be determined and compared with the behaviour of known compounds in a series of different solvent mixtures, each of which produces its own characteristic pattern of auxin movement. In this way we can get a fair idea of the chemical nature of the auxins present, although unequivocal identification is certainly not possible. The basic principles of this technique are illustrated in Fig. 4. The behaviour patterns of various auxins on paper chromatograms are recorded in Table II.

Although paper partition chromatography has so far been the most popular technique adopted for auxin fractionation, yet powder column adsorption chromatography has also proved useful, particularly when relatively large amounts of material are involved. Here separation depends on the relative degrees of adsorption of the components on to the solid particles of the columns. Thus Al_20_3 columns were first successfully used by Linser (1951). Cellulose powder or floc columns or even paper rolls have been used for bulk separations. Silica gel columns provide a quick and efficient second-stage process for the separation of components of acidic, basic and neutral fractions of extracts of auxins prior to more specific identifications by chemical and physico-chemical techniques (Powell, 1964).

Two other chromatographic methods that have recently been adopted for auxins are thin-layer chromatography (TLC) (Stahl & Kaldewy, 1961; Kaldewey, & Stahl, 1964) and gas/liquid chromatography (GLC) (Stowe & Schilke, 1964). In the former the solid phase is a uniformly-grained powder (e.g. silica gel) spread in a very thin layer (e.g. 0.25mm) on the surface of a

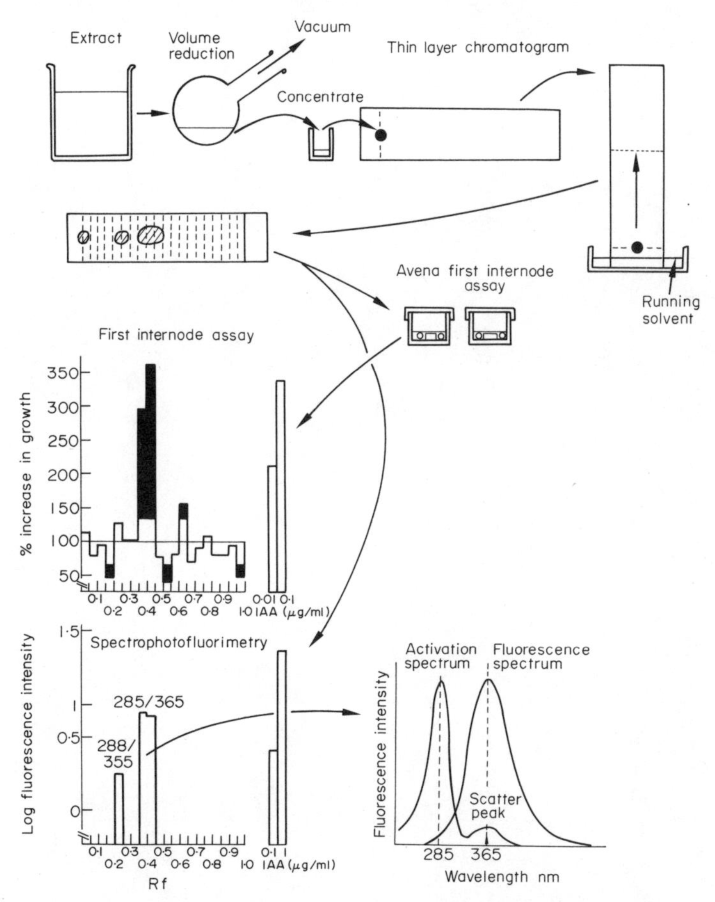

Fig. 4. Diagrams and graphs showing the methods and results of the chromatographic separation and assay of auxins. After extraction of a tissue in a solvent (e.g. ethyl ether) and reduction of volume under vacuum at near room temperature, the concentrate is spotted on a thin-layer chromatogram plate and run in a suitable solvent mixture. Subsequent to drying, the powder is scraped off the plate in transverse bands and each eluted in a suitable buffer, in which the plant material (*Avena* first internode segments) is incubated. A histogram of the response to each strip is plotted against Rf value, the fiducial limits of the assay being determined from an appropriate number of control samples of segments in buffer alone. In the illustration shown, of extracts from pea shoots (Nešković & Culafić, 1968), the black areas are for responses above (promotion) or below (inhibition) the 5 percent limit of probability. Responses to IAA standards are shown on the right-hand side. The fluorescence intensity of comparable aliquots of the eluates (with the wavelengths of the activation and fluorescent maxima noted in nm) are also plotted against Rf. Details of the activation and fluorescent spectra for the 'spot' at Rf 0.35 to 0.45 (characteristic of IAA) are shown on the right of the fluorescence chromatograms. (Fluorescence data also from Nešković & Culafić, 1968.)

TABLE II

RF VALUES OF A NUMBER OF INDOLE AUXINS ON PAPER CHROMATOGRAMS RUN WITH A NUMBER OF SOLVENTS

	Phenol-water	Butanol-propionic acid-water	Isopropanol-ammonia-water	Butanol-ethanol-water	Butanol-ethanol-ammonia	70 percent ethanol	Water	Pyridine-ammonia
Indol-3yl-acetic acid	0.8	0.95	0.25–0.35	0.7	0.8	0.8	0.9	0.6
Indol-3yl-carboxylic acid			0.2	0.8		0.8	0.9	0.6
Indol-3yl-acetaldehyde						0.45		
Ethyl indol-3yl-acetate	0.95	1.0	0.8–0.95	0.9	1.0	0.8	0.6	0.97
Tryptophan			0.2	0.3		0.4	0.6	0.45
Tryptamine			0.65–0.75	0.4	0.95	0.7	0.3	0.9
Indol-3yl-acetonitrile		0.95	0.75	0.95	1.0	0.85	0.4	0.95
Indol-3yl-pyruvic acid			0.1					
Indol-3yl-propionic acid	0.8	0.95	0.35–0.45	0.8	0.85	0.9	0.85	0.6
Indol-3yl-butyric acid	0.85	0.98	0.45–0.55	0.9	0.85	0.85	0.9	0.6

strip of glass. Mixtures of auxins are 'spotted' at one end and solvent is allowed to run as in paper partition chromatography (see Fig. 4). The advantages of this technique over paper partition chromatography are that it is much quicker, gives rather better separation of closely related compounds and is more convenient for the removal of the component 'spots' for further assay or analysis. In gas/liquid chromatography the moving phase is an inert gas (e.g. nitrogen, helium etc.) containing the vapours of the compound to be separated. The stationary phase is either a powder such as silica gel or a liquid of low volatility coating a suitable inert powder support. These are packed into narrow tubes down which the gas and vapour pass and the components of the vapour mixture, issuing in characteristic sequence at the other end, are detected by means of one or other physical property of the vapour. The complication of this latter method for auxins is that the substances to be analyzed have to be volatile without decomposition at the operating temperatures (e.g. 200–300°C). Thus acid indole auxins must be converted into their methyl or ethyl esters. But this method can be made extremely sensitive and quantities of IAA of the order of 2×10^{-3} μg can be detected (Stowe & Schilke, 1964; Grunwald *et al.*, 1967) through its use.

Another very useful technique for the separation of small quantities of auxins in mixtures is paper electrophoresis in which separation on paper strips soaked in appropriate buffer solutions is induced by an electrostatic potential difference. First used by von Denffer *et al.*, in 1952, it has now become standard practice with some workers and can be used for the separation of auxins which are resolved with difficulty by paper chromatography (Khalifah *et al.*, 1964).

(iii) Minimizing Losses

At all stages of extraction, purification and assay, a proportion of any labile auxin present may be lost and accurate quantitative estimates of tissue content become impossible. Not enough attention has been given to this problem in the past although some attempts to minimize losses have been made from time to time. Recent work on IAA extractions indicate the advantages of extraction under nitrogen to minimize oxidation, concentration at atmospheric pressure to reduce loss by vaporization, and the use of an antioxidant (1,2-dihydro-6-ethoxy-2,2,4-trimethylquinoline) during paper chromatography, similarly to reduce oxidation (Mann & Jaworski, 1970).

(iv) Identification – Chemical Tests

With the small amounts of auxins present in many extracts of plant materials, the only tests usually applicable are the very sensitive but, by chemical standards, inaccurate and extremely non-specific biological assays. However, concentrations of plant indoles in extracts do occasionally reach levels allowing determinations by chemical colour reactions and, in any case, biochemical studies with synthetic indole auxins demand the use of such specific and accurate tests. Some of the most important of these are discussed below.

Two of the most sensitive quantitative tests so far devized were first described by Mitchell & Brunstetter (1939), namely:

Nitrite Test. This is suitable for a concentration range of IAA from 0.01 to 0.15 mg per ml. To 50 ml of the acid solution are added 0.5 ml of 5 percent gum-arabic solution, 2.0 ml of a 0.5 percent solution of potassium nitrite, and 0.4 ml of concentrated nitric acid – in that order. The mixture is shaken and left to stand for 2 hours at room temperature when a red colour will develop if the auxin is present, the intensity of which can be estimated in a colorimeter against a standard IAA solution.

Ferric Chloride Test (*Salkowski reaction*). This is suitable for a concentration range of 0.02 to 0.1 mg per ml and can usually detect total quantities of the acid of the order of 2 μg. To 10 ml of syrupy sulphuric acid add 0.5 ml of M/10 ferric chloride. Then add 14.5 ml of distilled water, shake, and cool to room temperature. Next add 5 ml of the aqueous IAA acid solution. Shake again and cool to room temperature, when a red colour will develop if the auxin is present. This can be similarly estimated in a colorimeter.

A slight modification of this method (Tang & Bonner, 1947) is claimed to give a more accurate quantitative measure over the range 5 to 100 μg of IAA. Two ml of the aqueous solution are added to 8 ml of the reagent which has the following constitution: 15 ml of 0.5M.$FeCl_3$, 300 ml of H_2SO_4, sp. gr. 1.84, 500 ml of distilled water. After mixing, the colour is allowed to develop for a standard time of 30 minutes, as it fades slowly after the maximum has been reached.

These tests give various purplish and blue colours with several indole compounds and are not specific for IAA, although a modification of Tang & Bonner's ferric chloride reagent has been recently described and claimed to give a much better colour development and to be much more specific for IAA than for other indole compounds (Gordon & Weber, 1951). In this modification, perchloric acid is used in place of sulphuric acid. The solution can be sprayed on paper chromatograms when the development of a spot of colour demonstrates the presence of an indole compound if present in sufficient amounts. Since these solutions rapidly cause paper to disintegrate it is better to use a modification due to Jepson (1958) in which the paper is momentarily dipped into a 20 per cent solution of HCl in acetone to which a few drops of a saturated solution of potassium perchlorate has been added. Indole colour reactions appear on drying.

Ehrlich Reaction Test. In the Ehrlich reaction *p*-dimethylaminobenzaldehyde and HCl give a red-violet colour. It is useful for identification on paper chromatograms by dipping in a solution of 1 volume of 10 percent *p*-dimethylaminobenzaldehyde in concentrated HCl to 4 volumes of acetone (Jepson, 1958). This test will detect 1 μg of IAA on a chromatogram. A range of purple and blue colours are given with indoles although it is not completely specific for them. A test, similar to but much more sensitive than the original Ehrlich test, employs *p*-dimethylaminocinnamaldehyde made up as a one percent solution in a 50:50 mixture of ethanol and ^{6}N HCl (Harley-Mason & Archer, 1958). This, applied as a spray, gives a blue reaction in the presence of indoles and is claimed to detect as little as 0.1 μg IAA.

Other Tests. A dozen or so additional colour reaction tests have been used from time to time to detect indoles. The reader should refer to Bentley (1961*b*) for details.

(v) Identification – Use of physico-chemical properties.

It has already been seen that one aid to the identification of auxins is their relative rates of movement on paper chromatograms, a behaviour dependant on the physico-chemical properties of the molecule. The Rf values, i.e. the ratios of the distance moved by the molecules concerned to the distance moved by the solvent front, are known for a number of indole compounds in a number of solvents and extensive tables have been published by Sen & Leopold (1954), Stowe & Thimann (1954), Larsen (1955*b*), and Bentley (1961*a*). Table II records the approximate Rf values for some of the important indole compounds on paper chromatograms.

Absorption spectroscopy is of great value in the identification of pure compounds in solution since the wavelengths at which light is absorbed are characteristic of the compound concerned; each compound has its own characteristic absorption spectrum. In the case of the indole auxins there are no absorption bands in the visible spectrum and the main absorption peak occurs in the ultra-violet region. Measurement of the absorption of ultra-violet light can be carried out in dilute solutions in suitable solvents but needs relatively large quantities of material (e.g. 25 μg/ml). Ultra-violet spectrograms of purified fractions of extracts have been frequently used to help in the identification of auxins (Post, 1959; Vendrig, 1961; see also Pilet, 1961). However the broad, simple absorption bands are not very discriminating and much more positive identification is possible with infra-red absorption spectroscopy, where spectra are highly complex and highly specific. But here milligramme quantities of pure crystals are required. Nevertheless such spectrograms have been obtained in certain favourable cases (Post, 1959).

A method with an intermediate specificity but with a sensitivity equalling that of most biological assays involves the use of fluorometry. A great proportion of known organic compounds, after absorbing light at one wavelength, may re-emit it at a longer wavelength; most indole compounds show this phenomenon of fluorescence. The wavelengths most effective in exciting the compounds to fluoresce, and the wavelengths of the emitted light, are characteristic of the compound and can be accurately determined in an instrument called a spectrophotofluorometer. These characteristics are a considerable help in the identification of indole compounds in plant extracts (Burnett & Audus, 1964; Burnett *et al.*, 1965; Stowe & Schilke, 1964) (see Fig. 4). Changes in the fluorescence pattern with pH are a further guide to identity (Burnett & Audus, 1964). The intense fluorescence of most of the indole compounds means that the technique can be used to estimate very small concentrations of indoles in solution, providing no interfering compounds are present, since fluorescence intensity is proportional to concentration. Quantities of IAA of the order of 10^{-3} μg/ml can be measured by this method.

Fluorescence in the ultra-violet has long been used to detect compounds on chromatogram paper. By the use of appropriate filters and a photomultiplier tube, quantities of IAA of the order of 0.5 μg can be determined (Mavrodineanu *et al.*, 1955). This method has the great advantage that no chemical colour-developer is needed and the indole auxin is therefore not destroyed during the estimation.

The coloured products of the ferric chloride, nitrite and Ehrlich tests also fluoresce. For example indol-3yl-acetonitrile gives a characteristic violet product in the ferric chloride reaction and this product shows an intense light yellow-green fluorescence in ultra-violet light. These two properties taken together are highly specific to the nitrile (Linser & Kiermayer, 1956).

The most precise method available now is that of mass spectrometry; in this chemically most refined of techniques very small quantities (i.e. 1 μg) of an organic compound can be disintegrated by ion bombardment in a vacuum, the resulting charged molecular fragments having a wide range of different yet characteristic weights. A beam of these charged fragments, directed through a magnetic field will be 'analysed', the particles being deflected to different extents depending on their mass/charge ratios. The pattern of fragmentation is extremely characteristic of the compound and yields a kind of 'fingerprint' allowing precise identification of unknown substances (see examples for gibberellins in Fig. 23). Very small amounts of naturally occurring indoles, separated and purified by the methods already described, can be injected directly into the ionization chamber of the mass spectrometer and their 'fingerprints' obtained. They can then be identified by comparing these fingerprints with those of known indole compounds and quantitative determinations subsequently follow. An index of the fingerprints of twenty natural indoles has already been made (Jamieson & Hutzinger, 1970) but the application of the method to *in vivo* problems is still in its infancy.

The Nature and Origins of the Natural Auxins

(A) THE MYSTERY OF AUXINS A AND B

It has been seen that the first attempts at isolating the auxins in sufficient quantity and purity to allow investigations into their chemical structure resulted in the preparation of crystals purporting to be three distinct chemical substances, auxin A (auxentriolic acid) from urine, auxin B (auxenolonic acid) from malt and maize germ oil, and heteroauxin (indol-3yl-acetic acid) from urine. According to the formulae proposed, the first two substances were previously unknown to chemical science.

The third compound, indol-3yl-acetic acid (IAA) (Fig. 5 I), has quite a different structure and has long been known to the organic chemist, since it was first synthesized by Ellinger in 1904. It was also known to the biologist, since Hopkins & Cole in 1903 had shown that it was produced from the amino acid, tryptophan, by the activity of bacteria (*Escherichia communis*). Its presence in a higher plant (*Celtis reticulosa*) was even suggested as early as 1909 (Herter, 1909), although at that time there was no clue as to its physiological action.

After the first isolations of auxin A and heteroauxin had been made from urine, much attention was given in the early days to the identity of the naturally occurring auxins of plants. Owing to the extremely small quantities obtainable from the tissues studied (e.g. coleoptiles), direct chemical tests could not have been applied even if they had been available, as their sensitivities would have been far too low. Recourse was made therefore to indirect identification methods, involving (*a*) the acid-alkali stability relationships, and (*b*) determination of the diffusion constants in agar. Kögl had claimed that both auxins A and B were destroyed by hot alkali, whereas IAA was stable under the same conditions. On the other hand, hot acid decomposed IAA and auxin B, but not auxin A and early results of the application of this differentiation technique supported the view that auxin A was the hormone of the oat coleoptile and other seedling shoots (Went & Thimann, 1937). Subsequent attempts could not confirm them but these ideas of the key nature of auxins A and B persisted until after the second World War when the reliability of the method came under suspicion. Its uselessness was confirmed by the demonstration that a naturally-occurring indole compound

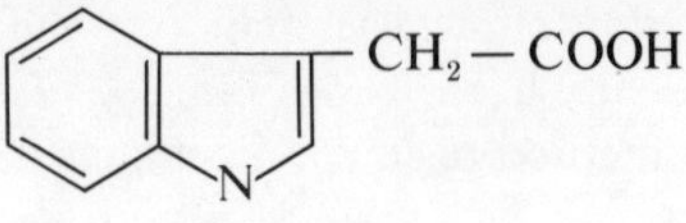

CH_2-COOH

I

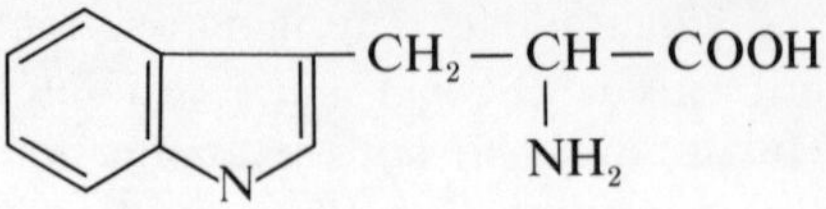

$CH_2-CH(NH_2)-COOH$

II

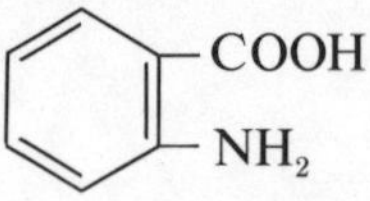

$COOH$, NH_2

III

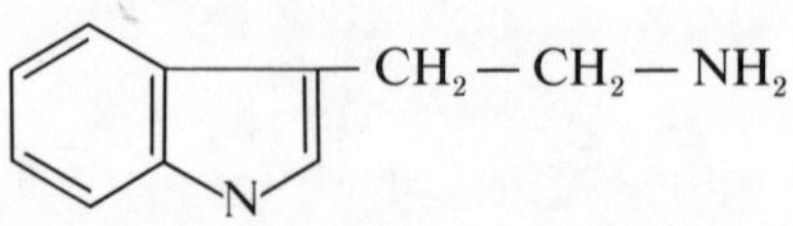

$CH_2-CH_2-NH_2$

IV

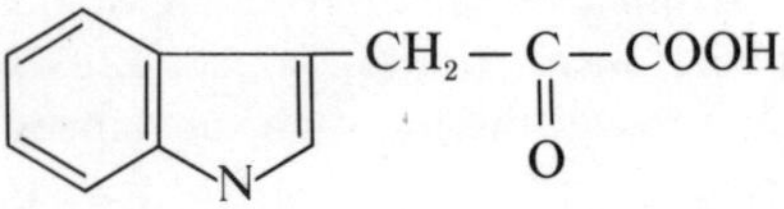

$CH_2-C(=O)-COOH$

V

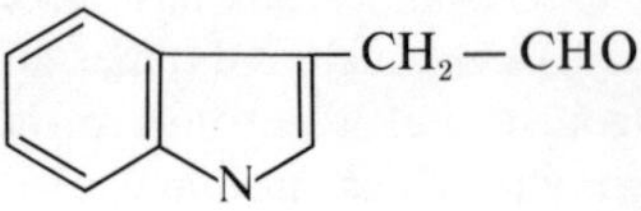

CH_2-CHO

VI

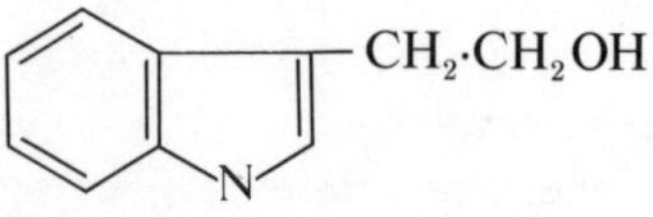

$CH_2{\cdot}CH_2OH$

VII

$CH_2-CH(OH)-COOH$

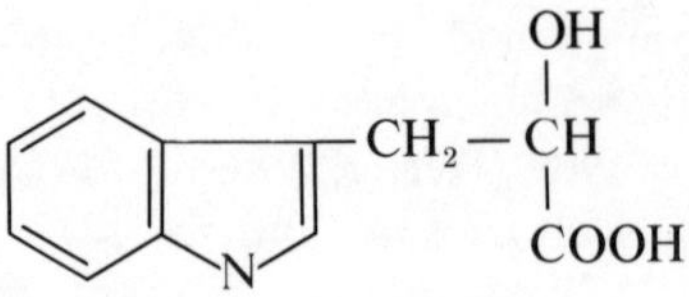

VIII

Fig. 5. The natural auxins and related compounds.

I	Indol-3yl-acetic acid (IAA)
II	Tryptophan α-amino-β-(indol-3yl)-propionic acid
III	Anthranilic acid
IV	Tryptamine β-(indol-3yl)-ethylamine
V	Indol-3yl-pyruvic acid (IPyA)
VI	Indol-3yl-acetaldehyde (IAAl)
VII	Tryptophol β-(indol-3yl)-ethanol
VIII	Indol-3yl-lactic acid

(indol-3yl-acetonitrile; Fig 6 IX), which can be detected by the *Avena* assays, is stable to hot acid and is converted into IAA by hot alkali (Henbest *et al.*, 1953), thus giving the appearance of an acid- and alkali-stable auxin.

Measurements of diffusion constants have been obtained by allowing diffusion to take place through a series of identical rectangular agar blocks arranged linearly and in close contact, followed by assay of the individual blocks. From the distribution curve thus obtained, the value of the diffusion constant can be calculated and thereby the molecular weight. Early observations on the auxin from the oat coleoptile (Went, 1928) and on pure auxin A (Kögl *et al.*, 1934) gave a molecular weight of 376, agreeing reasonably well with that for the proposed formula of auxin A (M = 328) and not with IAA (M = 175).

Later repetition of the early determinations (Wildman & Bonner, 1948) showed that auxin obtained by diffusion from the coleoptile tips into agar and then extracted from the agar with ether gave a quite different diffusion constant which was very close to that of IAA.

In spite of many attempts no subsequent isolations of auxins A and B have been made from plants. Furthermore a synthetic compound with the proposed auxin B formula has virtually no auxin activity (Nakamura *et al.*, 1966). Now they have receded from the botanists view and are mere ghostly mysteries of the past. They died slowly, since less than twenty years ago it was still thought that 'heteroauxin' (IAA) might owe its activity to a conversion to auxin A in plant cells (von Guttenberg *et al.*, 1953). The final death-blow came when modern techniques of chemical analysis (X-ray and mass-spectrometry) were applied to some authentic samples of the original auxins A and B preserved in the organic chemistry laboratories of the University of Utrecht (Vliegenthart & Vliegenthart, 1966). Auxin A was unambiguously identified as cholic acid (Fig. 7 XIX) whereas auxin B proved to be thiosemicarbazide. Neither of these two compounds has any growth promoting activity. The mystery of the activity of the fresh samples still remains. There is therefore little doubt now in the minds of plant physiologists that IAA is the major natural auxin; the epithet 'heteroauxin', once applied to it, is a most unfortunate misnomer and *should be dropped completely*. It was with this compound as a model that the chemists subsequently synthesized the very wide range of compounds of similar structure and growth-promoting activities which have proved to be of such immense value in practical agriculture and horticulture and which we now term the synthetic auxins (see Chap. III).

One last word must be said about the auxins of urine; is it not possible that the auxin A found there might not be a product peculiar to the metabolism of the animal? The first careful analysis of human urine which followed the work of Kögl could detect no auxins other than those of the indole group of compounds. Bennet-Clark and his colleagues (1952), using paper partition chromatography (see pp. 42–44), demonstrated that the greater part of the activity of urine was due either to IAA or to closely related compounds. Similar results were obtained by Wieland *et al.*, (1954), who carefully followed Kögl's purification techniques and analysed the products by paper partition chromatography. Only IAA and its methyl-ester (an artifact of one of the purification steps) could be demonstrated. However, the most recent work of Vendrig (1967*a*) has shown the presence in human urine of sterol-like compounds which have a definite auxin activity. As we shall see later (p. 65) similar compounds occur in plants. The mystery of the urine auxins still remains.

(B) THE OCCURRENCE AND DISTRIBUTION OF IAA

Since the advent of chromatographic methods for the separation and partial identification of auxins, a very wide range of plants and plant organs have been extracted and their auxins studied. From the position of active compounds (identified by bioassay) on chromatograms (Rf values), supplemented under favourable conditions of concentration by specific colour reactions, the presence of IAA has been indicated in a high proportion of these extracts. By 1954 (see review by Gordon) about a dozen species formed this list and included such organs as shoots of garden pea, roots of wheat, fruits of *Asparagus* and maize, and tubers of potato. By 1958 (see review by Bentley) a dozen or so more had been added. The liverwort *Lunularia* was one of them. Since then reports of IAA 'spots' on chromatograms have become so commonplace that few doubt its widespread occurrence throughout the whole of the vegetable kingdom, from the lowly algae and fungi to the largest angiosperm trees. However its ubiquity has been doubted, since a substantial number of investigations have failed to detect it and the list continues to grow. It is surprising that this list includes organs that are growing rapidly and in which high concentrations of IAA might therefore be expected (e.g. coleoptiles, potato sprouts, young black-current berries etc.) (see Bentley, 1958, Table I). Whether IAA is really absent from these tissues or whether deficiencies in technique have prevented its detection, are questions still to be answered. On the other hand it must be admitted that plant physiologists have often been a little optimistic in their acceptance of chromatographic behaviour and indole colour reactions as specific indications of IAA, since final unequivocal identification can come only from the isolation of pure crystals and rigorous chemical tests upon them. This has been done for very few plants indeed, e.g. for the immature grains of maize (Haagen Smit *et al.*, 1946), for cabbage leaves (Post, 1959) etc., probably because of the large amount of plant material required for IAA isolation in sufficient quantities. For example Post worked up a ton of cabbages over a period of three to five weeks to obtain 3 mg of IAA.

(C) THE ORIGIN OF IAA FROM TRYPTOPHAN: THE OCCURRENCE OF INTERMEDIATE COMPOUNDS

Chromatographic and electrophoretic studies have shown that a variety of indole compounds may occur in plant tissue. Some of them have been detected in the first instance by bioassay and there has been much debate as to their physiological significance. Thus we have always had with us the problem of whether activity at a particular region of a chromatogram or an electrophoretogram is due to an auxin in its own right, or to an inactive compound, convertible to an auxin (perhaps IAA) by enzymes in the plant organ used for the assay.

A close chemical relative of IAA is the amino-acid tryptophan (Fig. 5 II) and this has been know for a very long time to occur in all living plant cells, often free but usually as a constituent part of the proteins in the protoplasm. As far back as 1932 it was shown that the mould *Rhizopus suinus*, one of the earliest natural sources of auxin, developed a high activity by the oxidation of amino acids from peptone (J. Bonner, 1932). Later it was shown (Thimann, 1935*a*) that this mould could convert tryptophan to IAA (cf. Hopkins & Cole, 1903). When we turn to the higher plants we find that it is almost certain that a similar process may be taking place. The evidence for the existence of a compound (precursor) easily transformed into auxin by such organs as the coleoptile tip was first shown by Skoog (1937). A direct experiment with tryptophan, indicated that this compound could act as a precursor and was converted to an active auxin by coleoptile tips. Since that time, data have accumulated indicating that probably most green plants possess the capacity to convert tryptophan to IAA and that activity may reside in the cells of many different organs, e.g. leaves, stems, buds, seeds, roots, etc. (see Gordon, 1954, 1961). The greatest capacity is usually to be found in those tissues which are associated with auxin production, e.g. coleoptile tips, young buds, etc. An enzyme system complex which brings about the production of IAA from tryptophan *in vitro* was first isolated from spinach leaves and some of its characteristics worked out (Wildman *et al.*, 1947). Since then the enzyme system has been isolated from several other plants (see Gordon, 1961) and it might seem to be of widespread occurrence although the ability of intact tissues to bring about this conversion is much more difficult to demonstrate. Strong evidence that such an enzyme system is controlling auxin production, and thereby growth, in seedlings, came from studies using X-radiation. Irradiation of plants with very low doses (100 roentgens) reduced stem elongation and also reduced the concentration of native auxins. Similar low doses caused inactivation of the tryptophan-converting enzyme *in vitro* (Gordon, 1953). As auxin synthesis was not able to keep pace with auxin utilization during growth and other auxin-destroying processes (see pp. 333–343), the concentration rapidly became lower and growth declined.

It appears that some mineral deficiency diseases may be linked with this conversion. For example, zinc-deficient plants of tomato are very low in auxins (Skoog, 1940; Tsui, 1948). It seems that zinc is necessary for the synthesis of tryptophan from which the auxin is presumably derived.

One serious threat to these well-established theories comes from the demonstration that in bacteria-free preparations of organs of a number of species of plant (*Helianthus annuus, Cucumis sativus, Phaseolus vulgaris, Pisum sativum*, *Triticum vulgare*, *Zea mays*) tryptophan conversion to IAA is very meagre and that it is only in non-sterile material that IAA is actively produced (Libbert *et al.*, 1966; Libbert *et al.*, 1969; Wichner & Libbert, 1968; Kaiser, 1967). Earlier claims of IAA production from tryptophan may have been based largely on bacterial metabolism and the quantitative work on higher plants will need to be reappraised. A similar situation seems to hold for the *Avena* coleoptile which shows no growth response to tryptophan under sterile conditions (Thimann *et al.*, 1967); since there is a response to anthranilic acid (Fig. 5 III) under similar conditions, it has been suggested that this compound may be the precursor of IAA (Winter, 1966). On the other hand bacteria-free tissues of higher plants and also of many seaweeds (Schiewer, 1967) can convert tryptophan to IAA and there is convincing evidence that cell-free *sterile* extracts of plant tissue can also do so (Valdovinos, Ernest & Henry, 1967). Furthermore the required enzymes for tryptophan conversion to IAA (see later scheme) have been extracted and purified from some plants (e.g. mung bean by Wightman & Cohen, 1967); the threat to the well-established ideas of IAA biosynthesis is probably more apparent than real.

Further indirect evidence for tryptophan as the sole precursor of IAA in pea tissue has come from experiments in which tritium-labelled serine and ^{14}C-labelled indole have been fed to sterile stem tips (Erdmann & Schweiwer, 1971). Subsequently, tryptophan and IAA were extracted and the ^{3}H/^{14}C ratios determined in both compounds; they were virtually identical. Since serine and indole are known precursors of tryptophan, this is a very strong indication that IAA is derived solely from tryptophan and that there is no biosynthetic by-pass leading straight to IAA from indole.

The most probable biochemical sequences whereby IAA could be formed from tryptophan are illustrated below, and we will now briefly consider the experimental evidence relating to them. (In the formulae I represents the indol-3yl- ring system).

Starting with indol-3yl-acetaldehyde (Fig. 5 VI) it is now widely accepted that this is the most probable immediate precursor of IAA. Larsen (1944), who made the first demonstration of this conversion, showed that the

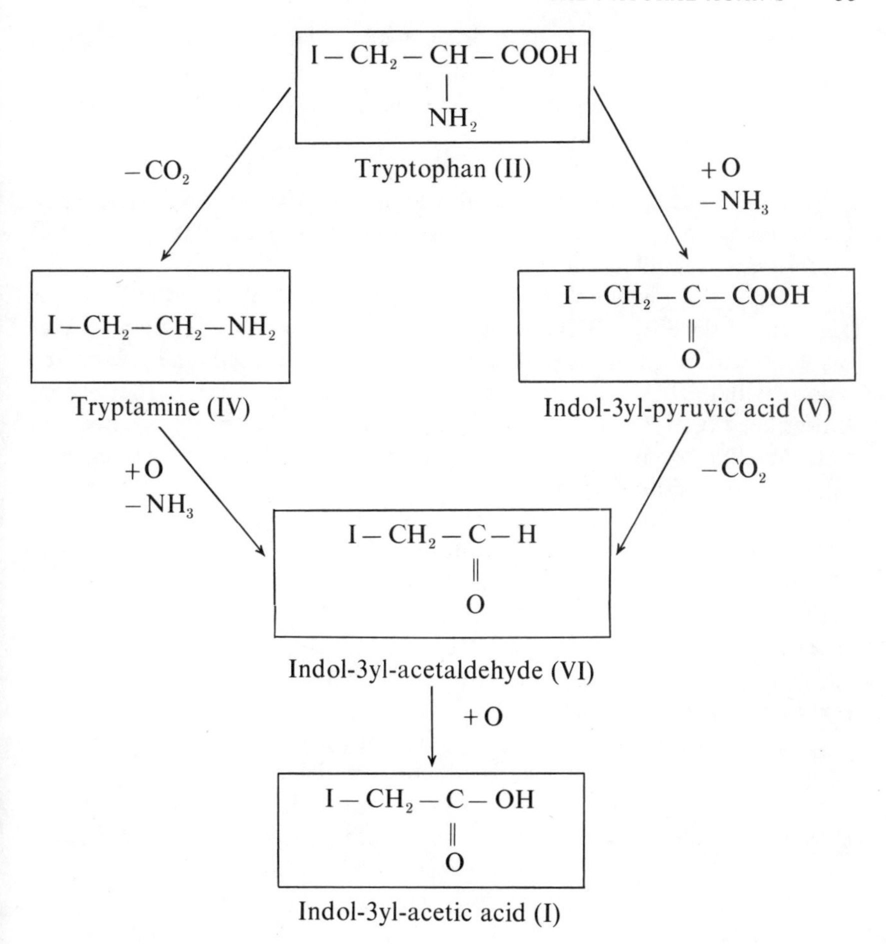

aldehyde can be easily oxidized to IAA by an enzyme system (aldehydrase) present in soil. Since then many plant tissues have been demonstrated to oxidize the synthetic aldehyde. Coleoptiles, coleoptile segments, and coleoptile juice will all form from it an acid auxin which is most probably IAA (Larsen, 1949; Bentley & Housley, 1952). It can also be oxidized by pineapple tissue, and by an enzyme which it contains. Furthermore, IAA production in the leaf is blocked by the aldehyde-specific reagent 'dimedon'. This is strong evidence for the suggestion that indol-3yl-acetaldehyde is the immediate precursor of IAA in the pineapple.

Owing to its extreme ease of conversion to IAA even *in vitro*, indol-3yl-acetaldehyde is extremely difficult to demonstrate in plant extracts; even the chromatographic behaviour of the synthetic compound has never been satisfactorily studied. Its occurrence was first demonstrated by the formation of the reversible bisulphite addition compound in pineapple (Gordon & Sanchez-Nieva, 1949). Its presence in mung bean extracts has recently been shown by the formation of its 2,4-dinitrophenylhydrazone derivative (Wightman & Cohen, 1968). It is not itself an auxin but owes its activity in various tests apparently to its conversion to IAA (Larsen & Rajagopal, 1964). Its normal level of concentration in plant tissues relative to IAA is therefore not likely to be high.

Evidence that the aldehyde is an intermediate in the biosynthesis of IAA comes from the effects of X-radiation, which, as we have seen before, reduces plant growth and lowers IAA levels in mung bean seedlings. Correlated closely with the decline in IAA is a rise in the indol-3yl-acetaldehyde concentration as radiation dose is increased, strongly indicating thereby, not only that the aldehyde is the immediate precursor of IAA but that the enzyme (aldehydrase) responsible for the conversion is the component of the total enzyme system sensitive to these radiations (Gordon 1956).

The oxidation of the aldehyde in cell-free preparations of mung bean seedlings has been shown to be catalysed by the enzyme aldehydedehydrogenase in the presence of the co-enzyme NAD (Wightman & Cohen, 1968). The fact that this reaction takes place in sterile cell-free systems eliminates the possibility that it does not take place in plants in the absence of bacteria (see also Rajagopal, 1968).

Larsen (1950) has earlier indicated that the conversion of the aldehyde to IAA was a dismutation process in which an equivalent quantity of β-(indol-3yl)-ethanol (tryptophol) (Fig 5 VII) was produced. Recent work suggests that this is unlikely in plant tissues and that oxidation to IAA and reduction to tryptophol are two separate processes catalysed by two separate enzyme systems (Wightman & Cohen, 1968; Rigaud, 1970), the balance of the two reactions depending, among other things, on the state of reduction of the available co-factors such as NAD, $NADH_2$ etc.

However, not all the evidence is clear cut in favour of the aldehyde as an intermediate, since in cell-free preparations of pea tissue the oxidation of tryptamine to IAA is not inhibited by dimedon (Muir & Lantican, 1968). What the pathway to IAA would be in this case remains obscure.

Most discussion of the proposed pathways outlined above has centred round the alternative routes *via* tryptamine (Fig. 5 IV) and indol-3yl-pyruvic acid (Fig. 5 V) respectively. Turning first to the keto acid we find matters somewhat obscured by the instability of the compound, which decomposes spontaneously into unidentified acid and neutral compounds (Bentley *et al.*, 1956; Grunwald *et al.*, 1968). Its occurrence in any quantity in plants is still in some doubt; a demonstration of its presence in acetone extracts of maize embryos, *Avena* coleoptiles, wheat roots, etc., was first claimed by Stowe &

Thimann (1954) on the basis of its Rf values on a paper chromatogram. Since then its presence has been suggested in other tissue (see Bentley, 1961*a*) and unequivocally demonstrated in mung beans from which it was isolated as the 2,4-dinitrophenylhydrazone derivative (Wightman & Cohen, 1968).

There is now reasonable evidence that plant tissues contain enzymes that will convert tryptophan to the pyruvic acid and then to IAA. There are two probable pathways of tryptophan conversion to indol-3yl-pyruvic acid. Oxidative deamination is one and transamination with α-oxoglutaric acid the other (see scheme below; I in the formulae represents the indol-3yl-ring system)

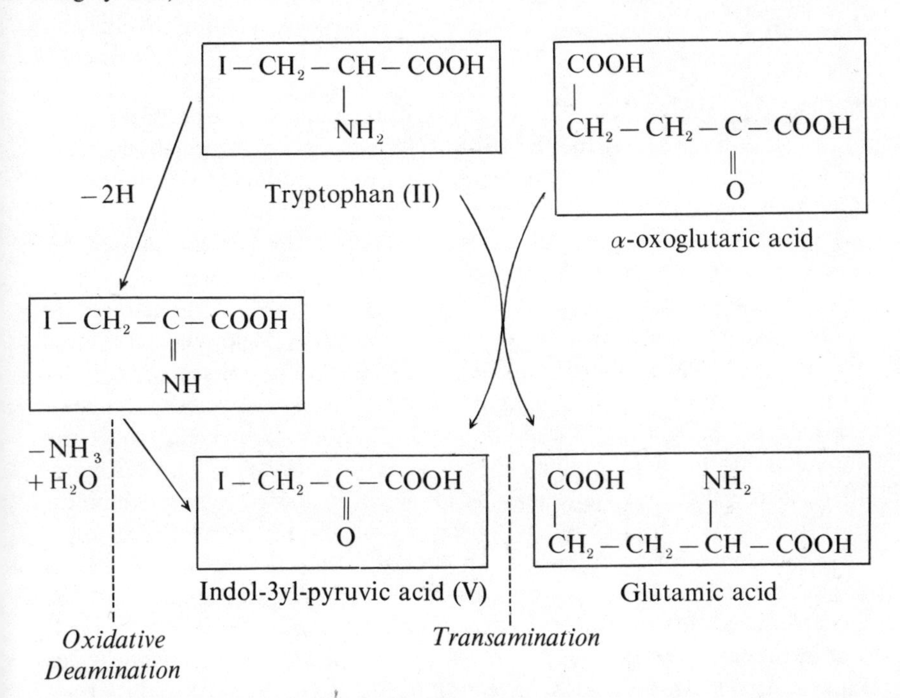

This transamination has been shown to occur in IAA synthesis from tryptophan in bacteria (Stowe, 1955) and is strongly indicated for higher plants (see Gordon, 1961). Indeed this reaction has been demonstrated in purified cell-free enzyme extracts from mung bean seedlings (Wightman & Cohen, 1967; Moore & Shaner, 1968). In the bacterium *Agrobacterium tumefaciens* in the absence of oxygen, tryptophan is converted to indol-3yl-lactic acid (Fig. 5 VIII) and β-(indol-3yl)- ethanol (Fig. 5 VII), both presumably formed by successive reductions of the indol-3yl-pyruvic acid. On the other hand the lactic acid has been shown to be decarboxylated to tryptophol, which is then oxidized to IAA in tomato shoots (Wightman, 1964); this, coupled with the demonstration that tryptophol occurs naturally in cucumber seedlings (Rayle & Purves, 1967*a* & *b*) suggests that it may be an important intermediate in at least some species of plant. Tryptophol can also be produced in mung bean by the reduction of the aldehyde (Wightman & Cohen, 1967) and this is accepted by Rayle & Purves (1968) as the most probable pathway of its formation in cucumber, where it forms a branch-point on the main route to IAA:

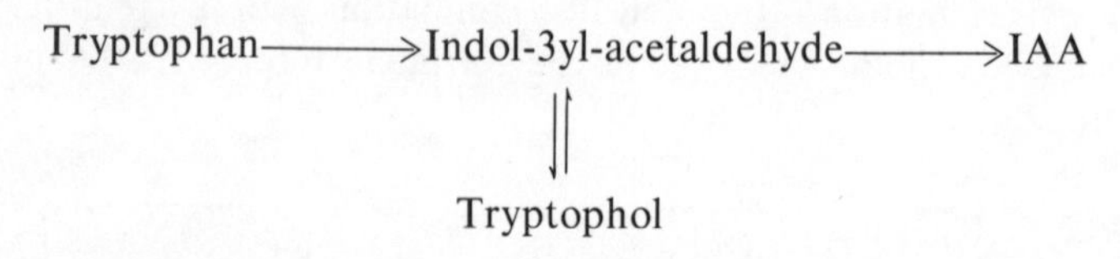

They suggest that the significance of this branch may be the control of flow in the main pathway. If the dehydrogenase converting the aldehyde to the acid were subjected to allosteric inhibition by the product (IAA), then such inhibition would divert the flow into tryptophol which would be a temporary storage material. A drop in IAA levels would release the dehydrogenase from inhibition and allow a rapid conversion of tryptophol to IAA.

The conversion of the pyruvic acid to IAA has been demonstrated by feeding experiments in tissues such as spinach leaves (Wildman *et al.*, 1947) and pineapple leaves (Gordon & Sanchez-Nieva, 1949) but its most probable pathway to IAA via indol-3yl-acetaldehyde is not yet fully established. For example in mung bean X-radiation, which blocks the oxidation of the aldehyde, has no effect on the conversion of the pyruvate to IAA (Gordon, 1956). More than one route may be involved, including a direct chemical breakdown.

The evidence for tryptamine as an intermediate is not so strong. So far it has been identified from a few species in easily estimated amounts, e.g. in the flowers and shoots of *Acacia floribunda* and *A. pruinosa* (E. P. White, 1944) and in the leaves of the American honey mesquite (*Prosopsis juliflora*) (see Larsen, 1967), where it may have nothing to do with auxin metabolism. Enzyme preparations from some plants will convert it to IAA but other preparations, e.g. from spinach leaves, will not. It has been produced in tomato and tobacco tissue by feeding with tryptophan (Wightman, 1964; Phelps & Sequeira, 1967). However, specific inhibitors of the enzyme which oxidizes tryptamine have no effect on the production of IAA in the fungus *Taphrina deformans*. (Perley & Stowe, 1966). Further, the decarboxylation of tryptophan in cell-free pea extracts produces no trace of tryptamine in the many products which are formed (Reed, 1968). It has been suggested (Gordon, 1961) that tryptamine is not on the main conversion route but may be formed by side-reactions from the aldehyde either directly or via the nitrile (see next section), since the aldehyde-specific inhibitor dimedon blocks the formation of tryptamine from tryptophan in plant tissues. On the other hand *sterile* tips of *Avena* coleoptiles show growth responses to tryptamine but not to tryptophan (Thimann & Grochowska, 1968). Furthermore, this response to tryptamine is blocked by inhibitors of mono-amine oxidase enzymes (e.g. semicarbazide). This suggests that the tryptamine is oxidized to IAA by amine oxidases and implies its role as a normal auxin precursor. In strong contrast to this, more direct evidence from experiments involving the incubation of sterile and non-sterile pea tissue with tryptophan, tryptamine and indol-3yl-pyruvic acid and identification of the products, indicates that tryptamine is an intermediate only in conversions by epiphytic bacteria, whereas in pea tissue the major pathway is through the pyruvate (Libbert *et al.*, 1968). On the other hand completely cell-free preparations from both sterile and non-sterile tissues of *Avena* and *Pisum* have been shown to convert both tryptophan and tryptamine to auxin although to different extents in the two tissues (Muir & Lantican, 1968). These conflicting results, even for one and the same tissue, underline the complexities of the situation and suggest that enzymes for both pathways, i.e. through tryptamine and indol-3yl-pyruvic acid respectively, may be present in all tissues but that the predominance of one or other pathway in normal metabolism will depend on growth conditions which will determine the balance of the operative enzymes in the tissue.

(D) OTHER INDOLE COMPOUNDS

A considerable number of additional indole compounds whose physiological importance still remains to be discovered, have been isolated from plant tissues. One of the first of these to be characterised was the neutral compound indol-3yl-acetonitrile (IAN, Fig. 6 IX) prepared in crystalline form after purification by column chromatography (E. R. H. Jones *et al.*, 1952).

This compound, which is highly active in many of the main tests for auxins, was extracted in amounts greatly in excess of the acid auxin components which are negligible in quantity; in fact, at least in Brussels sprouts, virtually all the auxin 'potential'seemed to reside in this one neutral molecule. An interesting feature of the nitrile is that, like the hypothetical auxin A, it is alkali-labile and somewhat acid-stable. Extraction of alkali-treated material with ether will not remove the IAA thus produced until the mixture is re-acidified (Bentley & Housley, 1952). It is therefore conceivable that this nitrile may have been responsible for some of the reports of alkali-labile 'auxin A'.

IAN is not an auxin but owes its activity in certain biological assays to the fact that the tissue concerned can convert it to IAA enzymatically. Insensitive tissues do not possess the enzyme. The evidence for this is discussed later. Since its first isolation, IAN has been reported as present in many

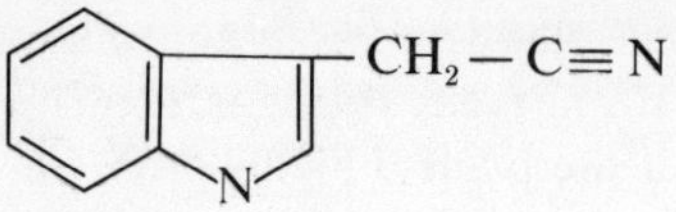

$CH_2-C\equiv N$

N

IX

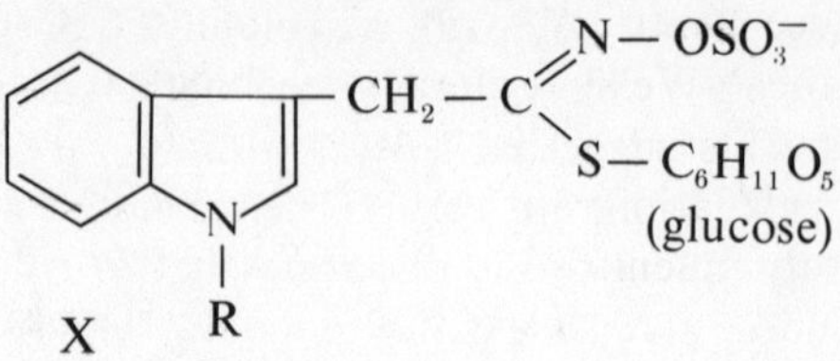

$N-OSO_3^-$

CH_2-C

$S-C_6H_{11}O_5$

(glucose)

N

R

X

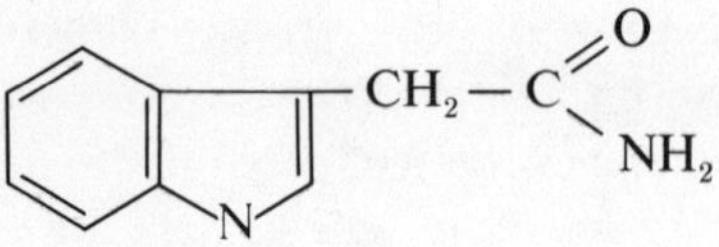

O

CH_2-C

NH_2

N

XI

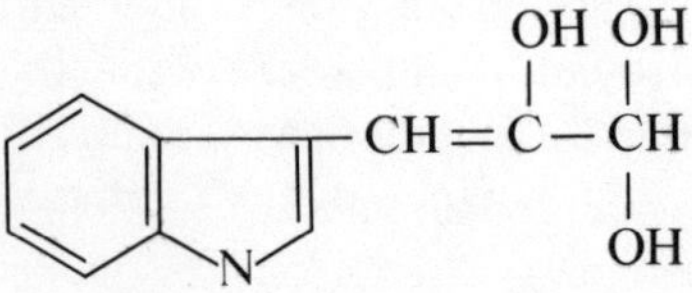

CH_2

O O

N

OH OH

$CH=C-CH$

OH

CH_2-CH =O

O

OH OH

N

XII

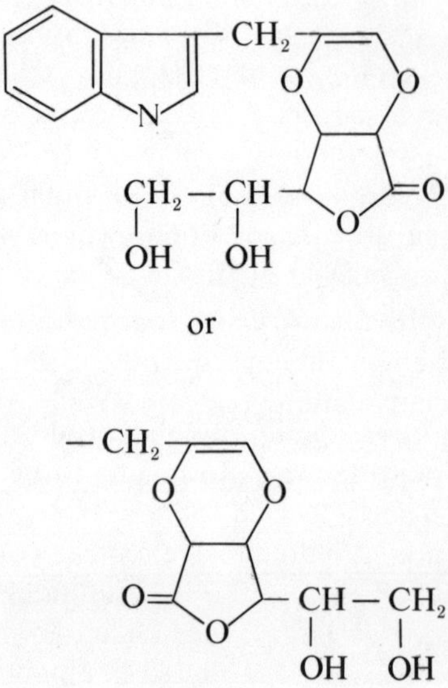

or

CH_2-OH

$-CH_2$

O O

N

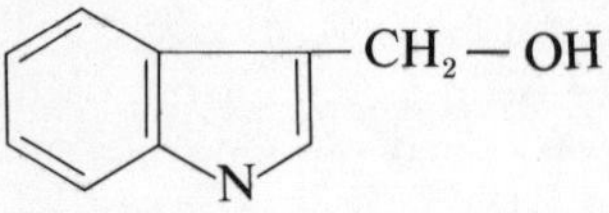

XIV

O= $CH-CH_2$

O

OH OH

XIII

Fig. 6. The natural auxins and related compounds (continued).

IX Indol-3yl-acetonitrile (IAN)

X R = -H: Glucobrassicin

R = $-OCH_3$: Neoglucobrassicin

XI Indol-3yl-acetamide

XII Skatylglyoxal hydrate

XIII Ascorbigen

XIV 3-Hydroxymethylindole. Indol-3yl-methanol.

different tissues but the identifications have been based merely on chromatographic behaviour and colour reactions and can only be accepted as tentative (see Bentley, 1961*a*). Its role in the growth of the plant is highly problematical. We have already seen that it is unlikely to be a direct precursor of IAA and its suggested role as a temporary storer of auxin 'potential' is entire speculation. Its very existence in the undamaged cell is now in question subsequent to the discovery that *Brassica* spp. contain large amounts of an indole glucoside called glucobrassicin (Fig. 6 X) (Gmelin & Virtanen, 1961). At pH3 this substance is broken down by enzymes in the tissue 'brei' and considerable amounts of IAN are formed. Since no IAN could be detected in extracts made with boiling methanol, where enzymatic attack was prevented, it was suggested that the IAN reported in other work with *Brassica* spp., where such precautions were not taken, might be an artefact of preparation. The possible widespread occurrence of glucobrassicin and its derivative neoglucobrassicin (Fig. 6 X) in a number of families of flowering plants (Schraudolf, 1965) suggests that the earlier work on IAN occurrence should be checked.

The physiological significance of glucobrassicin is not clear. The fact that it occurs in particularly large amounts in the growth zones of plants suggests that it is manufactured there and acts as the main component of an auxin-precursor pool. On the other hand the morphological peculiarities of the cultivated *Brassica* spp. indicate an unusual auxin situation and glucobrassicin might consequently be a secondary metabolite of the nature of an IAA-detoxication product (Gmelin & Virtanen, 1961) and restricted furthermore to a few exceptional families, Cruciferae, Resedaceae, Capparidaceae, and Tovariaceae (Schraudolf, 1965; Kutáček, 1967). It has been pointed out that it is unlikely to be an auxin precursor in these plants since neither IAN nor any compound with auxin activity can be derived from it at pH values above 5.2, whereas the pH of the cell sap is much nearer neutrality (Schraudolf & Weber, 1969).

Recent work involving the incubation of *Brassica* tissue with ^{14}C-labelled tryptophan suggests that the major pathway of metabolism of this amino acid in this genus goes through glucobrassicin and leads to ascorbigen (see later) and IAN; very small amounts of IAA are formed (about 0.25 percent of the tryptophan supplied) and this comes partly from the IAN (Kutáček & Kefeli, 1968). These authors suggest that glucobrassicin could serve as a 'reserve precursor' of IAA. In this genus a significant amount of the tryptophan is metabolized to indol-3yl-acetamide (Fig. 6 XI) (Kutáček & Kefeli, 1968) which could be produced by oxidative decarboxylation catalysed by peroxidase (Riddle & Mazelis, 1964); this might be another pathway contributing to IAA synthesis in this rather exceptional genus.

Another indole compound also isolated from the genus *Brassica* has created considerable interest (Procházka & Kořístek, 1951; Procházka *et al.*, 1957). This is ascorbigen (Fig 6 XIII), an indole-ascorbic acid complex which liberates ascorbic acid on acid hydrolysis, and IAA with other indole compounds on alkaline hydrolysis. Like glucobrassicin it occurs predominantly in growth zones and was originally regarded by its discoverers as being involved in IAA biosynthesis from tryptophan. There are indications however that, like glucobrassicin, it may be a secondary by-product of auxin metabolism or, like IAN, an artefact of preparation. For example Post (1959) has isolated from cabbage considerable quantities of skatylglyoxal hydrate (Fig. 6 XII), which is very active in auxin assays, but whether as an auxin in its own right or after conversion to an auxin, is not yet clear. Ascorbigen is not active in these tests and Post regards it as arising from skatylglyoxal by a side reaction, not directly related to IAA biogenesis. Again glucobrassicin at pH7 forms 3-hydroxymethylindole (Fig. 6 XIV) as a hydrolysis product from which ascorbigen could be formed by combination with ascorbic acid (Gmelin, 1964). Thus ascorbigen could arise during extraction as an artefact, a view now shared by its discoverers (Kutáček & Procházka, 1964). Discussion of this compound has been included because it illustrates the pitfalls inherent in studies of native plant hormones and the dangers of applying conclusions from observations on damaged tissue to situations in intact plants.

Other indole compounds, which may be auxins, precursors of auxins or metabolic products of auxins have been isolated on chromatograms and tentatively identified from their speeds of migration and their colour reactions with indole reagents. They include the methyl and ethyl esters of IAA, indol-3yl-propionic (Fig. 8 XXI), -butyric (Fig. 8 XXII), -glycollic (Fig. 12 LXIII) and -carboxylic (Fig. 13 LXVII) acids, 3-indolaldehyde and various derivatives of tryptophan and tryptamine. In view of the extreme rarity of chlorine-containing organic compounds in nature, it is interesting that the 4-chloro derivative of IAA (Fig. 7 XV) has been isolated as the methyl ester in methanol extracts of immature pea seeds and shown to be active in auxin tests (Marumo *et al.*, 1968). It is apparently as active as IAA in auxin tests but its biological role remains obscure. D-4-chlorotryptophan has also been isolated from the same source (the first record of a chlorine-containing amino acid occurring in nature) and may be a precursor of the 4-chloroindol-3yl-acetic acid (Marumo & Hattori, 1970).

Other active compounds giving an indole reaction but otherwise eluding identification are regularly being reported. IAA, when fed to plant tissue is known to conjugate with certain normal metabolites such as aspartic acid and glucose. These substances will be discussed in a later chapter (Chap. XI).

Although it has been widely assumed since the earliest days of auxin study that IAA is in normal control of the growth of roots as well as shoots, this is by no means unequivocally established; this situation, together with the somewhat anomalous behaviour of IAA *vis-à-vis* roots (see Chap. VII pp. 183–6 and Chap. XI pp. 327–30) has suggested to a number of physiologists that the natural auxin of roots is not IAA but quite a different substance. Direct evidence for this is scanty and unsatisfactory although substantial enough not to be ignored. Thus in *Lens culinaris* roots the main auxin is claimed to be 3-(indol-3yl)-acrylic acid (Hofinger *et al.*, 1970). However, excised wheat roots growing in sterile culture excrete into the medium a substance which is extremely active as a root growth inhibitor (Woodruffe *et al.*, 1970); although, like the indole auxins, it gives a positive Ehrlich colour reaction, there is no clear evidence as yet that it is an indole. On the other hand in *Vicia faba* roots the inhibiting substance, which is also undoubtedly an active auxin in the standard tests, has fluorescent properties quite different from those of IAA and resembling very closely those of the non-indole 'citrus auxin' discussed in the next section (Burnett *et al.*, 1965). The real nature of the auxin in control of root growth is still a mystery and a challenge.

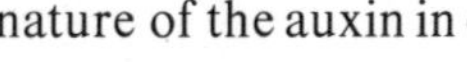

(E) NON-INDOLE AUXINS

Ever since the marriage of chromatography with auxin assays there has been a steady trickle of reported isolations of unknown active compounds not giving indole colour reactions. Bentley in 1958 already listed over a score of such reports. However, caution is needed when considering such claims because it has now become clear that some of the auxin assays in use are not specific for auxins but can also show responses to a very different group of hormones, the gibberellins (see Chapter IV). But a few closer identifications have been claimed, indicating that plants may indeed contain true auxins that are not indolic compounds.

For example Vlitos and his colleagues (Crosby & Vlitos, 1959; Vlitos & Crosby, 1959) have extracted from Maryland Mammoth tobacco leaves a material with auxin activity, closely resembling the long-chain primary alcohol l-docosanol (Fig. 7 XVI) in its infra-red absorption spectrum. Later, waxes scraped from sugar-cane stems were also claimed to have auxin properties (Vlitos & Cutler, 1960). These claims have puzzled plant physiologists in view of the extreme insolubility of such substances in water and have been rejected by one (Kefford, 1962b) who could not repeat the observations. Further study is obviously needed. The saponins, which are plant products and, like the alcohols mentioned above, are highly surface-active, have also been shown to have marked auxin activity (Vendrig, 1964).

Derivatives of phenylacetic acid (see Chapter III) may also occur as natural auxins e.g. phenylacetamide (Fig. 7 XVII) which has been isolated

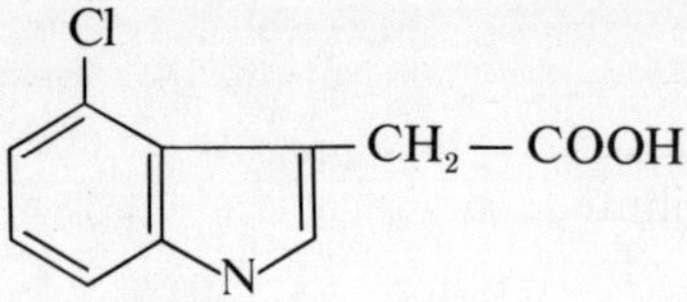

Cl
CH_2-COOH
N
XV

$CH_3-(CH_2)_{21}-OH$
XVI

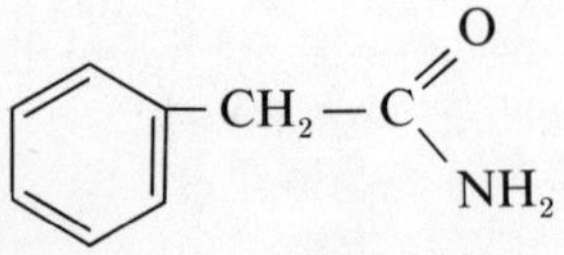

O
CH_2-C
NH_2
XVII

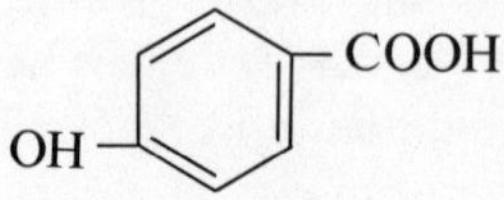

COOH
OH
XVIII

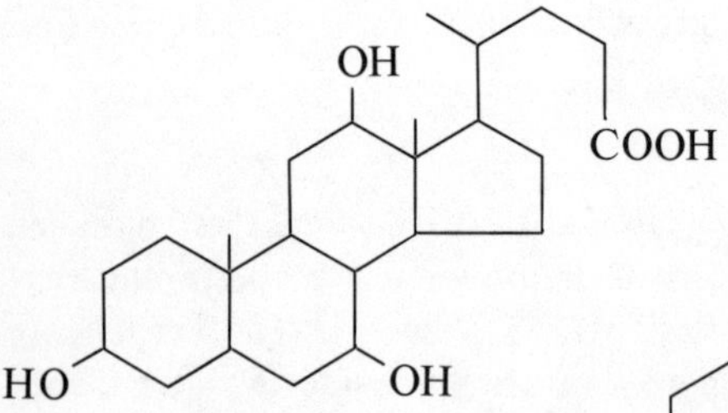

OH
COOH
HO
OH
XIX

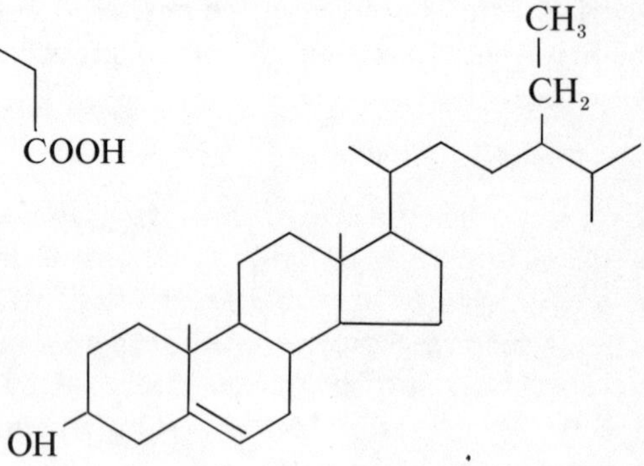

CH_3
CH_2
OH
XX

Fig. 7. The natural auxins and related compounds (continued).

XV	4-Chloroindol-3yl-acetic acid
XVI	1-Docosanol
XVII	Phenylacetamide
XVIII	*p*-Hydroxybenzoic acid
XIX	Cholic acid
XX	β-Sitosterol

from seedlings of *Phaseolus* sp. (Isogai *et al.*, 1964, 1967) and *p*-hydroxybenzoic acid (Fig. 7 XVIII) from *Ribes rubrum* (Vieitez *et al.*, 1966).

The use of spectrophotofluorometry (see p. 48) has facilitated the separation of a possible non-indole auxin from IAA in plant extracts (Khalifah *et al.*, 1963, 1965). It has been called '*Citrus* auxin' because it was first isolated from young *Citrus* fruit. Its fluorescence behaviour is very different from that of any known indole compound but resembles that of β-naphthol. Its infrared absorption spectrum indicates that it is a carboxylic acid with an aromatic ring system. It seems to be widely distributed, occurring in at least six unrelated families (Lewis *et al.*, 1965). Finally from *Coleus* spp. a system of interconvertable auxins has been isolated and partially characterized as sterols, very closely resembling β-sitosterol (Fig. 7 XX) and campesterol; they have a very high activity in the *Avena* curvature test (Vendrig, 1967*b*). Their relationship to cholic acid (Fig. 7 XIX) (cf. recent analysis of the original sample of auxin A – see p. 51) is provocative of speculation.

The absolute dominance of the indole group of auxins is again being challenged.

Chapter III

The Synthetic Auxins

THE STRUCTURAL REQUIREMENTS FOR AUXINS

Naturally after the isolation of IAA as an active plant growth hormone, plant physiologists looked around for other compounds of similar chemical constitution which might have similar physiological activities. In the year 1935, several workers announced the discovery of a number of such compounds, which included other indole carboxylic acids such as indol-3yl-propionic acid (Fig. 8 XXI), γ-(indol-3yl)-butyric acid (Fig. 8 XXII) (Zimmerman *et al.*, 1936) and indol-3yl-pyruvic acid (Fig. 5 V) (Haagen Smit & Went, 1935), and also naphth-lyl- and naphth-2yl-acetic acids (Fig. 8 XXIII and XXIV), phenylacetic acid (Fig. 8 XXV) and anthrylacetic acid (Fig. 8 XXVI) (Zimmerman *et al.*, 1936). Later on naphth-2yloxy-acetic acid (Fig. 8 XXVIII) was added to the list (Irvine, 1938) and still later the phenoxyacetic acid (Fig. 8 XXVII) series (Zimmerman & Hitchcock, 1942).

It became apparent very early, however, that although such compounds closely resembled one another in possessing the properties of auxins, they differed widely amongst themselves as regards specific activity; accordingly, estimates of that activity varied with the nature of the assay used in the determinations. This was shown in the work of Thimann (1935*b*) who investigated the activities of indol-3yl- (Fig. 51), inden-3yl- (Fig. 8 XXIX), and benzofuran-2yl-acetic acids (Fig. 8 XXX). Inden-3yl-acetic acid showed very small, and benzofuran-2yl-acetic acid no activity in the *Avena* curvature test, whereas a considerable activity was shown both in the *Avena* coleoptile straight growth test and in the pea test. These two latter compounds, therefore, although possessing auxin activity, were prevented by some secondary property of the molecule from showing this activity in the *Avena* curvature test. The small curvatures produced were very largely confined to the upper portion of the coleoptile, and suggested that impeded transport (see Chapter

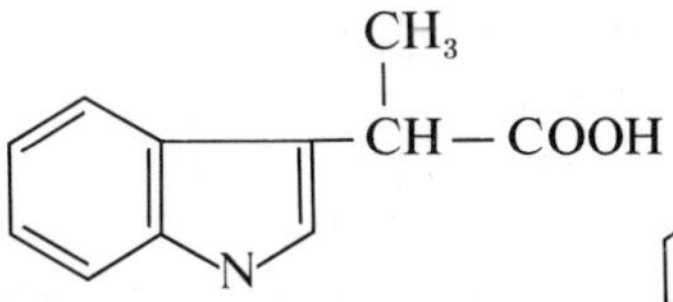

CH_3 | CH—COOH (indol-3-yl)

XXI

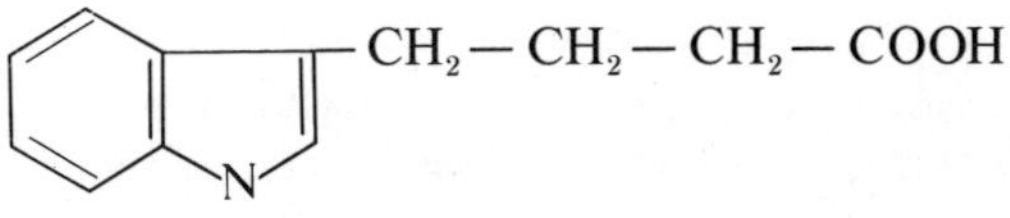

CH₂—CH₂—CH₂—COOH (indol-3-yl)

XXII

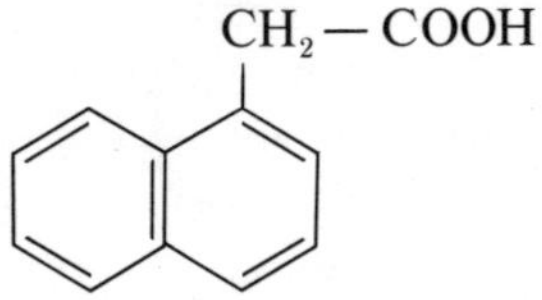

CH_2—COOH (naphth-1-yl)

XXIII

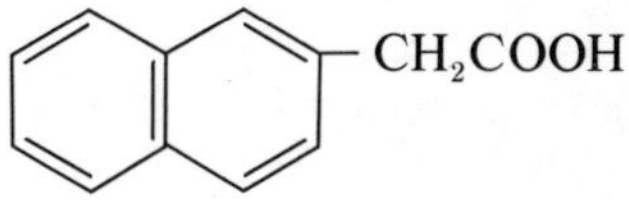

CH_2COOH (naphth-2-yl)

XXIV

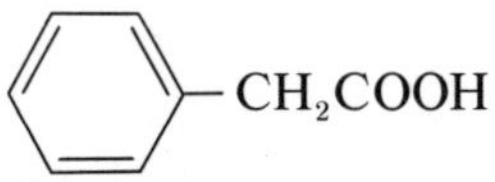

CH_2COOH (phenyl)

XXV

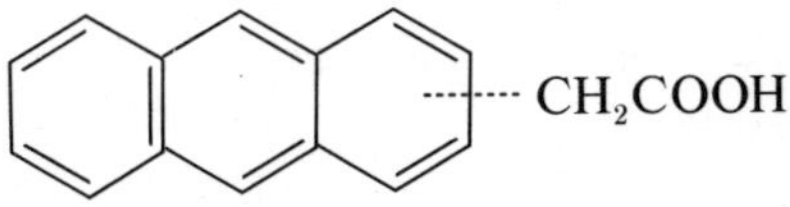

CH_2COOH (anthryl)

XXVI

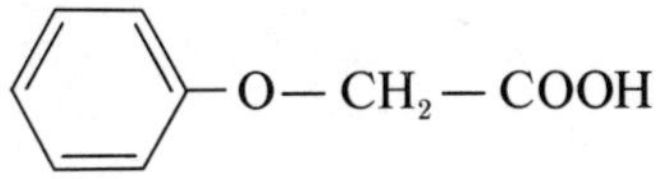

O—CH_2—COOH (phenyl)

XXVII

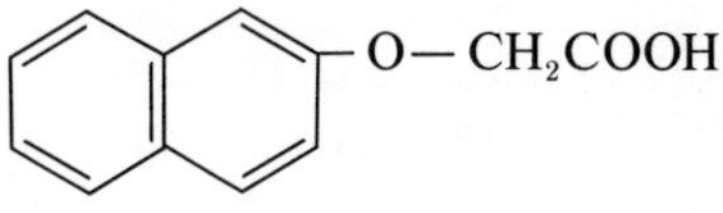

O—CH_2COOH (naphth-2-yl)

XXVIII

CH_2COOH (inden-3-yl)

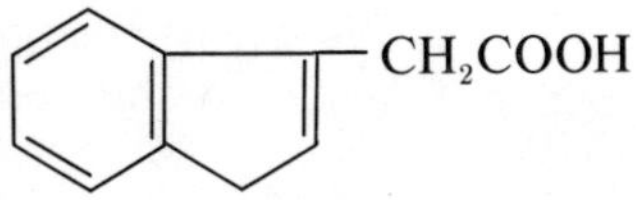

XXIX

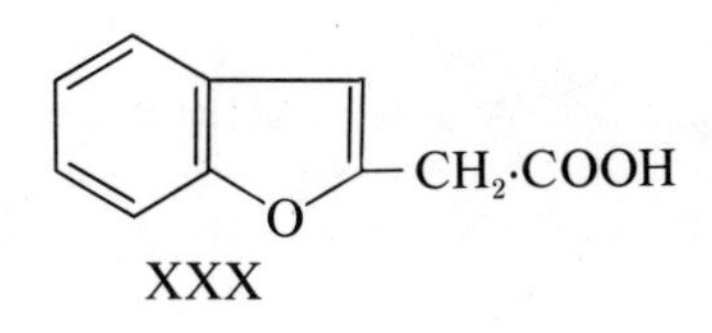

XXX

Fig. 8. The synthetic auxins.

XXI	α-(Indol-3yl)-propionic acid
XXII	γ-(Indol-3yl)-butyric acid
XXIII	Naphth-lyl-acetic acid (NAA)
XXIV	Naphth-2yl-acetic acid.
XXV	Phenylacetic acid
XXVI	Anthrylacetic acid
XXVII	Phenoxyacetic acid
XXVIII	Naphth-2yloxy-acetic acid
XXIX	Inden-3yl-acetic acid
XXX	Benzofuran-2yl-acetic acid

XI) of these two latter compounds accounted for their very low activity. Thus it is evident that the relative activities of molecules in any one test may be no precise measure of their relative activities at the growth centres. Secondary properties of the molecule affecting its availability at the site of its action, e.g. the rate of penetration into cells and tissues, the rate of transport from the site of application, and the rate of inactivation or destruction in the tissues of the plant, will all greatly affect the ultimate response (see Chapter XI).

The operation of such secondary molecular properties may be of the greatest importance when we came to the application of compounds of this type in growth control of plants in agriculture and horticulture. Another possible source of confusion is the fact that auxins usually provoke other growth responses which may not be closely related quantitatively to the main auxin activity (stimulation of extension-growth of stems). The employment of assay tests involving such responses may easily lead to conflicting generalizations concerning structure–activity relationships.

In spite of these difficulties, which are unavoidable in the methods at present available for the estimation of auxin activities, a massive amount of work has been and is still being done on the interrelationships of chemical structure and activity in a very wide range of synthetic compounds. Before going on to consider the results of such investigations in detail, it will be useful to set down the general structural requirements for auxin activity drawn up from the early work on this subject. These generalizations, which were first put forward by Koepfli *et al.* (1938) and were based on the pea test (primary activity), run as follows:

The active molecule was one which had:

1. A ring-system as a nucleus.
2. At least one double bond in this ring.
3. A side-chain possessing a carboxyl group (or a group easily convertible into a carboxyl group).
4. At least one carbon atom between the ring and the carboxyl group in the chain.
5. A particular spatial relationship between the ring-system and the carboxyl group.

It will be seen that all the compounds so far mentioned, conform to these requirements. Went & Thimann (1937) also enunciated secondary properties which were supposed to affect only transport, inactivation rates, etc.,

They were:

1. Length of the side-chain.
2. Nature and degree of substitution in the nucleus and the side-chain.
3. The basic structure of the nucleus.

Let us now see how far these requirements are still borne out by the mass of data on structure–activity relationships accumulated over the intervening thirty odd years.

Effects Of Variation In Structure On Auxin Activity

(A) THE NATURE OF THE RING

The nature of the ring has a great effect on activity. Five-membered ring compounds are without activity. Again, the substitution of carbon and of oxygen for nitrogen in the heterocyclic ring of indole greatly reduces the primary growth activity of the corresponding auxins. On the other hand, naphthalene and anthracene give compounds about as active as IAA, although in the former the position of the side-chain greatly affects the activity. The probable reasons for these differing activities will be considered later, but the number of double bonds in the ring may be important. This has been demonstrated in the case of naphth-1yl-acetic acid and phenylacetic acid, in which progressive hydrogenation of the double bonds in the nucleus results in a progressive activity loss which is complete in the saturated ring compounds, cyclohexylacetic and decahydronaphthylacetic acids. No compound with a completely saturated ring has yet been found to have auxin activity. A considerable number of ring systems, monocyclic, polycyclic, aromatic and heterocyclic, are now known to bestow auxin activity on molecules, provided suitable substituent groups are present. A range of such systems, additional to those already mentioned, is shown in Fig. 9.

There has been much inconclusive debate on the significance of the ring and its double bonds – features which had come to be accepted as absolute essentials for auxin activity. The situation now threatens to become completely fluid owing to the discovery of some active 'auxins' which do not possess such a ring-system. These were the carboxymethyl-*N*:*N*-dialkyldithiocarbamates (Fig. 10 XL) (van der Kerk *et al.*, 1955). These substances, particularly those with lower alkyl substituent groups (R_1 and R_2), have a low but distinct auxin activity in the Went pea test and in the *Avena* cylinder test. The unexpected activity of these molecules might be explained by an effect on the status of the native auxin in the assay tissue, for example by mobilising it for action from a bound inactive state at some site in the cell (see Chapter XI). This seems improbable from studies of the exchange of ^{14}C-labelled IAA in *Avena* coleoptiles treated with one of those compounds (Gordon & Moss, 1958). That they are auxins in their own right is suggested by the proposed interaction of N with the neighbouring S atom in the molecule producing a flat pseudo-ring structure (Fig. 10 XL) which has the

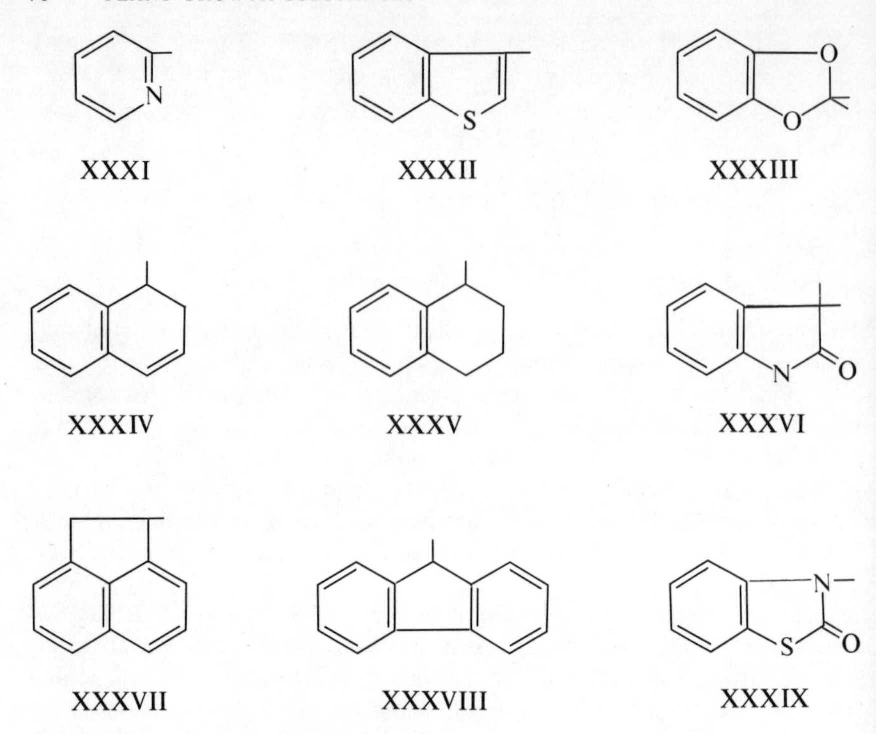

Fig. 9. Synthetic auxins. Range of ring structure.

XXXI	Pyridyl-2-
XXXII	Benzothien-3yl-
XXXIII	Benzodioxolyl-
XXXIV	1,2,-dihydronaphth-lyl-
XXXV	1,2,3,4-tetrahydronaphth-lyl-
XXXVI	2-oxindol-3yl-
XXXVII	Acenaphthen-lyl-
XXXVIII	Fluoren-9yl-
XXXIX	2-Oxobenzothiazolin-3yl-

attributes necessary for auxin action (Pluijgers & van der Kerk, 1961). The auxin activity of carboxymethyl-*o*-alkylxanthates (Fig. 10 XLI) (Fawcett, Wain & Wightman, 1956) and of *S*-methyl-*N*-methyl-*N*-carboxymethyldithiocarbamates (Fig. 10 XLII) (Rothwell & Wain, 1963) subsequently discovered are not inconsistent with a similar explanation.

However, it has been pointed out that these dithiocarbamates are powerful chelating agents (Heath & Clark, 1956) and that many such agents have auxin properties, even though they are of widely different chemical constitution and unrelated structurally to the natural auxins. Chelating agents cause auxin-like growth-responses in the oat and wheat coleoptile cylinder

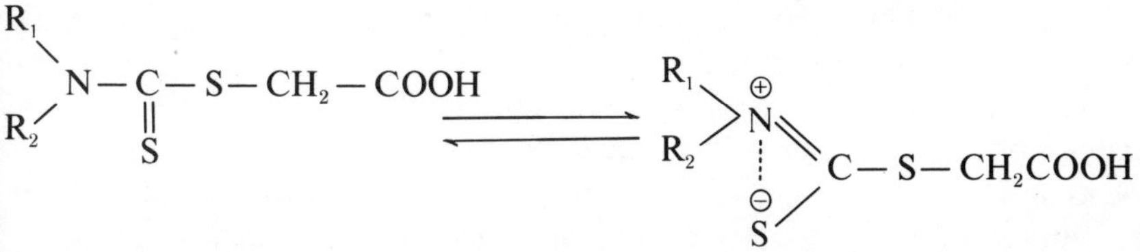

$$\begin{array}{l}R_1 \\ \quad \diagdown \\ \quad\; N - \underset{\underset{S}{\|}}{C} - S - CH_2 - COOH \\ \quad \diagup \\ R_2\end{array} \rightleftharpoons \begin{array}{l}R_1 \diagdown \;\overset{\oplus}{N} \\ R_2 \diagup \;\; \vdots \;\; \diagdown C - S - CH_2COOH \\ \qquad\;\; \overset{\ominus}{S} \diagup \end{array}$$

XL

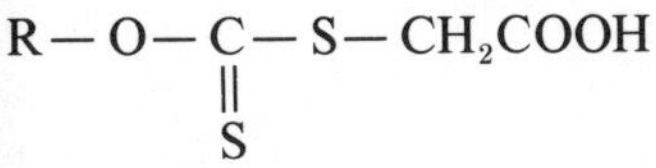

$$R - O - \underset{\underset{S}{\|}}{C} - S - CH_2COOH$$

XLI

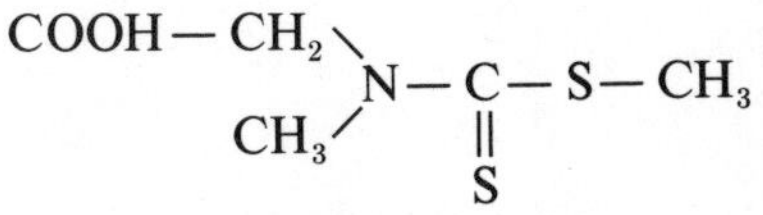

$$\begin{array}{l}COOH - CH_2 \diagdown \\ \qquad\qquad\quad N - \underset{\underset{S}{\|}}{C} - S - CH_3 \\ \qquad\quad CH_3 \diagup\end{array}$$

XLII

$$CH_3 - \underset{S \cdot CH_2 \cdot COOH}{\overset{S \cdot CH_2 \cdot COOH}{C}} - CH_2 \cdot CO \cdot CH_3 \longrightarrow \begin{array}{l}CH_3 \diagdown \\ \quad C - CH \\ O \diagup\!\!\diagup \qquad \diagdown\!\!\diagdown C - S - CH_2COOH \\ \quad H - CH_2 \diagup\end{array}$$

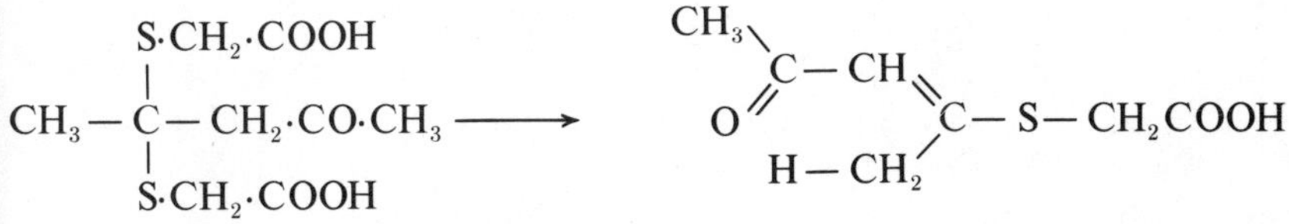

XLIII XLIV

Fig. 10 Synthetic auxins. Non-ring compounds.

XL	Carboxymethyl-*N*,*N*-dialkyldithiocarbamates
XLI	Carboxymethyl-*O*-alkylxanthates
XLII	*S*-methyl-*N*-methyl-*N*-carboxymethyldithiocarbamates
XLIII	2,2-Di(carboxymethylthio)-pentan-4-one
XLIV	2-Carboxymethylthio-2-pentene-4-one

test (Bennet-Clark, 1956; Heath & Clark, 1956) (see Chapter XII). They include ethylendiaminetetra-acetic acid (EDTA), nitrilotriacetic acid, iminodiacetic acid and 8-hydroxyquinoline. Concentrations as low as 10^{-10} M of EDTA have a measurable stimulation of wheat coleoptile growth. This suggests that the auxins may be chelating agents and that their physiological actions may be based on such properties. More attention will be given to this theory later in the book (Chapter XII) but it should be mentioned that auxin activity has not been confirmed for all these chelating agents (Fawcett, Wain & Wightman, 1956) and that the dithiocarbamates may still conform to the classical structure/activity rules. Other compounds which show some auxin activity (e.g. in the wheat coleoptile segment test) and which may owe their activities to the formation of a derivative with a planar pseudo-ring, are thioglycollic acid derivatives of β-diketones (Barrett & George, 1969). Thus the compound 2,2-di(carboxymethylthio)-pentan-4-one (Fig. 10 XLIII) dissociates in water to form 2-carboxymethylthio-2-pentene-4-one (Fig. 10 XLIV) and thioglycollic acid. A weak bond between a terminal hydrogen and the = 0 group of compound XLIV could ensure a ring arrangement and thus auxin activity.

(B) THE NATURE, LENGTH, AND POSITION OF THE SIDE-CHAIN

According to requirement (3) of Koepfli *et al.*, indicated before, the carboxyl group must be separated from the nucleus by at least one carbon atom. It has, however, been shown by later work that this minimum requirement is not valid. Thus considerable auxin activity has been reported in many compounds (Zimmerman & Hitchcock, 1944; Bentley, 1950*b*; Veldstra & van der Westeringh, 1952; Heacock *et al.*, 1958) where the carboxyl group is attached directly to a ring (Fig. 11 XLV to LI). In fact one of the most active auxins discovered in recent years, 4-amino-3,5,6-trichloropicolinic acid (Fig. 11 LI) is lacking such a bridging carbon atom (Kefford & Kaso, 1966). The implications of these differences will be considered later.

Nevertheless in the active series with those side-chains, compounds having one bridging carbon atom between nucleus and carboxyl group usually show the highest activity. When the number of these bridging atoms is increased there is generally a fall in activity, but this fall does not run parallel with the increase in the number of carbon atoms. Thus in the higher homologues of naphth-1yl-acetic acid (Fig. 8 XXIII) (Grace, 1939), IAA (Thimann & Bonner, 1938), and 2,4-dichlorophenoxyacetic acid (Fig. 11 LII) (Synerholm & Zimmerman, 1947), great activity is shown by those compounds having an odd number of bridging carbon atoms in the side-chain whereas those having an even number have very low or no activity (see Plate 7). Synerholm & Zimmerman (1947) suggested that this alternation of activity might be due to the β-oxidation of the side-chain in the manner first proposed by Knoop (1904) and now well established for fatty acids in both animals and plants (see Davies *et al.*, 1964). The progressive removal of two carbon units from the side chain would yield the corresponding inactive phenol or active acetic acid for even and odd numbers of bridging carbon atoms respectively. Direct experimental evidence for this was first obtained by treating flax plants with a range of homologous ω-phenoxy-alkylcarboxylic acids (general formula C_6H_5–O–$(CH_2)_n$–COOH) (Wain, 1951). After absorption by the roots, considerable quantities of the phenol were shown to be formed only in those plants that had been treated with acids having an even number of carbon

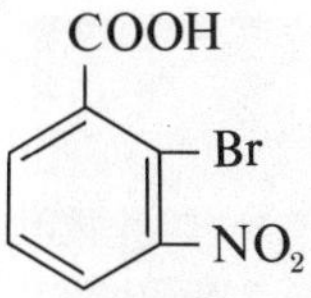

XLV

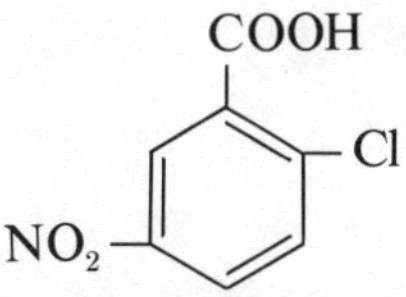

XLVI

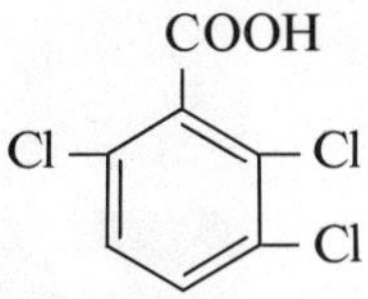

XLVII

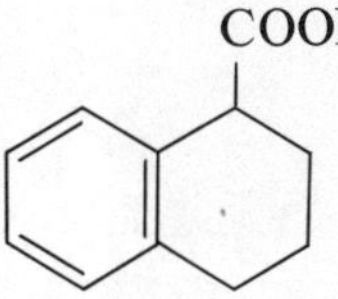

XLVIII

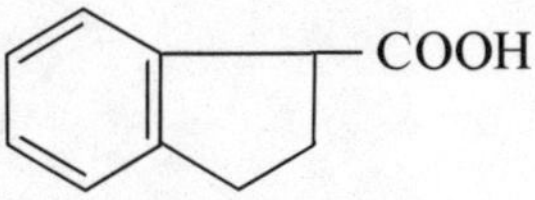

XLIX

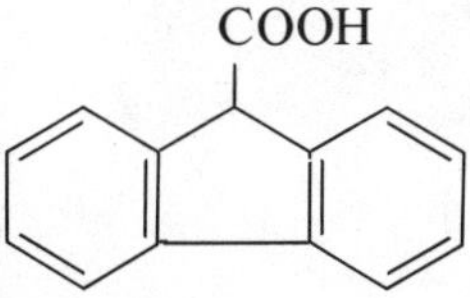

L

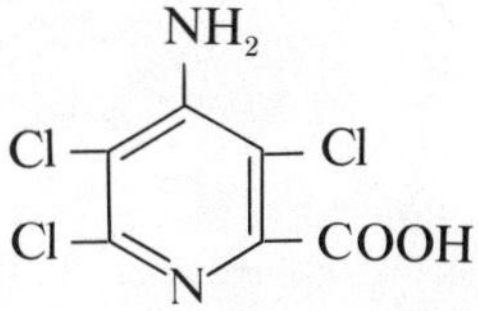

LI

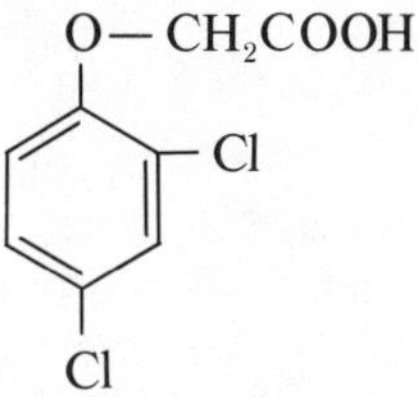

LII

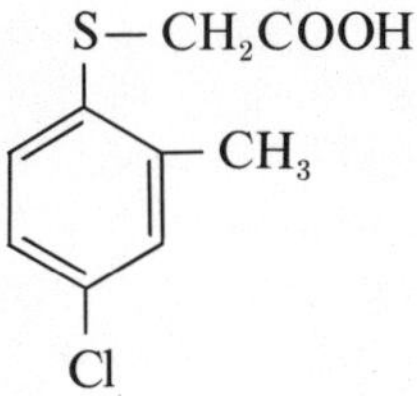

LIII

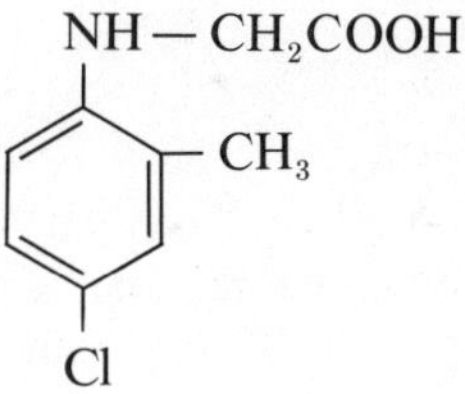

LIV

Fig. 11. Synthetic auxins. Side-chain effects.

XLV	2-Bromo-3-nitrobenzoic acid
XLVI	2-Chloro-5-nitrobenzoic acid
XLVII	2,3,6-Trichlorobenzoic acid
XLVIII	1,2,3,4-Tetrahydronaphthoic acid
XLIX	Indan-lyl-carboxylic acid
L	Floren-9yl-carboxylic acid
LI	4-Amino-3,5,6-trichloropicolinic acid
LII	2,4-Dichlorophenoxyacetic acid (2,4-D)
LIII	4-Chloro-2-methylphenylthioacetic acid
LIV	*N*-(4-chloro-2-methylphenyl)-glycine

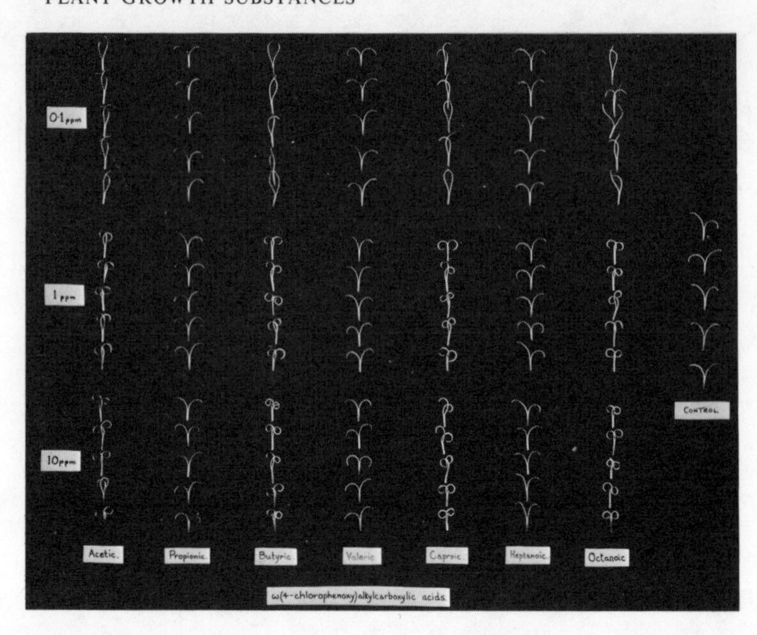

Plate 7. Alternation of auxin activity with increasing side-chain length in the ω-(4-chlorophenoxy)-alkanecarboxylic acid series tested in the split pea stem curvature assay. Concentrations applied: top row, 0.1 ppm; middle row, 1.0 ppm; bottom row, 10 ppm. The members of the homologous series in sequence in columns from left to right are; acetic, propionic, butyric, valeric, caproic, heptanoic and octanoic respectively. Extreme right: control in water. (Photograph by kind permission of Prof. R. L. Wain, Wye College, University of London.)

atoms between carboxyl groups and ring. Subsequently investigations were extended by analyzing ether extracts of treated material by paper chromatography. In the homologous 2,4-dichlorophenoxy series a strict alternation was observed in the wheat cylinder test, the Went pea curvature test, and the tomato leaf epinasty test (Wain & Wightman, 1954). Soaking wheat coleoptile cylinders or pea stem segments in these acids showed that all acids having an odd number of bridging carbon atoms in the side-chain were converted by the tissue to the acetic homologue (Fawcett, Taylor *et al.*, 1956). Further experiments on other homologous series show that ability to bring about β-oxidation to the acetic acid and the phenol may depend both on the tissue and on the nature of the ring-system (Wain & Wightman, 1954; Fawcett, Pascall *et al.*, 1959). We will return to this subject later.

The position of the side-chain may also have a very great effect in asymmetric ring-systems. The most striking are the indolylacetics themselves. IAA (indol-3yl-acetic) is the most active, indol-lyl-acetic (Jönsson, 1955) and indol-2yl-acetic (Schindler, 1958) are less active but indol-4yl-acetic (Thimann, 1958) has very low activity indeed. Equally puzzling are the naphth-lyl- and naphth-2yl-acetic acids: the former has an activity in various auxin assay tests as high as that of IAA, whereas the activity of the 2-substituted-compound is very slight. It is possible that this suprising difference may be due to secondary properties of the molecule, although theories based on molecular shape have been advanced to account for it in terms of true auxin activity (see p. 91).

The substitution of other elements to link the side-chain to the nucleus can also greatly modify the activity. Oxygen so introduced in phenylacetic acid, to form phenoxyacetic acid, merely eliminates the small activity of that compound, whereas in the naphthylacetic acids the effect is more marked. With naphth-lyl-acetic acid (Fig. 8 XXIII), which has high activity, its introduction to form naphth-lyloxy-acetic acid causes a very great reduction in activity, but with the slightly active naphth-2yl-acetic acid (Fig. 8 XXIV) the reverse is true. Thus the effects of position of the side-chain on the resultant activity of the molecule is seen to be closely dependent on the nature of the ring, and one is tempted to look for an explanation of these effects in the stereo-chemistry of the molecule. Such considerations will be dealt with in more detail later on, when more data have been presented. Although oxygen as a linkage element in the side-chain can bring about this greatly enhanced activity, other elements or groups can have the reverse effect. Thus, 2,4-dichlorophenoxyacetic acid (Fig. 11 LII) (2,4-D) is a compound as active as IAA in certain assay tests (e.g. *Avena* coleoptile cylinder straight growth test). Replacement of this oxygen by sulphur lowers the activity considerably in similar compounds (Fig. 11 LIII) (Wain, 1949; Wilske & Burström, 1950; Fawcett *et al.*, 1955). The activity is reduced to very low values or is destroyed by substituting the -S- group by an -SO- or an -SO_2- group (Wilske & Burström, 1950), and it is suggested (Erdtman & Nilsson, 1949) that this is caused by the removal of the two lone electrons of the -S- group, activity being dependent on the presence of these electrons at a suitable distance from the carboxyl group. Such considerations are also supposed to determine the activity of the other auxins and of IAA and 2,4-D. Substitution of an -S- group at an intermediate point in the side-chain destroys auxin activity and turns the molecule into an anti-auxin, e.g. NMSP (Fig. 44 CXLVIII) (see later). The activity arising from the substitution of a linking -NH- group to form the phenylglycines (Fig. 11 LIV) varies with the tissue and the nature of the substitution in the benzene ring. Thus high activity in the pea test has been claimed for the 2,4-D homologue (Veldstra & Booij; 1949; Fawcett *et al.*, 1955), while the corresponding homologue of naphth-2yloxy-acetic acid has negligible activity in the tomato parthenocarpy test (Wain, 1949). A systematic comparison of ten chloromethylphenylglycines with their corresponding phenoxyacetic acid homologues made by Clarke & Wain (1963) using the wheat coleoptile cylinder, the pea segment, and the pea curvature tests, showed that the phenylglycines were, on the whole, less active than the phenoxyacetic acids, although the effects of ring-substitution (see later) followed the same pattern in both series. There was one notable exception in that 3-chloro-4-methylphenylglycine was more active than the corresponding phenoxyacetic acid.

Substitution on a methylene carbon of the side-chain can have remarkable effects on auxin activity particularly in the acetic acids. The replacement of one hydrogen atom, say by a -CH_3 group, produced an asymmetric carbon atom and hence two optical isomers which show widely differing activities (see section on Stereochemical considerations, p. 89). Replacement of both hydrogen atoms by such groups usually produces a molecule completely inactive as an auxin but capable of opposing auxin action in growth, i.e. an anti-auxin (see later, Chapter VII). Thus in the naphthyloxy series, naphth-2yloxy-acetic acid (Fig. 12 XXVIII) and α-(naphth-2yloxy)-propionic acid (Fig. 12 LV) are highly active while naphth-2yloxy-isobutyric acid (Fig. 12 LVI) possesses negligible activity. The same phenomenon was shown

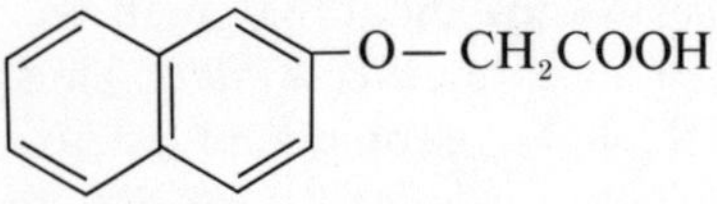

O—CH_2COOH

XXVIII

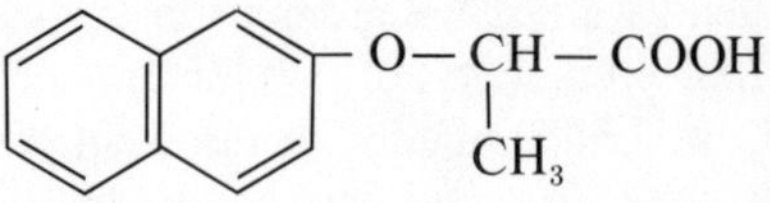

O—CH—COOH
|
CH_3

LV

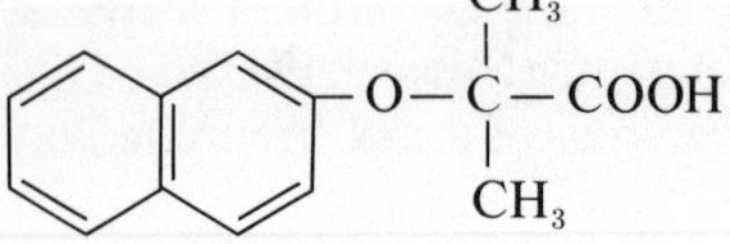

CH_3
|
O—C—COOH
|
CH_3

LVI

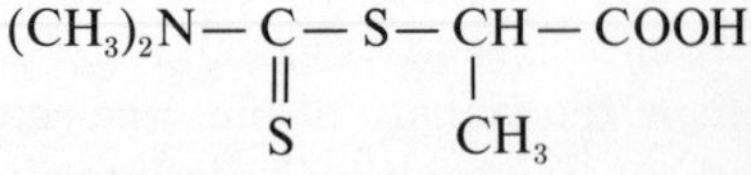

$(CH_3)_2$N—C—S—CH—COOH
‖ |
S CH_3

LVII

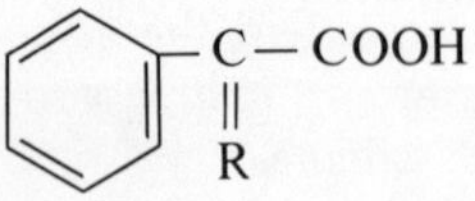

C—COOH
‖
R

LIX

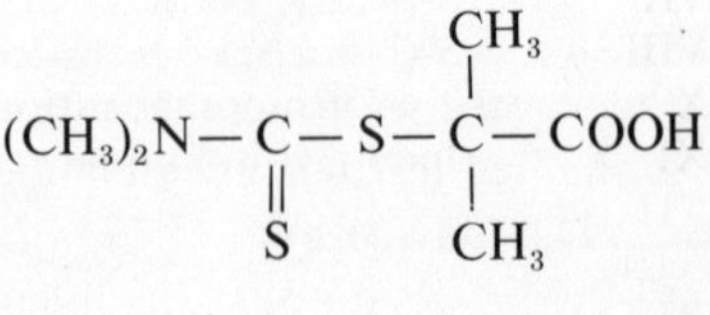

CH_3
|
$(CH_3)_2$N—C—S—C—COOH
‖ |
S CH_3

LVIII

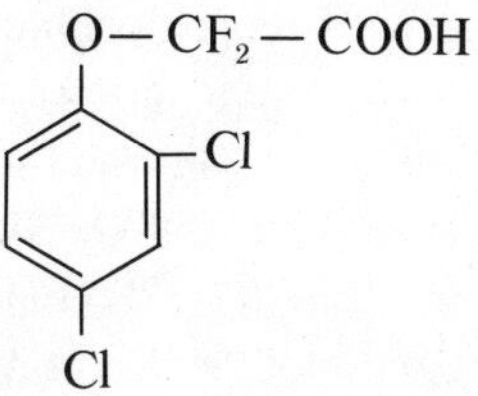

LX

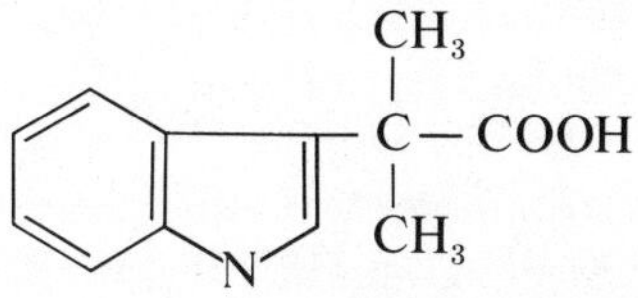

LXI

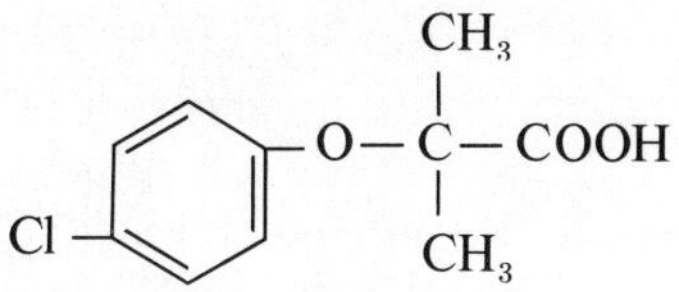

LXII

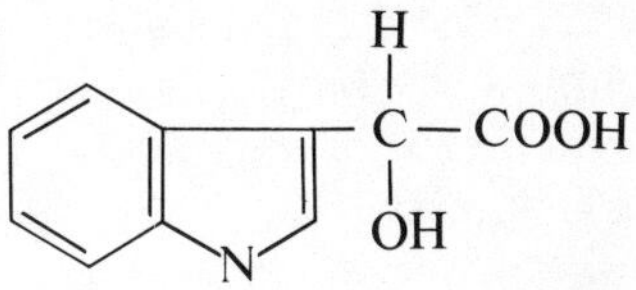

LXIII

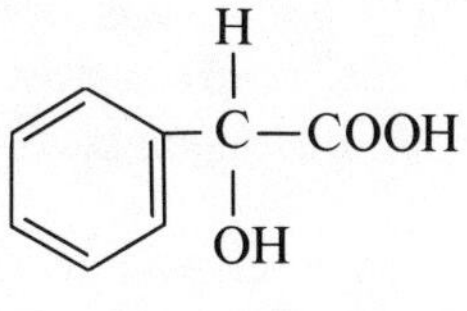

LXIV

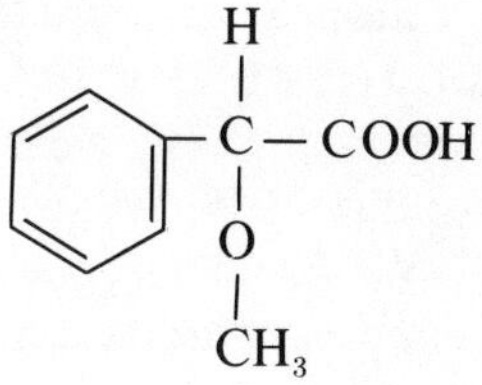

LXV

Fig. 12. Synthetic auxins. Side-chain effects (continued)

XXVIII	Naphth-2yloxy-acetic acid
LV	α-(Naphth-2yloxy)-propionic acid
LVI	Naphth-2yloxy-isobutyric acid
LVII	*S*-(l-carboxyethyl)-*N*,*N*-dimethyldithiocarbamate
LVIII	*S*-(l-carboxy-l-methylethyl)-*N*,*N*-dimethyldithiocarbamate
LX	2,4-Dichlorophenoxy-difluoroacetic acid
LXI	Indol-3yl-isobutyric acid
LXII	4-chlorophenoxy-isobutyric acid (PCIB)
LXIII	Indol-3yl-glycollic acid
LXIV	Phenylglycollic acid: mandelic acid
LXV	Phenylmethoxyacetic acid

to occur with several series of phenoxy acids (Wain, 1951) and phenyl-thio-acids (Fawcett *et al.*, 1955) in three different tests—the wheat cylinder test, the tomato epinasty test, and the pea curvature test. Even with the dithiocarbamates the same pattern is observed (Pluijgers & van der Kerk, 1961), the compound LVII (Fig. 12) being active but the compound LVIII (Fig. 12) being inactive. Wain (1949) suggests that the hydrogen atom is essential for the fundamental biochemical processes in which auxins are involved, including not only the promotion of stretching growth in these tests but also in other growth phenomena e.g. induction of seedless fruits in the tomato, initiation of rooting in cuttings, etc. (Osborne & Wain, 1951). A theory of action based on this assumption is discussed later.

In the mono-substituted series small substituents (e.g. methyl, ethyl or even propyl groups) give activity similar to that of the parent acetic molecule, while bulkier, straight or branched alkyl group substituents reduce activity e.g. in the indol-3yl- (Schlender *et al.*, 1966) and phenyl- (Fujita, Kawazu, Mitsui & Kato, 1966) series. A phenyl substituent may completely destroy activity (Veldstra, 1953).

However there are important exceptions to this α-hydrogen rule. Thus acids of formula LIX (Fig. 12), where R is $=CH_2$ or $=C(CH_3)_2$ are very active (Veldstra, 1955). The α-difluoro-derivative of 2,4-D (Fig. 12 LX), although inactive in the *Avena* coleoptile segment test, has definite activity in the pea segment test (Greenham, 1958) while indol-3yl-isobutyric acid (Fig. 12 LXI) shows weak activity in the *Avena* segment test but high activity in the pea curvature test (Thimann, 1958). Even 4-chlorophenoxy-*iso*butyric acid (PCIB) (Fig. 12 LXII), which has for a long time been used as a powerful antiauxin, has a weak auxin action on *Avena* first internode segments (Ng & Audus, 1964, 1965).

Wain and his colleagues (Fawcett *et al.*, 1955) have themselves shown that some iso-butyric acids have a low but definite auxin activity (e.g. 2,3,4-trichlorophenoxy-iso-butyric acid, and 2,4,5-trichlorophenoxy-iso-butyric acid in the wheat cylinder test, and some others in the pea curvature test). The suggestion that this activity might be the result of partial metabolic conversion in the tissues to the corresponding α-substituted propionic acid has not been confirmed experimentally. These discrepancies are further discussed later (pp. 181).

If an -OH group is a substituent on the α-carbon then even mono-substitution seems to reduce activity to very low values, e.g. indol-3yl-glycollic acid (Fig. 12 LXIII) (Thimann, 1958), phenylglycollic acid (Fig. 12 LXIV) (Takeda, 1959). A previous report of the complete inactivity of phenylglycollic (mandelic) acid (see Thimann, 1951*b*) is probably erroneous; while the racemic mixture is inactive, resolution shows the L(+) form to have a significant action on *Avena* coleoptile segments while the D(−) form is inactive and a strong antiauxin. In the racemic mixture, therefore, the activity of the L(+) form is completely blocked by the D(−) form, making the mixture inactive as a whole (Åberg, 1963). This illustrated clearly the dangers of drawing conclusions from tests on such racemic mixtures and points to the need for further studies on similar compounds.

Turning to the effect of the α-substituted -OH group, a change in the physical properties of the molecule and a greatly enhanced water solubility due to the hydrophilic -OH group, may, however, be playing a part in this lowered activity. This suggestion is supported by the fact that the substitution of a -OCH_3 group (e.g. on the α-carbon of the phenoxyacetic acids) gives compounds (Fig. 12 LXV) which are much more active (Mitchell *et al.*, 1959; Takeda, 1959).

(C) THE CARBOXYL GROUP

In the majority of instances, active compounds have shown a terminal carboxyl group or a group easily convertible into one. With the esters of IAA the activity decreases with the increasing size of the esterifying alcohol and probably runs parallel with the ease of enzymatic hydrolysis of the ester in the plant (Kögl & Kostermans, 1935). The activities of some esters have been reported to exceed those of the corresponding acids (Zimmerman & Hitchcock, 1937), but this again may be merely a secondary property of the molecule bringing about a more rapid tissue penetration. From time to time there have been indications in the literature that the sodium salts of the auxins have a greater specific activity than the corresponding acids. It seems most likely that such observed differences may result either from secondary properties of the salt (e.g. ease of penetration to the site of action) or from a specific stimulation effect of alkali ions on auxin activity (Thimann & Schneider, 1938*a*).

There are, however, a few compounds which do not possess a terminal carboxyl group, but which nevertheless have definite activities in some tests. In many instances the activity of these compounds is held to be caused by their conversion in the plant to the corresponding acid. Both indol-3yl-acetaldehyde and indol-3yl-ethylamine (tryptamine) have already been discussed in relation to auxin precursors. Other aldehydes (i.e. naphth-1yl-acetaldehyde and phenylacetaldehyde) have also been shown to be easily converted to the corresponding acid by enzymes present in *Avena* coleoptile tissue, two molecules of the aldehyde giving one of the acid: the conversion is therefore probably a dismutation (Larsen, 1951); the presence of an enzyme (aldehyde mutase) catalysing this reaction has been demonstrated in plant tissues.

From this one would expect the activities of the aldehydes in the *Avena* test to be exactly half those of the corresponding acids, whereas the actual ratio is of the order of 0.08. This discrepancy may be the reflection of a slow penetration or translocation of the aldehyde or its sluggish conversion to the acid in the coleoptile tissue. On the other hand it has been reported (Jones *et al.*, 1951) that 2,3,6-trichlorobenzaldehyde has a higher auxin activity than the corresponding acid in the root growth inhibition test, suggesting a much more rapid penetration into the tissues and a considerable degree of conversion into the effective acid. There seems little doubt that different tissues possess, to different degrees, the ability to oxidize aldehydes of this kind directly to the corresponding acids and even further. For example, an enzyme system capable of oxidizing phenylacetaldehyde has been isolated from tissues of a variety of plants (Kenten, 1953). The conversion goes almost entirely to benzoic and formic acids. The presence of such a system could easily account for the low yields of IAA from indol-3yl-acetaldehyde in the investigations described above. Furthermore, different tissues show markedly differing abilities to carry out this more complete oxidation. Pea root is extremely active, while dandelion root has virtually no activity. In spite of these various disturbing influences, it seems reasonably safe to conclude that the activity of the acetaldehydes as auxins depends entirely on their conversion to the homologous acetic acids. The activity, though slight, of such alcohols as tryptophol (β-(indol-3yl)-ethanol) (Fig. 5 VII) and β-(naphth-1yl)-ethanol (Veldstra, 1953) suggests that these two are oxidized, probably via the aldehydes to the corresponding acids in the plant.

Amides are similarly converted to the corresponding acids, and it has been shown (Jones *et al.*, 1949) that a very close correlation exists between the activities of the amides of a wide range of synthetic auxins (e.g. 2,4-dichlorophenoxyacetic acid, 4-chloro-2-methylphenoxyacetic acid, 2,4,5-trichlorophenoxyacetic acid, and naphth-2yloxy-acetic acid) and the activity of their corresponding acids. This has been firmly established by chromatographic analysis of extracts of plant tissues treated with the amides of a homologous series of ω-(2,4-dichlorophenoxy)-aryl-amides (Fawcett, Taylor *et al.*, 1956). Only those compounds possessing an odd number of bridging carbon atoms in the unbranched side-chain formed the homologous acetic acid. This shows that conversion of the amides to the corresponding acids must have preceded β-oxidation to the acetic acid. Similar feeding experiments with indole-3-acetamide have demonstrated the varying ability of different plant tissues to carry out the hydroloysis (Seeley *et al.*, 1956).

One dissenting voice has been raised against this comfortable explanation of amide activity. Johnston (1962) has studied the activities of eighteen *N*-substituted amides of 2,4-D of formula LXVI (Fig. 13), where R_1 = H, aryl or alkyl group and R_2 = aryl or alkyl group. The relative toxicities to plants (supposed to parallel auxin activity) of these homologues were compared with their rates of hydrolysis by KOH. No correlation could be found between these two molecular properties, which led Johnston to deduce that the amides were active as such and that the differing activities of his various compounds could be related to the electronic status of the -C(:O)-N- portion of the molecule. In other words the auxin activity of a molecule is not obligatately dependant on a carboxyl group but on the existence of a suitable donor group or nucleophilic centre in the side-chain. Such conclusions may be premature, since toxicity at high concentrations does not necessarily reflect auxin activity and, furthermore, N- substitution is likely to influence differentially the ease of access of the applied substance to the site of action in the plant. Studies of hydrolysis rates *in the plant* are needed for these compounds as well.

Indol-3yl-acetonitrile (IAN) is another molecule that is converted to IAA and this we might expect to account for its high auxin activity. This molecule, however, possesses special properties which put it into a category different from that of indol-3yl-acetaldehyde. It shows, for example, a higher auxin activity than IAA itself when subjected to the *Avena* coleoptile segment straight growth test. In Went's *Avena* curvature test their activities are comparable in low concentrations, but IAA is relatively more active in high concentrations. It also exerts a greater inhibition in high concentrations in the coleoptile straight growth test (Bentley & Housley, 1952). On the other hand, the nitrile is inactive in Went's pea stem curvature test except at high concentrations (Bentley & Bickle, 1952). Such confusing discrepancies demonstrate the inadequacy of any one assay used by itself for estimating auxin activity, and meanwhile exemplify the innate difficulties of biological testing. This particular confusion has been cleared up by demonstrations of the varying abilities of different plant tissues to convert the nitrile into the acid. This was first demonstrated by soaking coleoptile segments in IAN solutions and then testing those solutions by the pea stem curvature test. A response was obtained corresponding to a 25 to 50 percent conversion of the nitrile to IAA (Thimann, 1953).

A subsequent survey of 21 families of flowering plants showed that the enzyme responsible for the conversion (nitrilase) is not widely distributed, being demonstrable in only three families: the grass family (Gramineae), the cabbage family (Cruciferae) and the banana family (Musaceae) (Thimann & Mahadevan, 1964). IAN, then, is not active *per se* but must first be converted by the nitrilase to IAA. The same situation holds for the other nitriles. Fawcett, Taylor *et al.*, (1956) have studied a homologous series of ω-(2,4-dichlorophenoxy)-alkyl carboxylic acids and -alkyl-nitriles, using the wheat cylinder and split pea stem curvature tests. In the wheat cylinder test the typical alternation of activity with chain length was noted in both acids and

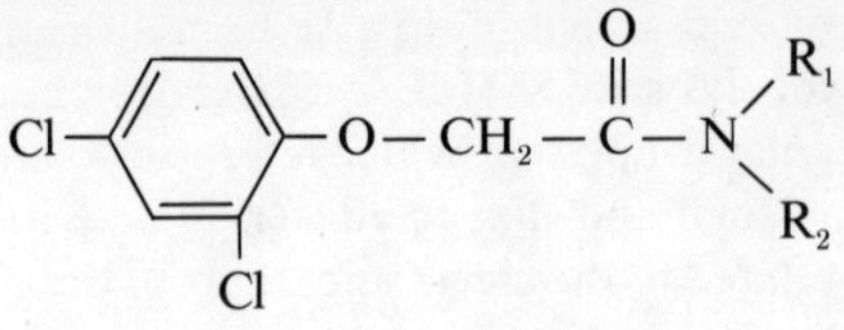

LXVI

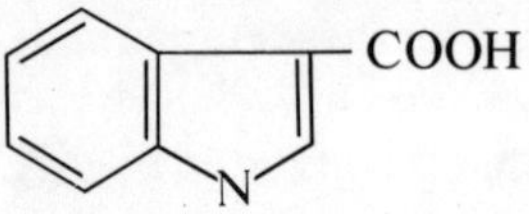

LXVII

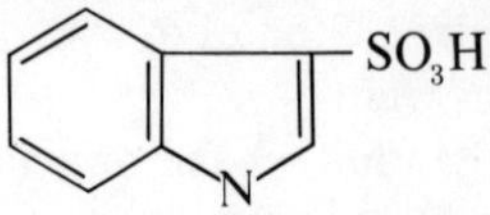

LXVIII

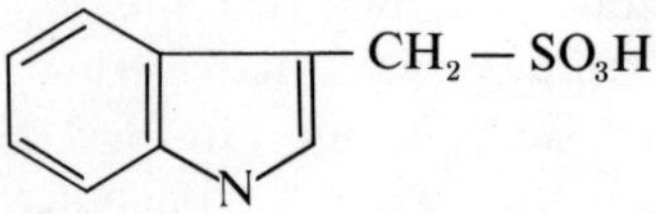

LXIX

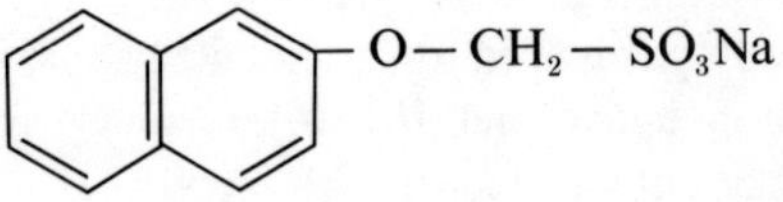

LXX

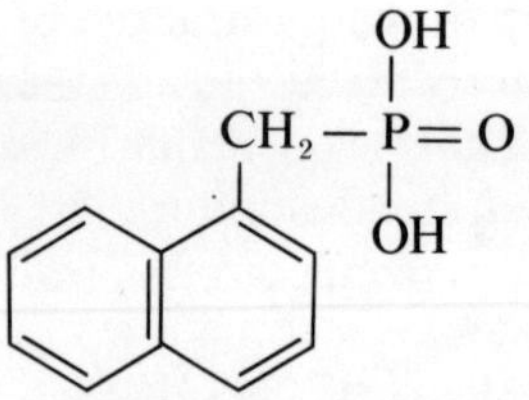

LXXI

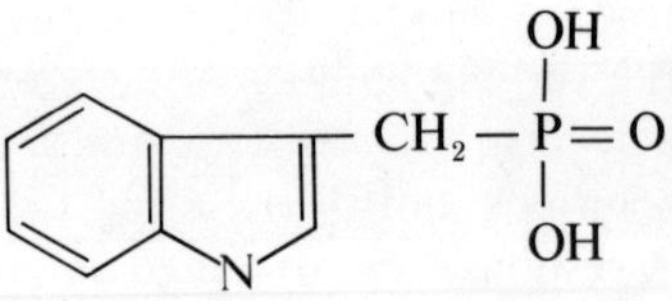

LXXII

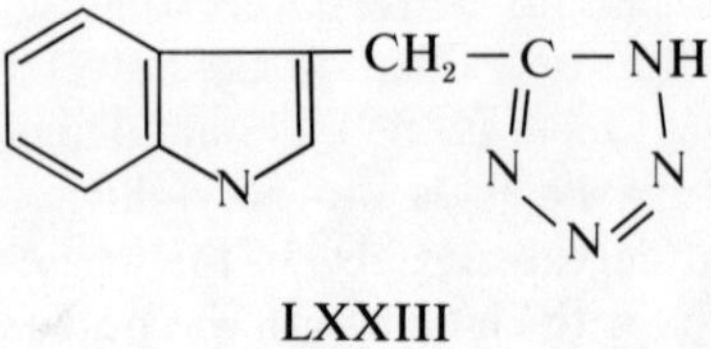

LXXIII

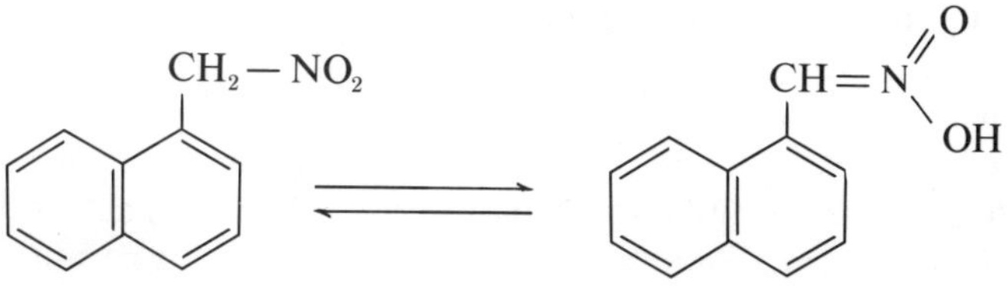

LXXIV

'aci-form'

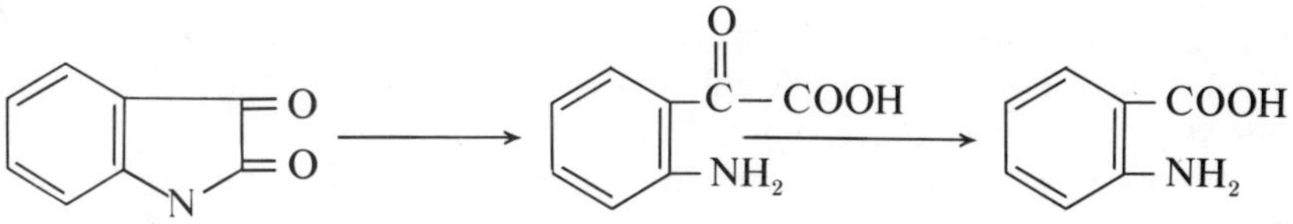

LXXV

LXXVI

III

Fig. 13. Synthetic auxins. Terminal acid groups.

LXVI	N-substituted 2,4-dichlorophenoxyacetamides
LXVII	Indol-3yl-carboxylic acid
LXVIII	Indican
LXIX	Indol-3yl-methanesulphonic acid
LXX	Naphth-2yloxy-methanesulphonic acid
LXXI	Naphth-lyl-methanephosphonic acid
LXXII	Indol-3yl-methanephosphonic acid
LXXIII	5-(Indol-3yl)-methyltetrazole
LXXIV	Naphth-lyl-nitromethane
LXXV	Isatin
LXXVI	Isatic acid: 2-aminophenylglyoxylic acid
III	Anthranilic acid.

nitriles. This behaviour ran precisely parallel to the production of 2,4-D by the tissues. In the pea test however only the acetonitrile was active, an unexpected finding as pea tissue does not contain the nitrilase. Subsequent study has indicated that some non-enzymatic hydrolysis may take place in that tissue (Mahadevan & Thimann, 1964).

While clearing up one source of confusion, however, these experiments introduced yet another, namely that all the nitriles showed some slight activity in the wheat cylinder test, irrespective of chain length, and this was accompanied by a corresponding slight production of 2,4-D. It has been suggested that, in this tissue at least, an 'α-oxidation' of a $-CH_2-CN$ group to a carboxyl group can take place to reduce the number of bridging carbon atoms by one, and allow β-oxidation of the carboxylic acid to follow. Support is given to this suggestion by the detection of indol-3yl-carboxylic acid (Fig.

13 XVII) in germinating peas (Cartwright *et al.*, 1956; Seeley *et al.*, 1956). This compound is most probably produced by the α-oxidation of IAN such as has been shown to occur in feeding experiments with wheat coleoptile tissue (Seeley *et al.*, 1956), thus adding one further complication to our comparisons of nitrile and acid activities in different tissues. The higher activity of the nitrile in the *Avena* straight growth test could be explained by its more rapid entry into the tissue and its conversion therein into IAA. A much more logical explanation arises from the observation that IAA activity in the pea test is greatly increased by the presence of the relatively inactive nitrile. The nitrile may therefore enter the coleoptile tissues and be partially converted into the acid. The unconverted nitrile could then potentiate the action of this acid to an extent sufficient to give the apparently superior activity reported for the nitrile in this tissue (Osborne, 1952). The relatively low activity in high concentrations in the *Avena* curvature test is also explainable in terms of such rapid transverse movement of the nitrile across the coleoptile that concentration differences, and therefore growth-rate differences, between the two sides cannot be maintained. This is to some extent supported by indications of a more rapid polar transport of the nitrile in coleoptile tissue. Clearly, at the moment, there is no real evidence that the nitrile is active *per se*.

There are other compounds, however, for which the experimental evidence suggests that activity is not the result of partial conversion to compounds with a terminal carboxyl group. One clear exception is indican (Fig. 13 LXVIII), a constituent of urine, with an activity about 5 percent that of IAA (Veldstra, 1944).

Indol-3yl-methanesulphonic acid (Fig. 13 LXIX), which is non-toxic and can be applied in high concentrations, can be clearly shown to be active at these high levels in the pea test. So can the homologous 2,3,6-trichlorophenyl homologue (Veldstra *et al.*, 1954). On the other hand compounds of the type of formula LXX (Fig. 13) have been shown to have negligible activity (Wain, 1949). Both the α-naphthyl- (Fig. 13 LXXI) and indol-3yl-methanephosphonic acids, which also have a low toxicity, show distinct auxin activities in the pea test in high concentrations (Veldstra *et al.*, 1954; Veldstra, 1955). The phenoxythioacetic acids, of general formula R-O-CH_2-CO-SH, where R may be various substituted phenyl groups, show growth-regulating activities as high as or sometimes even surpassing those of their phenoxyacetic acid homologues (Burström *et al.*, 1956). The tetrazole ring is also acidic and the compound 5-(indol-3yl)-methyltetrazole (Fig. 13) LXXIII) has a very high activity in both the pea curvature and the *Avena* cylinder tests (van der Westeringh, 1957). It would seem, therefore, that activity may be imparted to an otherwise suitable molecule by an acid group other than a carboxyl. This is one explanation which has been evoked for the small but definite activity of the compound naphth-1yl-nitromethane (Fig.

13 LXXIV). It has been suggested (Veldstra, 1944) that this compound is present in solution as an equilibrium mixture of two tautomeric forms and that it is the 'aci-form' of the molecule which possess the activity (see formulae LXXIV, Fig. 13).

Phenolic compounds also appear to have auxin activity. Phenol itself is inactive although naphth-l-ol has been claimed to act as a weak antiauxin (Åberg, 1950) (see later section). Di-ortho-substitution by halogens, however, introduces activity in four different tests, the 2,6-dibromo- and 2,6-di-iodophenols having very high activity (Wain & Taylor, 1965). This effect of substitution is reminiscent of the effect of similar substitution in the benzoic acids, where also the requirement for a side-chain (see section (B)) has been obviated (see also section (D)). Some misgiving is justified concerning the activities of the phenols as auxins *per se*. This is because there is very strong evidence that, with many phenols, action on growth is due to their effects on the activity of the enzyme catalysing the oxidation of the endogenous IAA (IAA-oxidase; see Chapter XI). Thus monophenols promote the oxidation and inhibit growth while diphenols and polyphenols inhibit the oxidation and promote growth (see for example Åberg & Johansson, 1969; Nitsch & Nitsch, 1962*b*). However, the suggestion that an influence of these monophenols on the IAA-oxidase activity of the test tissue could account for their growth-promoting action is strongly disputed by Harper & Wain (1969) who point out that the effects of ring substitution (see later, section (D)) on activity are quite different for the two actions.

An even more puzzling situation arises with the compound isatin (Fig. 13 LXXV) which has been shown to have auxin activity in pea stem segments, although at fairly high concentrations. If this substance is really an auxin in its own right, further revision of our ideas of structure/activity laws will be called for; but in view of its close chemical relationship with IAA, this naturally-occurring compound may owe its growth-promoting action to a disturbance of the natural auxin balance, either by acting as a precursor of IAA or by preventing its destruction (Galston & Chen, 1965). A study of the metabolism of ^{14}C-labelled isatin in pea tissues (Kutaček & Galston, 1968) showed that it was mainly converted to anthranilic acid (Fig. 5 III) which could produce small amount of tryptophan and therefore presumbaly IAA. Such a metabolic conversion might account for the activity of the isatin molecule. However, doubts on this explanation have come from the work of Wain and his colleagues (see Wain, 1968) which suggests that the activity is due to conversion to isatic acid (2-aminophenyl-glyoxylic acid Fig. 13 LXXVI). Comparisons of the activities of a range of substituted isatic acids and the corresponding anthranilic acids (Fig. 13 III) point to isatic acid having activity *per se* and *not* by virtue of its conversion to anthranilic acid. Both investigations suggest that isatin is not itself an auxin.

(D) SUBSTITUTION IN THE NUCLEUS

Early work on substitution in the pyrrole ring of IAA showed in all cases a reduction in activity (Kögl & Kostermans, 1935). This reduction was least for a CH_3– group, greater for C_2H_5–, and activity was completely destroyed by a CH_3O– group in any position. Substitution on the nitrogen atom brought about the greatest reduction in activity. On the other hand, mono-substitution of halogens or a methyl group in the phenylene moiety gives an increased activity (pea curvature test), the greatest effect being obtained in the 4- or 6-position (Hoffman, *et al.*, 1952; Porter & Thimann, 1965). A similar effect of substitution has been shown on the capacity to induce fruit-set (Sell *et al.*, 1953).

Marked effects of substitution are also seen in all the synthetic auxins, and the nature of the effect depends upon the nucleus and the type of side chain.

We will take first the series derived from phenoxyacetic acid which itself shows weak auxin activity. Here substitution of halogen or lower alkyl groups into the ring may greatly increase the activity, which in some compounds (e.g. 2,4-dichlorophenoxyacetic acid) and in some tests (i.e. *Avena* cylinder test) equals that of IAA itself. So far it has not been possible to formulate any rigid laws relating activity to nature and degree of substitution. The following broad generalizations can, however, be drawn. Effectiveness of substitution decreases in the following sequence, from left to right:

Cl Br $-CH_3$ I

Substitution in the 2- and 4-positions is usually more effective than in the 3-position. Single substitutions usually give less active compounds than double substitutions, but further increase to and beyond tri-substituted compounds gives no further improvement and may reduce the activity considerably. However, the effect of the substituent is closely dependent on the position of the substitution. Thus the small fluorine atom substituted in the 2-position reduces activity to almost zero whereas in the 4-position it gives reasonably high activity. In this case the differences are probably due to the electron-attracting properties of the fluorine substituent. In the 3-position a methyl-substitution produces a very weak auxin whereas the corresponding methylthio-substituted acid is 6 to 25 times stronger by virtue of the additional -S- atom (Åberg, 1967), presumably an effect of group size. We now have a wealth of data of this kind but the details are beyond the scope of this book. A general discussion of the implications of these effects will follow (p. 98–102).

There are very strong indications that halogen substitution in both ortho positions in phenoxyacetic and naphth-2yloxy-acetic acid completely destroys activity (Muir *et al.*, 1949; Hansch & Muir, 1950; Seely & Wain, 1950; Luckwill & Woodcock, 1956), suggesting that essential chemical reactions in the plant, in which the auxins take part, involve a combination at this carbon atom. The implications of this situation will be discussed in a later section. Another generalization concerning the effects of ring substitution has come from a study of the complete range of chlorine-substituted acids (Leaper & Bishop, 1951). It was found in tomato epinasty tests that, for high activity, two carbon atoms *para* to each other in the ring must be left unsubstituted. A theory of action involving quinone formation was built on these observations. The same general trend has been observed in the *Avena* cylinder test and in the West pea test, but there are too many exceptions for comfort. Thus in the pea test both the 2,3- and the 2,3,4- substituted compounds have a high activity and in the *Avena* coleoptile segment test the 2,3,4-substituted compound is nearly as active as the mono-substituted 4-chlorophenoxyacetic acid (Wain & Wightman, 1953). A further complication is the suggestion that complete inactivity is conferred in both *meta* positions (Toothill *et al.*, 1956).

The pattern is quite different in the benzoic and phenylacetic acid series. Benzoic acid itself is not an auxin but here *ortho*-substitution (by a halogen or a methyl group) *introduces* auxin activity. Activation increases with further substitution in the 3- or 6-position and reaches a maximum activity in 2,3,6-trichlorobenzoic acid (Fig. 11 XLVII). This is the reverse of the situation just noted for the phenoxyacetic acids where di-*ortho*-substitution destroys activity (Pybus *et al.*, 1959). Again the dimensions of the substituting group are of importance, iodine greatly reducing the activity *vis-à-vis* the corresponding chloro- or bromo- compounds. For example 2,5-di-iodobenzoic acid is almost completely inactive, as compared with the highly active 2,5-dichlorobenzoic acid (James & Wain, 1969).

Phenylacetic acid is a weak auxin and, in closely similar fashion, 2-, 3-, and 6-substitution of a halogen greatly enhances activity; values equalling that of IAA in a variety of tests have been obtained with 2,3-dichloro- and 2,3,6-trichlorophenylacetic acids (Jönsson, 1955; Pybus *et al.*, 1958; Gander & Nitsch, 1959). However the effects of substitution are quite different from those in the phenoxyacetic acid series. To quote just a few examples, 2-chlorophenylacetic is about 100 times more active then 2-chlorophenoxyacetic acid whereas with a 4-chloro-substitution the phenoxyacetic acid is about 10 times stronger than the corresponding phenyl acid. Again 2,6-dichlorophenylacetic acid is a medium-strength auxin whereas 2,6-dichlorophenoxyacetic acid has no auxin activity; indeed it acts as an antiauxin (Åberg, 1969) (see later, Chapter VII).

In the case of the phenols high auxin activity requires a substituent with high electron-attracting properties in the 2- and 6- positions. *Meta*-substitution reduces activity and *para*-substitution abolishes it (Harper & Wain, 1969). Possible explanations of all these differences will be put forward later (see pp. 101–2).

In most instances the substitution of other groups such as hydroxy-, alkyloxy- and nitro- reduce or destroy activity in whatever position they are introduced. For obvious reasons such compounds have received very little attention.

A different effect of substitution is its influence on the metabolism of the side-chain by plant enzymes (Wain, 1955*a*, 1955*b*). We have already seen (p. 72–4) that ω-aryl-alkylcarboxylic acids may show alternation of auxin activity with increasing length of the side-chain; this is due to β-oxidation resulting in either active aryl-acetic acid or inactive phenol. Extension of the early observations to a number of homologous series of these ω-substituted acids, and the employment of different assay tests, have revealed that the ability of any particular plant tissue to carry out the oxidation depends on the nature of the terminal aryl-group (see also Luckwill & Woodcock, 1956). Table III taken from the data of Wain (1955*a*) and Fawcett, Pascall *et al.* (1959) shows the effect of this phenomenon in tests with three different types of plant.

TABLE III

ALTERNATION OF AUXIN ACTIVITY IN HOMOLOGOUS SERIES OF CHLORINE-SUBSTITUTED ω-PHENOXY-ALKYLCARBOXYLIC ACIDS

(+ = alternation; O = no activity beyond the acetic acid)

Substitution position				Test		
2	3	4	5	Wheat coleoptile	Pea curvature	Tomato epinasty
*					O	
	*				+	
		*		+	+	+
*	*				O	
*		*		+	+	+
*			*	+	O	O
	*	*		+	+	+
*	*	*			O	
*		*	*	+	O	O

Clearly there are plant-group differences, the wheat enzymes seeming indifferent to the nature of the ring-substitution whereas pea and tomato enzymes are ineffective with substituents in certain positions. Where such insensitivity occurs it seems that the 2-substitution usually blocks the oxidation, the only exception being the 2,4-dichloro- series. The 4-chloro-2-methyl- series behaves similarly. These discoveries have very important implications for the practical use of these synthetic auxins.

(E) STEREOCONSIDERATIONS

We have already noted that a marked difference in activity exists between isomers of naphthylacetic acid and naphthyloxyacetic acid, and we have suggested that the explanation for this might lie in the difference in shape of the molecules. This phenomenon was first indicated in 1937 by Kögl, who showed that for the optically active α-(indol-3yl)-propionic acid (Fig. 14 XXI) the dextro-rotatory isomer was about thirty times more active in the *Avena* test then the laevo-rotatory form. However, it was decided from later work (Kögl & Verkaaik, 1944), that both compounds were equally active in the segment test and that the difference in the *Avena* test was due to a secondary property of the molecule resulting in differential basipetal transport rates in the coleoptile.

There is now overwhelming evidence that the primary action of the auxins is determined largely by the configuration of the terminal polar carboxyl group in relation to the ring-system. The first observation pointing to this was of the high activity of *cis*-cinnamic acid (Fig. 14 LXXVIII) and some of its derivatives in the cylinder test and the inactivity of the *trans*-isomer (Fig. 14 LXXVII).

Such facts led Veldstra (1944) to postulate that for the molecule to show activity the carboxyl group should not lie in the plane of the ring and that the maximum activity will arise when the dipole is perpendicular to that plane. He was led to this conclusion by contemplations of Stuart models of the above compounds. It will be seen that in both compounds the side-chain is free to rotate around the –C–C– bond attaching it to the ring. In the *trans*-form the models show that this rotation would take place through 360° and that therefore the average orientation of the carboxyl group would be in the plane of the ring. In the *cis*-form, however, complete rotation through 360° was not possible, and movement of the side-chain around this bond was restricted to a lateral oscillation on one side. In this case the average orientation of the carboxyl dipole would be at a definite angle to the plane of the ring and have a component perpendicular to it. According to Veldstra, this latter component is responsible for the activity of the *cis*-isomer. Similar considerations were shown to apply to the active *cis*- and inactive *trans*-forms of 2-phenyl-cycloprop-lyl-carboxylic acid (Veldstra & van der Westeringh, 1951). This principle was supported by observations on the *cis*- and *trans*-isomers of 1,2,3,4-tetrahydronaphthylidene-1-acetic acid (Fig. 14 LXXIX and LXXX). Stuart models show that the *trans*-form, which is physiologically inactive, has the carboxyl dipole in the plane of the ring, whereas the *cis*-form cannot be constructed without some strain in the hydrogenated part of the nucleus and the side-chain therefore does not lie in that plane. There is, therefore, a component of the carboxyl dipole perpen-

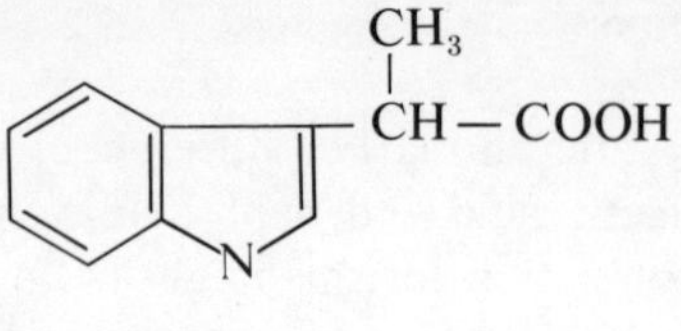

XXI

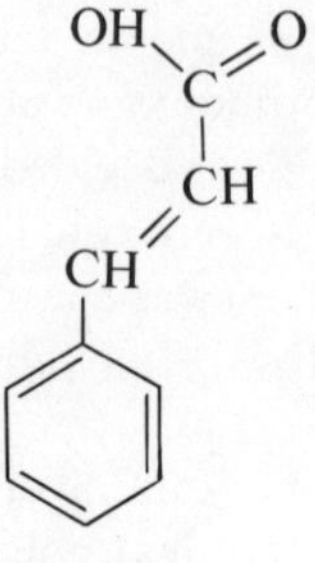

LXXVII 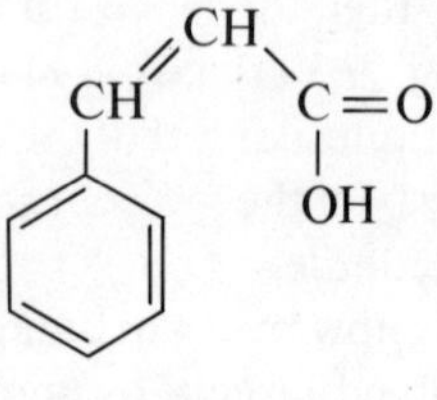 LXXVIII

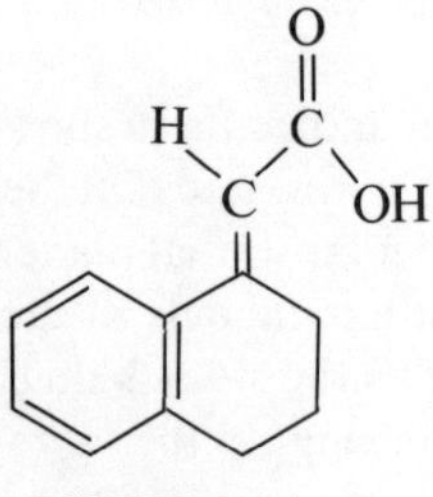

LXXIX (*cis*) 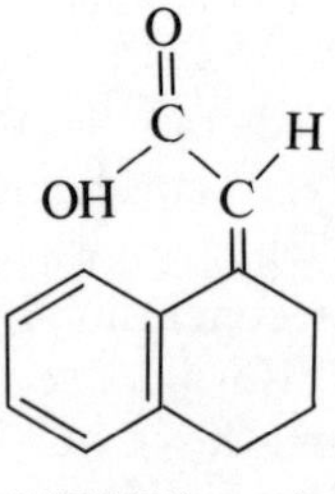 LXXX (*trans*)

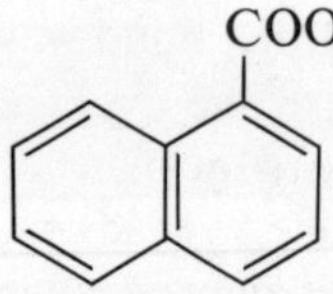

LXXXI 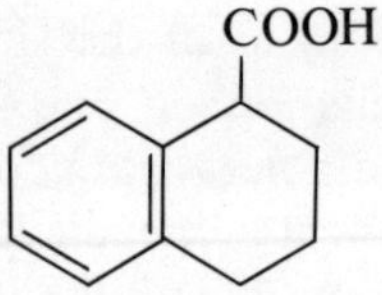 XLVIII 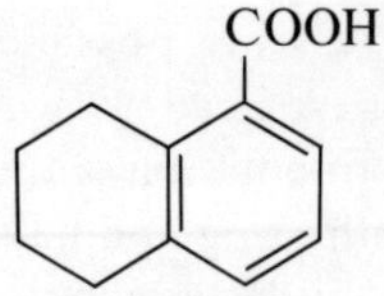 LXXXII

Fig. 14. Synthetic auxins. Stereo considerations.

XXI	α-(Indol-3yl)-propionic acid
LXXVII	*Trans*-cinnamic acid
LXXVIII	*Cis*-cinnamic acid
LXXIX	*Cis*-1, 2, 3, 4-tetrahydronaphthylidene-1-acetic acid
LXXX	*Trans*-1, 2, 3, 4-tetrahydronaphthylidene-1-acetic acid
LXXXI	1-Naphthoic acid
XLVIII	1,2,3,4-Tetrahydro-l-naphthoic acid
LXXXII	5,6,7,8-Tetrahydro-l-naphthoic acid

dicular to the plane of the ring, and this acid is physiologically active (Veldstra, 1944; Havinga & Nivard, 1948). Veldstra has, with some success, applied such theoretical considerations to the explanation of the activity differences of the napthyl- and napthyloxy-acetic acid isomers and also to account for the unexpected activity of the naphthoic acids and the substituted benzoic acids (see p. 72) (Veldstra, 1949). Among the former compounds l-naphthoic acid (Fig. 14 LXXXI) shows very small activity in the pea test. Hydrogenation of the ring to form 1,2,3,4-tetrahydro-l-naphthoic acid (Fig. 14 XLVIII) greatly enhances the activity. This hydrogenation has produced an asymmetric carbon atom at the position where the carboxyl is attached, and this necessarily brings the carboxyl group out of the plane of the ring and, on Veldstra's theory, increases activity. Hydrogenation to form the 5,6,7,8-tetrahydro-derivative (Fig. 14 LXXXII) retains the low activity (Veldstra, 1953, p. 167).

Similar explanations could be given for the activity of such compounds as 2,3-dihydrobenzothien-3yl-carboxylic acid (Fig. 15 LXXXIII) (Burström & Hansen, 1956) and indan-1yl-carboxylic acid (Fig. 15 XLIX) (Heacock *et al.*, 1958). Furthermore, the considerable enhancement of the activity of l-naphthoic acid by chlorine, bromine, iodine or methyl substitution in the 8-position (Fig. 15 LXXXIV) is consistent with the forcing of the carboxyl out of the plane of the ring by neighbouring substituents (Veldstra, 1956; Koshimizu *et al.*, 1960). A similar view is proposed for the high activity of 2,6-substituted benzoic acids (Veldstra, 1952). An interesting situation arises if the number of carbon atoms in the hydrogenated ring of 1,2,3,4-tetrahydro-l-naphthoic acid is increased by one or two to form 1,2-benzo-l-cycloheptene- and 1,2-benzo-l-cyclooctene carboxylic acids (Fig. 15 LXXXVIII). In these compounds the polar carboxyl group is virtually in the plane of the ring and the compounds have no auxin activity (Fujita, Kawazu, Mitsui, Katsumi & Kato, 1966).

But there are important exceptions to these generalizations. Thus Veldstra himself has shown that both 2-naphthoic acid and 1,2,3,4-tetrahydro-2-naphthoic acid (Fig. 15 LXXXV) are devoid of activity, whereas, on his theory, it is difficult to see how the latter compound could remain inactive. Two further exceptions are the inactive indan-2yl-carboxylic acid (Fig. 15 LXXXVI) and xanthen-9yl-carboxylic acid (Fig. 15 LXXXVII) (Heacock *et al.*, 1958).

A clue to the solution of these problems came with the recognition that the two optical isomers in a racemic mixture may have very different auxin activities. Thus Veldstra has shown that in 1,2,3,4-tetrahydro-l-naphthoic acid the activity was entirely due to the (−) form, the (+) form being completely inactive (Veldstra & van der Westeringh, 1951). Later Smith & Wain, (1952) succeeded in resolving a racemic mixture of the compound α-(2-naphthyloxy)-propionic acid. Here the (+) isomer has high auxin activity,

COOH
S

LXXXIII

COOH

XLIX

R COOH

LXXXIV

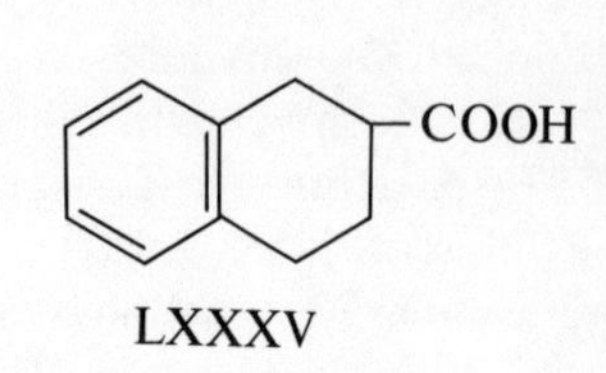

LXXXV

COOH

LXXXVI

COOH
O

LXXXVII

COOH

LXXXVIII

Fig. 15. Synthetic auxins. Stereo considerations (continued).

LXXXIII	2, 3-Dihydrobenzothien-3yl-carboxylic acid
XLIX	Indan-1yl-carboxylic acid
LXXXIV	8-Substituted l-naphthoic acid
LXXXV	1,2,3,4-Tetrahydro-2-naphthoic acid
LXXXVI	Indan-2yl-carboxylic acid
LXXXVII	Xanthen-9yl-carboxylic acid
LXXXVIII	1,2-Benzo-l-cyclooctene carboxylic acid

not only in the pea curvature test but also in five other auxin assays, whereas the (−) isomer is inactive.

At the present time we have data on the comparative activities of a considerable number of isomeric pairs of plant growth regulators. The situation has been reviewed by Jönsson (1961) and more recently by Fredga & Åberg (1965). The compounds listed exhibit a range of structure almost as wide as that covered so far in this chapter (see following generalized formulae A and B).

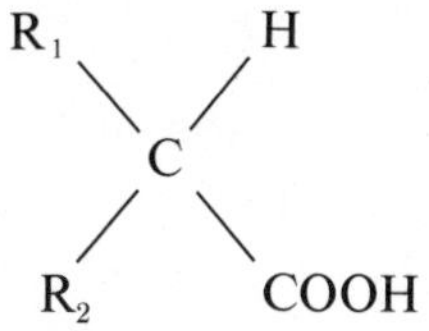

R_1 H C R_2 COOH

A

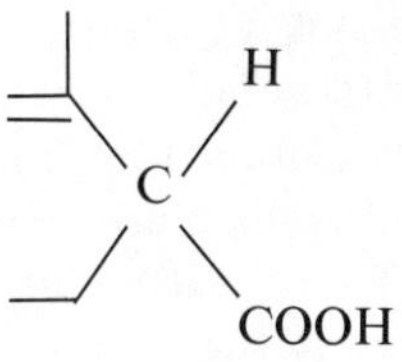

H C COOH

B

They include compounds where the ring-group R_1 is an indolyl-, a phenyl-, a phenoxy- (variously substituted), an anilino-, a naphthyl- (also substituted) or a naphthyloxy-group, and even the ring-like dithiocarbamyl-group. The other substituent (R_2) may be methyl- (α-propionic acids), ethyl- (α-butyric acids) or other alkyl- group (e.g. allyl- in α-allylphenylacetic acid). It may be a hydroxy- group, as we have seen in mandelic acid. Compounds where the asymmetric carbon atom is part of the ring (e.g. dihydro- and tetrahydronaphthoic acids and indanylcarboxylic acid; formula B above) are also included.

It is of particular interest that the majority but not all of the active isomers belong to the D configuration series (i.e. they are related sterically to D(−)lactic acid and are mostly dextroratatory (+). The L(−) forms are usually inactive as auxins but may actively antagonise auxin actions (see later section). In racemic mixtures therefore they will not only dilute the activity of the D(+) isomer but may positively suppress it completely. The lack of activity reported earlier for α-propionic acids may have been the result of testing racemic mixtures of this kind. The few exceptions to the above rule include L(−)α-(indol-3yl)-propionic acid, L(−)α-(naphth-lyl)-propionic acid and L(−)α-(benzothien-3yl)-propionic acid. The most puzzling is the indolyl- acid, where activity depends on the test employed, both forms being equally active in the *Avena* segment test and the L(−) form being slightly more active than the D(+) form in the flax root inhibition test (Åberg, 1958). In view of their relationships with native auxin (IAA) one wonders whether metabolic interconversions may not enter and confuse the picture in these tests. The L(−) form of the other two isomeric pairs have a definite auxin activity although it is lower than that of the D(+) form. The explanation of these departures from a fairly well-established rule still eludes us.

Theoretical Considerations of Structure/Activity Relationships

In the search which is constantly going on for new synthetic growth regulators it is clear that an exact knowledge of the mechanism in the plant could be of great service. To this end therefore much attention has been paid to structure/activity relationships and to physical properties of active molecules which are similarly correlated with activity.

One of the earliest observations was that extension growth of many organs is much greater in acidic than in neutral or alkaline media. The optimum pH usually falls between 4 and 5. The native auxins were soon involved to explain this phenomenon by assuming that they are active only in the undissociated state. Being weak acids, they would be almost completely undissociated and hence active at pH 4 to 5. At higher pH they would be more dissociated and thus less active. Support for this hypothesis came from observations that the activity of synthetic auxins in various tests varied inversely with the degree of ionic dissociation as calculated from pK values. Furthermore, in comparing a number of synthetic auxins, it was noted that their activities, calculated per unit of undissociated acid, were all of the same magnitude (D. M. Bonner, 1938). Thus cinnamic acid, a weak auxin, on a total concentration basis, was now as strong as IAA. It was therefore supposed that the main differences in activity of the synthetic auxins resulted from this ionic dissociation in aqueous solution. A quarter of a century of subsequent research has shown that this is not so. It does not explain the different activities of the naphthylacetic acids and their higher homologues (Veldstra, 1944), or of phenoxyacetic acid and its mono-halogen derivatives (Muir *et al.*, 1949). Ionic dissociation of the acid undoubtedly modifies the effectiveness of the molecule at any given total concentration, by regulating penetration rates through the cell membrane (Wedding & Erickson, 1957); other overriding properties determine intrinsic activities.

Another line of enquiry was stimulated by the obligate double bond in the nucleus. On the analogy of the mode of action of certain known enzyme systems, it was supposed that the auxins might be playing the part of organic catalysts in some metablic process essential for extension-growth and that the catalytic action might depend on a reversible oxidation-reduction of this double bond. As a result of experiments designed to check this possibility, Veldstra (1944) stumbled upon a property of the auxin molecule which shows a considerable degree of correlation with the growth-promoting activity. This was the tendency of active compounds to be adsorbed at a water-mercury interface. Thus compounds of high auxin activity were strongly adsorbed whilst those of low activity were not. This was shown to hold for the naphthylacetic acids and for the chlorinated phenoxyacetic acids. Exceptions were found, e.g. in *cis*- and *trans*-cinnamic acids (see p. 89) which showed identical surface activities but markedly different auxin actions. Veldstra therefore modified the structural requirements postulated by Koepfli *et al.* (1938), adding that the ring must have a high surface activity, and, as we have already seen, the carboxyl dipole must have a component perpendicular to the plane of the ring. Veldstra visualized the action of the auxins as depending on adsorption to the lipoid constituents of the membranes bounding the cell protoplast. By the solution of the lipophilic end of the auxin molecule in the membrane a powerful 'turgescent' action could be exerted, increasing permeability to dissolved substances. In this way the food supply and the growth of the cell could be regulated. Using the rate of outward diffusion of the red pigment from beetroot slices as a measure of this permeability, Veldstra (1949) compared surface activities and effects on permeability of several members of a homologous series of halogenated phenoxyacetic acids. He found that the two ran closely parallel, but the auxin activity in the series was not similarly correlated. Indeed, it rose to a maximum in compounds having an intermediate value of surface activity. Further modifications of his theory were therefore required; it was suggested that a correct balance should be maintained between the lipophilic activity of the ring and the hydrophilic activity of the carboxyl (or similar) group terminating the side-chain. Much further evidence has been produced in support of these contentions. For example, the introduction of a second carboxyl group either on the same side-chain (e.g. the aryloxysuccinic acids, Fig. 16 LXXXIX) or as part of another ring-substituent (e.g. on the $-CH_3$ group of MCPA, Fig. 16 XCI and XCII) (Julia & Baillargé, 1953, 1954; Thimann, 1958) destroys activity, presumably by making the molecule too hydrophilic.

A theory based on similar requirements has been put forward by van Overbeek (1961) who thinks that membrane-bound enzyme systems rather than membrane permeability are regulated by the adsorbed auxins. The nature of the ring substituents, the length and orientation of the side chain etc., determine the degree to which the molecule will be drawn into the membrane and the extent to which the hydrophilic (carboxyl) group will project and interact with the surface-borne network of enzymes where auxins exert their primary regulatory action. These theories, however, still remain too generalized to be of any great practical importance. Furthermore, the dangers of over-generalization have been stressed by Brian (1967) who

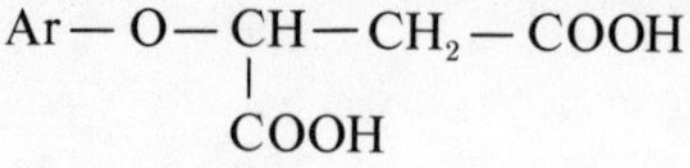

$Ar-O-CH(COOH)-CH_2-COOH$

LXXXIX

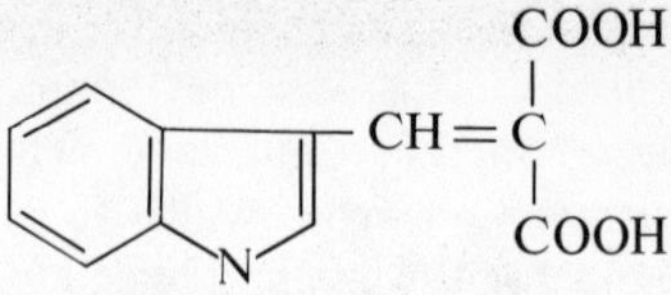

$CH=C(COOH)_2$

XC

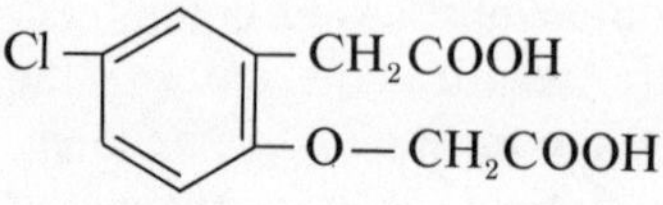

CH_2COOH, $O-CH_2COOH$, Cl

XCI

CH_3, $O-CH_2COOH$, Cl

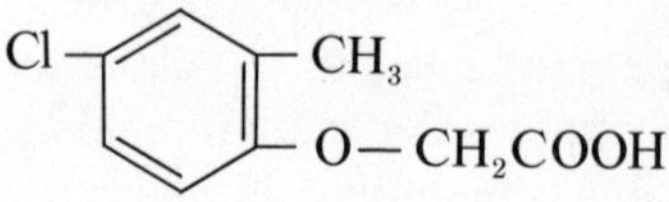

XCII

Fig. 16. Synthetic auxin analogues.

LXXXIX	Aryloxysuccinic acids
XC	Indol-3yl-methylenemalonic acid
XCI	2-Carboxymethyl-4-chlorophenoxyacetic acid
XCII	4-Chloro-2-methylphenoxyacetic acid (MCPA)

has made a study of the adsorption of a large number of ring-substituted aromatic acid auxins to protein monolayers from oat. He could find no direct relationship between adsorption and activity to embrace all his compounds and pointed out that any such relationship which did exist might be masked by different adsorption patterns, at sites not related to growth centres, where the auxin would be immobilized and rendered ineffective.

We have already seen that Wain regarded the possession of an α-hydrogen atom as essential for auxin activity (p. 78). He suggested that activity depends primarily on three essential units of structure, i.e. an unsaturated ring-system, a carboxyl group, and at least one α-hydrogen atom, all having a definite spatial relationship with each other, so that the auxin molecule can 'fit on' at the centre of its action in the growing cell (Smith & Wain, 1952). This 'fitting' was visualized as involving a three-point contact of these essential structural units, each with its own specific reaction point in the growth-centre. Lack of any one of these three units, or an unsuitable configuration of them, would result in no activity. This is illustrated in Fig. 17. What is more, one would expect inactive compounds, possessing groups that fit at two points only, to compete with active molecules for the growth centres and thereby reduce their activity. This was confirmed by demonstrating the very strong competitive inhibition of the auxin action of the D(+)

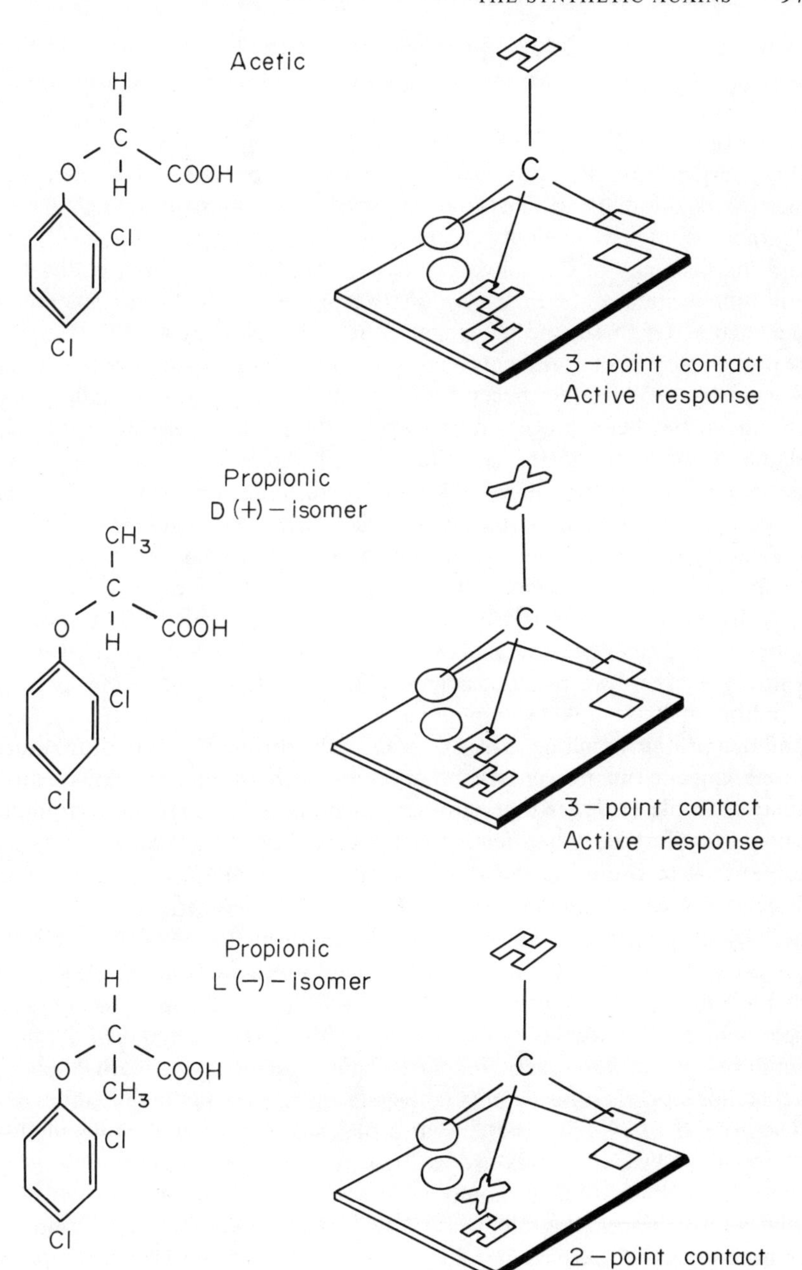

Fig. 17. Diagrammatic representation of Wain's 'three-point-attachment' theory, to account for the relationship between structure and activity in the auxins.

isomer of α-(naphth-2yloxy)-propionic acid by the L(−) isomer in the *Avena* segment test. As we have seen a large number of other racemic acids have now been resolved. In almost all, the inactive isomer blocks the action of the active one (Smith *et al.*, 1952; Matell, 1953; Åberg, 1953; Fredga & Åberg, 1965). Again complete absence of activity has been demonstrated for a number of *a-iso*butyric acids (Fawcett *et al.*, 1953; Jönsson, 1961). All these characteristics give weighty support to the 'three-point-contact' theory and the necessity of the α-hydrogen atom for true auxin activity. But there still remain some apparent exceptions that have not so far been moulded into the theory. Of these, the halogenated benzoic acids are perhaps the most exploited in potent arguments against the three-point contact theory (Muir & Hansch, 1953). Other exceptions have already been pointed out on page 78, and it has been suggested that activity depends on general molecular shape rather than on the substitution and configuration of groups on a particular carbon atom in the auxin molecule. Indeed it has been suggested (Jönsson, 1961) that it is not the *presence* of an α-hydrogen atom which is obligatory for activity but the *absence* of a large atom or group preventing the proper contact of the rest of the molecule with the active sites in the cell.

A theory resembling the above and based on the inactivity of the di-*ortho*-substituted phenoxyacetic acids was put forward in 1951 (Hansch *et al.*). Known as the 'two-point-attachment' theory, it supposed the auxin to combine at its carboxyl group and at a free *ortho*-position with a cysteinyl unit of a protein as in Fig. 18 (A). Di-*ortho*-substitution blocks the attachment to the ring and thus prevents the compound from acting as an auxin. Although there is ample evidence that auxins may be 'bound' on to plant proteins, yet this particular theory has been much criticized on the grounds of the many exceptions to the proposed di-*ortho*-substitution rule. Thus, for example, Thimann (1952) claimed that 2,6-dichlorophenoxyacetic acid has some activity in the pea test, and this has been confirmed by Wain & Wightman (1953) and extended to include 2,3,6-trichlorophenoxyacetic acid, α(2,6-dichlorophenoxy)-propionic acid, and α(2,6-dichlorophenoxy)-butyric acid, which are all active in the wheat cylinder test (Fawcett *et al.*, 1955). A similar series of 2,6-dimethylphenoxy- and 2,4-dichloro-6-methylphenoxy-acids show considerable activity in the pea curvature and *Avena* cylinder tests (Osborne *et al.*, 1955). The activities of 2,4-dichloro- and 2,4-dibromo-6-fluoro-phenoxyacetic acids are as high as is that of 2,4-dichlorophenoxy-acetic acid, whereas they should be no greater than the very low values of their 2,6-dichloro- homologues (Wain, 1953). On the other hand the complete lack of activity in *n*-butyl-5-chloropyridinoxy-2-acetate (Fig. 19 XCIII) (Crosby & Vlitos, 1958) is strong support for the theory since there is no hydrogen to replace on either side at the two −N = atoms in the ring.

The clearest exceptions to the theory as it stands (see section D) are the

(A) Cl — C_6H_3(Cl) — O — CH_2 — COOH + HS — CH_2 — CH(NH_2) — C=O — NH — PROTEIN ⟶ Cl — C_6H_2(Cl)(O — CH_2 — CO — NH —) — S — CH_2 — CH — C=O — NH — PROTEIN

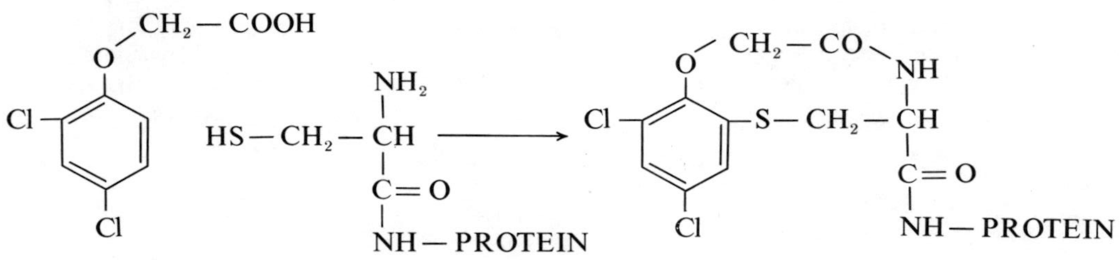

(*After* Hansch, Muir & Metzenberg, 1951)

Fig. 18. Diagrams showing the characteristic features of 'two-point-attachment' theories.

(A) Theory of Hansch, Muir & Metzenberg (1951), involving covalent linkage of carboxyl and an *ortho* carbon of the auxin with amino and sulphydryl groups respectively of a protein molecule.

(B)

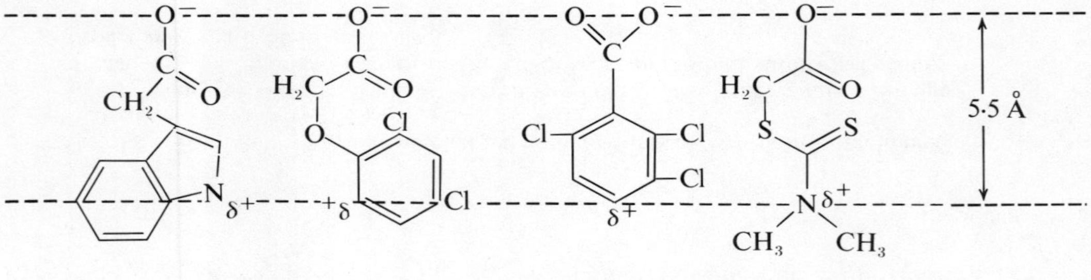

5·5 Å

IAA — 2,4-D — 2,3,6-TBA — Dithiocarbamates

(*After* Porter & Thimann, 1959)

Fig. 18. continued.

(B) Theory of Thimann showing the 5.5 Å space separation of charges on the molecule of four active auxins, indol-3yl-acetic acid (IAA), 2,4-dichlorophenoxy-acetic acid (2,4-D), 2,3,6-trichlorobenzoic acid (2,3,6-TBA), and the dithiocarbamates.

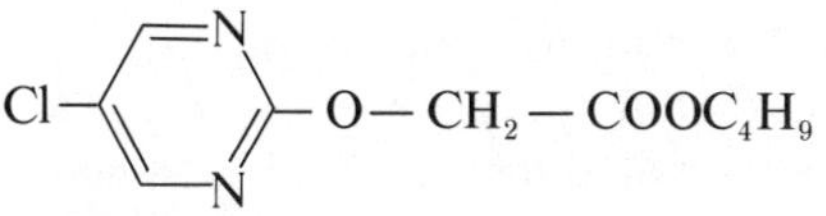

Cl— N N —O—CH_2—$COOC_4H_9$

XCIII

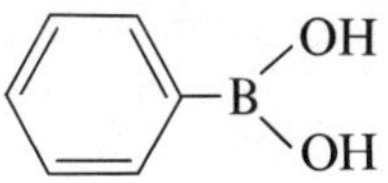

—B(OH)(OH) XCIV

Fig. 19. Synthetic auxins.
XCIII *n*-butyl-5-chloropyridinoxy-2-acetate.
XCIV Phenylboric acid.

substituted benzoic acids and phenylacetic acids. The proponents of the two-point-attachment theory have tried to explain these exceptions on chemical grounds which cannot be discussed here, by suggesting that *ortho* chlorine atoms are easily displaced by the nucleophilic substrate groups of plant proteins during the relevant growth reaction. A production of ionic chlorine from these molecules during response has also been claimed, but weighing strongly against this suggestion is the high activity of 2-6-dimethyl-substituted acids; for example, 3-chloro-2,6-dimethyl-, 3-bromo-2,6-dimethyl- and 3-iodo-2,6-dimethylbenzoic acids are just as active in the pea test as is the trichloro-homologue (Veldstra & van der Westeringh, 1952). It seems extremely unlikely that methyl groups could be displaced from the benzene ring as chlorine is supposed to be.

The particular difficulty of the benzoic acids was resolved when it was pointed out (Thimann & Leopold, 1955) that attachment should involve two points on the receptor molecule which should be a specific, fixed distance apart. As there is no linking carbon atom between ring and carboxyl in the benzoic acids, the combining point corresponding to the *ortho* position of the phenoxy acids would have to be the *meta* position in the benzoic acids. Blocking of the *ortho* positions with substituents would not therefore be expected to have any effect on activity.

Thimann has since modified and extended this theory to explain the activity of the major groups of the auxins. Considerations of electron distributions in the molecule in relation to substitution etc., led him to conclude (Porter & Thimann, 1959; Thimann, 1963) that active molecules all possess a fractional positive charge at some point in the molecule at a distance of about 5.5 Å units from the negative charge of the dissociated carboxyl (or similar) group. This is illustrated in Fig. 18 (B) for four widely different auxins.

This specific charge separation of the dipolar auxins allows them to be bound to the active surfaces in the cell which have a pattern of similarly spaced complementary charges. Such an association of an auxin with its site of action does not involve a chemical reaction as in the two-point-attachment theory of Hansch & Muir. We can therefore explain the activity of the tetrazoles (see p. 84) which on that theory would have to form somehow a covalent bond with the $-NH_2$ group of a protein, a process virtually impossible to imagine. On the other hand the tetrazole group does dissociate a H^+ ion and become negatively charged. Thimann's theory also explains why di-*ortho*-substitution destroys activity in the phenoxyacetic acids but enhances it in the benzoic and phenylacetic acids. Thimann also claims that it can explain the long-standing puzzle of inactivity in the di-*meta*-substituted acids in all three series since such substitution would confer positivity on the 1– position of the ring, giving a charge-separation completely unsuitable for activity. A check on this theory has been made (Porter & Thimann, 1965) by studying the effects of halogen and methyl substitution on the auxin activity of IAA (pea curvature test) and on the fractional positive charge at the -NH- position estimated from infra-red absorption data. Chlorination in the 2- and 5- positions gave greatly increased activity while methylation in the same positions decreased it. The spectroscopic data indicated that the fractional positive charge was reasonably correlated with auxin activity in these compounds and IAA. One exception, the 5,7-dichloro- derivative, having a low auxin activity and an apparently high fractional positive charge, points to the need for further work. A further difficulty arises with the substituted phenols which show auxin activity, since the charge distribution on the molecule does not fit Thimann's theory, and Harper & Wain (1969) bring forward arguments that characteristics of these active phenols fit more properly into a modified 'two-point-attachment' theory.

Auxin Synergists

From time to time several substances, inactive either in themselves or at the concentrations employed, have been claimed to augment the growth action of auxins. The word synergism (Gk. *synergos* = working together), borrowed from similar phenomena of drug action in man, has been applied to these processes. The compounds exhibiting these properties have been called auxin synergists. One such example is TIBA (Fig. 43 CXLII). In low non-inhibiting concentrations (0·1 mg/l) it will considerably increase the effectiveness of IAA in the *Avena* curvature test (Thimann & Bonner, 1948). Another comparable compound is indole; this is claimed to augment the action of auxin in promoting the rooting of *Ageratum* cuttings (van Raalte, 1951). Many other examples could be quoted. Their chief characteristic is their chemical diversity, while the mechanisms of their actions may be equally diverse. Since most of these actions are probably of an indirect nature, operating metabolically or by some other means to raise the level of native or applied auxins at the site of action in the cell, a consideration of them will be left to a more appropriate chapter (Chapter XI). So far their study has helped very little to advance our knowledge of structure/activity relationships in the auxins and related compounds.

General Considerations

The mind of the reader will no doubt be confused by the plethora of schemes proposed to explain structure/activity relationships in substances active in auxin tests. He will be dismayed by the frequency with which exceptions occur to every proposed rule. The search for a consistent underlying theme of structure or property still goes largely unrewarded. It is of course possible that growth responses may be released, directly or indirectly, in a variety of ways, and that each may require a different molecular structure or property. This is clearly indicated by the study of the activity of chelating agents (Heath & Clark, 1956). Interaction studies with IAA indicate that they are not behaving in the same way as IAA does when it promotes extension growth. To give an example from another type of regulator, phenylboric acid (Fig. 19 XCIV) and various ring-substituted derivatives have a remarkable stimulating action on root extension and therefore have a claim to be grouped with the auxin antagonists (see Chapter VII). But the relatively small effects of ring-substitution on activity (Torssell, 1956), and other physiological considerations (Odhnoff, 1961), indicate that their action has nothing in common with that of the auxin antagonists (root auxins) described in Chapter VII. May not such considerations also apply to some of the exceptions to structure/activity rules discussed previously? There is a great need to extend the exploration of these rules by probing deeper into the growing cell, to the molecular level of growth-regulator operation. It seems to the author that the onus is now on the biologist to devise techniques for doing this, thereby giving a new impetus to this important aspect of the study of growth substances.

Chapter IV

The Gibberellins

The Original Isolation of the Gibberellins and their Establishment as Plant Hormones

In view of the dominant position that the gibberellins have taken up as plant hormones in the last decade, it is unfortunate that their recognition as such was so long delayed. We have seen in the first chapter that the eventual discovery of these growth-regulating substances stemmed from studies of the *bakanae* disease of rice, which had long been recognised (Ito, 1932) as one of the three most serious diseases of this vital crop. The disease is world-wide and has been the subject of scientific articles by plant pathologists (see Stodola, 1958). As early as 1921 the plant pathologist Sawada had recognised the growth-stimulating action of the fungal mycelium and a few years later his student, Kurosawa (1926), showed conclusively that the stimulating principle was a heat-stable chemical secreted by the fungus. This discovery prompted a series of investigations into the active material but it was not until another ten years had passed and a reliable bioassay using rice seedlings had been perfected that a group in the University of Tokyo succeeded in preparing a crystalline sample of it (Yabuta, 1935). Yabuta christened it 'gibberellin'. Finally in 1938 the group announced the isolation of two crystalline compounds which they called gibberellin A and gibberellin B (Yabuta & Sumiki, 1938). The latter compound, later called gibberellin A_1 (Fig. 20 XCVI) was extremely active in the rice seedling assay. The other, (Fig. 20 XCVIII) was weakly active and could be produced from gibberellin A by heating (50 – 70°C) with dilute acid. Degradation studies established both these compounds as derivatives of the tricyclic hydrocarbon fluorene (Yabuta *et al.*, 1941).

Language barriers and a world war prevented the news of these extremely important discoveries from spreading to the western world and it was not

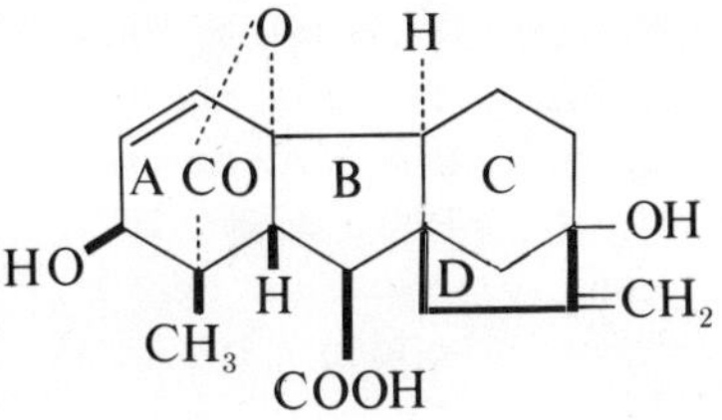

XCV 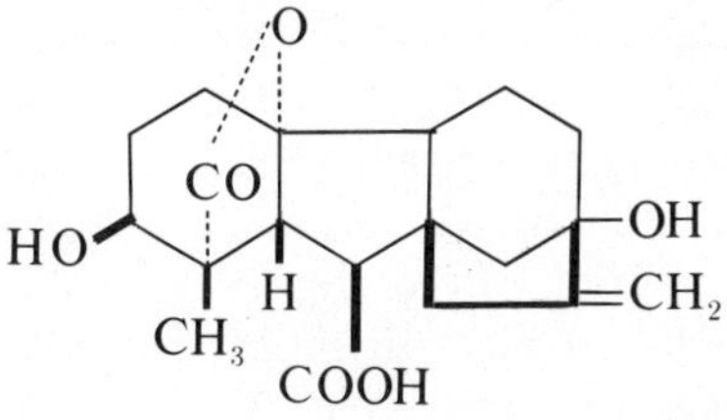 XCVI

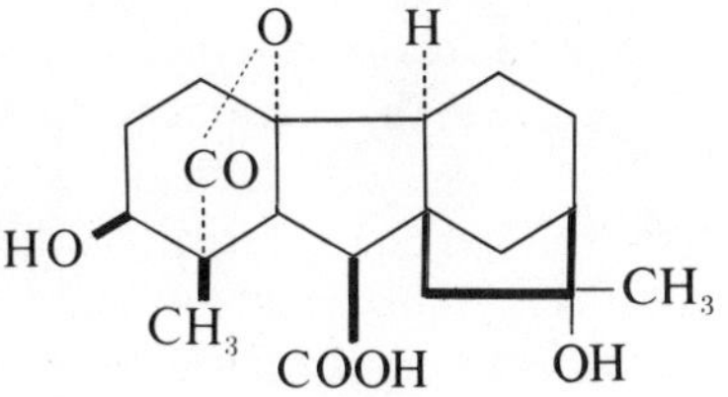

XCVII 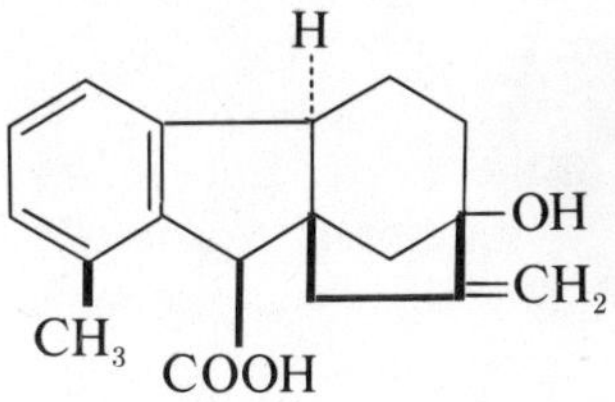 XCVIII

Fig. 20. The first gibberellins.

XCV	Gibberellic acid. Gibberellin A_3
XCVI	Gibberellin A_1
XCVII	Gibberellin A_2
XCVIII	Gibberellin B. Allogibberic acid

until ten years later that the intensive investigations on the gibberellins started in England under the leadership of Dr. P. W. Brian at Imperial Chemical Industries and in the United States Department of Agriculture under Dr. F. H. Stodola. Within a year or so both groups had isolated the same highly active compound from the culture medium of *Gibberella fujikuroi.* The English group called it gibberellic acid (Curtis & Cross, 1954) and the American group named it gibberellin X (Stodola *et al.*, 1955). It was quite different from the Japanese gibberellin A in its optical rotation and the melting point of its esters. Compared with IAA, the molecules of the gibberellins are very complex and it was some time before the formula (Fig. 20 XCV) now generally accepted for gibberellic acid, was worked out (Cross *et al.*, 1959).

The American workers isolated a second substance very closely resembling gibberellin A (Stodola *et al.*, 1955), which was subsequently shown to differ from gibberellic acid in the absence of a double bond in ring A. This compound is now called gibberellin A_1. Prompted by these studies the

Japanese workers made a re-examination of the crude gibberellin produced by their particular strains of fungus and eventually isolated from it three components, gibberellic acid, (now called gibberellin A_3), gibberellin A_1 and gibberellin A_2 (Fig. 20 XCVII) (Takahashi *et al.*, 1955). The exact constitution of the original gibberellin A still remains obscure. Gibberellin B, when subject to intensive purification was shown to be identical with allogibberic acid (Fig. 20 XCVIII) (Cross, 1954) and appeared to have no growth-promoting properties, suggesting that the original sample had contained active gibberellins as impurities. However more recent work (see Paleg, 1965; Sembdtner *et al.*, 1965) has vindicated the early Japanese claims and shown that it may promote some growth responses on its own. In a highly purified form it still has activity on dwarf maize mutants (Brian *et al.*, 1967). A comprehensive account of the work on the characterization of these early gibberellins has been written by Brian *et al.*, (1960).

The dramatic promotion of stem extension which is characteristic of the gibberellins was thought by the Japanese pioneers to be the result purely of an augmented cell elongation, this closely resembling the action of the auxins. They also showed that, like auxin, gibberellins seemed to affect other growth processes such as the development of the shape of leaves of several plants, the elongation and initiation of roots and the time of flowering (Yabuta & Hayashi, 1939). But it was only when gibberellic acid became obtainable commercially in America and Great Britain that we were made aware of the amazing range of its growth-regulating properties. Thus before 1955 attempts to explain the slow and stunted growth of dwarf varieties of cultivated plant in terms either of low auxin content or of low auxin sensitivity had largely met with failure (von Abrams, 1953). Then Brian and Hemming (1955) showed that a number of dwarf and intermediate varieties of garden pea (*Pisum sativum*) were so responsive to gibberellic acid that they could be made to grow as fast as the tallest varieties by applications of small doses of this substance (see Fig. 21 and Plate 8). A year later the same sensitivity was shown in a number of single-gene dwarf mutants of maize (*Zea mays*) (Phinney, 1956). Clearly the gibberellins were growth regulators different from the auxins and might even be higher plant hormones. Then an effect of gibberellin in causing the 'bolting' of certain plants with a rosette habit was shown by two independent workers (Lona, 1956; Lang, 1956*a*) and this response was in large part due to a greatly augmented cell division. In the same year there came from several different laboratories the announcement that gibberellic acid would induce the formation of flowers on various species of plant under conditions otherwise unfavourable to flowering (Lang, 1956*b*; Lona, 1956; Curry & Wassink, 1956; Harder & Bünsow, 1956). Equally surprising were the discoveries of its action in breaking the dormancy of buds of potato tuber (Brian *et al.*, 1955) and in

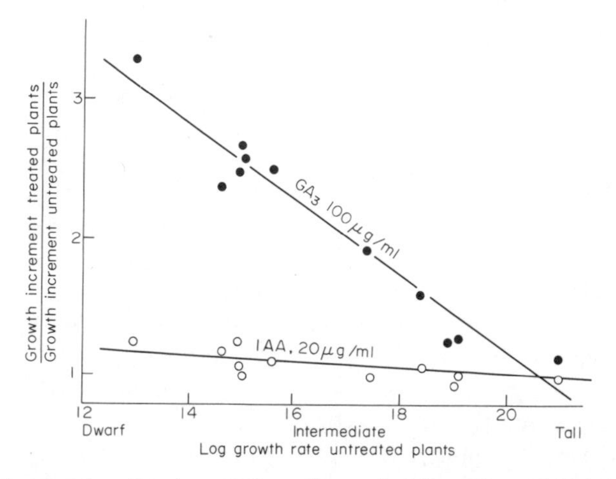

Fig. 21. Graphs showing the relative effects of gibberellic acid (GA_3) and indol-3yl-acetic acid (IAA) in stimulating the extension growth of dwarf, tall and intermediate varieties of garden peas. Whereas the small stimulation by IAA is virtually the same for all varieties, the stimulation by GA_3 increases progressively as one passes from tall to dwarf varieties (from Brian & Hemming, 1955).

Plate 8. Effects of the application of gibberellic acid to Meteor pea seedlings. Photograph taken 26 days after the application of alcohol solutions to the first true leaf. Treatments from right to left; 1, Control, untreated; 2, Control, treated with alcohol solvent alone; 3, Treated with 0.01 µg gibberellic acid in alcohol. Thence doses increase in two-fold steps up to the last (13, extreme left) dosed with 10.24 µg gibberellic acid. (Photograph by kind permission of Drs. P. W. Brian and H. G. Hemming).

promoting the germination of lettuce seeds (Lona, 1956). Such a wide spectrum of action, equalling, although differing considerably in detail from that of IAA, strongly suggested that the gibberellins, as well as being the accidental by-products of fungal metabolism, might very well be important natural hormones of higher plants. The scene was set for the search for them in higher plant tissue.

Already in 1951 a substance had been obtained from immature seeds of bean (*Phaseolus*) which, when applied to dwarf bean test plants, almost trebled their height (Mitchell *et al.*, 1951). Although not then recognised as such, this substance was clearly of the nature of a gibberellin. A direct search soon showed the presence of substances with gibberellin-like activities in other seeds, the liquid endosperm of the wild cucumber (*Echinocystis macrocarpa*) being extremely rich (West & Phinney, 1956). In pea shoots a substance behaving like gibberellic acid on paper chromatograms was recorded by Radley (1956). Final proof came with the isolation from immature seeds of *Phaseolus multiflorus* of pure crystals of gibberellin A_1 (MacMillan & Suter, 1958). Since then many others have been isolated and characterised (see later section) and the gibberellins are clearly as well established as IAA as natural hormones of higher plants.

At the time of writing twenty-nine natural gibberellins are known. They have been given serial numbers and are called gibberellins A_1 to A_{29}.

Bioassay Methods

Just as Went's *Avena* curvature test was crucial for the early chemical work on the auxins, so too did the isolation and characterisation of the gibberellins in Japan depend on the rice seedling bioassay developed by Kurosawa. The technique was simple since it involved nothing more than applying the test solution to the sterile sand in which rice seeds had been sown and an increase in height of the seedlings indicated the presence of gibberellin. The seedlings are insensitive to auxins. Since then many more bioassay methods have been developed. These will now be briefly surveyed.

(A) SEEDLING TESTS

On the whole intact seedlings are much more sensitive to applications of gibberellins than are isolated organs and so have been to a large extent the favoured material for bioassays. Furthermore the dramatic reactions of dwarf varieties of certain cultivated plants soon singled out these plants as especially suitable. Two major tests use them.

The dwarf pea (*Pisum sativum*) test was first introduced by Brian & Hemming (1955) (see Fig. 22A). A suitable variety (Meteor has been very popular) is grown in sand or vermiculite in normal daylight. About ten days after planting, when the seedlings have two fully expanded leaves, one or more leaflets are treated with gibberellin by applying very small (a few μl) droplets of solution. The subsequent accelerated growth of selected internodes gives a measure of the gibberellin applied. The response is directly proportional to the logarithm of the dose (total quantity applied) over the range 10^{-4} to 10 μg of gibberellic acid. The advantages of this test are its simplicity and the ready availability of genetically very uniform seeds of suitable dwarf varieties.

The dwarf maize (*Zea mays*) test (Fig. 22B) was developed by Neely & Phinney (1957). Grains of certain dwarf mutants (in particular one designated d-l) are sown in a suitable medium and allowed to germinate for 6 to 7 days, when the first leaf is just unfolding from the coleoptile. The test solution, containing a trace of a surface-active compound (to make the solution spread over the leaf surface and promote optimal penetration of the gibberellin) is dropped into the conical cavity made by the base of the emerging leaf blade. The response is an accelerated extension of the leaf sheath, which is measured after 7 to 10 days more. In this test the logarithm of the response is proportional to the logarithm of the dose over the range 0.001 to 1.0 μg per plant. The great disadvantage of this test is the difficulty associated with obtaining the mutant grain, which is not obtainable from commercial sources.

A method specifically for use in paper chromatography was developed by Frankland & Wareing (1960) using two-day-old seedlings of lettuce (*Lactuca sativa*) (Fig. 22C). These can be planted directly on filter-paper containing the gibberellins, Increased hypocotyl growth is recorded after a further 5 days. Response is proportional to the logarithm of the gibberellin concentration in the growth solution over a range of 0.1 to 10 μg. The method is much less sensitive than the preceding two but has the great convenience of simplicity.

Seedlings of several other species have been used. They include dwarf seedlings of *Pharbitis nil* (epicotyl elongation) (Hirono *et al.*, 1960) and cucumber (hypocotyl elongation) (Lockhart & Deal, 1960). The latter species has rather widely differing sensitivities to the different gibberellins (see p. 121) and this promises certain practical advantages as an aid to identification in the analysis of plant extracts. Similar differential responses are shown by a number of dwarf cultivars of rice which have been used in a very sensitive bioassay involving the application of micro-drops of extracts to coleoptiles and measuring the subsequent response of the leaf sheath (Murakami, 1968).

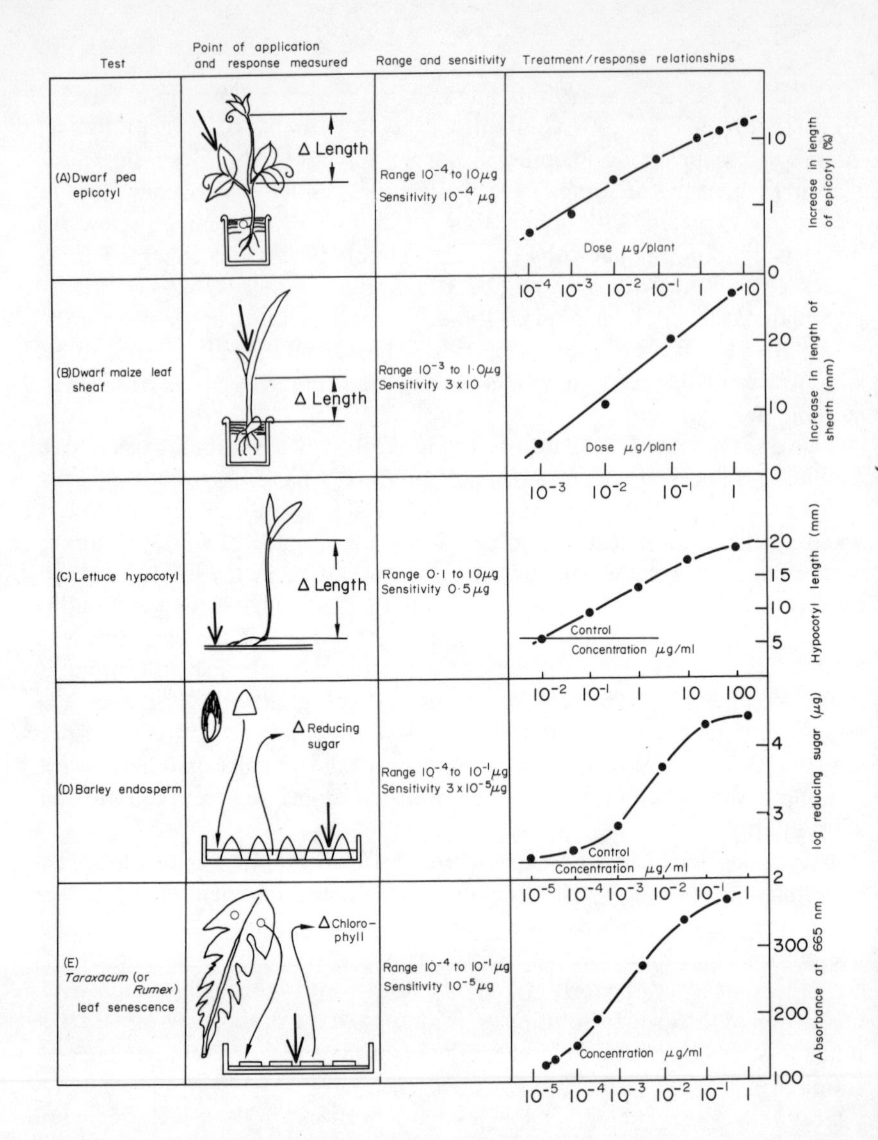

Fig. 22. Diagram illustrating the salient features of five of the main gibberellin assay techniques, (A) dwarf pea epicotyl, (B) dwarf maize leaf sheath, (C) lettuce hypocotyl, (D) barley endosperm α-amylase induction and (E) *Taraxacum* leaf senescence. In the first column are illustrated the major steps in the preparation of the material for the assay. The black arrows show the points of application of the gibberellins in each assay. The second column shows the range of dose over which a quantitative determination is possible and also the limit of sensitivity (i.e. the lowest amount that can be detected by the assay). The third column shows the quantitative relationships between the response and the dose or concentration of the gibberellin (gibberellic acid in all these graphs). The data were taken from the following sources; (A) McComb & Carr, 1958, (B) Brian *et al.*, 1964, (C) Frankland & Wareing, 1960. (D) Dr. Alan Crozier (personal communication), (E) Whyte & Luckwill, 1966.

(B) LEAF GROWTH

All the seedling tests suffer from the disadvantage that a number of days are needed after gibberellin applications before final measurements can be made. To reduce this waiting time recourse has been made to the use of segments of the immature first leaf base of certain cereals (i.e. *Avena*, *Hordeum* and *Triticum*), whose growth in length is greatly promoted by the gibberellins. The assays are preformed in a way similar to that of the *Avena* coleoptile segment test for auxins. The growth of these segments may also be inhibited by IAA (and possibly other auxins) (Radley, 1958) which may lead to complications in the assay of plant extracts (van Overbeek & Dowding, 1961). It is important to mention at this point that certain standard auxin assays involving segments, e.g. *Avena* coleoptile and first internode (mesocotyl) tests may also be sensitive to gibberellins (Recalde *et al.*, 1960; Ng & Audus 1964) so that many of the reports of unidentified auxins in plant extracts need re-investigation. It is a little perturbing that gibberellic acid is also slightly active in the *Avena* curvature test (Saebø, 1960). The growth of immature leaves of broad-leafed plants can also be promoted by gibberellins, and disks cut from the young leaves of *Phaseolus* and floated on the solutions have been used as a specific qualitative test (see Brian *et al.*, 1960; Wheeler, 1961).

(C) BARLEY ENDOSPERM TEST

One of the most striking actions of the gibberellins is that they promote and sometimes induce the germination of seeds. Barley (*Hordeum vulgare*) is one such sensitive species. One of the first happenings in this process is the appearance of the enzyme α-amylase, which attacks the reserves of starch in the endosperm and releases reducing sugars. This enzyme, which is produced in the outermost layer of the endosperm (aleurone) can be caused to appear by gibberellin treatment of extremely low dosage. This has been made the basis of a bioassay, which is one of the most sensitive so far devized (Nicholls & Paleg, 1963) and is completely insensitive to IAA or other hormones. The technique consists of taking samples of barley grain, cutting them in half transversely (see Fig. 22 D) and placing the distal embryo-less half with its cut surface in the test solution, which contains an antibiotic (streptomycin) to prevent the growth of bacteria. After two days of incubation at 30°C the amount of reducing sugar released from the endosperm into the solution can be determined by a sensitive chemical estimation (Somogyi, 1952); the quantity produced is roughly proportional to the concentrations of the gibberellin in the medium, e.g. for gibberellic acid, over a range 10^{-4} to 10^{-1} μg/ml. Apart from its greater sensitivity, this test has the advantage of speed, since answers can be obtained a day or two after applying the test solution. Its disadvantages are that reducing substances other than glucose from the hydrolysis of starch and present in the extracts may give a positive reading, necessitating the use of blanks with no endosperm for each test. A modification of the assay using soluble starch, the hydrolysis of which is measured directly by iodine-staining, avoids this complication (Jones & Varner, 1967)

As a parallel measure of gibberellin action the protein content of the solution can also be measured since this too is augmented in proportion to the gibberellin concentration. A further test involving rice instead of barley endosperm has the advantage of giving virtually no activity in the absence of gibberellin in the test solution (Murakami, 1966).

(D) LEAF SENESCENCE TEST

Another phenomenon controlled by various hormones is the yellowing of leaves (senescence) which is due to the loss of the pigment chlorophyll consequent upon changes in protein content in the ageing cells. At low concentrations gibberellic acid has been shown to retard senescence and hence loss of chlorophyll in isolated disks of dandelion (*Taraxacum officinale*) (Fletcher & Osborn 1966) or dock (*Rumex* spp.) (Whyte & Luckwill, 1966) leaves, and this can be used for a very simple and sensitive assay for gibberellins. Disks about 7 mm in diameter are incubated for 4 to 5 days at 25°C in gibberellin solution in the dark, i.e. until most or all of the green chlorophyll has disappeared from the control (untreated) disks (see Fig. 22 E). The chlorophyll is then extracted in hot acetone or ethanol and estimated in a spectrometer at 665 nm (absorption peak of chlorophyll *a*). The pigment concentration in the disks is roughly proportional to gibberellin concentration over the range 10^{-4} to 10^{-1} μg/1. Aliquots of solution can be as small as 0.3 ml thus giving a test sensitivity of 10^{-5} μg.

The attractiveness of this test is its extreme simplicity, but its disadvantages are that auxins and cytokinins (see Chapter V) also retard senescence, although, in these particular species, only when present in much higher concentrations.

(E) SUGAR-CANE SPINDLE BIOASSAY

A very recent test with unusual characteristics has made use of the 'spindle', i.e. the young stem apex, of sugar-cane, *Saccharum officinarum* (Most, 1968). Apical portions of 4–8 month-old canes are starved for 28 hours in darkness at 26–28°C. Then 5 cm portions of these spindles are placed with their cut ends in sterile distilled water and the extract to be assayed applied in solution containing a wetting agent (0.05 percent Tween-20) as 10 μl droplets to the upper cut surfaces. Subsequent growth takes place in the light and the growth increment is measured at 24 and 48 hours. The material is insensitive to IAA and to cytokinins (See Chapter V) but can detect quantities of gibberellic acid somewhat less than 10^{-2} μg per spindle.

One striking property of this assay is that it is very sensitive to gibberellin A_8 and can detect quantities as low as 10^{-3} μg per spindle. As we shall see later gibberellin A_8 is virtually inactive on the vast majority of plants so far tested.

Extraction and Purification

The extraction of gibberellins from the culture filtrate of *Gibberella fujikuroi* is by adsorption on activated charcoal. Adsorbed substances are then eluted by ethanolic ammonia or by acetone. Because the gibberellins are all weak acids extracts can be purified still further by methods similar to those used for IAA; they can be extracted from aqueous extracts (e.g. *Gibberella* culture media) at pH3 by organic solvents such as ethyl ether or ethyl acetate.

Higher plants pose different problems since the amounts of gibberellin present are very low (of the order of 0.5 mg/kg fresh weight) and, in addition, more refined steps for concentration and purification are needed. The very large amount of solvent and associated reagents needed to extract and purify great quantities of plant material for such small yields of active gibberellins have their attendant dangers. Small amounts of impurity can be concentrated to high levels in small quantities of final extract used for separation and assay and these may greatly distort the estimates made. For example in one series of experiments blank runs involving no plant material indicated the presence of two substances with gibberellin activity in the final preparation and these were traced to impurities in a sample of sodium bicarbonate used in the fractionation procedure (Hartley *et al.*, 1969). This should be taken as a warning of possible artefacts whenever a high degree of concentration of bulky extracts are involved in the study of naturally-occurring hormones.

Fresh material can be extracted by acetone/water or methanol/water mixtures or the material can be freeze-dried and extracted with ethyl acetate. Aqueous solutions are then subjected to first-stage purification by adsorption on charcoal, charcoal/Celite or silicic acid columns and are then subjected to gradient elution by increasing concentrations of acetone in water or of ethyl acetate in chloroform respectively.

Owing to the very small concentrations of gibberellins in higher-plant tissue very large amounts of material have to be extracted and this means that crude extracts contain impurities far in excess of the gibberellins present. Very considerable 'clean-up' is therefore needed before the more sophisticated chromatographic-separation techniques described below can be applied and unambiguous chemical identification made. For such bulk preparations very efficient initial purifications can be obtained by a sequence involving first a countercurrent distribution technique between two immiscible solvents (e.g. ethyl acetate and aqueous buffer at pH 5·5) followed by 'molecular sieving' on a column of Sephadex G-10 and silicic acid column chromatography (Crozier *et al.*, 1969, 1971).

Active fractions, detected by bioassay, can be fractionated by paper chromatography or by thin-layer chromatography (TLC) on silica gel (Cavell *et al.*, 1967). This latter technique is in principle the same as paper chromatography, with the substitution of a very thin (250 μm) uniform layer of a suitable powder (e.g. silica gel) as the adsorbant spread on a glass plate. The unknown mixture is spotted at one end of the adsorbant film and suitable solvent mixtures are run from that end. The great advantage of this method is the flexibility introduced by a wide range of absorbant powders, the speed of running (1 to 2 hours) and the greater precision of separation. A considerable number of solvent systems have already been investigated and details of the behaviour of the first seventeen gibberellins have been tabulated by Paleg (1965) and Cavell *et al.* (1967). The most effective separation is obtained with the methyl esters and trimethylsilyl ethers of the methyl esters by gas-liquid chromatography, in which the mixtures in vapour form are passed along a tube containing a suitable adsorbant. The separated components as they emerge in characteristic sequence at the other end may be detected by a variety of physical devices (Cavell *et al.*, 1967) (see Fig. 23).

Chemical Detection

One happy characteristic of the gibberellins is that when heated with sulphuric acid, substances are produced which fluoresce with characteristic colours in ultra-violet light. By spraying thin-layer chromatograms with ethanol/concentrated H_2SO_4 (95/5), heating in an oven at 120°C and examining under a dark u.v. lamp, the gibberellins can be partly identified from their fluorescence colours (Jones *et al.*, 1963). A further aid to identification is the relative amount of heating necessary to produce fluorescence, the development of colour occurring in a few minutes or so with some gibberellins but requiring up to 20 minutes with others. Details are shown in Table IV. By using marker spots identified in this way the gibberellin zones on thin-layer chromatograms of plant extracts can be located; then the powder can be scraped off the plate and gibberellins eluted from it. The eluate can than be heated in 50 percent ethanolic sulphuric acid (25 min

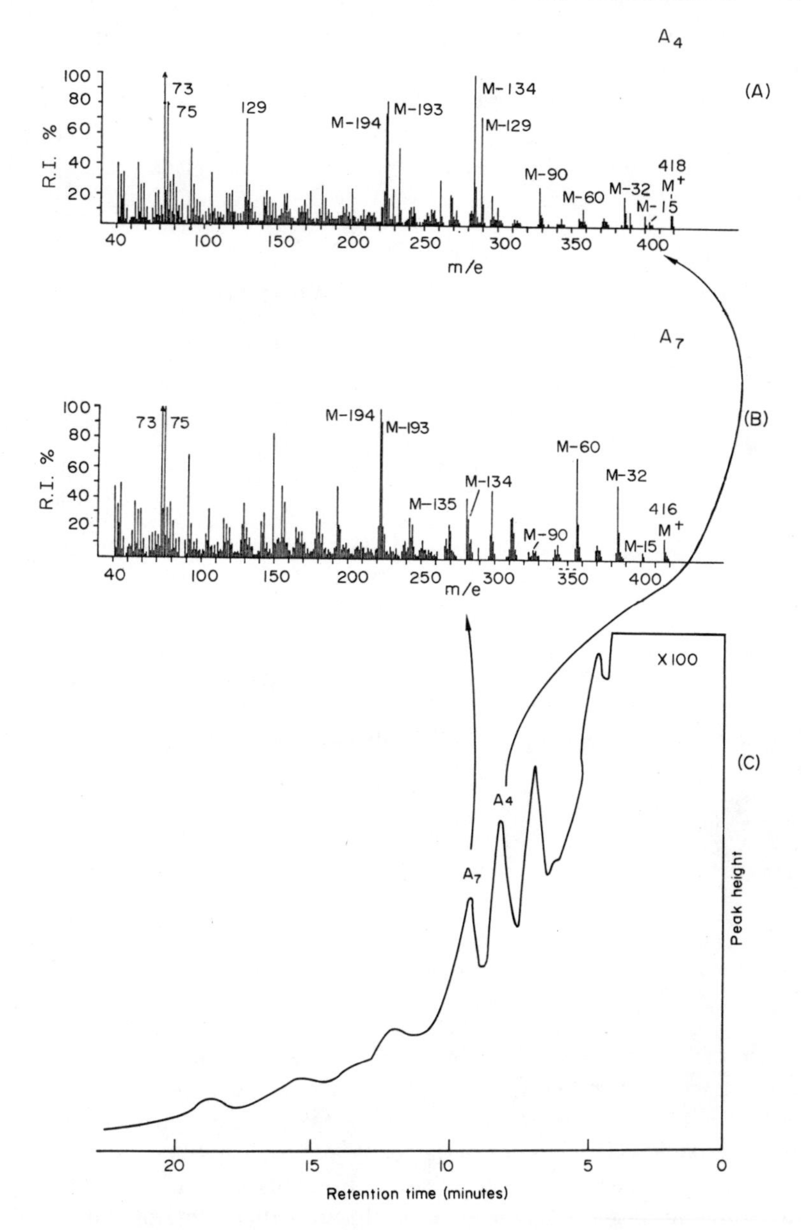

Fig. 23. Mass spectra of the methyl ester trimethylsilyl ethers of gibberellin A_4 (A) and gibberellin A_7 (B) together with the gas chromatogram of a methylated and (C) from MacMillan (1968); with kind permission of the authors.)
(C) showing peaks of these two compounds. ((A) and (B) from Binks *et al.* (1969) and (C) from MacMillan (1968); with kind permission of the authors.)

at either 100°C or 50°C) and the fluorescence which then develops can be determined and used as an accurate measure of gibberellin concentration. The several gibberellins studied have characteristic excitation and emission spectra (cf. fluorescence of indoles in Chapter II) and these too can be very valuable guides to identification (Elson *et al.*, 1964) (see Table IV).

TABLE IV

GIBBERELLIN FLUORESCENCE CHARACTERISTICS

Gibberellin	TLC fluorescent colours*		Fluorescence in solution†		
	Induction period (min)	Colour of spot		Excitation maximum(nm)	Fluorescence maximum(nm)
A_1	30–40	Blue	(a)	417	463
A_2	4–8	Purple	(b)	{390 440	420 468
A_3	1–3	Greenish blue to blue on prolonged heating	(b)	417	463
A_4	4–8	Purple	(b)	{390 440	420 468
A_5	10–20	Blue	(a)	417	463
A_6	10–20	Blue	(a)	417	463
A_7	1–2	Bright yellow to pale yellow on prolonged heating	(b)	{390 455	420 473
A_8	10–20	Blue	(a)	417	463
A_9	4–8	Purple	(b)	{390 440	420 468

* Sprayed with ethanolic H_2SO_4 (95/5) and heated at 120°C

† Heated 25 min in 50 percent ethanolic H_2SO_4 at (a) 100°C and (b) 50°C

As in the case of the natural indole auxins already mentioned (p. 48–9) the most precise method now available is that of mass spectrometry. The combination of this method of characterization with gas chromatography brings a completely new order of precision into the identification and measurement of the components of mixtures of gibberellins, even in relatively crude preparations. The mass spectra of the methyl esters of twenty-four gibberellins and of the trimethylsilyl ethers of the methyl esters of the hydroxylated gibberellins (see later section) have been published by Binks *et al.*, (1969). Furthermore the sensitivity of this combined technique is very high since quantities of the order of 10 ng can be detected with it.

Structure and Occurrence

Since the first identification of a gibberellin (A_1) from a higher plant (*Phaseolus multiflorus* seeds) (MacMillan & Suter, 1958) there has been a great surge of research on the occurrence of these hormones throughout the plant kingdom. Two gibberellins, first called Bean Factors I and II, were next isolated by West and Phinney (1959). Bean Factor I was subsequently shown to be a new gibberellin and was given the serial number A_5 (MacMillan *et al.*, 1959) (see Fig. 24). The extremely high activity of the liquid endosperm of *Echinocystis macrocarpa* was then shown to be due to the presence of four gibberellins (A_1, A_3, A_4 and A_7), the last two being previously known only from cultures of the fungus *Gibberella fujikuroi* (Elson *et al.*, 1964). Gibberellic acid (A_3) has now been identified from the seeds of quite a few plants. Furthermore many new gibberellins have been discovered in other immature seeds. Thus in 1964 two new gibberellins (A_6 and A_8) were isolated from *Phaseolus multiflorus* seeds (Jones, 1964). A further gibberellin (A_{10}) was identified in *Echinocystis* (Cross *et al.*, 1964). Two additional gibberellins were shown to be present in bamboo (Tamura *et al.*, 1966) and in lupin (Koshimizu *et al.*, 1966) and these were later given the serial numbers A_{19} and A_{18} respectively. Later a second new gibberellin (A_{23}) was isolated from lupin seed (Koshimizu, Fukui *et al.*, 1968). The seeds of Japanese Morning Glory (*Pharbitis nil*) yielded gibberellin A_{20} (Takahashi, Murofushi, Yokota and Tamura, 1967) and those of *Canavalia gladiata* (Sword Bean) were shown to contain two others (Takahashi Murofushi, Yokota, Tamura, Kato and Shiotani, 1967) later named A_{21} and A_{22}. To date the total number in the seeds of *Phaseolus multiflorus* has been brought up eight by the inclusion of A_{19} (Pryce *et al.*, 1967), A_{20} (Takahashi, Murofushi, Yokota and Tamura, 1967) and a new member A_{17} (Pryce & MacMillan, 1967).

Fig. 24 shows the full range of natural gibberellins isolated and characterized at the time of writing (October, 1970). There are twenty-nine so far, fifteen of which are produced by the fungus *Gibberella* and nineteen found in higher plants. Only five are common to both fungus and higher plant.

The unequivocal chemical identifications of gibberellins from higher plants as detailed above has relied heavily on the newly-applied techniques of gas/liquid chromatography – mass spectrometry but in addition there has been a very much greater list of partial identifications using specific bioassays and the determination of Rf values of active compounds located on various types of chromatogram. This is not the place to try to document such a mass of information but some indication of the relative ubiquity of gibberellins (or more accurately gibberellin-like substances) throughout the plant kingdom should be of interest. It should also be realised that the brief survey which follows is by no means exhaustive. In addition to the immature

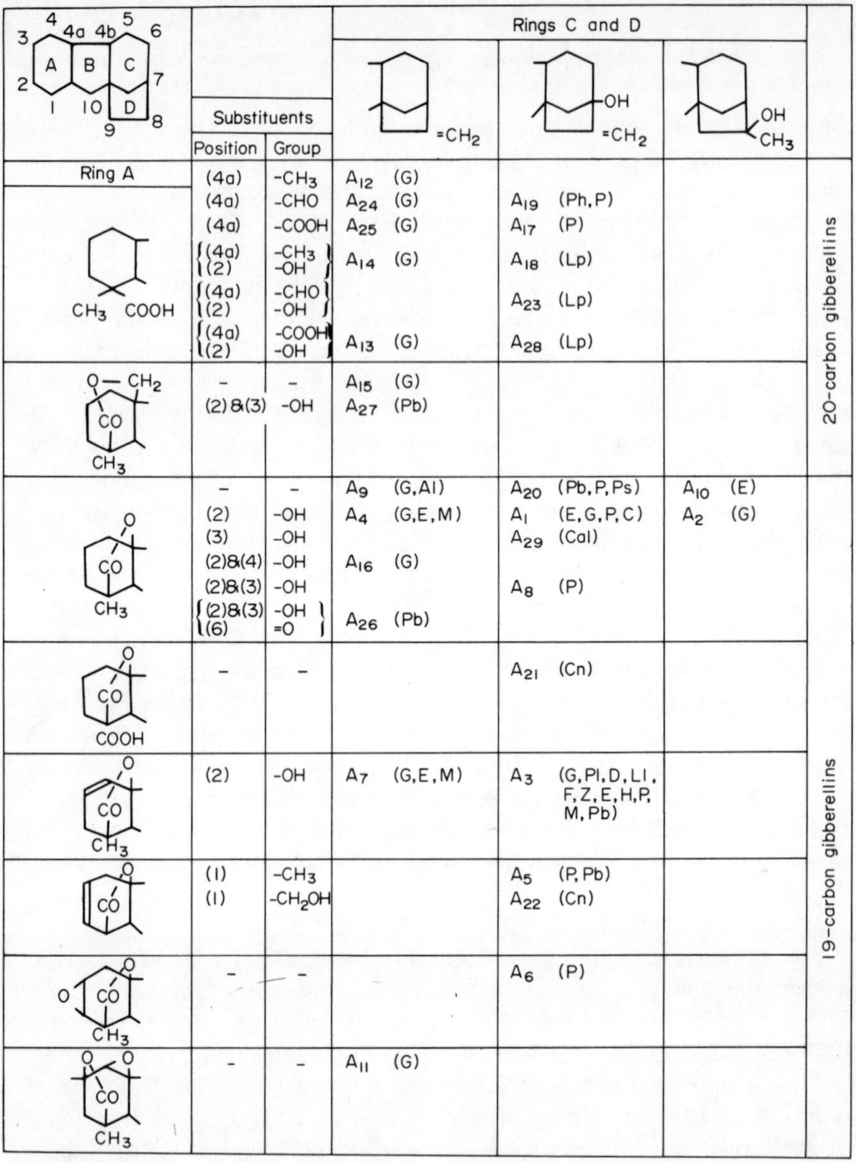

Ring A	Substituents: Position	Substituents: Group	Rings C and D: =CH₂	Rings C and D: -OH, =CH₂	Rings C and D: OH, CH₃	
Ring A	(4a)	$-CH_3$	A_{12} (G)			20-carbon gibberellins
CH_3 COOH	(4a)	-CHO	A_{24} (G)	A_{19} (Ph, P)		
	(4a)	-COOH	A_{25} (G)	A_{17} (P)		
	(4a) (2)	$-CH_3$ -OH	A_{14} (G)	A_{18} (Lp)		
	(4a) (2)	-CHO -OH		A_{23} (Lp)		
	(4a) (2)	-COOH -OH	A_{13} (G)	A_{28} (Lp)		
O— CH_2, CO, CH_3	–	–	A_{15} (G)			
	(2)&(3)	-OH	A_{27} (Pb)			
O, CO, CH_3	–	–	A_9 (G, Al)	A_{20} (Pb, P, Ps)	A_{10} (E)	19-carbon gibberellins
	(2)	-OH	A_4 (G, E, M)	A_1 (E, G, P, C)	A_2 (G)	
	(3)	-OH		A_{29} (Cal)		
	(2)&(4)	-OH	A_{16} (G)			
	(2)&(3)	-OH		A_8 (P)		
	(2)&(3) (6)	-OH =O	A_{26} (Pb)			
O, CO, COOH	–	–		A_{21} (Cn)		
O, CO, CH_3	(2)	-OH	A_7 (G, E, M)	A_3 (G, Pl, D, Ll, F, Z, E, H, P, M, Pb)		
O, CO	(1)	$-CH_3$		A_5 (P, Pb)		
	(1)	$-CH_2OH$		A_{22} (Cn)		
O, O, CO, CH_3	–	–		A_6 (P)		
O, O, CO, CH_3	–	–	A_{11} (G)			

Fig. 24. Scheme showing the range of chemical structure and the chemical relationships in the first twenty-nine gibberellins to be isolated and characterized. Although the details are not complete, the letters in brackets after each gibberellin record the plant from which it has been isolated.

Al = *Althaea rosea*; C = *Citrus*; Cal = *Calonyction*; Cn = *Canavalia*; D = *Dactylis*; E = *Echinocystis*; F = *Festuca*; G = *Gibberella*; H = *Hordeum* Ll = *Lolium*; Lp = *Lupinus*; M = *Malus*; P = *Phaseolus*; Pb = *Pharbitis*; Pl = Phleum; Ps = *Pisum*; Ph = *Phyllostachys*; Z = *Zea*.

seeds recorded in Fig. 24. lettuce, plum and horsechestnut seeds show gibberellin activity when extracted. Activity is also found in fruits (e.g. tobacco, *Datura. Citrus*). Although more difficult to detect, gibberellins undoubtedly occur in vegetative shoots. Their wide distribution is shown by the following examples:- bamboo, beet, rhubarb, pea, bean, potato, lettuce, rice, wheat, maize and strawberry. Roots may be a main source of gibberellins which have been demonstrated in the sap which bleeds from the stump of sunflower plants when the shoot is cut off (Phillips & Jones, 1964).

They can also be demonstrated in gymnosperms and ferns (Kato *et al.*, 1962), brown and green algae (Mowat, 1963), soil yeasts (Chailakhyan *et al.*, 1958) and soil bacteria (*Arthrobacter globiforme*) (Katznelson *et al.*, 1962). The significance of their occurrence in these colourless, heterotrophic plants is obscure. It is possible that here, as in the case of *Gibberella* itself, they are the result of some kind of kink in the metabolic system of the organisms and have no effect, good or bad, on their growth.

Since new demonstrations of gibberellin occurrence seem to have become almost daily happenings, it would not be surprising if the gibberellins did not prove to be characteristic of the metabolism of most, if not all plants. But gibberellin-like substances are not confined to plant tissues since a wide variety of animal tissues (e.g. earthworm bodies, silkworm pupae, chick embryo, codfish sperm) have been shown to contain substances which promote α-amylase production in rice endosperm (see bioassay section) (Ogawa, 1965). Even human saliva and urine possess them. Whether these are gibberellins extracted by the animal from their plant foods or whether they are natural animal products that are not true gibberellins but have some of their characteristic properties (see ecdysone in following section) have still to be decided.

STRUCTURE-ACTIVITY RELATIONSHIPS

In striking contrast to the auxins, the gibberellins seem to have a very closely prescribed basic structure; the gibbane ring system seems obligate for activity, as is the carboxyl group on position 10 of ring B. Furthermore activity seems to be associated with certain specific substituent groups, particularly at positions 1, 2, 7 and 8.

As will be seen from Fig. 24, the gibberellins fall naturally into three groups, depending on the nature of these substituent groups at positions 7 and 8 on ring D. The gibberellins of the first group (no hydroxyl on carbon 7) are predominantly fungal in origin whereas the higher plant gibberellins are found mainly in the second group (hydroxyl on carbon 7). The full biological significance of this remains to be seen, but it probably indicates that the biosynthetic pathways for gibberellins are different in the fungus and in the higher plant.

During the last decade several series of studies have been made on the comparative activities of the first nine gibberellins in a considerable variety of growth responses. In the last few years these have been extended to all but one of the twenty-nine gibberellins known to date (October, 1970) in a series of standard tests (Brian *et al.*, 1967; Crozier *et al.*, 1970). A broad analysis of the comprehensive data of Crozier *et al.* (1970) allows us to distinguish five categories of activity as follows:

Very high	A_3, A_7
High	A_1, A_4, A_5
Intermediate	A_2, A_6, A_9, A_{18}, A_{20}, A_{22}, A_{23}
Low	A_{10}, A_{15}, A_{19}, A_{24}
Very low or zero	A_8, A_{11}, A_{12}, A_{14}, A_{17}, A_{21}, A_{25}, A_{26}, A_{27}

However it is clear from the more extensive studies on the first nine gibberellins that activity is highly species-dependent and may also vary from one type of physiological response to another. In order to illustrate this in a way which may be more easily grasped by the reader, an analysis has been carried out on the data from fifteen independent surveys. Because of the diversity of the original data presentation, only four broad categories of activity have been made, i.e. high, medium, low and zero and these have been recorded on a scale of shading in Figs. 25 and 26 for five types of response (i.e. extension growth of seedling stems, coleoptiles, leaves etc., the induction of parthenocarpic fruit, the induction of flowering, the acceleration of abscission of leaf petioles and the induction of α-amylase in the aleurone layer of barley grain).

The order of the gibberellins has been so arranged that those with the highest average activity are on the left, starting with A_7, and the lowest are on the right, ending with A_8. The three classes of activity, high, medium and low, are more coarsely drawn than the finer comparisons of most of the relevant surveys, but even these simplified patterns are too complex for any but the broadest generalizations. To arrange the nine compounds in the order shown the activities have been scored as high = 3, medium = 2, low = 1 and no activity = 0 and the mean score per observation has been calculated for each of the nine gibberellins. This score is recorded for extension growth responses in Fig. 25.

It seems reasonably clear that in these responses the gibberellins fall into five groups. Gibberellin A_7 has a high or medium activity in all of the tests with the exception of two (*Avena* first leaf test and one report of the d-3 dwarf maize test). Next comes a group of three (A_4, A_3 and A_1) which are virtually

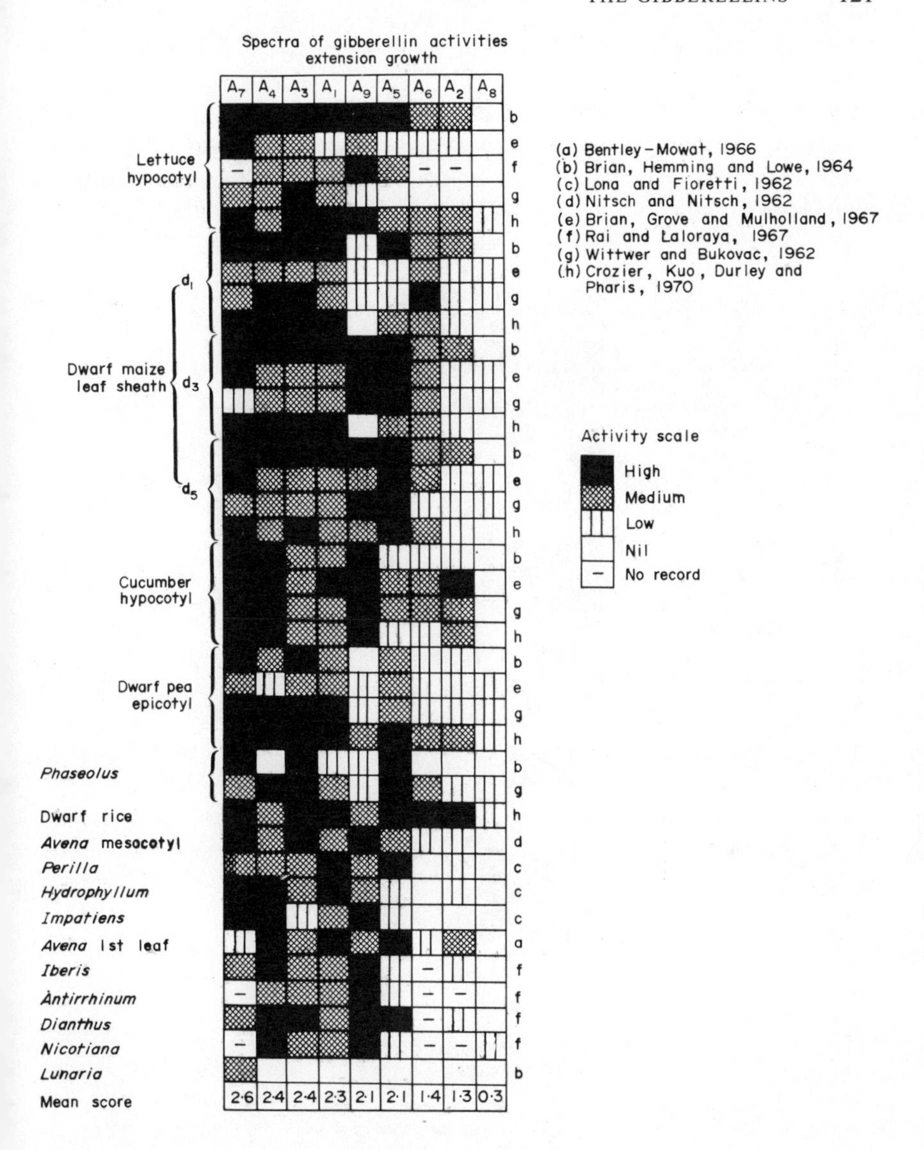

Fig. 25. Diagram showing the spectra of activity of the first nine gibberellins in promoting the extension growth of various organs of a number of plant species. The gibberellins (indicated in the top row of the diagram) have been ranked in descending order of mean activity (bottom row) based on an arbitrary numerical score (see text) corresponding with the activity scale of shading on the right of the main diagram. The small letters on the right-hand side of each row indicate the sources of the data; a key to these sources is given in the top right-hand corner.

indistinguishable. On the whole they show high activity, although there are some records of low activity. The third group (A_9 and A_5) show a considerable range of activities while the next (A_6 and A_2) have low activities in almost all tests. The last group (A_8 alone) is virtually without activity in all tests. A similar rough order of activity is followed in the other four response types (Fig. 26).

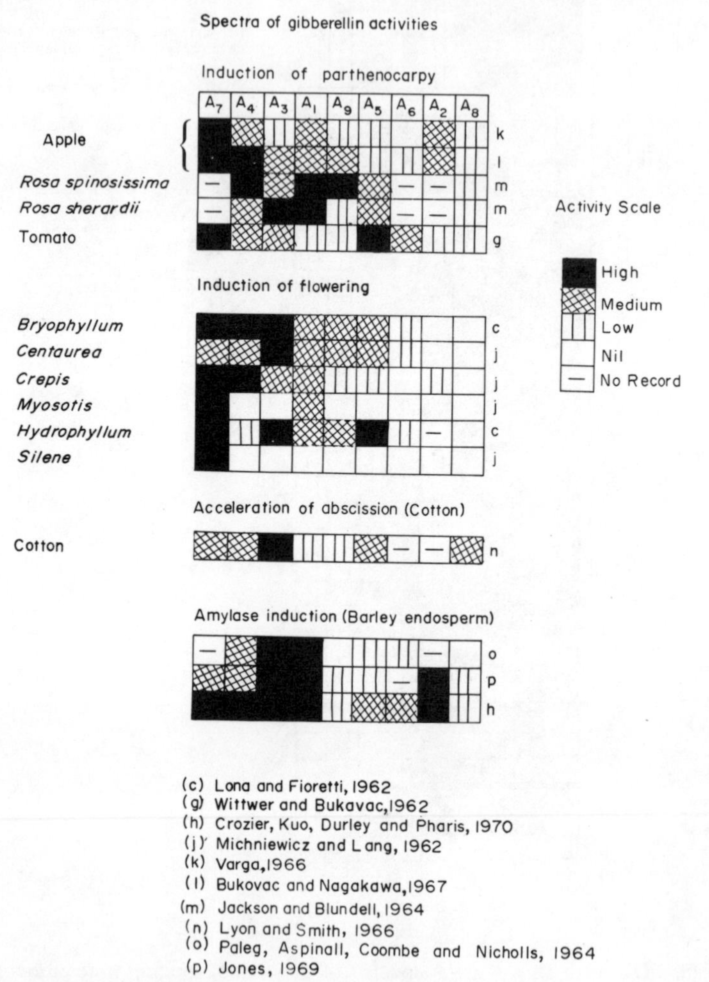

Fig. 26. Diagram showing the spectra of activity of the first nine gibberellins in four responses *other than* extension growth, i.e. induction of parthenocarpy, induction of flowering, acceleration of abscission and induction of amylase in barley endosperm. The gibberellins have been ranked in the same order as in Fig. 25. The small letters on the right-hand side of each row indicate the sources of the data. A key to these sources is given at the bottom of the diagram.

No strictly consistent pattern of relative activity seems to emerge with regard to species differences or type of response. Such a pattern is perhaps too much to expect in view of the discrepancies which appear in the data from different groups of workers, although too much should not perhaps be made of most of the smaller discrepancies in view of the broad interpretation of activity scale I have had to make in drawing up the diagram. No doubt some of them arise from significant differences in testing techniques and the attendant complications of relative uptake and movement of the applied compounds. Another factor will undoubtedly be the relative purities of the compounds used by different workers; small traces of an active gibberellin in a sample of an inactive one could drastically distort the patterns. One or two surprising exceptions to the general pattern might however be pin-pointed; for example the low activity of A_7 in the growth of the first leaf of *Avena*, the lack of activity of A_4 in *Phaseolus* and the low activity of A_3 in *Impatiens*. The absence of activity in all but A_7 in the stem growth of *Lunaria* and in flower induction in *Silene* is also very striking. The high activity of A_2 in the barley endosperm bioassay is also worthy of particular note. Another striking departure from the general pattern is one not included in these diagrams. This is the very high activity of A_8 in the sugar-cane spindle test (see p. 112).

The significance of these differences, both in overall activity of the various gibberellins and in the specific activities of the individual gibberellins between species and tests, still eludes us. One possibility is that only one gibberellin is truly active at the control centres of the cell; all of the others are precursors that must first be converted to this key gibberellin before response is elicited. The activity patterns could then be reflections of the relative ease with which the particular plant or organ could bring about their conversions. One is reminded of the case of IAN, which is an inactive precursor of this kind for IAA. This explanation was first put forward by Phinney & West (1960) to explain specificity patterns in dwarf maize mutants. It is clear from the complexities of Figs. 25 and 26 that no such simple situation will suffice to explain all the nuances of the observed patterns. The most likely explanation is that there are a number of active gibberellins which have to fit on a specific receptive surface in the cell, and that the growth activity is related to the goodness of fit. If the shape of this molecular receptor varies somewhat from species to species, then the goodness of fit of any one gibberellin would similarly vary, bringing with it differences in activity. A good example of such highly specific stereo requirements is pseudogibberellin A, where the -OH group on the A ring of gibberellin A_1 (see Fig. 20 CVI) is changed from the β- to the α-position, i.e. to the other side of the ring. This compound is normally an inhibitor, antagonizing A_3, but shows gibberellin activity in the cucumber hypocotyl and dwarf maize tests (Kato & Katsumi, 1967). The pros and cons of both theories are discussed by Brian (1966) who concludes that both mechanisms are probably in operation to a greater or lesser extent in different species.

However certain broad structure/activity relationships can be made out, particularly now that additional information is available for gibberellins beyond A_9 and a greatly increased number of related molecules (Brian *et al.*, 1967).

An intact gibbane ring seems to be essential (but see however discussion on helminthosporol later) for activity in all assays. The fission of ring A or ring D destroys activity. The possession of a carboxyl group at position 10 on ring B is also essential. Ring A can be saturated or have one unsaturated double bond but aromatization of this ring gives complete inactivity, with one exception, i.e. that of allogibberic acid (see p. 105). The lactone ring across ring A is not essential for activity although gibberellins A_{12}, A_{13} and A_{14} which do not possess it, show low or no activity in all assays so far; but even this activity might be due to active compounds produced by lactonization in the test tissue. The nature of the substituents at positions 1, 2, 4a, 7 and 8 seem to decide relative activities in the several plant responses. On position 3 the introduction of an -OH group, as in A_8, seems to reduce activity to low or zero value. These structure/activity relationships, in a group of molecules so closely related, afford the most exciting prospects for our future understanding of the mechanism of these molecular 'keys' and stand in such strong contrast to the still nebulous relationships seen in the chemically much more diverse auxins.

Gibberellin Biosynthesis

The close relationship that is apparent between the chemical structure of the gibberellins and that of the widely distributed and important class of plant products the terpenes, led to early speculation that their respective pathways of biosynthesis might also be closely related. The terpenes are of universal occurrence in plants. They are unsaturated hydrocarbons with a basic structure of two or more linked 5-carbon units arranged as in isoprene ($CH_2{=}C(CH_3){-}CH{=}CH_2$). They may form linked chains or they may cyclize into polycyclic molecules. Closely related compounds are the steroids the yellow pigments carotenes and xanthophylls and natural rubber (for further details of the biochemistry of these compounds see Davies *et al.*, 1964). It is now reasonably established that the starting point for the biosynthesis of this wide range of substances is acetic acid which can be linked, three molecules at a time, to form the intermediate mevalonic acid (see Fig. 27). This then loses carbon dioxide to form the five-carbon isoprene-like

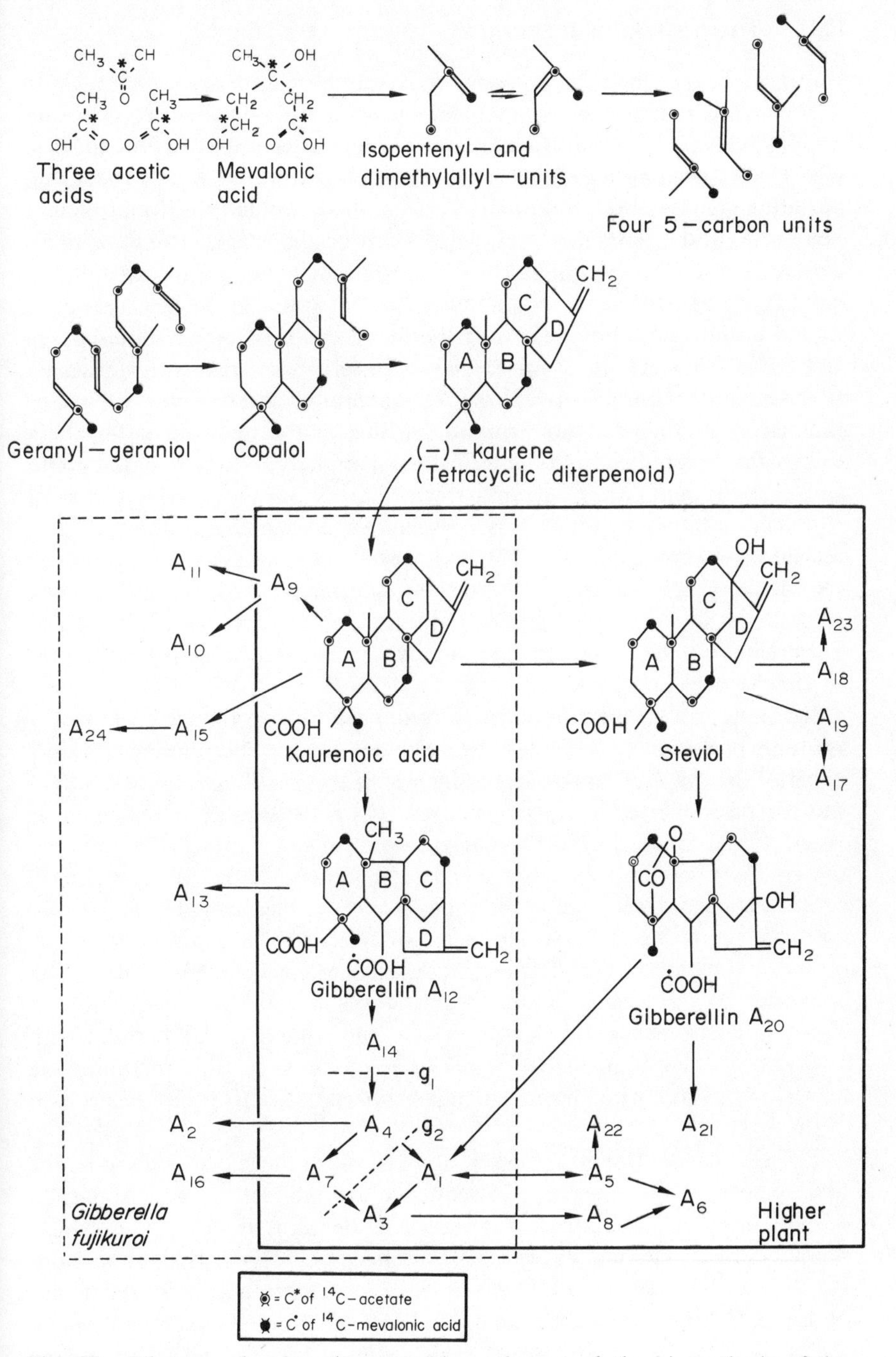

Fig. 27. Diagram showing the probable pathways of the biosynthesis of the gibberellins. The lower half of this diagram has been adapted from Lang (1970, Fig. 4.).

building bricks, which are subsequently assembled into terpenes etc.

The use of acetate and mevalonic acid (as mevalonic lactone) containing radioactive ^{14}C in the culture medium on which *Gibberella* was growing (see Fig. 27 for labelling positions in the molecules) resulted in the production of radioactive gibberellin A_3 (Birch *et al.*, 1958), thus making it reasonably certain that the gibberellins are synthesized by the fungus along a closely similar pathway. Subsequent chemical analyses of the radioactive gibberellin A_3 to determine the positions of the ^{14}C atoms in the molecule gave results which were consistent with the biosynthesis scheme outlined in the first half of Fig. 27, in which a tricyclic diterpenoid was suggested as an intermediate on the direct pathway to gibberellin. But this was still 'paper chemistry' and needed confirmation, which first came after painstaking search for other metabolites of *Gibberella* under conditions which might favour the accumulation of such intermediates (Cross *et al.*, 1964). A most important compound which was isolated and characterized was the diterpenoid (–)–kaurene (see Fig. 27), a tetracyclic compound, not far removed chemically from the tricyclic diterpenoid already proposed as an intermediate. ^{14}C–labelled (–)–kaurene fed to *Gibberella* subsequently produced gibberellin A_3 labelled with ^{14}C in such a way as to establish (–)-kaurene as an unequivocal precursor of the gibberellins.

Subsequent work of a similar nature has established the pathway to (–)-kaurene proposed in Fig. 27. Thus the open-chain diterpenoid geranylgeraniol (as the pyrophosphate) is formed from mevalonate by the fungus and can be converted by it to (–)-kaurene (Upper & West, 1967; Shechter & West, 1969). Similarly the bicyclic diterpenoid copalol (see Fig. 27), was shown to be an intermediate (Shechter & West, 1969), which obviously comes between geranylgeraniol and (–)-kaurene. The next step in the biosynthetic sequence is the oxidation of (–)-kaurene to kaurenoic acid, which is a normal metabolite of *Gibberella* (Cavell & MacMillan, 1967) and can be converted by the fungus to gibberellins (Hanson & White, 1969). The final conversion of kaurenoic acid, with its six-carbon ring B, to a gibberellin, with a five-carbon ring B, involves a remarkable series of reactions including the extrusion of one carbon atom which is oxidized to the 10-position carboxyl of the gibberellins.

The simplest of the gibberellins, and therefore the most likely to be the first formed in the biosynthetic sequence are gibberellin A_{12} (20-carbon gibberellin) and gibberellin A_9 (19-carbon gibberellin). For the latter, experimental data are lacking, but there is evidence in *Gibberella* that gibberellins A_{12}, A_{14} and A_3 may arise in that sequence from kaurenoic acid (see review by Lang, 1970). Evidence for the central role of gibberellin A_4 in the biosynthesis of many of the 19-carbon gibberellins comes from similar feeding studies with radioactive kaurenoic acid. In the initial stages of incubation most of the label appears as gibberellin A_4 and A_7 and subsequently in gibberellins A_1 and then A_3 (Geissman *et al.*, 1966). The solution of the interrelationships of these four gibberellins came from an approach via the biochemical genetics of *Gibberella* (Phinney & Spector, 1967). It is now a basic tenet of biology that the production of an enzyme is controlled by its own particular gene. The use of fungal mutants, in which one or more of these genes are missing so that particular biochemical processes cannot go on, has been a major technique in the elucidation of biosynthetic pathways (for an exposition of the rationale and methodology of such studies the reader should refer to suitable textbooks such as Srb, Owen & Edgar, 1965). *Gibberella* is the name given to the sexual (perfect) stage of the fungus; the asexual (imperfect) stage is known under the name *Fusarium moniliforme*. Early work on the strains then available was unproductive because of the great difficulties involved in inducing the formation of the perfect (sexual) stage necessary for all genetic work. Finally from a world-wide collection of 121 strains made by Dr. W. E. Gordon, Phinney & Spector (1967) were able to select 12, in which they could cause sexual reproduction and which showed different patterns of the production of gibberellins A_1, A_3, A_4 and A_7. Suitable crosses between strains were thus possible and from an analysis of the patterns of gibberellin production in the progeny, Spector & Phinney (1968) were able to recognize the operation of two genes controlling the biosynthesis of these four gibberellins. One, g_1, controlled the biosynthesis of all four, while the other, g_2, controlled the synthesis of gibberellins A_1 and A_3 only. It seemed reasonable to suppose therefore that gene g_1 controls some step preceding the formation of gibberellin A_4 while gene g_2 controls the hydroxylation of ring C in position 7, whereby A_4 is converted to A_1 and A_7 to A_3 (see pathways in Fig. 27).

In the last few years remarkable strides have been made in the elucidation of the biosynthetic pathways of gibberellin synthesis in higher plants and there is now every indication that they are basically the same as those in *Gibberella*. Enzymically-controlled steps have been studied in cell-free preparations from a number of higher plant tissues and considerable progress has been made in the characterisation of the enzymes concerned. The endosperm of *Echinocystis macrocarpa*, a rich source of gibberellins as we have already seen, has been shown to contain the enzymes which catalyse the biosynthetic sequence of Fig. 27 from acetic to kaurenoic acids (West *et al.*, 1968). Similar production of kaurene from mevalonate has been shown in enzyme preparations from *Cucurbita pepo* endosperm, from the seeds and fruits of *Pisum sativum* and from *Ricinus communis* seedlings (see review by Lang, 1970). The formation of the intermediate copalol was also shown in *Echinocystis* endosperm preparations (Shechter & West, 1969). From feeding studies it is clear that gibberellins are formed, not only in cell-free preparations but also in growing plant tissue. For example kaurenoic acid causes growth responses in dwarf (d-2) maize and dwarf rice seedlings and it is also active in the α-amylase test on barley endosperm (for references see Lang, 1970). Details of the sequence steps however remain to be worked out.

A possible alternative pathway from kaurenoic acid involves a closely related diterpenoid steviol (Fig. 28 XCIX), which occurs in fair quantity in the leaves of the South American plant *Stevia rebaudiana*. It was first shown to have gibberellin-like activity in the dwarf maize assay (Ruddat *et al.*, 1963) and to be converted to a gibberellin-like substance (not gibberellin A_3) by *Gibberella* (Ruddat *et al.*, 1965*a*), suggesting that it could also be a procursor of the gibberellins. More direct support comes from experiments in which ^{14}C-mevalonic acid has been fed to cut stems of *Stevia rebaudiana*.

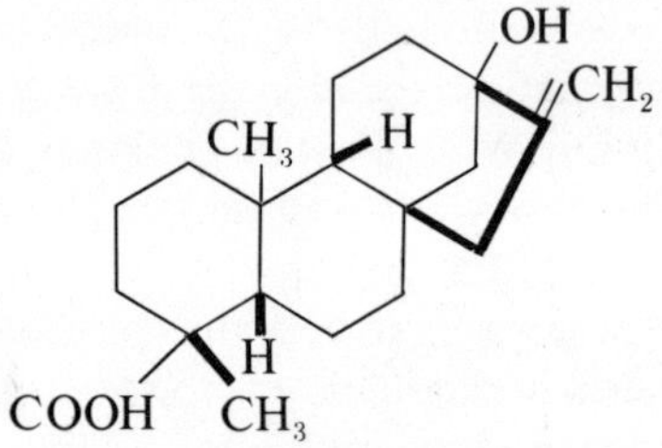

Fig. 28. DCIX Steviol.

Radioactive kaurene, kaurenoic acid and steviol were subsequently isolated and the pattern of labelling suggested that kaurene and kaurenoic acid were the immediate precursors of steviol (Hanson & White, 1968). The fact that steviol could be derived from kaurenoic acid by hydroxylation of the C ring and that gibberellins with such a hydroxyl group are unique to the higher plants, suggests a possible unique pathway through steviol to gibberellin A_{20}, and thence to the A_1-A_3-A_4-A_7 nexus (see Fig. 27). However steviol has so far been shown to occur naturally only in *Stevia* and so the importance of this proposed pathway has still to be proved.

Other Substances with Gibberellin-like Activities

Although the chemical structures allowing activity in the gibberellins so far isolated is of apparently limited range, yet there are other substances of different chemical constitution, that share some of their characteristic properties.

The first example is another metabolite from a fungus, *Helminthosporium sativum.* This metabolite was isolated by Tamura *et al.*, in 1963. It has been given the name *helminthosporol* and has the formula C of Fig. 29. It stimulated the growth of both tall and dwarf varieties of rice, but to a lesser extent than gibberellic acid. However, in striking contrast, helminthosporol has no effect on the growth of any of the three dwarf maize mutants (d-1, d-2 or d-5) and it also seems to have few typically auxin-like properties (Kato *et al.*, 1964). It also shares with the gibberellins one of their most characteristic properties, namely the induction of α-amylase synthesis in the endosperm of barley and of rice (Briggs, 1966; Mori *et al.*, 1965). Closely related compounds e.g. helminthosporic acid (Fig. 29 CI) and dihydrohelminthosporic acid (Fig. 29 CII) also show some activity, e.g. in the lettuce hypocotyl and rice seedling tests (Sakurai & Tamura, 1966). Both helminthosporol and helminthosporic acid share with gibberellins and auxins the capacity to promote the growth of oat coleoptile and mesocotyl segments but are, like the gibberellins, inactive in the oat curvature and split pea epicotyl curvature tests (Hashimoto & Tamura, 1967*a*). Like the gibberellins but unlike the auxins they break seed dormancy (Hashimoto & Tamura, 1967*b*) but unlike both they stimulate primary root growth in lettuce (Hashimoto *et al.*, 1967).

Since these compounds bear a close resemblance to that part of the gibberellin molecule represented by rings C and D, Briggs (1966) has suggested that these rings represent the 'effector' part of the molecule. The data already reviewed on p. 123–4 suggest that this is unlikely and that other parts of the gibberellin molecule are equally important (Brian *et al.*, 1967) for example the carboxyl group on ring B.

The most puzzling new development has been the isolation from *Phaseolus* seeds of a compound with distinct gibberellin-like properties but quite unrelated in structure (Fig. 29 CIII). It has been called phaseolic acid by its discoverers (Redemann *et al.*, 1968). It stimulates the growth of dwarf pea and maize varieties, it induces α-amylase synthesis in barley and it retards senescence in barley leaves. If this proved to be a true natural hormone the implications for the structure/activity relationships of the true gibberellins are very considerable.

A more intriguing recent discovery is that a steroid hormone from locusts (*Locusta migratoria* and *Schistocerca gregaria*) has typical gibberellin activity in the dwarf pea and dwarf maize tests. This hormone, called λ-ecdysone seems to control moulting in these insects during the change from the larval to the adult form. What is even more interesting is that gibberellic acid, when applied to the larvae, can also accelerate the rate of moulting (Carlisle *et al.*, 1963). This is the first unequivocal case of reciprocal activity of two different hormones, one plant and one animal, and is the first hint of a possible similarity of action. Ecdysone has even been reported as occurring

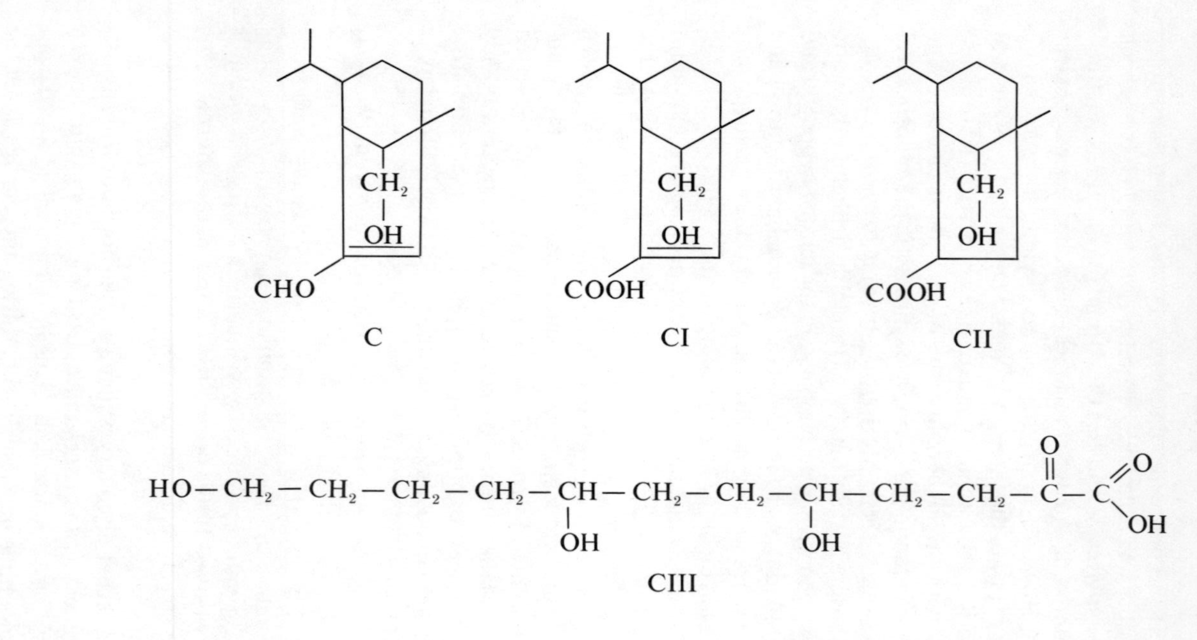

Fig. 29. Other compounds with gibberellin-like activities.

C	Helminthosporol
CI	Helminthosporic acid
CII	Dihydrohelminthosporic acid
CIII	Phaseolic acid
CIV	Ethylene-bis-nitrourethane

in the rhizomes of the fern, *Polypodium vulgare* (Heinrich & Hoffmeister, 1967).

A group of synthetic substances stimulating plant growth in a way superficially resembling that of the gibberellins is represented by ethylene-*bis*-nitrourethane (EBNU) (Fig. 29 CIV), and certain related compounds. EBNU treatment increases the stem length of *intact plants* particularly peas, while inhibiting the growth of roots (Mees, 1965). However its activity is much less than that of gibberellic acid and the patterns of response of various pea varieties are also very different with the two substances. For example, while varieties Meteor and Onward are the most responsive to gibberellic acid, Meteor is the most responsive and Onward completely unresponsive to EBNU. Another difference arises in seed responses; gibberellic acid will promote the germination of dormant lettuce seed while EBNU has no effect (Morgan, 1965). Clearly then this compound cannot be classified with the gibberellins. There is some suggestion that it may augment final organ length by prolonging the period of cell extension, perhaps by retarding the natural ageing processes in the cells.

A survey of range of related compounds (Borer *et al.*, 1966) shows that this kind of growth-regulating activity is confined to compounds of the general structure R-N(NO_2)-X-(NO_2)N-R where X is an alkylene or a substituted alkylene chain. Furthermore the evidence indicated that activity is associated with a conversion in the plant to the dinitramines (R = H) but no real clue to the nature of this action has yet emerged. They are clearly neither gibberellins or auxins.

Chapter V

The Cell-Division Hormones

Introduction

So far the plant hormones we have considered have been those whose most characteristic action seems to be the control of cell expansion, which accounts for most of the size increase in plants and, to some extent, contributes to the determination of the three-dimensional shape of organs. But the pattern of organ production, their nature, frequency, position and shape are largely the outcome of the spatial pattern of cell multiplication (division) in the apical and lateral meristems (see Chapter 1), and there is now no doubt that this is also under hormonal control. Auxins and gibberellins can exert far-reaching influences on cell multiplication processes but there are other plant hormones which seem to be more specifically concerned with cell division and are on the whole much more effective in regulating it. These cell division hormones are the subject of this chapter.

THE TECHNIQUE OF TISSUE AND ORGAN CULTURE

Most of our knowledge of the cell division hormones has come from the exploitation of a relatively new and revolutionary technique in the study of plant growth problems. In this technique, individual organs or tissues removed from the parent plant are grown on sterile nutrient media, usually in the dark, so that they are isolated from all nutrient or hormonic influences which would normally come to them from other organs and tissues. These 'tissue culture' and 'organ culture' methods depend for their success on the maintenance of absolutely sterile conditions and the continued supply of fresh culture media of the correct constitution. Such conditions are usually obtained by transferring pieces of the growing organ or tissue at frequent intervals from one sterile culture vessel to another containing fresh medium: in this way indefinite growth can be obtained over long periods of time. This indefinite growth, in which the participating cells may traverse the whole cycle from meristematic division to full maturation and differentiation, is therefore quite distinct from the limited extension-growth of such portions of organs as oat coleoptile sections and pea stem sections, etc., as discussed in previous chapters. The difference arises in that, in tissue cultures, some part of the tissue isolated is either already meristematic, or is induced to become so by the applied cultural conditions, whereas, in the organ section, used for the several auxin and gibberellin assay techniques, few such dividing cells are to be found. The perfection of this tissue culture technique has enabled plant physiologists to study the precise requirements of a number of different plant organs and tissues, and in particular to isolate and characterise the cell-division hormones themselves.

Although the technique is fairly new, the idea goes back to the famous German botanist, Haberlandt (1913–14), who first considered the theoretical possibilities of it. Unfortunately, all his early attempts to realize such culture met with failure because the nutrients employed were not suitable. After twenty years of unsuccessful experimentation, success was finally achieved by two independent workers, Kotte in Germany (1922) and Robbins in America (1922). They realized that tissues most likely to be induced to grow *in vitro* were those still capable of active cell division at the time of separation from the parent plant (i.e. meristematic tissues), and that organic growth factors, other than simple nutrients, might also be required. Applying these ideas, both workers succeeded in growing roots from isolated root tip meristems in artificial nutrient media containing sugar, inorganic salts and yeast extract. Cell multiplication in the tip meristem was strictly limited however, and it was not until another decade had passed that unlimited growth was obtained. After a series of detailed experiments on the optimum concentrations of each of the components of the nutrient medium, a combination was eventually evolved which gave this unlimited growth when the medium was regularly and frequently renewed (White, 1934).

EARLY STUDIES ON KNOWN CELL DIVISION FACTORS

Very much research has also been done on isolated tissue fragments and the growth, by rapid cell multiplication, of an undifferentiated tissue mass, or callus as it is called. The indefinite growth of this tissue is isolation was realized almost simultaneously by three independent workers, White (1939) in America, Gautheret (1939) in Paris, and Nobécourt (1939) in Grenoble. Callus can be induced to form from a number of different plant tissues. White obtained his from the natural callus or tumour tissue which occasionally forms at a graft union in tobacco, and also from the tumour tissue caused in certain plants by the activities of the crown-gall disease organism, *Agrobacterium tumefaciens.** Gautheret has obtained his from many sources, viz. the cambial tissue of many woody species such as goat willow (*Salix caprea*), the stems of herbaceous plants such as tobacco (*Nicotiana tabacum*) and *Antirrhinum*, and the storage organs of certain vegetables such as carrot, chicory, *Scorzonera*, etc.; Nobécourt obtained his from carrot (see also callus growth on Virginia Creeper in Plate 9). The tissue formed is usually compact and made up of regular undifferentiated cells in which occasional isolated conducting elements are found. Under some growth conditions it may develop into a loose, friable, cancerous mass, in which the cells are either swollen and rounded or grown out into filaments like those of a mould fungus.

In all these successful growth media, not only were sugars and mineral salts necessary as normal growth nutrients, but other organic growth factors had also to be present in very small amounts. In addition to auxin (either natural indole axin or synthetic homologues such as 2, 4-D) there were often requirements for certain vitamins, the chief being thiamin (Fig. 30 CV) (Vitamin B_1), pyridoxine (Fig. 30 CVI) (Vitamin B_6) and niacin (Fig. 30 CVII). A range of amino acids is usually required and this can be most conveniently provided by the mixture of amino acids produced by hydrolysing milk proteins (casein hydrolysate). One fundamental deduction that we can make from these facts is that such tissues and organs are incapable of making their own accessory growth factors, at least in sufficient quantities, from the simple sugars and mineral salt nutrients provided. This is carried out in the normally growing plant in organs possessing the appropriate biochemical machinery i.e. in most cases in the leaves.

It is not surprising that tissues from different organs have both qualitatively and quantitatively differing growth-substance requirements. These differences are probably most marked in respect of auxin requirements. For example, although most dicotyledonous plant tissues demand a supply of auxin in the culture medium before they will grow, a few plants can do without it. This is so in the tuber tissue of some varieties of Jerusalem artichoke (*Helianthus tuberosus*) at certain stages of development (Gautheret, 1942), and also in bramble (*Rubus* spp.), and willow (*Salix* spp.) (See Gautheret, 1955*b*.) The experimental evidence indicates that these tissues possess the capacity of manufacturing sufficient of their own auxins, and so do not need an external supply. The same seems to be true of tumour tissue produced by

*This tumorous tissue was grown free of the causative organism and shown to be capable of more rapid growth than normal tissue, thus demonstrating for the first time the possibility of true plant cancers. See review by Gautheret (1950).

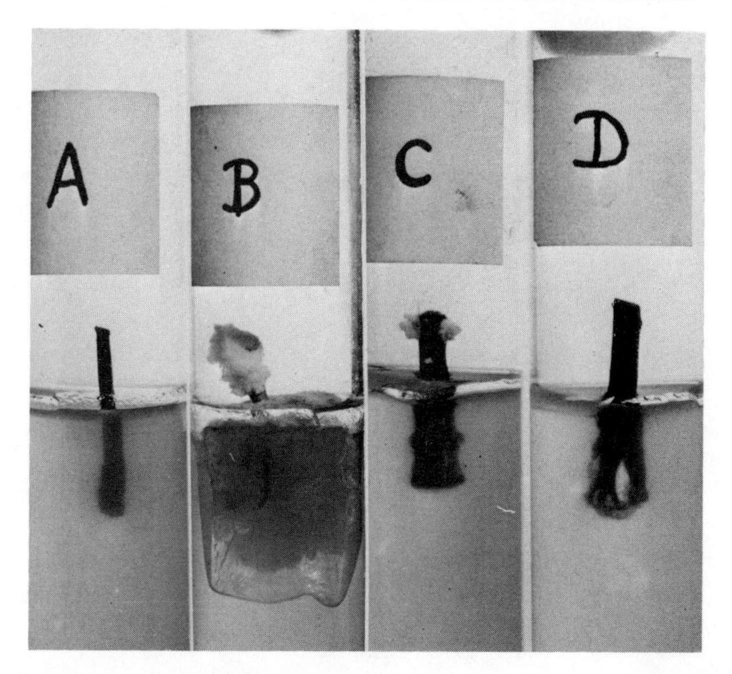

Plate 9. Tissue cultures of Virginia creeper stem.

(A) *Parthenocissus tricuspidata.* Medium without IAA. There is no appreciable callus formation.

(B) *Parthenocissus tricuspidata*. Medium containing 1.0 ppm. IAA. This species shown no particular polarity of auxin movement and hence there is a voluminous formation of callus tissue at both the upper and the lower ends of the segment.

(C) *Parthenocissus quinquefolia.* Culture medium without IAA. This species shows a marked basipetal polarity of auxin movement. The segment of stem is inverted in agar (i.e. apical end downwards in the agar) and appreciable amounts of callus has been formed at the basal end above the agar, due to the basal accumulation of the endogenous auxin.

(D) *Parthenocissus quinquefolia.* As for (C) but with the basal end downwards in the agar and showing callus formation. This demonstrates that callus formation in (C) cannot have been due to a better aeration above the agar medium.

(Photographed from cultures generously supplied by Dr. R. Heller, University of the Sorbonne, Paris.)

parasitic organisms, e.g. crown-gall on *Scorzonera* caused by the bacterium *Agrobacterium tumefaciens* (Gautheret, 1955*b*), and the tumour on sorrel (*Rumex acetosa*) roots caused by the virus *Aureogenus magnivena* (Black, 1945).

There are several interesting phenomena associated with the auxin relations of tissues in culture. One is the phenomenon of 'sensitization', in which a series of passages of tissues through auxin-containing media pro-

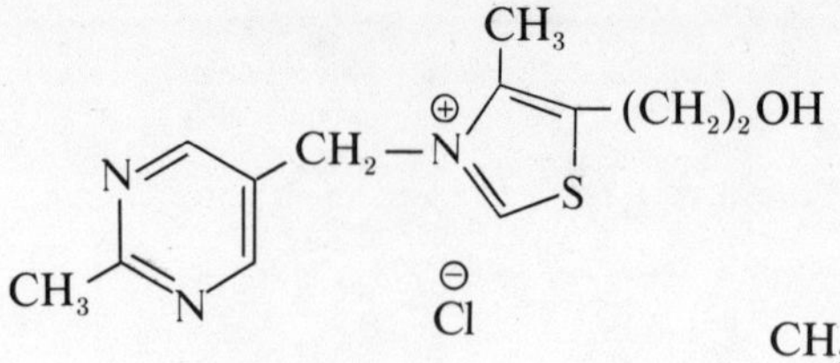

CV

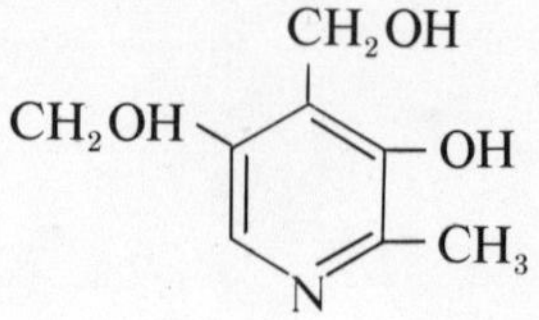

CVI

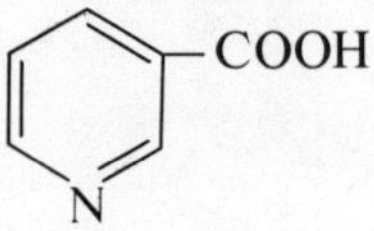

CVII

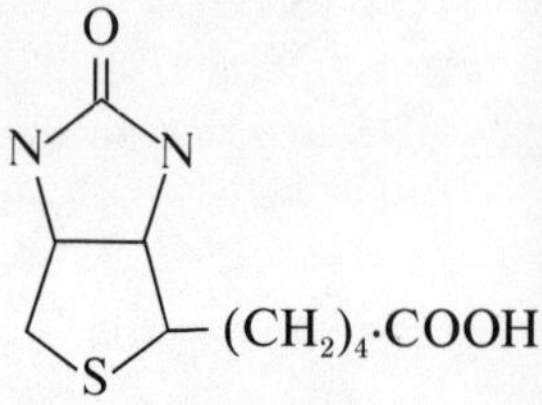

CVIII

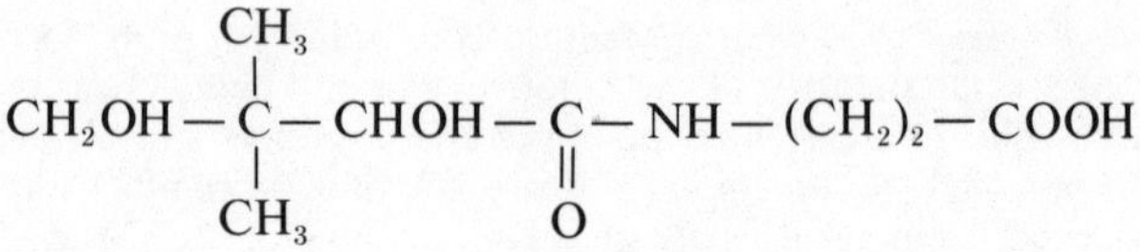

$$CH_2OH-C(CH_3)_2-CHOH-CO-NH-(CH_2)_2-COOH$$

CIX

Fig. 30 Growth factors for tissue and organ culture.
CV Thiamin: Vitamin B_1
CVI Pyridoxine: Vitamin B_6
CVII Niacin
CVIII Biotin
CIX Pantothenic acid

gressively increases the sensitivity of the new tissue to auxins, so that optimum concentrations get smaller (Gautheret, 1955*a*). Jerusalem artichoke and grapevine (*Vitis* sp.) tissue show this response.

A related but much more fundamental change in auxin requirements sometimes occurs spontaneously in tissue cultured in auxins. This is the phenomenon which Gautheret has called 'accoutumance' (English equivalent is 'habituation') or, more recently *anergy* (Gautheret, 1955*a*). This change is announced by the loss of the faculty to form roots in high auxin concentrations, and by the otherwise compact and opaque tissue becoming friable and translucent. This altered appearance is caused by many small loosely adhering nodules in the tissue. The most important feature of the change is that such tissue has acquired the capacity for continued growth without an external auxin supply. This is correlated with a much higher endogenous auxin level in the tissue (Kulescha & Gautheret, 1948; Gautheret, 1955*a*). Undoubtedly there has been some radical shift in the biochemical equilibrium in the tissue cells, resulting in a higher auxin level, but whether this latter is due to a more rapid auxin synthesis or to a slower auxin destruction is still a moot point (Gautheret, 1955*a*). So far, the causes which underlie this switch in growth requirements remain obscure.

The vitamin requirements of tissues in culture may show similar variations. Thiamin, although it may stimulate proliferation in some tissues (e.g. of *Scorzonera*), is not a necessity in the growth medium, as adequate amounts are manufactured in the tissue itself. Pantothenic acid (Fig. 30 CIX), which belongs to the vitamin B complex, also stimulates the growth of some tissues e.g. of hawthorn (*Crataegus* sp.) (Morel, 1946), whereas there is no response with carrot (*Daucus carota*), grape (*Vitis*) and Virginia creeper (*Parthenocissus*) tissue (Gautheret, 1955*b*). A number of other vitamins behave similarly. The phenomenon of *anergy* has also been noted in willow tissue with thiamin and other vitamins, biotin (Fig. 30 CVIII) and pantothenic acid (Fig. 30 CIX) (Gautheret, 1955*b*).

Cell Division Factors From Endosperm Tissue

The search for cell division factors in plant extracts and fluids of natural origin has revealed the presence of a whole complex of growth substances in the nutritive material (endosperm) surrounding young embryos in seeds. One of the most active sources is the fluid which more or less fills the internal cavity in the nutritive flesh of the coconut (*Cocos nucifera*). Small quantities of this coconut milk will greatly promote the growth in liquid culture of such tissues as the phloem of carrot roots, the stimulation being many times greater than that produced in other tissues by IAA. Furthermore, in contrast to the action of auxins, the stimulating effects are almost entirely on cell division and not on cell extension (Steward & Caplin, 1954*a*). Their action may, however, be closely associated with auxins, as some tissues (e.g. potato tuber tissue), whose proliferation is not promoted by coconut milk alone, can be made very sensitive to that material if small traces of a synthetic auxin (e.g. 2, 4-D) are present as well (Steward & Caplin, 1951). One suggested explanation of this is that a natural inhibitor of the action of the coconut milk factor, shown to be present in the potato tuber, is put out of action by the auxin (Steward & Caplin, 1954*b*). It does not require a great effort of the imagination to visualize the control of cell division by the balance of three such factors. Similar division-promoting activity has been shown in extracts of the reproductive structures of quite different plants. It is present for example in immature grains of maize (*Zea mays*) in what is called the 'milk stage'; it is also to be found in young fruits of both an American species of horse-chestnut (*Aesculus woerlitzensis*) and the unrelated gymnospermous maidenhair tree (*Ginkgo biloba*). The development of these fruits in relation to the production of cell-division substances has been fully described by Steward & Shantz (1959). This activity of extracts declines as the 'seed' matures. Presumably the growth factors which are characteristic of such nutritive tissue, are brought under control as the cells begin to differentiate into compact endosperm (Steward & Caplin, 1954*b*).

Much attention has been given to the isolation and identification of these growth factors. The first attempts were made by Shantz & Steward (1952), who extracted 2,500 litres of coconut milk and separated a number of active fractions therefrom by solvent partitioning and chromatography on columns of pure cellulose and of synthetic resins. These were first allocated to two distinct categories, specific cell-proliferation substances and amino acids, upon whose presence the high activity of the unknown growth substances seemed to depend. Some of these unknown substances have been separated and identified by a new technique of plant tissue culture allowing observations on the growth of free-floating cells. One was identified unequivocally as *N*,*N'*-diphenylurea (Fig. 31 CX) (Shantz & Steward, 1955). A little later in the same laboratory activity in a neutral (non-acid) fraction was shown to be due to certain sugar alcohols, predominantly sorbitol (Fig. 31 CXI) with lesser amounts of *myo*- (Fig. 31 CXII) and *scyllo*-inositol (Fig. 31 CXIII) (Pollard *et al.*, 1961).

Great difficulties and frustrations beset workers who attempt the task of identifying specific cell-division factors in coconut milk and other similar materials, firstly because of the great diversity of active substances present and secondly because no one component is sufficient by itself, a considerable number of them needing to be present in proper proportions in any synthetic growth medium to bring activity up to that of pure coconut milk. Thus apart from the amino acids and the growth substances already mentioned there are present vitamins (Vandenbelt, 1945), auxin (Paris & Duhamet, 1953), gibberellins (Radley & Dear, 1958), nucleic acids (Tulecke *et al.*, 1961) and the N^6-substituted adenines (see later). The exact biochemical rôles of each, apparently essential, component still largely elude us. Doubts have been expressed as to whether some of the so-called specific cell-division factors

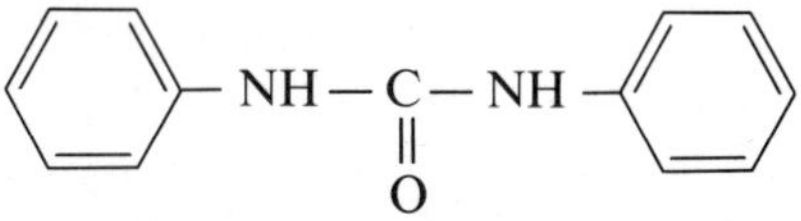

CX

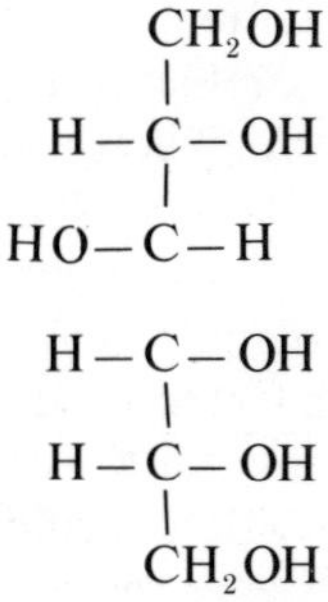

CXI

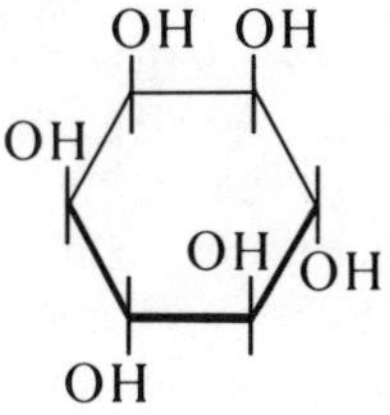

CXII

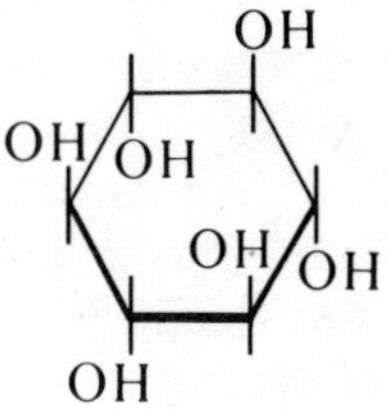

CXIII

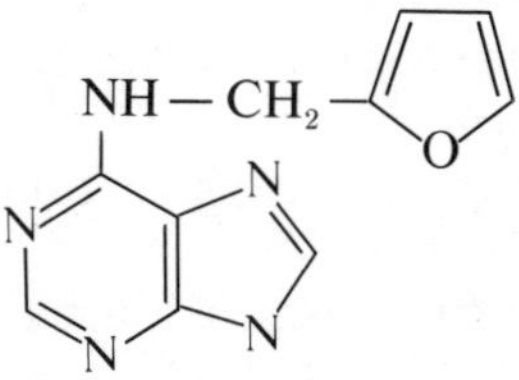

CXIV

Fig. 31. Cell division factors
CX *N,N*-diphenylurea
CXI Sorbitol
CXII *myo*-inositol
CXIII *scyllo*-inositol
CXIV Kinetin. 6-furfurylaminopurine.

are really acting as cell-division hormones. For example it has been suggested that the diphenylureas may be acting merely as sources of the side chains in the synthesis of the N^6-substituted adenines (see later p. 146–8) which are undoubted cell-division factors. Even more doubt exists for *myo*-inositol, for which there is much evidence that it is an important key intermediate in the synthesis of a number of basic polyuronides and pentosans of the cell wall and in this way would influence growth processes in tissues (see review by Lamport, 1970).

The Cytokinins

In studies of the growth in sterile culture of pith tissue from the stem of tobacco, Jablonski & Skoog (1945) observed that, when auxin was the only growth substance present in the culture medium, considerable expansion of existing cells was the only growth response to be observed. If however the pith contained small amounts of differentiated vascular (food-conducting) tissue or if the piece of pith were placed in contact with a small fragment of such vascular tissue, then active division took place in the pith cells. This discovery indicated the presence of a soluble cell-division factor in vascular tissue, a suggestion subsequently verified by experiments with extracts which, in combination with auxin, induced active cell division in cultured pith tissue. The exploration of other plant sources disclosed a remarkable activity in extracts of yeast and this led ultimately to the richest source of this new factor – deoxyribonuclei acid (Miller *et al.* 1955). Purification of extracts of the samples of the acid from herring sperm and from calf thymus gland resulted in a pure crystalline preparation of the new cell-division factor, which was christened *kinetin*. It was identified as 6-furfurylamino-purine (Fig. 31 CXIV) and was subsequently synthesized (Miller *et al.* 1956). The full history of kinetin discovery is described in a review by Miller (1955).

Many other aminopurine derivatives with other groups replacing the furfuryl group were then synthesized and shown to have activity (Skoog & Miller, 1957). A whole range of such active compounds are now known and these are discussed in a later section.

Although it is now widely recognized that kinetin was an artefact of the nucleic-acid-extraction procedure, yet many subsequent searches have revealed the presence of kinetin-like substances in many plant tissues. They are, therefore, like the auxins and gibberellins, native growth substances of plants and their nature and occurrence will be described in a later section (p. 146). They form a group quite distinct from the auxins and gibberellins in their physiological actions and for this reason have been given the name *cytokinin* by Skoog *et al.* (1965). This term is now applied, not only to the naturally-occurring substances, but also to purely synthetic analogues which produce the same physiological reactions.

The Biological Assay of the Cytokinins

As we have already seen for the auxins and gibberellins, all detection, purification and thus the ultimate chemical characterisation of the cytokinins have hinged on efficient, sensitive and specific biological assays. We will now consider the most important of these in some detail.

(A) CALLUS TISSUE GROWTH

Naturally the first and still a most widely used assay was based on induced cell proliferation in homogeneous parenchymatous tissues from the stem pith of tobacco (*Nicotiana tabacum*). As first used (Miller *et al.*, 1956), small pieces of callus tissue, preformed in culture from such sterile pith fragments, were grown in a medium composed of a balanced solution of mineral salts, vitamins (thiamin, nicotinic acid and pyridoxine), auxin (IAA), the amino acid glycine, sucrose (as an energy and carbon source) and agar (to solidify the medium for surface culture of the tissue). On such a medium callus tissue will grow and respond to cytokinins and reach its maximum bulk in three to four weeks (see Plate 10 and Fig. 32 A). The increase in volume (wet weight) can then be determined and is the parameter used as a measure of the response to cytokinin. Unfortunately, as we have seen, many factors may affect callus growth and thus much empirical research has gone into the design of an assay medium of optimal constitution such that all factors other than cytokinins in the plant extract under study will have their minimum effect and the assay thus attain its maximum sensitivity to and specificity for cytokinins. Such a medium has been described by Murashige & Skoog (1962) and contains the same constituents as the original medium, except in different proportions, and also the sugar alcohol *myo*-inositol, which we have already seen was identified as a growth factor from coconut milk. While giving minimal growth in the absence of cytokinins, this medium gives a response to cytokinins which is linearly related to concentration over the range 1 – 15μg/1 (Rogozinska *et al.*, 1964); higher concentrations progressively depress growth from its maximum. It is an extremely sensitive assay and will detect natural cytokinins (e.g. zeatin – see later) at concentrations as low as 5×10^{-11} M (Skoog & Armstrong, 1970).

Although this method measures only the increase in fresh weight, yet on this medium such growth is closely correlated with the increase in total cell numbers and so can be taken as a measure of cell division intensity. Even so some workers have preferred to make counts of the total cell numbers per callus isolate, a somewhat tedious process involving the separation of cells by chemical methods (2 percent aqueous chromic acid) and counting them in known volumes of a sample suspension under the microscope (Bottomley *et al.*, 1963).

Naturally tobacco is not the only species to yield suitably responsive callus and various other sources are now used. Callus derived from the cotyledons of sterile soybean (*Glycine max*) was introduced for assay by Fox & Miller (1959). Phloem tissue from the carrot root has been used by Letham in all his researches with increases in wet weight and cell number as measures of response (see Letham, 1963*b*). This carrot tissue is more sensitive to cyto-

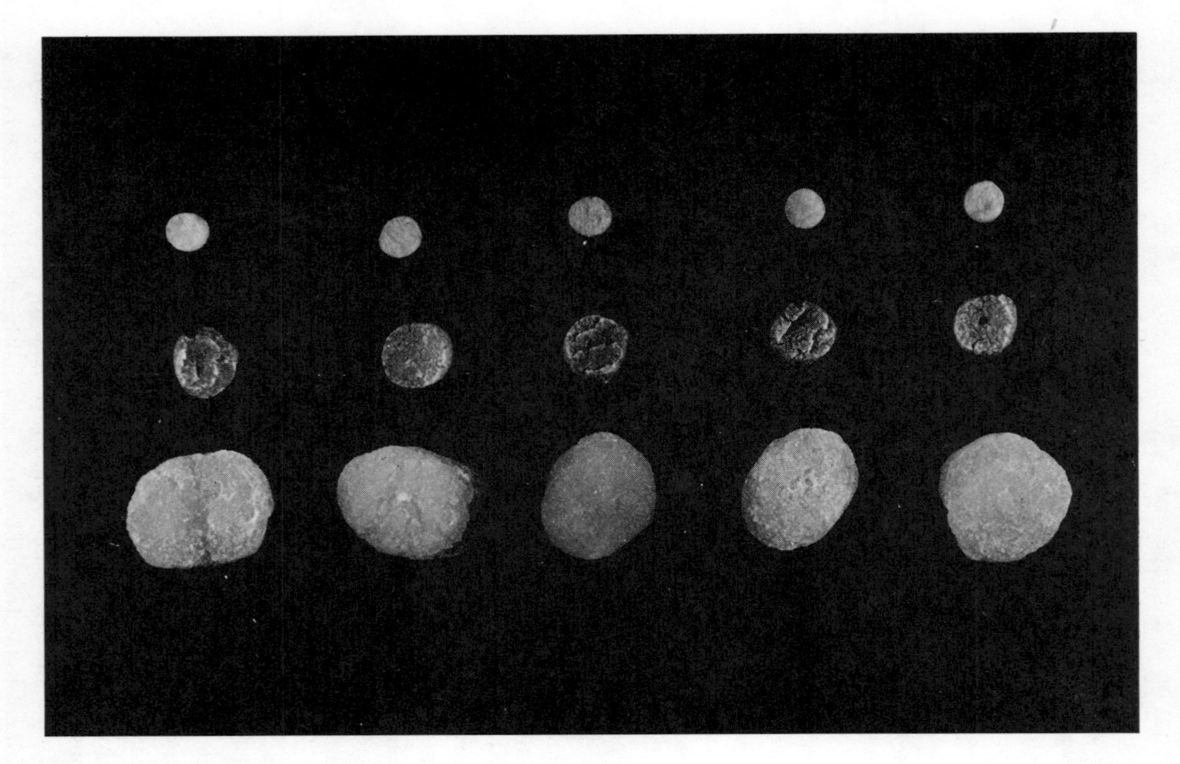

Plate 10. Carrot phloem callus assay for cytokinins.
Tow row. Carrot phloem explants after excision. *Middle row*. Explants after growth in basal medium for 21 days. *Bottom row*. Explants after growth for 21 days in a medium containing zeatin (5×10^{-10} M). Magnification $\times$ 4. (Photograph by kind permission of Dr. D. S. Letham, National University of Australia.)

kinins if certain vitamins present in Murashige & Skoog's medium are omitted (Letham, 1967*b*).

(B) BUD FORMATION

Assays which depend on callus growth, which is slow, are necessarily of long duration and hence involves the risk of loss of active material from the culture medium during the growth period. A relatively rapid assay of high sensitivity and specificity has been based on the promotion of bud formation in the protonema of the moss *Funaria* (Szweykowska, 1962). However it lacks the precision of the tobacco callus assay.

(C) LEAF EXPANSION

The growth of leaves is also controlled by hormones. Auxins act as inhibitors and cytokinins are stimulants. This stimulating action has been used for quantitative assay for the cytokinins. Partially expanded leaves from plants grown in the dark (etiolated) are taken and from the blade small disks are cut with a suitable tool (e.g. a cork-borer). Samples are then floated on a buffer solution containing the suspected cytokinin and left for one to three days at a constant temperature (*c.* 25°C) in the light. In the absence of cytokinins there is little leaf expansion (measured by increase in diameter or wet weight of the disk) and the degree of promotion produced by the cytokinins is proportional to the logarithm of the applied concentration over the range 0.003 to 1.0 μg/ml. Suitable species have been found to be dwarf bean (*Phaseolus vulgaris* L.) (Miller, 1961*b*; Letham, 1967*b*) and radish (*Raphanus sativus*) (Kuraishi, 1959).

A rapid bioassay has been developed by Letham (1968) using radish cotyledons excised from the seed immediately after their emergence from the testa during germination. Their expansion in the light is considerably increased by cytokinins, a respons which is sensitive to a concentration of 10 μg/1. The response is specific to cytokinins, with the exception of a small stimulation produced by gibberellic acid (see Fig. 32 B).

Various members of the duckweed family (Lemnaceae) provide very convenient material for cytokinin assay. Species of *Lemna* and the closely related genus *Spirodela* are tiny plants consisting of one or two very small green fronds that grow floating on the surface of ponds. Loeffler & van Overbeek (1964) used the increased rate of frond expansion of *Lemna minor* induced by cytokinins in the culture medium as a basis of a quantitative assay, and a similar method has been employed by Letham (1967*a*) using

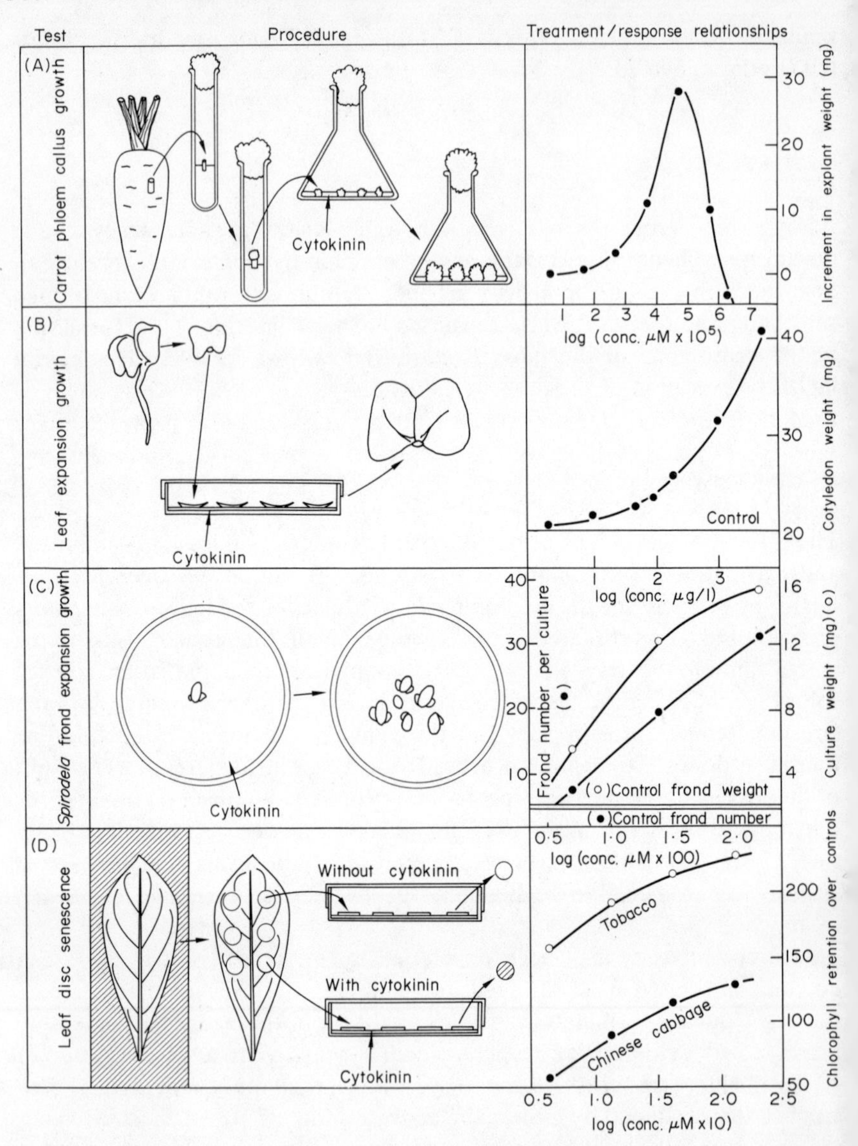

Fig. 32. Diagram showing the salient features of four of the main assays for cytokinins. (A) carrot phloem callus growth, (B) radish leaf expansion (C) *Spirodela* frond expansion growth and (D) leaf disk senescence test. In the first column are illustrated the major steps in the preparation of the material for assay. The second column shown the quantitative relationships between the response and the concentration of applied cytokinin (kinetin in all these graphs). Data of (A), (C) and (D) from Letham (1967*a*). Data of (B) from Letham (1968).

Spirodela oligorhiza (see Fig. 32 C). The response is related in an approximately linear fashion to the logarithm of the cytokinin concentration from about 0.05 μM. Increases in frond number can also be used as a basis for assay.

(D) LEAF YELLOWING TEST. (DELAY OF SENESCENCE)

One of the most striking effects of the cytokinins is that they will delay the yellowing (loss of chlorophyll) in ageing leaves of certain species if applied to the surface in solution. This was first shown by Richmond & Lang (1957) and the phenomenon was soon exploited for a sensitive bioassay for cytokinins (Osborne & McCalla, 1961). In this original test leaves were harvested from the American cocklebur (*Xanthium pennsylvanicum*) and stored under humid conditions in dim light until the yellowing process was just beginning. Disks were then cut from the blades and samples floated on moist filter-paper in darkness for 48 hours (see Fig. 32 D). While the control (no cytokinins) disks went completely yellow those treated with the hormone retained their chlorophyll and remained green. The test is put on a quantitative basis by extracting the chlorophyll pigments with a suitable solvent (e.g. 80% ethanol) and estimating it in a colorimeter or spectrometer. There is a linear relationship between the chlorophyll retention and the logarithm of the cytokinin concentrations from about 0.5 to 50 μM.

Since then other workers have found other species more convenient. Kende (1964) has used segments of barley leaves, Seth & Wareing (1965) oats and bean and Letham (1967*a*) tobacco, Chinese cabbage (*Brassica pekinensis*) and radish disks.

(E) BETACYANIN SYNTHESIS

Certain species of the genus *Amaranthus* produce the red pigment betacyanin in the hypocotyl in the light but not in the dark. Synthesis in the dark is promoted by cytokinins, especially if the amino acid tyrosine is supplied as a substrate. This response has been made the basis of a relatively quick and sensitive assay for cytokinins by Bigot (1968) using *Amaranthus caudatus*. Its sensitivity reaches down to concentrations of about 3×10^{-7}M and is highly specific for cytokinins.

Other assays have been used (see review by Miller (1961*b*)) but (A), (C) and (D) above are those now regarded as standard. Of these the callus culture assay is the most sensitive but takes 3 to 5 weeks to complete. Leaf yellowing inhibition is the least sensitive but can be completed rapidly (2 days). However it is not specific for cytokinins. A table of data for all current assays has been compiled by Letham (1967*b*).

Extraction, Purification and Separation of Natural Cytokinins

Extraction of cytokinins from plant tissue is usually carried out with aqueous ethanol (70–95 percent). The tissue is macerated with the solvent, allowed to extract for a standard time, then centrifuged and the residue re-extracted one or more times. The supernatant extracts are recombined, evaporated to dryness under vacuum below 40°C and the residue resuspended in water for immediate assay. Alternatively the aqueous extract of this residue can be re-extracted with ethyl acetate and further evaporated to dryness before making the final assay solution (see Letham & Williams, 1969). Further purification can make use of suitable ion-exchange resins to remove unwanted materials (see Klämbt *et al.*, 1966).

The separation of components of the final purified extract makes use of chromatographic methods, most commonly paper chromatography and thin-layer chromatography on silica gel or alumina. Such separations are normally followed by bioassay of segments of the chromatogram to identify active materials with the methods already described above. Separation and preliminary identification of the cytokinin bases and their ribosides (see later p. 147) can be carried out with considerable precision by gas chromatography of the trimethylsilyl derivatives (Most *et al.*, 1968) (cf. separation of the gibberellins p. 115). The method is extremely sensitive, the limit of detection with pure compounds being 0.005 μg. Positive identification by mass spectrometry is the next logical step.

The Naturally-Occurring Cytokinins of Higher Plants

After the initial isolation and identification of kinetin, the question whether it, or related compounds, were naturally occurring hormones of higher plants, remained for a few years an open one. The first clue to the answer came from the work of Goldacre & Bottomley (1959) who made extracts of small immature apples two weeks after pollination at a time when growth was mainly by cell multiplication and when they might be expected to contain optimal concentrations of cell-division factors. Assay by tobacco callus tests revealed the presence of an active cell-division factor which could not be characterised chemically. Shortly afterwards Miller (1961*a*) made extracts of immature maize grain in the 'milk' stage of development and obtained evidence for the presence of more than one cell-division factor, one of which showed chemical properties very similar to those of kinetin. Working independently Letham (1963*a*) isolated an apparently identical compound from the same source and gave it the name *zeatin*. A year later came chemical identification (Letham *et al.*, 1964) as 6-(4-hydroxy-3-methylbut-2-

enyl)-aminopurine (Fig. 33 CXV) and proof of its very close structural affinities with the original kinetin. Its identity with the kinetin-like substance first isolated by Miller followed a year later (Letham & Miller, 1965). Zeatin is much more active than kinetin, being effective in the tobacco callus test at one twentieth to one thirtieth the concentration (Skoog *et al.*, 1967); there are similar differences in the carrot phloem callus test and in the *Spirodela* test but zeatin is much less active than kinetin in the leaf senescence test (Letham, 1967*a*).

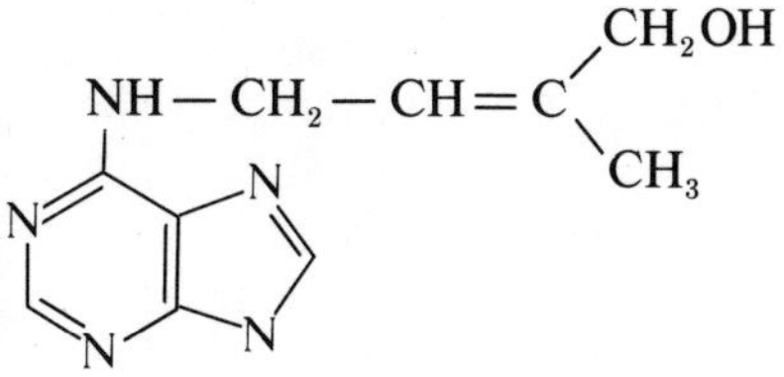

$NH-CH_2-CH=C(CH_2OH)(CH_3)$ (purine ring)

CXV

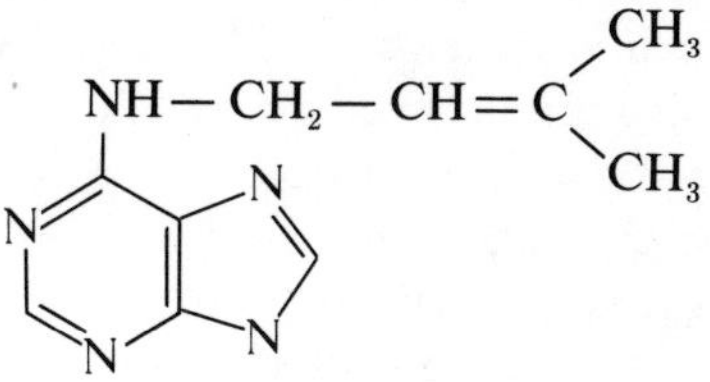

$NH-CH_2-CH=C(CH_3)(CH_3)$ (purine ring)

CXVI

$NH-CH_2-CH_2-CH(CH_3)(CH_2OH)$ (purine ring)

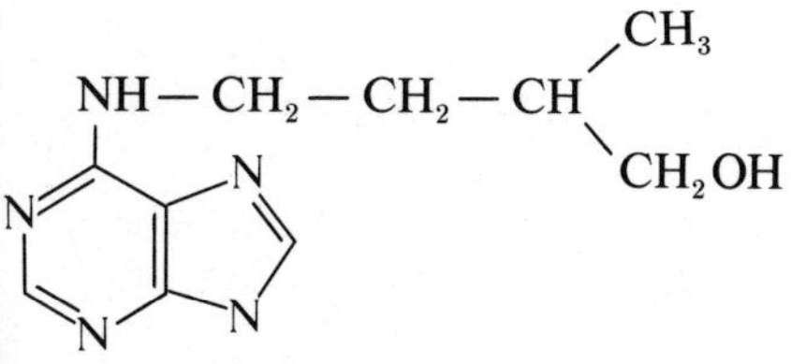

CXVII

Fig. 33. Natural cytokinins.
CXV Zeatin. 6-(4-hydroxy-3-methylbut-2-enyl)-aminopurine.
CXVI Isopentenyladenine (IPA). 6-(γ,γ-dimethylallyl)-aminopurine
CXVII Dihydro-zeatin

Since then other cytokinins have been shown to accompany zeatin; one is a compound of zeatin with a pentose sugar, β-D-ribofuranose (i.e. zeatin riboside) and another a phosphate ester of the riboside (Letham, 1966*a* & *b*). It would also appear that zeatin is not a cytokinin specific to maize kernels since work on immature fruitlets of plum, which, like apple fruitlets expand mainly by cell multiplication, has shown the presence therein of cell-division factors apparently identical with zeatin (Letham, 1964). Zeatin in the form of its riboside now appears to be one of the cell-division factors present in coconut milk (Letham, 1968) but is not identical with the cytokinin of apple fruitlets, which nevertheless may be a zeatin derivative (Zwar & Bruce,

1970). Zeatin and its riboside have also been reported from *Helianthus annuus* leaves and xylem sap (Klämbt, 1968) and also as a metabolic product in cultures of the mycorrhizal fungus *Rhizopogon roseolus* (Miller, 1967). It promises therefore to have a wide distribution in plants.

Corynebacterium fascians is an organism which induces multiple bud production and witches broom in certain trees, a phenomenon one might expect to be due to the production by the parasite of bud-inducing cytokinin-like substances. From cultures of this bacterium a cytokinin has been isolated and shown to be a close relative of zeatin (Helgeson & Leonard, 1966). It is 6-(γ,γ-dimethylallyl)-aminopurine, or, to use a more popular name, *iso*-pentenyladenine (IPA) (Fig. 33 CXVI). Two *Exobasidium* species, which also produce witches broom symptoms in host plants, have also been shown to synthesise cytokinins in culture. Another tumour-producing micro-organism, *Agrobacterium tumefaciens*, is also claimed to produce the same cytokinin (Klämbt, 1967) which may play a contributory part, in the promotion of cell proliferation in the tumour tissue. A parallel situation is seen in the nitrogen-fixing bacterium *Rhizobium*, which invades the roots of leguminous plants and induces the formation of root nodules. A cytokinin has been shown to be produced by *R. japonicum* when grown in pure culture (Phillips & Torrey, 1970) and the cell proliferation in the cortex of the roots which largely accounts for nodule formation may result from the action of the cytokinin which this nodule organism produces.

Yet another close relative of zeatin has been isolated from immature *Lupinus luteus* seeds (Koshimuzu, 1967) and this is dihydrozeatin (Fig. 33 CXVII).

An isomer of IPA, triacanthine, is present in the leaves of *Gleditsia triacantha, Holarrhena floribunda* and *Chidlovia sanguinea*, the isopentyl side chain being attached to carbon 3 of the adenine nucleus instead of to the 6-amino group. (Cavé *et al.*, 1962). This compound has been reported to have a small cytokinin activity, but this is likely to be due to a partial transfer of the side chain from the 3- to the 6-amino position during autoclaving, giving rise to small amounts of IPA (Rogozinska *et al.*, 1964).

The list of uncharacterized cytokinins, detected by one or other standard bioassay, is rapidly lengthening. They have been reported as present in the cambia of trees, in xylem sap, in root apices, in potato skin, in tomato juice, in crown-gall tissue etc. Detailed references to these researches can be found in a review by Shantz (1966) and in another by Nitsch (1967*a*).

Cytokinins have also been shown to occur as components of transfer RNA in a number of bacteria, in yeast, in wheat germ and in tissues of rat and monkey (see Skoog & Leonard, 1968 for references). The significance of these occurrences in RNA will be considered in relation to the mechanism of cytokinin action in Chapter XII.

SYNTHETIC CYTOKININS AND STRUCTURE/ACTIVITY RELATIONSHIPS

Since the isolation and identification of kinetin, many related compounds have been synthesized and their cytokinin activities tested in a variety of assays. From such studies ideas of the molecular structure necessary for action are beginning to emerge and will now be briefly discussed.

(A) 6-SUBSTITUTED AMINOPURINES

These very close homologues of kinetin have been the most extensively studied. We have already seen that one (zeatin) is much more active. A considerable number of substituents other than the furfuryl group produce substances with activities equal to that of kinetin. A list of them is drawn up in Table V. They show a wide range of chemical structure with nothing obvious by way of a common feature or features. However the size of this substituent seems to be important since if it is too small (e.g. $-H$ (adenine itself), $-CH_3$ or $-C_2H_5$) or too large (e.g. a decyl group) activity is considerably

TABLE V

ACTIVITIES OF 6-SUBSTITUTED AMINOPURINES IN THE TOBACCO CALLUS TEST RELATIVE TO THE ACTIVITY OF KINETIN

Estimates based on concentration (μM) giving approximately half maximum response.

Name of substituent group	Formula of substituent group	Relative activity	
γ, γ-dimethylallyl	$-CH_2-CH=C(CH_3)_2$	10.0	1
n-pentyl	$-(CH_2)_4-CH_3$	1.6	1
isoamyl	$-(CH_2)_2-CH(CH_3)_2$	1.3	1
benzyl	$-CH_2-$	1.3	1
2-thenyl-	S	1.0	1
2-pyridylmethyl	N CH_3-	0.7	1
phenyl		0.4	1
n-butyl	$-(CH_2)_3-CH_3$	0.4	1
n-hexyl	$-(CH_2)_5-CH_3$	0.25	1
β-ethoxyethyl	$-(CH_2)_2-O-CH_2-CH_3$	0.1	2
β-propoxyethyl	$-(CH_2)_2-O-(CH_2)_2-CH_3$	0.1	2

(1) Skoog *et al.*, 1967.
(2) Rothwell & Wright, 1967.

reduced (Skoog *et al.*, 1967). This is clearly demonstrated in Fig. 34. Again a double substitution may increase or decrease activity depending on the final size of the side chains. Thus in the induction of buds on the protonema of the moss *Tortella caespitosa*, another characteristic of cytokinins, 6-(di-*n*-propyl)-aminopurine is more active than 6-(*n*-propyl)-aminopurine while 6-(di-*n*-butyl)-aminopurine is less active than the 6-(*n*-butyl)- homologue (Gorton *et al.*, 1957). Where the N^6-substituent is a benzyl group, further substitution in the phenyl ring usually reduces activity, *para*-substituents doing so more than *ortho*- or *meta*-substituents. Only in *o*-chloro- or *o*-methyl-substitution is a slight increase in activity recorded (Kuraishi, 1959). Such relationships give rise to the suggestions that the biological activities of these 6-substituted aminopurines might be related to their hydrophilic/lipophilic characteristics (cf. similar considerations for the synthetic auxins in Chapter III) and this was tested in a range of compounds by measuring oil/water partition coefficients (Rothwell & Wright, 1967);

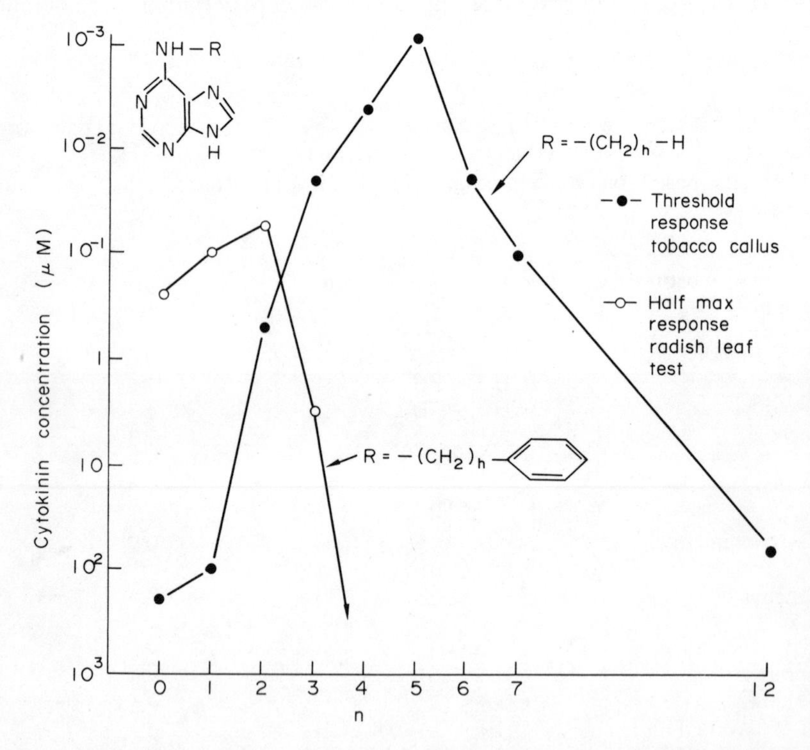

Fig. 34. Graph showing the relationship between side-chain length and biological activity in two homologous series of cytokinins. (Constructed from the data of Skoog *et al.*, (1967)).

no such correlation was disclosed. Aspects of substituent structure other than mere size or lipoid solubility are more important. Thus the presence of an unsaturated double bond greatly increases activity, e.g. the γ, γ-dimethyl-allyl-group confers much greater activity than the corresponding saturated isoamyl-group. Similarly the phenyl group confers more activity than the cyclohexyl group (Skoog *et al.*, 1967). Hydroxylation of the side-chain also affects activity. Thus with the isopentyl side-chain, a hydroxyl introduced in position 4 enhances activity but similar substitution in positions 2 and 3 reduce it (Leonard *et al.*, 1969).

A basic characteristic of such side-chains which is important for cytokinin activity may be its planarity. This is indicated from a comparison of the activities of a number of close relatives of N^6-isopentenyladenines with slightly different side-chains. Thus the *trans* form of 6-(3-chloro-2-butenyl-amino)-purine is about three times as active as the *cis* form of this molecule. Again the activity of 6-(2-bromo-3-methyl-2-butenylamino)-purine has only about one percent of the activity of its non-halogenated homologue. In both these pairs the least active compound has the side chain with the most disturbed planarity (Hecht *et al.*, 1970*b*).

Substitution in the ring of N^6-isopentenyladenine on the whole reduced physiological activity. As would be expected the reduction of activity is a function of the substituent group and for those groups studied runs in the order:

$$=\mathrm{O} \doteqdot \mathrm{Cl} < \mathrm{NH_2}- \; = \; -\mathrm{S}-\mathrm{CH_3} < \mathrm{HO}-$$

(Hecht *et al.*, 1970*a*). A similar situation holds for the hydroxy isopentenyl (zeatin) series. Esters of zeatin formed by elimination of water between the hydroxyl group of the side chain and the carboxyl group of formic, propionic and indol-3yl-acetic acids all have cytokinin activities slightly greater than that of zeatin itself, a situation which has been explained in terms of either a more rapid penetration into the assay tissue or to a stabilization or the molecule by esterification (Schmitz *et al.*, 1971). In contrast to the partial deactivation of the unesterified molecule by ring substitution, the insertion of chlorine into the 2 position of the zeatin ring system doubles the activity of the acetic and propionic acid esters; it is suggested that this may be related to the ultimate functioning of these cytokinins in t-RNA and its improved binding on the ribosomes (Schmitz *et al.*, 1971) (see Chap. XII, p. 400).

(B) OTHER ADENINE DERIVATIVES

Substitution in positions other than N^6 have been reported in the earlier literature to yield active cytokinins, but more recent and critical work has suggested that the activities recorded were due to N^6-substituted derivatives produced by conversions during sterilization prior to bioassay (Skoog *et al.*, 1967). The situation seems to be that only the N^6-substituted adenines have activity. The addition of a second *N*-substituent group to an active N^6-substituted adenine greatly reduces its activity, the N^6-methyl derivatives of kinetin and 6-benzylaminopurine for example having about one tenth the activity of the parent compounds. Substitution in the 1- or 3-positions to give for example 1-methyl-6-benzylaminopurine results in virtually complete loss of activity. (Skoog *et al.*, 1967).

(C) OTHER PURINE DERIVATIVES

Derivatives of purine other than adenine have also been investigated. The replacement of N^6 with a sulphur atom (Fig. 35 CXVIII) still allows some activity (Skinner *et al.*, 1957) but this tends to be very low; for example 6-(γ, γ-dimethylally)thiopurine has an activity only 0.01 to 0.05 that of 6-(γ, γ-dimethylallyl) aminopurine (Skoog *et al.*, 1967).

(D) CLOSELY-RELATED NON-PURINES

The substitution of a nitrogen atom for carbon atom 8 in the purine ring of kinetin yields azakinetin (Fig. 35 CXIX). This still has activity which is about 0.05 that of kinetin itself (Skoog *et al.*, 1967). Similar reductions in activity are seen when an oxygen or a sulphur atom are substituted for nitrogen in the N^6 position in the adenine nucleus (Skoog *et al.*, 1967). In leaf senescence tests with rice leaves weak cytokinin-like activity has been demonstrated in imidazole (Fig. 35 CXX) and benzimidazole (Fig. 35 CXXI) (Yamada *et al.*, 1964). These authors conclude that it is the imidazole ring which is essential for activity. However the leaf senescence test is relatively unspecific and confirmation is needed from cell-division tests. Nevertheless some support for these contentions comes from studies with 4-substituted pteridines (Fig. 35 CXXII), which seem to have no cytokinin activity (Lloyd *et al.*, 1967).

(E) COMPOUNDS NOT RELATED TO PURINES

We have already seen that in studies of the cell division factors from coconut milk, Shantz & Steward (1955) isolated a compound *N,N'*-diphenylurea (Fig. 31 CX) which was active in their carrot explant assay. The activity of this compound was verified in the tobacco pith test (closely related to the tobacco callus test) by Bruce & Zwar (1966) who tested many hundreds of compounds including substituted ureas (Fig. 35 CXXIII) and thioureas (Fig. 35 CXXIV) in an attempt to derive structure/activity relationships in these groups of compounds.

Firstly they established that substituted thioureas showed activity which was on the whole smaller than that of the corresponding ureas. For example

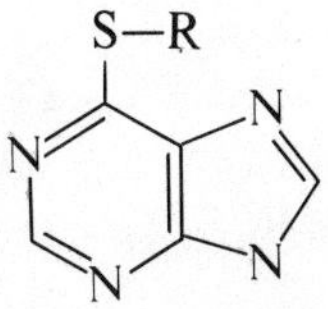

CXVIII

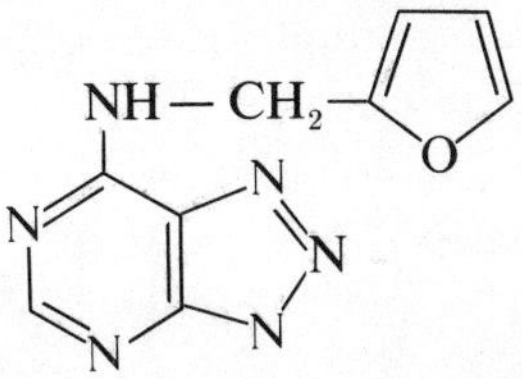

CXIX

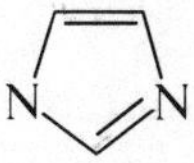

CXX

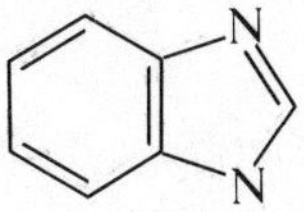

CXXI

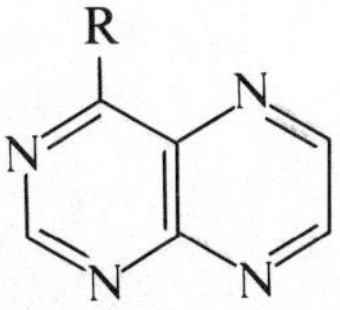

CXXII

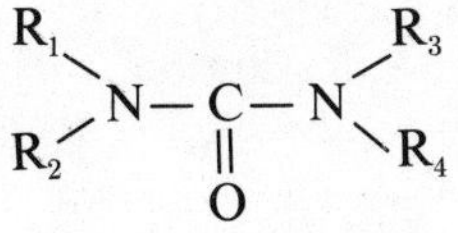

CXXIII

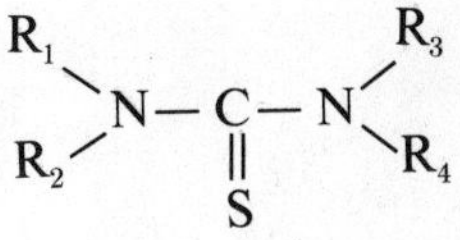

CXXIV

Fig. 35. Synthetic cell division factors.

CXVIII	6-substituted thiopurines.
CXIX	Aza-kinetin
CXX	Imidazole
CXXI	Benzimidazole
CXXII	4-substituted pteridines
CXXIII	*N*-substituted ureas
CXXIV	*N*-substituted thioureas.

with $R_1 = R_2 = R_3 = H$ and R_4 = one of a dozen or so substituted phenyl groups, the ureas were from about twice to hundreds of times as active as the corresponding thioureas.

At least one hydrogen on one of the nitrogen atoms is essential for activity, e.g. *N,N'*-diphenyl-*N,N'*-dimethylurea ($R_1 = R_3$ = phenyl; $R_2 = R_4 =$

methyl is completely inactive. Di-substitution on only one nitrogen leads to greatly reduced activity, e.g. *N,N′*-diphenylurea ($R_1 = R_3$ = phenyl; $R_2 = R_4$ = H) is highly active whereas *N,N′*-diphenyl-*N*-*N*-methyl urea ($R_1 = R_3$ = phenyl; R_2 = methyl; R_4 = H) is completely inactive and *N*-4-chloro-phenyl-*N,N′*-dimethylurea (R_1 = 4-chlrophenyl; R_2 = H; $R_3 = R_4$ = methyl) has a low but definite activity.

In mono-substituted ureas ($R_1 = R_2 = R_3$ = H; R_4 = a phenyl ring) activity increases with ring substitution, the highest activity being associated with *meta* and the lowest with *ortho* substitution. Pyridyl substituents also confer activity.

In di-substituted ureas ($R_1 = R_3$ = H; $R_2 = R_4$ = phenyl or substituted phenyl group), the highest activity is associated with those compounds having one substituted phenyl ring.

Whether these substituted ureas induce cell division by the same biochemical mechanisms as the N^6-substituted aminopurines is still a moot point. It has been suggested that they might act as side chain donors, forming active cytokinins by reactions with adenine in the cell, but direct checks to test this were not conclusive, and it was concluded that the adenine and phenylurea derivatives belong to two distinct classes of growth factor, functioning at different sites in the cell (Skoog & Armstrong, 1970). These contentions have been further explored by the synthesis and biological testing of a rang of 6-ureidopurines, with a side chain of general formula —NH——CO—NH—R. With R = phenyl, i.e. the substitution of a purine for a phenyl in *N,N′*-diphenylurea, a cytokinin was obtained with an activity in the tobacco callus test about a fifth that of kinetin but twenty times that of *N,N′*-diphenylurea. Furthermore the effect of monosubstitution of Cl in the 2, 3 or 4 position of the phenyl ring caused a pattern of activity change different from that recorded for similar substitutions on one phenyl ring of *N,N′*-diphenylurea by Bruce and Zwar (1966). However its activity in the promotion of bud formation on tobacco callus cultures was very much greater than that of kinetin (McDonald *et al.*, 1971). The differences in the chlorine substitution effects suggest that these new compounds are acting in a manner different from the diphenylureas whereas their higher bud-inducing activities suggest that they are not acting in the same way as kinetin. A wider series of biological tests on these new compounds is required.

Cell-division factors isolated from the crown gall tissue of *Vinca rosea* have been partially identified as derivatives of nicotinamide, containing sulphur, glucose and a straight-chain fatty-acid (Wood *et al.*, 1969). They induce cytokinin responses in the tobacco callus assay but not in all tests and it was suggested by the discoverers that the N^6-substituted adenines might owe their activity to their effects on the biosynthesis of these new nicotinamide derivatives. So far there is no direct evidence to support this suggestion.

Chapter VI

Natural Plant Growth Inhibitors

INTRODUCTION

So far, in our considerations of the normal role of plant growth substances, we have been concerned mainly with direct growth promotion. However, just as a motor car needs a brake as well as an accelerator pedal if we are to exercise proper control over its progress, so too one would expect the progress of growth to require both accelerators (auxins and gibberellins) and brakes (growth inhibitors) to afford the precision of control that obviously obtains in normal cells and tissues. There are indeed some growth phenomena in plants where, at relatively high concentrations, auxins seem to operate the braking system themselves, for example the growth dominance of apical buds, root growth inhibition, and possibly the inhibition of flowering; but it is not yet established by unequivocal evidence that direct auxin inhibitions are an integral part of normal growth control in the plant. Whatever the future may decide on this point, there is no doubt that growth-inhibiting substances are of very wide occurrence in plants, and that they play their part in regulating the initiation and growth of plant organs. The idea of a dual balanced control is by no means new, since it had already been formulated at the turn of the century by Czapek (1903), the famous Czech botanist, who gave the name of 'negative catalysts' or 'anti-enzymes' to these hypothetical inhibiting substances; but it is only in the last few years that such substances have been isolated, characterized chemically and implicated with the auxins and other growth hormones as full partners in the normal control of growth processes. In this Chapter we will survey current knowledge of the nature of these substances leaving it to later chapters to consider their modes of action and their roles in growth and development. But first we must survey briefly those phenomena in which inhibitors of one kind or another seem to be directly involved.

Growth Phenomena Involving Inhibitors

(A) GERMINATION

The most clear-cut case of growth suppression by natural inhibitors is in the maintenance of a dormant state in the propagules (seeds, spores etc) of the majority of plants, pending the advent of conditions favourable for the growth and development of the seedling (or sporeling).

Thus a phenomenon of almost universal occurrence in the plant kingdom is the inability of such reproductive structures to germinate and grow into new plants while still associated with the organs in which they are borne on the mother plant. With the exception of a few species of higher plants that are confined to the salt swamps (mangroves) of tropical estuaries and sea-coasts, and which produce 'viviparous' seedlings, the seeds of higher plants will not generally germinate while still within the fruit or pod, even though they may be fully mature and capable of immediate germination when shed. In many plants, this state of affairs is induced by a germination inhibitor which is present in some part of the fruit. In the pear, apple, and fig, for instance, it is present in the fleshy part of the fruit. Indeed, inhibitors in fruit flesh are of very wide occurrence, one survey revealing that they could be demonstrated in 33 genera of 16 families from 16 widely spaced orders of plant (Moewus *et al.*, 1951). In the majority of instances, they can be found also in the juice of the fruit, as in the tomato, the orange, the berry of mountain ash, etc. On the other hand, in very many seeds an inhibitor is resident in the seed coat itself, as in cabbage, lettuce, clover, coffee, etc. In many plants, where the 'seed' of commerce is really a fruit, e.g. wheat, buckwheat, sunflower and beet, the inhibitor is in the fruit 'coat.' As we shall see, all these inhibitors have little direct chemical relationship to the auxins; although diverse in nature, they have been christened 'blastocholines' by a pioneer in their study (Köckemann, 1934) by combining the two Greek words *blastanêin* (germination) and *chōlyēin* (prevent).

In the lower plants, the same phenomenon is widely observed. Spores of ferns and mosses will not germinate when still within their containers (sporangia, capsules, etc.), owing in all probability to the presence of inhibitors (Froeschel, 1954). In the green, ribbon-like liverwort *Marchantia*, which grows on the surface of damp soil in shady places, reproduction may take place by the development of minute, detachable green 'buds' (gemmae) that are formed inside special cup-like containers on the upper surface of the ribbon. These, too, will not develop into new liverwort plants until they have been detached and separated from the parent plant. In this, the controlling agent is a specific inhibitor produced by the parent tissue. In the microscopic green water plant *Chlamydomonas* (see p. 297), the resting body produced after sexual fusion also contains a germination inhibitor. In one variety this can be washed away by flowing water, but in another it cannot escape through the thick coat until the spore has been subjected to a certain minimum temperature. Subsequently the washing away of the inhibitor allows the spore to germinate when more favourable temperature conditions return (Moewus, 1950*b*).

In seed plants the main function of these inhibitors, when present in fruit tissue, would seem to be the prevention of premature germination before adequate dispersal is attained. A correlated effect in some plants is the spreading out of germination over a long period, amounting to years in some cases. Thus, in charlock (*Sinapis arvensis*), some seeds may be retained in the pod when it falls. Those that are freed can germinate immediately, while the ones retained do not; these latter seeds will germinate only when the pod has rotted or the inhibitor has been washed out by prolonged exposure to rain. In this way germination may be spread over at least two years (Evenari, 1949). This phenomenon may be of widespread occurrence. Mechanically cracking, scratching or otherwise rupturing the seed coat – a necessary prelude to germination in many seeds – may act purely by allowing the inhibitor to escape, and this may be just as important for the germination process as facilitating the entry of water.

The most dramatic example of this control by rain-leaching is seen in the 'rain gauges' found in the seeds and fruits of short-lived (ephemeral) desert plants. Many such species appear to contain, in their dispersal units, germination inhibitors nicely adjusted to quantities which can be washed out by just that amount of rain necessary to allow subsequent growth to total maturation (see Fig. 36). Rain, insufficient for the production of a new crop of seed, might, in the absence of such rain gauges, induce all dormant seeds to germinate and denude the whole area of that species, probably exterminating it. Even different varieties of the same species may show differences in inhibitor content related to the rainfall patterns of their particular local habitats (for further details see Koller, 1959; Mayer & Poljakoff-Mayber, 1963).

Inhibitors present in the seeds themselves may play an all-important rôle in those species that require special external conditions for their germination. This is particularly apparent in seeds that are sensitive to light. Lettuce is an example of the type of plant which, under certain conditions, will not germinate unless the seeds are exposed for a short time to light at the red end of the visible spectrum (650 – 680 nm).

A likely explanation of this phenomenon is that light triggers the destruction of a germination inhibitor, present in lettuce seed, which was maintaining the seed in a dormant state in spite of otherwise favourable conditions. This theory is supported by a number of experimental facts. For example,

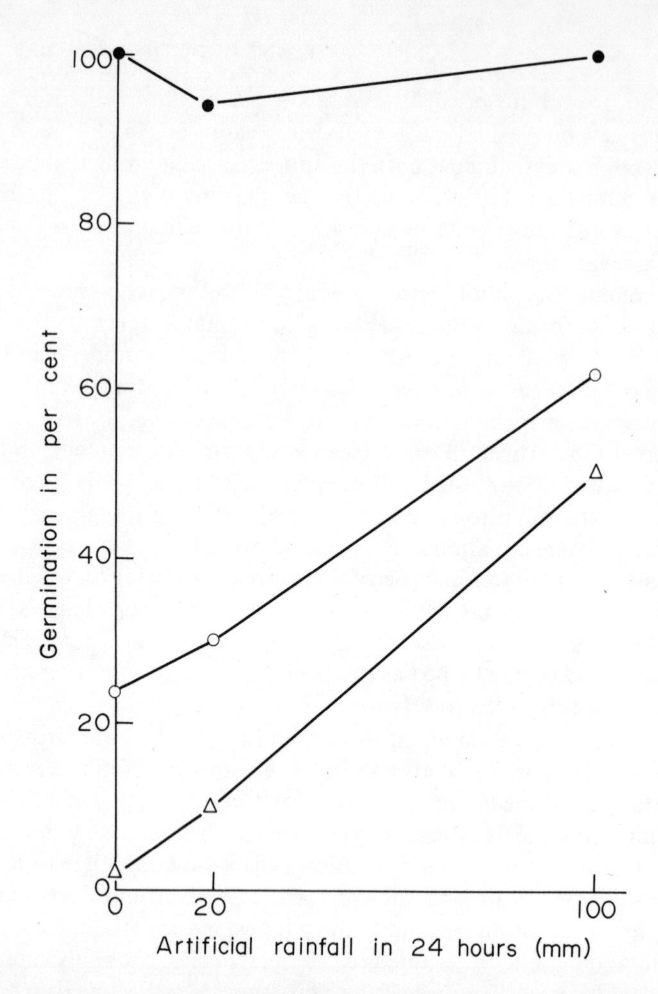

Fig. 36. Rain gauge plants. Graphs showing the results of experiments in which the seeds of desert plants were subjected to artificial rainfall of two different intensities. The seeds were then kept in moist soil and their germination recorded. *Anastatica hierochuntina* seeds (—●—) germinated completely even in moist soil and did not need additional rainfall. With the other two, *Erucaria boveana* (—○—) and *Carrichtera annua* (—△—) there was very little germination in moist soil alone but a progressive increase with increasing amounts of artificial rain, which leached out a chemical germination inhibitor (from the data of Koller, 1959).

the seed does contain a blastocholine; furthermore, non-dormant seeds in the dark can be made dormant by treatment with a number of known blastocholines including the widely occurring coumarin; again induced dormancy can be very largely counteracted by exposure to light (Nutile, 1945; Weintraub, 1948).

The oxidative destruction of an inhibitor seems to be responsible for the breaking of dormancy in *Xanthium* seeds by exposure to pure oxygen (Wareing & Foda, 1956, 1957) (see Fig. 37 A). The promotion of germination by this means is accompanied by a significant reduction in the content of a growth-inhibiting substance in aqueous extracts of the treated seeds.

In other types of seed, germination is inhibited by light, and here we would need to postulate either the production of an inhibitor in the light, or the photosensitization of one already present, a situation suggested by several experiments in which blastocholines extracted from beet seed, from the seeds of the dark-germinating *Phacelia tanacetifolia*, and from red kidney bean (*Phaseolus vulgaris*) seeds (Siegel, 1950), are much more effective in the light. (See Evenari, 1949, pp. 178–9.)

We must not lose sight of the fact, however, that these natural inhibitors, in spite of their obvious importance, are not the only factors controlling seed dormancy. The necessity for caution in our approach is shown by apple seeds, which require an exposure to a 60-day 'after-ripening' ('stratification') period under cold moist conditions before they will germinate. One of the contributing causes of the subsequent breaking of dormancy seems to be the disappearance of a natural inhibitor from the embryo itself, and the leaching of a similar inhibitor from the seed coat (Luckwill, 1952). But embryos excised from seeds not exposed to 'after-ripening' conditions will not germinate, even though they contain no inhibitor. This germination could be governed by the production of a growth-stimulating hormone, triggered off by the cold temperature treatment. In the seed of the lentil (*Lens culinaris*), temperatures of 2°C. bring about a fall in free auxins but a considerable increase in that extractable by mild alkaline hydrolysis. Subsequent germination at favourable temperatures results in a rapid accumulation of free auxin to a level well above that in normal seeds (Pilet, 1954). From the biochemical standpoint, seed germination is a very complicated affair which we are just beginning to understand.

In recent years it has become clear that the gibberellins may be much more intimately concerned with the regulation of seed germination than the auxins. In many cases their application to light-requiring seeds will trigger germination in the dark, e.g. in lettuce, suggesting that native gibberellins may play a vital role in the mechanism of light control. They may promote the germination of dormant light-insensitive seeds and fruits; good examples are barley and other cereal grain. Gibberellic acid treatment also replaces the low-temperature requirements of certain seeds which require 'stratification' and it is possible that natural gibberellins are induced to form in such seeds, whose dormancy they break. This gibberellin effect has been demonstrated in apple, beech, hazel and peach seeds. Fig. 37 B shows the effects of chilling of ash seeds on their content of growth substances (probably gibberellins), which stimulate the growth of excised unchilled embryos of the

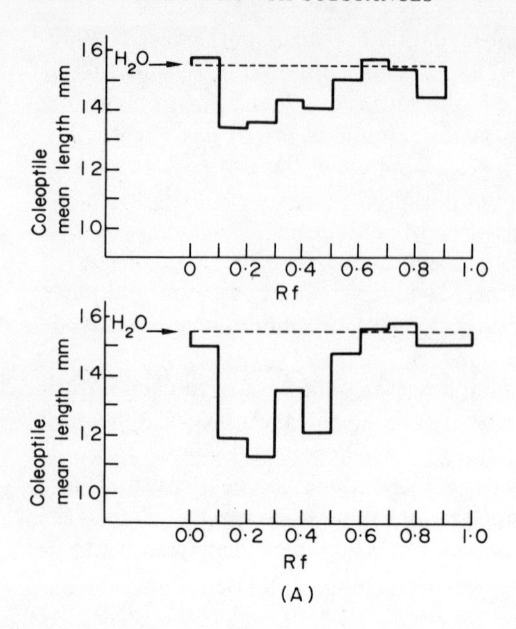

(A) Chromatograms of aqueous extracts of the upper seeds of *Xanthium* maintained in pure oxygen (upper chromatogram) or air (lower chromatogram). The seeds were extracted 30 hours after planting. It will be seen that the seeds which have been stimulated to germinate by high oxygen tensions (upper curve) have a much lower content of a growth inhibitor (wheat coleoptile segment assay).

(A) from Wareing & Foda, 1957.

(B) chromatograms of extracts of unchilled (upper chromatogram) and chilled (lower chromatogram) embryos of *Fraxinus excelsior.* The presence of active substances on the chromatograms were detected by their effects on the germination of unchilled embryos of the same species. It will be seen that chilling has caused the appearance of a germination stimulant which is not present in unchilled embryos.

(B) from Wareing & Villiers, 1961.

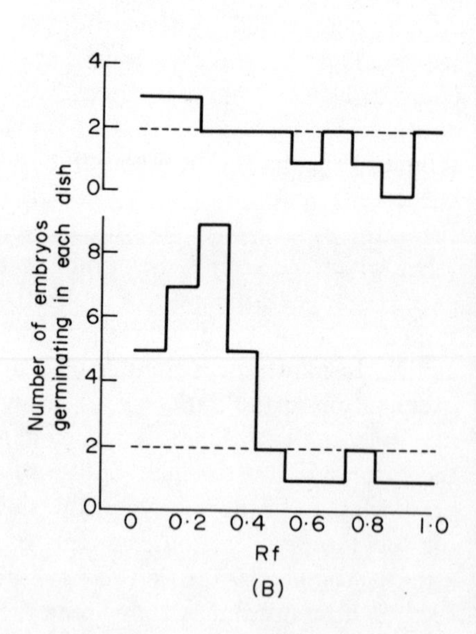

Fig. 37. The effect of conditions controlling seed germination on the growth substance content of treated seeds.

same species. Chilling has apparently promoted the production of these growth promoters. A considerable literature on these gibberellin effects has accumulated, and detailed references should be sought in the review by Vegis (1964).

Propagules other than fruits and seed can show dormancy phenomena. Thus the young potato is at first dormant and cannot be induced to sprout by environmental conditions optimal for growth. This is correlated with a high content of a growth inhibitor, which disappears when the tuber matures and acquires a capacity to grow.

The rôle of inhibitors, present in seeds, in the struggle for space in natural plant communities has already been hinted at. The precise importance of this factor is very difficult to assess, and very little systematic work has been done on that aspect of ecology. A number of clear demonstrations of this chemical antagonism between seeds are known. Corn cockle (*Agrostemma githago*) is seldom found in beet (*Beta*) fields and experiments have demonstrated the inhibition of germination of its seeds in mixed plantings with beet. Seeds of beet will also inhibit the growth of rye grass and campion (*Melandrium* sp.) when placed in the same pot of soil. Similarly, wheat and rye-grass will suppress the germination of certain weed seeds, e.g. corn chamomile (*Anthemis arvensis*) and scentless mayweed (*Matricaria maritima* subspecies *inodora*). Red kidney bean seeds will inhibit the germination and growth of flax and wheat, and seeds of *Viola* will also inhibit the germination of wheat. The growth of seedling and maturing plants may also be affected by seed inhibitors applied to their roots (Froeschel, 1956). Wheat, barley, pea and corn cockle, when grown in solutions of such inhibitors, remained stunted in leaf, root and stem, and fewer organs were produced over a prolonged growth period. An interaction in which stimulators rather than inhibitors of germination are involved is found in certain highly specialized plants which parasitise the roots of other (host) plants. These are the hemiparasites *Striga hermontica* and *Orobanche minor*, which grow on the roots of tropical grasses (e.g. maize) and of legumes respectively. The seeds of these plants may lie dormant for long periods, but germination is initiated when the soil is invaded by the roots of a suitable host plant. It has been clearly demonstrated that this phenomenon is due to the exudation from the host roots of a highly active substance which promotes the germination of these seeds. A haustorium (food-absorbing organ) then emerges and penetrates the tissue of the host root. Although the *Striga* substance has been collected and subjected to a high degree of purification, its chemical constitution has not yet been completely established.

These few examples stress the complexity of possible seed interactions in a natural plant community. This subject will be dealt with in more detail in a later chapter.

(B) DORMANCY OF BUDS

Just as a whole plant goes through a succession of discrete and well marked stages in its development from seed to full maturity, so we can distinguish a similar set of stages in the formation and development of a bud. Buds are indeed small embryonic shoots and are composed of stem, leaves and, in flowering buds, a complete flower or collection of flowers (inflorescence) in embryo. Such a bud starts its life either from an apical meristem (apical bud) or from a small lateral meristem in the angle that a leaf makes with the main stem (lateral or axillary bud). (See Fig. 2.) Initiation takes place at a very early stage, e.g. when the branch bearing them may itself be a bud. Indeed, it is often possible by dissection to make out the meristem in the axil of an embryonic leaf in an unopened bud. The development of the second generation bud from this meristem commences as soon as the parent bud bursts and starts its growth into a new branch. It finally reaches its mature size and development when the new branch is fully grown.

When buds have reached the mature stage, further development usually stops and they become dormant. Many factors seem to contribute to this state of dormancy. They may be of internal or external origin. Thus dormancy may be maintained in a bud by the physiological activities of a neighbouring organ and this reaches its most dramatic expression in the phenomenon of *apical dominance*, where growing apical buds induce a state of almost complete inactivity in neighbouring lateral buds. This phenomenon has been supposed to be mediated by an auxin mechanism. The factors external to the plant which induce bud dormancy are many and varied, and are usually associated with the onset of conditions unfavourable to bud growth. Thus temperature, light quality, length of day, the supply of water and mineral nutrients and combinations of these factors can all be determinants of dormancy in buds; changes in one or more of these conditions may result in dormancy 'break' and subsequent bud growth (for full discussion see Wareing (1956, 1969) and Vegis (1964)).

Dormancy break and the internal processes leading up to it, are usually associated with parallel changes in the levels, not only of growth promoting hormones but particularly of chemical inhibitors in the tissues concerned. At one time it was thought that a high level of native auxin was the main factor controlling bud dormancy, because high concentrations applied to growing buds inhibited their growth. Certainly the observed decline in total auxin content of dormant pear and apple buds as the end of the rest period approaches supported these contentions (see Samish, 1954), but the content of 'free' diffusible auxin, which is usually taken to be the component active in growth control, shows the reverse trend. For example, in pear buds there is a steady increase in diffusible auxin towards the end of dormancy (Bennet & Skoog, 1938) and Nitsch & Nitsch (1959) have shown that the native auxin content of stem tips of *Rhus typhina* decreased when plants were moved from summer day-lengths (long days, short nights) to winter day-lengths (short days, long nights), conditions which normally induce bud dormancy. Such facts throw doubt on the control of dormancy solely by auxins and this doubt is strengthened by the many observations that the dormant condition of buds is very closely correlated with a greatly increasing content of growth-inhibiting compounds. This is true of *Rhus typhina* (Nitsch & Nitsch, 1959) and of *Acer pseudoplatanus* (Phillips & Wareing, 1959) under short day conditions. In *Betula pubescens* the inhibitor is manufactured by the leaves under short day conditions (Eagles & Wareing, 1963). In blackcurrant dormancy-inducing short-day conditions not only bring about the decline in gibberellin content in the buds but also causes an increase in the levels of a growth inhibitor (see Fig. 39 C). Again natural bud dormancy in the potato and in the ash (*Fraxinus excelsior* var. *pendula*) is closely correlated with a high content of special inhibitory substance (tested on the growth of oat coleoptile), while the auxin content is low, and is not related in any obvious way to the state of dormancy (Hemberg, 1949*a* & *b*). In the grapevine, two neutral inhibitors accumulate as dormancy advances and then slowly disappear when the bud is about to break, while the auxin content remains stationary throughout the rest-period (Spiegel, 1954). An inhibitor, 5,7,4-trihydroxyflavanone (narengenin, Fig. 41 CXXXVI), similarly accumulates and disappears in peach flower buds (Henderschott & Bailey, 1955, Henderschott & Walker, 1959*a* & *b*) (see Fig.

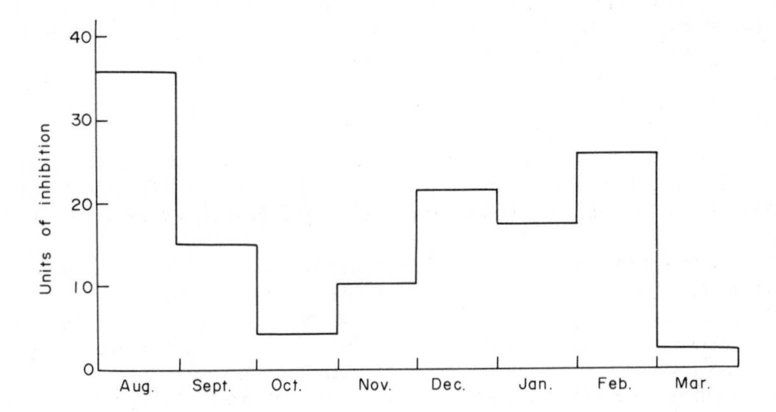

Fig. 38. Graph showing the seasonal variation in the concentration of a growth-inhibiting substance in the dormant flower buds of peach. Inhibitor assayed by the wheat coleoptile segment test. (Data from Henderschott & Walker, 1959*b*.) Note the dramatic drop in concentration of the inhibitor when dormancy is beginning to be broken.

38) while the auxin content shows no closely correlated trends except to rise rapidly as dormancy breaks (Blommaert, 1955).

In more recent years the gibberellins have firmly established themselves in the picture. Thus gibberellin application to tree buds can break dormancy in *Weigela*, *Citrus*, peach, blackcurrent and many others (For a full survey see Vegis, 1961). In other species the effects of gibberellin are obtained only when application is accompanied by partial chilling, e.g. in pear (Brown *et al.*, 1960). Again in *Rhus typhina* the long-day requirement for bud-dormancy break can be replaced by gibberellic acid treatment (Nitsch, 1957). However all species are not responsive and there are instances, e.g. in the swamp cypress (*Taxodium distichum*) where gibberellin treatment may even hasten the onset of dormancy (Brian *et al.*, 1959). As in the case of seeds, chilling can bring about dormancy break in buds and this is usually accompanied by a loss of an endogenous inhibitor and the accumulation of endogenous gibberellins. This is beautifully shown in Fig. 39 A & B for dormant buds of blackcurrant exposed to natural seasonal and artificial chilling. There can be little doubt that bud dormancy and its control is mediated by the interaction of growth inhibiting and growth-promoting factors, which may include both auxins and gibberellins. Closely associated with bud break is the initiation of cambial activity, which would seem to be induced by hormones emanating from the growing buds.

Fig. 39. Graphs showing the effects of conditions inducing dormancy break in buds of blackcurrant on the growth-substance levels in extracts of those buds.

(A) Shoots collected from our-of-doors in January and then stored at 2°C. Bud tissue sampled in January, February and March for extraction and essay. This series shows the effects of artificial chilling conditions.

(B) Buds sampled from bushes growing out-of-doors on January, February and March for extraction and assay. This series shows the effects of natural chilling conditions.

(C) Shoots subjected to different daylengths before bud sampling for extraction and assay. Left (LD) shoots maintained in continuous long days (18 hours); Centre (5 SD) shoots exposed to 5 short days; Right (10 SD) shoots exposed to 10 short days.

Assays of chromatogram segments were carried out with the dwarf maize leaf segment test for gibberellins. The black areas of the chromatograms represent regions where responses were above or below the sampling error (fiducial limits) of the assay and correspond therefore to the presence of gibberellins or growth inhibitors respectively. It will be seen that both artificial chilling (A) and natural chilling during the winter months (B), both of which break dormancy, cause the appearance of a gibberellin and the disappearance of an inhibitor from bud extracts. Similarly short days (C) which induce dormancy in the buds, cause a disappearance of a gibberellin, which has the same Rf as gibberellic acid (GA) and the appearance of apparently the same inhibitor in bud extracts (data from Wareing, 1969).

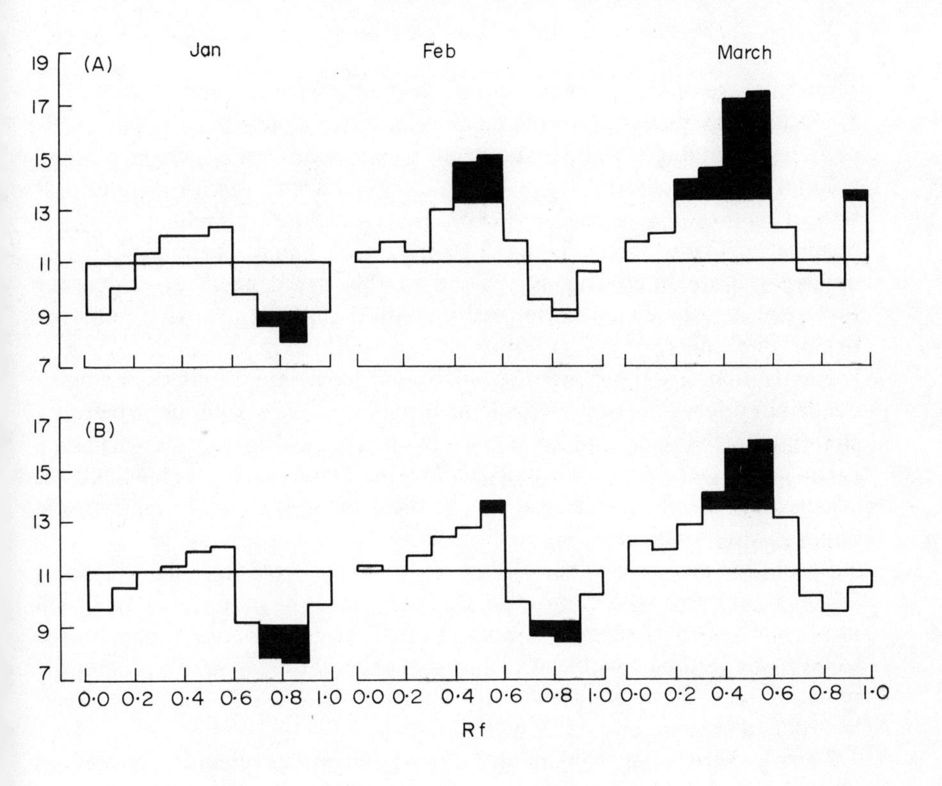
(A)
Jan
Feb
March
19
17
15
13
11
9
7
(B)
17
15
13
11
9
7
0·0 0·2 0·4 0·6 0·8 1·0
Rf

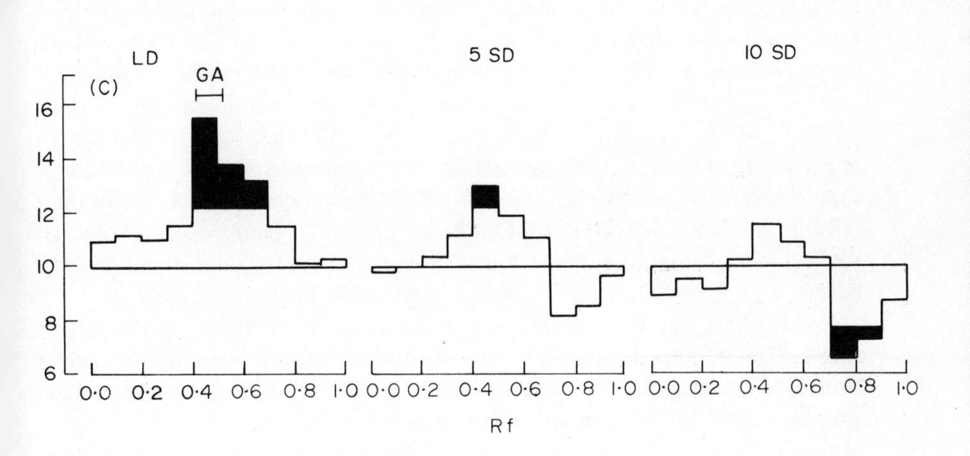
(C)
LD
GA
5 SD
10 SD
16
14
12
10
8
6
0·0 0·2 0·4 0·6 0·8 1·0
Rf

ISOLATION AND ASSAY

As in the case of the native growth-promoting hormones, growth inhibitors are usually extracted from plant material by suitable solvents. The crude extract so obtained must be subjected to purification and here paper and column chromatography have been much favoured. Again the identification of regions of growth-inhibitory activity, a necessary prelude to further purification and ultimate chemical identification, can be accomplished only by appropriate biological assay, and in this the dangers of misleading responses are somewhat more predominant than with the auxins, gibberellins or cytokinins.

Firstly there is the possibility of non-specific inhibitions of the assay material by very high non-physiological concentrations of common cell constituents. Such inhibitions can be purely osmotic, as has been shown in seed germination (see Mayer & Poljakoff-Mayber, 1963) and these mechanisms are usually made apparent after chemical identification, or appropriate dilution experiments.

Secondly there is the necessity to chose actively growing plant material for assay in order to obtain optimal responses to the inhibitor. When such material consists of seeds or seedlings only, the growth environment needs to be made optimal; with the more usual excised organs such as coleoptile or first internode segments of oat, it may be necessary to apply hormones in order to achieve optimal growth.

Thirdly there is the problem of the great diversity of chemical compound that can exert growth inhibitions, bringing with it a parallel range of plant specificities and the risk either of failure to 'spot' significant compounds or of exaggerating the importance of others if an inappropriate plant organ or tissue is used in the assay. It is always good practice, whenever possible, to chose, for assay, the plant from which the suspected inhibitor has been extracted (in the case of endogenous inhibitors). For example excised embryos of birch (*Betula*) have been used to assay inhibitors extracted from birch seed (Wareing, 1959) although there is always the inherent danger that the accuracy and sensitivity of the assay may be offset by residual inhibitors in the material itself. On the other hand if inappropriate material is chosen, for example *Avena* coleoptile segments for seed inhibitors, one may record activities which are irrelevent to the phenomena studied.

Lastly there are the ambiguities that can be introduced by the auxins themselves which can, under certain conditions and in certain concentrations, inhibit growth in some assaying organs. The use of the seedling root growth test or the root segment test involves such risks.

In conclusion it should be stressed that it is not possible to devise one or more general and standard assays for plant growth inhibitors of the kind set up for auxins, gibberellins and cytokinins; each problem may demand the devizing of a specific bioassay.

Extension-Growth Inhibitors

There is now a large and somewhat confusing literature reporting the occurrence of inhibitors of plant growth (as detected in auxin assays) in a wide variety of plant organs including leaves, stems, roots, underground storage organs, fruits and seeds. For details the reader should consult reviews by Hemberg (1961) and Shantz (1966). The vast majority remain as unidentified 'spots' on chromatograms and their significance and role in plant growth control are therefore equally obscure.

One such 'spot', christened inhibitor-β by its discoverers (Bennet-Clark *et al.*, 1952) occurs regularly as a weakly acid component in many plant extracts made in the study of auxins. It inhibits the growth of coleoptiles and roots alike but there is no evidence that it does so by interfering directly with hormone action. One chemical analysis which was carried out (Housley & Taylor, 1958) suggested that it was a mixture of at least three inhibitors, azelaic acid (Fig. 40 CXXV) an unsaturated lactone scopoletin (Fig. 40 CXXVI) and an unsaturated polyhydroxy fatty acid. Material from another source (Köves, 1957) seemed to be a mixture of salicylic acid (Fig. 40 CXXVII) and a derivative of cinnamic acid. However the most recent studies (Robinson & Wareing, 1964; Cornforth *et al.*, 1965; Milborrow, 1967) have shown that the most active component of the inhibitor-β complex from *Acer pseudoplatanus* leaves is abscisic acid, the inhibitor which is mainly responsible for bud dormancy (see p. 173); in the potato complex, abscisic acid is accompanied by a second very strong inhibitor which has not yet been characterized (Holst, 1971).

The group of organic compounds known as the unsaturated lactones are inhibitors of extension growth, particularly in roots. One of the best-known of these naturally-occurring lactones is coumarin (Fig. 40 CXXIX), a potent germination inhibitor (see next section). Scopoletin (Fig. 40 CXXVI), a derivative of coumarin, occurs widely in the plant kingdom and is found in the free state as well as in the form of its glucoside. It is regarded by some plant physiologists as an important component in the growth-regulating system, since it shows significant interactions with auxins and cytokinins in the control of some growth processes. A general antagonism of auxin action in several growth responses has been demonstrated by Libbert & Lübke, (1960) while Skoog & Montaldi (1961) have shown that high concentrations of applied IAA can bring about the release of large amounts of scopoletin from tobacco tissue cultured *in vitro* (see Chapter V) presumably from the

$COOH-(CH_2)_7-COOH$

CXXV

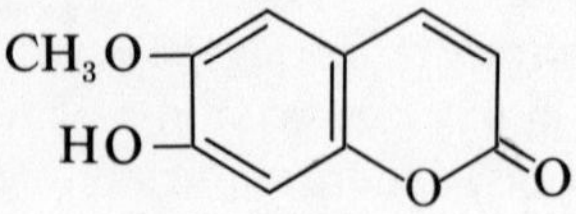

CH_3O, HO, O, O

CXXVI

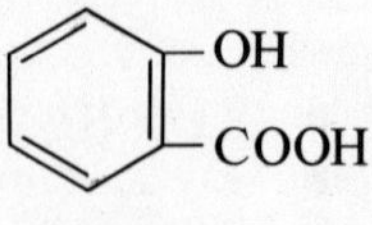

OH, COOH

CXXVII

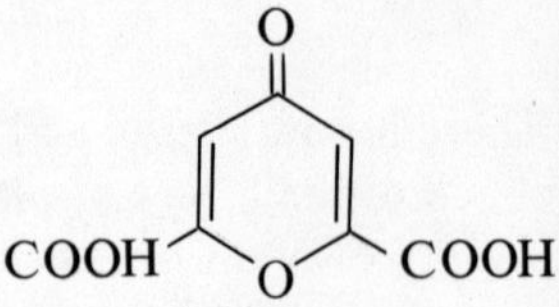

O, COOH, O, COOH

CXXVIII

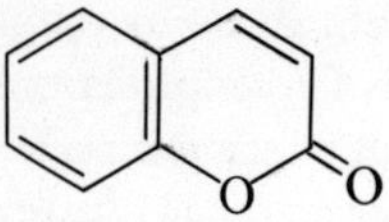

O, O

CXXIX

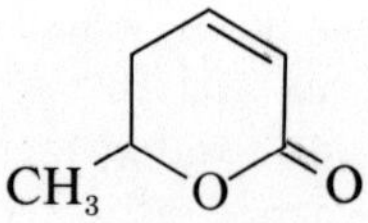

CH_3, O, O

CXXX

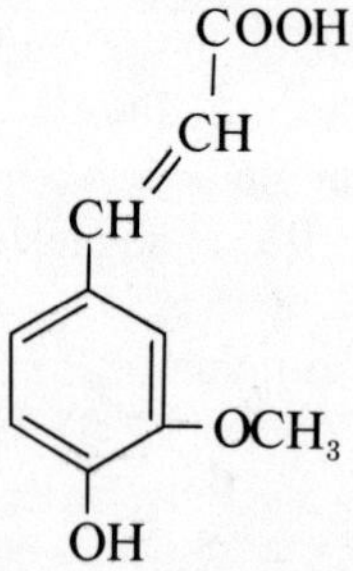

COOH, CH, CH, OCH_3, OH

CXXXI

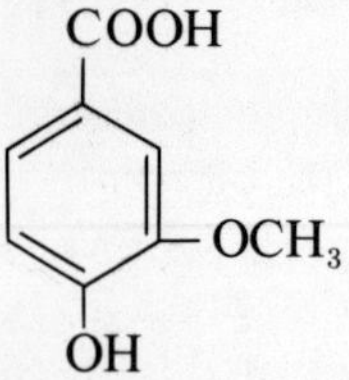

COOH, OCH_3, OH

CXXXII

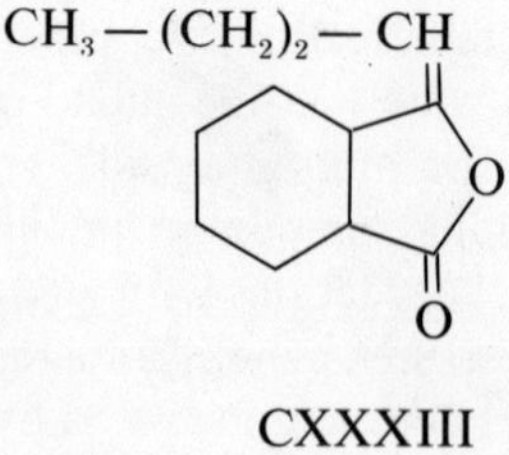

$CH_3-(CH_2)_2-CH$, O, O

CXXXIII

O, N, O

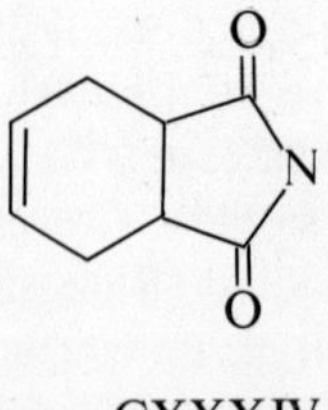

CXXXIV

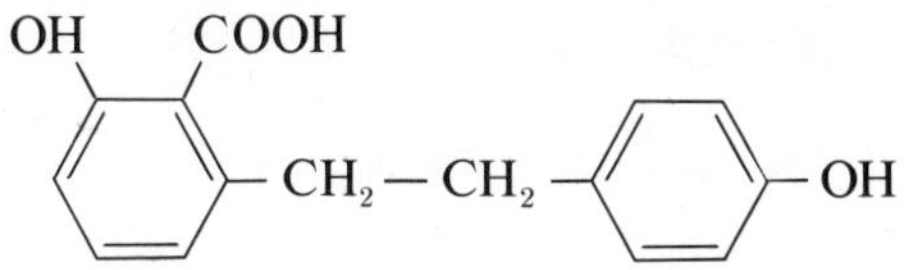

OH COOH

CH_2-CH_2 OH

CXXXV

Fig. 40. Inhibitors of extension growth and germination.

CXXV	Azelaic acid
CXXVI	Scopoletin
CXXVII	Salycilic acid
CXXVIII	Chelidonic acid
CXXIX	Coumarin
CXXX	Parasorbic acid
CXXXI	Ferulic acid
CXXXII	Vanillic acid
CXXXIII	*n*-Butylidene-hexahydro-phthalide
CXXXIV	*Cis*-4-cyclohexine-1, 2-dicarboximide
CXXXV	Dihydrohydrangeic acid. Lunularic acid.

glycoside. Indeed it has been suggested that this accumulation of scopoletin caused by the application of the synthetic auxin 2, 4-D may contribute to the herbicide action of that substance (Fults & Johnson, 1950). Another coumarin derivative, seselin, occurs in *Citrus* roots and is a potent inhibitor of seedling root extension (Goren and Tomer, 1971). It antagonizes IAA action by apparently promoting its enzymatic oxidation (see Chapter XI) suggesting that it may be playing an important role in the regulation of root growth in *Citrus*.

Another aromatic acid which occurs widely in a number of families of flowering plant is chelidonic acid (Fig. 40 CXXVIII), sometimes in quite large quantities (Ramstad, 1953). It is an active growth inhibitor in the pea segment assay for auxins, where it counteracts auxin action (Leopold *et al.*, 1952). However its functioning as a growth regulant in those species in which it occurs naturally has so far not been demonstrated.

Many phenolic compounds occurring in plants have been shown to inhibit growth in auxin-type assays. However their actions may be indirect in that they promote the destruction of native auxins in the tissue; a consideration of these compounds will therefore be left to a later chapter dealing with the regulation of hormone levels (Chapter XI). Other substances which, among other actions, inhibit extension growth (e.g. naringenin, abscisic acid, etc.) will be considered in a later section of this chapter.

Germination Inhibitors

The natural germination inhibitors are exceedingly numerous and correspondingly varied in their chemical structures. The vast majority are simple organic compounds, although, in a number of plants, inorganic products liberated by enzyme action from plant constituents can be effective. Two important examples of the latter are ammonia, produced from organic nitrogenous materials, and prussic acid, coming from the cyanogenetic glucoside reserves in seeds of certain groups of plants – in particular, those to which the plum and apple belong.

Of the organic compounds, the unsaturated lactones are among the most important, since the original 'blastocholine', parasorbic acid (Fig. 40 CXXX), isolated from the fruit of mountain ash (*Sorbus aucuparia*) belongs to this group. Coumarin (Fig. 40 CXXIX), a widely distributed substance responsible for the scent of woodruff (*Asperula odorata*) and new mown hay, also belongs here and is highly effective as a germination inhibitor. It has been identified as the natural germination inhibitor in the dispersal units of *Zygophyllum dumosum* and *Trigonella arabica* (Lerner *et al.*, 1959). Note that scopoletin discussed in the previous section is a derivative of coumarin. Many other coumarin derivatives are widely distributed in the plant kingdom (Späth, 1937) and may participate in growth-regulating processes. We have already seen that a coumarin may be involved in the light-sensitivity of lettuce seeds. In an attempt to clarify the mechanism concerned, the relative activities of a large number of coumarin derivatives, as inducers of light sensitivity in lettuce seeds, have been studied by Berrie *et al.*, (1968). On a basis of these comparisons they concluded that coumarin acts as an anti-gibberellin, presumably by competition related to their respective lactone bridges, at an active centre associated with germination processes. This needs further study since the lactone bridge is not present in all gibberellin molecules (see Chapter IV). Coumarin is the internal lactone, and a close relative, of *ortho*-hydroxy-*cis*-cinnamic acid. We have already seen that *cis*-cinnamic acid has the property of an auxin and that *trans*-cinnamic acid can be a strong growth inhibitor. It has therefore been thought that a study of the biochemical behaviour of these closely related compounds might lead to the solution of the mechanisms of auxin and blastocholine actions. *Trans*-cinnamic acid itself may be the blastocholine of the resting spore of the green alga *Chlamydomonas* (Moewus & Banerjee, 1951). The fruit of sugar beet contains at least four compounds, two of which are derivatives of cinnamic acid, i.e. 4-hydroxycinnamic acid and 4-hydroxy-3-methoxy-cinnamic acid (ferulic acid) (Fig. 40 CXXXI). The other two closely related compounds are 4-hydroxy-benzoic acid and 4-hydroxy-3-methoxy-benzoic and (vanillic acid) (Fig. 40 CXXXII) (Roubaix & Lazar., 1957).

One or more of these compounds are contained in the flesh or fruits such as strawberry, lemon and apricot (Varga, 1957), tomato (Akkerman & Veldstra, 1947) and the chaff of oats (Köves, 1957) and it is likely that they are natural germination inhibitors just as important as the coumarins themselves.

A closely related group of compounds, apparently the germination inhibitors of many seeds of the family Umbelliferae, are the phthalides, the most active of such compound so far isolated being *n*-butylidene-hexahydrophthalide (Moewus & Schader, 1951). It has the formula CXXXIII (Fig. 40). The very potent germination inhibitor of sugar-beet fruit has been identified (Mitchell & Tolbert, 1968) as *cis*-4-cyclohexine-1,2-dicarboximide (Fig. 40 CXXXIV), which is structurally comparable to the synthetic inhibitors maleic hydrazide and the maleimides (see Chapter VII).

A group of compounds specific to one family of plants (Cruciferae) are the mustard oils. These are volatile substances present not only in seeds (e.g. the dormancy-inducing substances of *Sinapis arvensis*), but also in the roots of such species as radish, horse-radish, etc. Essential oils from a range of other plant families can also be effective germination inhibitors e.g. those from the skin of the orange and lemon, the flower of the clove, leaves of peppermint, thyme, rosemary and *Eucalyptus* and fruits of fennel. These compounds are very heterogeneous from a chemical standpoint. Some of their vapours are effective in very small quantities. For example, the emanations from certain types of resinous wood (e.g. spruce, poplar, etc.) can inhibit seed germination (Weintraub & Price, 1948). Wood is obviously a material to be used with care in germination tests in seed testing institutions. Certain organic acids are also very effective germination inhibitors, and are often the active agents in tissues of fleshy fruits. As an example may be quoted malic and citric acids in the apple, (Akkerman & Veldstra, 1947). The steeping liquors, which arise as a byproduct of the malting of barley grain, contain many germination inhibitors. They consist mainly of simple carboxylic acids such as acetic, isobutyric and phenylacetic acids (Cook & Pollock, 1952), and phenol acids such as vanillic acid (Cook & Pollock, 1954). The last important group of compounds in this connection are the alkaloids, many of which are of value pharmaceutically. Not all alkaloids have these properties, and only a few are strong inhibitors: the most important are cocaine, physostigmin, caffein and quinine.

An inhibitor of growth and development, although not strictly a germination inhibitor, is that which prevents the development of the propagules (gemmae) of certain thallose liverworts while they still remain in their gemma cups. Such an inhibitor has recently been isolated and identified from the thallus of *Lunularia cruciata*, where it seems to be the main mediator of the control of dormancy in gemmae by the operation of environmental factors such as day-length (see Chapter X). The inhibitor, like some of the germination inhibitors already described, is a phenolic acid, dihydrohydrangeic acid (Fig. 40 CXXXV) for which the trivial name lunularic acid has been coined (Valio & Schwabe, 1970). It has a wide distribution in liverworts, which do not appear to contain abscisic acid (W. W. Schwabe, personal communication).

One significant point which emerges from this survey of the chemical nature of inhibitors is the large variety of chemical structure which they exhibit. It is obvious, therefore, that the mechanisms of their several actions are correspondingly varied, and that the germinative capacity of the seed or the growth of the plant may be attacked through one or more of its complex of essential enzyme systems.

Dormancy Inhibitors

Apart from the dormant seed, there are two other organs in which characteristic dormancy behavior is shown; these are underground storage organs such as potato tubers and the dormant buds of aerial shoots. Both these latter have been subject to many studies in attempts to isolate and identify their native inhibitors.

One of the first to be implicated was the 'inhibitor-β complex' already described. This has been shown to be present in the outermost tissues of the potato tuber (Housley & Taylor, 1958), where it could maintain the dormancy of the resident buds, and in the dormant buds of sugar maple (*Acer saccharinum*) (Lane & Bailey, 1964). However these complexes are not of identical chemical composition in different plant species and their exact physiological role remains obscure.

Coumarin, which is a very active inhibitor of seed germination, has also come under consideration for other organs. Thus it has shown considerable promise as a dormancy promoter in potato tubers in storage. Dipping of the tubers before storage in a 0.1 percent aqueous solution resulted in a 48 percent inhibition of sprouting, while treatment with a solution of similar concentration in lanolin prevented sprouting completely (Moewus & Schader, 1951). On the other hand Vegis (1964) tried unsuccessfully to induce dormancy in the buds of certain water-plants by immersing them in coumarin solutions. In view of its very restricted distribution it seems a very doubtful candidate for the general role of dormancy inhibitor.

An inhibitor of quite a different type has been isolated from the dormant buds of peach (Henderschott & Walker, 1959*a*); it was the yellow flavanone pigment narengenin (Fig. 41 CXXXVI). It suggested action as a native dormancy-inducer in buds is supported by observations that the dormancy-breaking action of applied gibberellins in both lettuce seeds and peach buds can be suppressed by the concommittant application of naringenin (Phillips, 1962).

The most extensive and profitable work on bud dormancy-inhibitors was initiated by Eagles & Wareing (1963), who extracted an inhibitor from birch leaves and showed that when it was applied to seedling leaves it completely prevented the growth of apical buds. Furthermore the concentration of inhibitor increased in these leaves under winter day-length conditions (i.e. short days, long nights) during which buds are naturally dormant (Eagles & Wareing, 1964). These workers coined the name 'dormin' for this specific inhibitor. Subsequently a concerted attack was made on highly purified extracts of sycamore leaves, with silicic acid column and paper chromatography and this eventually resulted in the preparation of a few crystals of a pure compound which was identified as 3-methyl-5-(1-hydroxy-4-oxo-2,6,6,-trimethyl-2-cyclohexene-1-yl)*cis,trans*-2,4-pentadienoic acid (Fig. 41 CXXXVII) (Robinson & Wareing, 1964; Cornforth *et al.*, 1965). By a strange coincidence this substance had been independantly isolated from young cotton fruits only a very short time before (Ohkuma *et al.*, 1963) and shown to be a potent accelerant of the process of abscission and had therefore been christened abscisin. A third independant isolation of a growth inhibitor from the pods of yellow lupin (Rothwell & Wain, 1964) has subsequently been identified with dormin. The name now agreed for this substance is abscisic acid (Addicott *et al.*, 1968).

Abscisic acid, abbreviated ABA, like the auxins and the gibberellins, occurs in very small quantities in plant tissue and so the isolation and purification of this substance requires the use of large quantities of suitable solvents and extensive fractionation of extracts. Aqueous organic solvents are normally used and extracts are subjected to acid/base fractionation, followed by paper, column or thin-layer chromatography. Successive steps in the purification are most conveniently followed by a bioassay, using such responses as the acceleration of abscission, the inhibition of extension growth in coleoptiles, the inhibition of seed germination or of the development of embryos excised from germinating seed. However these can only be indicators since they are completely unspecific and, as we have seen, a wide range of unrelated chemicals can cause such responses, possibly by quite different physiological mechanisms.

It is however very fortunate that there exists a highly specific and sensitive physico-chemical method for estimating abscisic acid. This is based on the almost unique optical rotatory dispersion of ABA which has a very intense 'Cotton effect'. In solutions of optically-active substances showing the Cotton effect, the rotation of the plane of polarized light is sensitive to

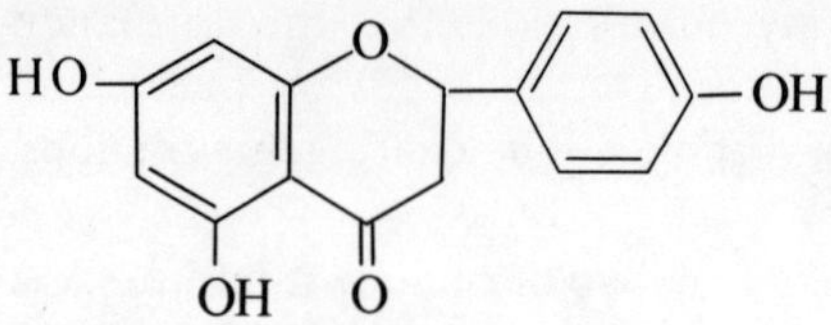

CXXXVI

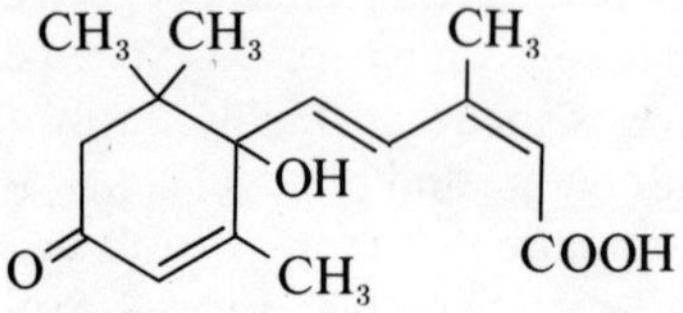

CXXXVII

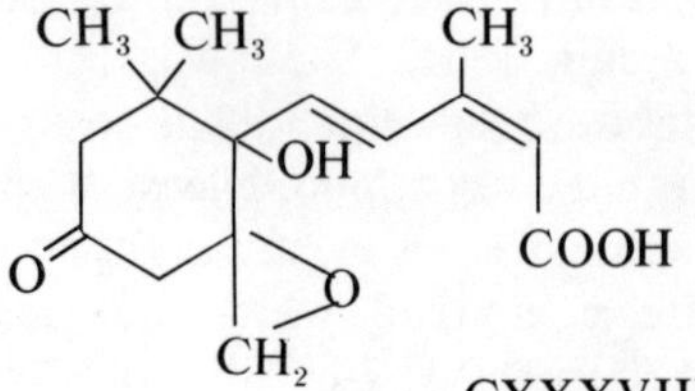

CXXXVIII

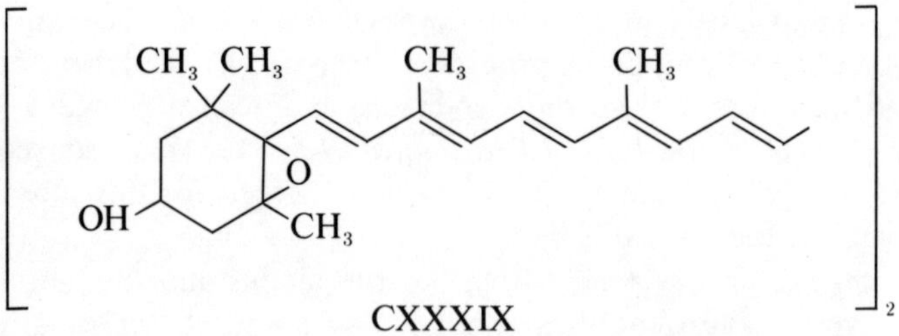

CXXXIX

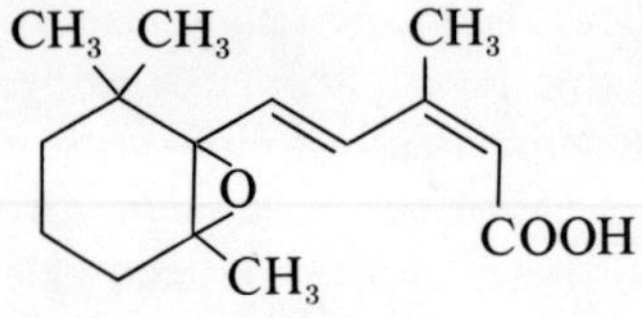

CXL

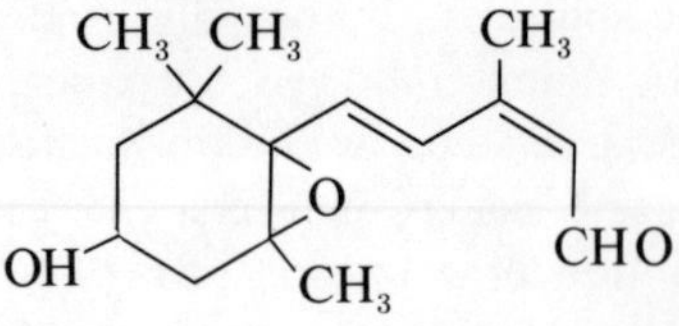

CXLI

Fig. 41. Dormancy inhibitors.

CXXXVI	Naringenin
CXXXVII	Abscisic acid
CXXXVIII	Phaseic acid
CXXXIX	Violaxanthin
CXL	5-(1, 2-epoxy-2, 6, 6-trimethyl-1-cyclohexyl)-3-methyl-*cis*, *trans*-2, 4-pentadienoic acid
CXLI	Xanthoxin

wavelength. As the point of maximum light absorption is approached the optical rotation rises to a maximum. At the absorption maximum the rotation drops to zero and then rises to a second maximum of reversed sign as the wavelength moves to the other side. The wavelength where reversal takes place is highly specific for ABA and can be used for its identification in very dilute solutions, provided there are no optically-active contaminants. The rotation 'peaks' are unusually high in ABA and thus quantitative estimates of concentrations can be made with a sensitivity as low as 0.3 μg/ml (Milborrow, 1968). An even more sensitive chemical method is that of gas/liquid chromatography. The ABA is converted to its trimethylsilyl derivative before it is run and the peak on the chromatogram can be located and estimated quantitatively by comparison with peaks of known amounts of authentic ABA (Davis *et al.*, 1968).

Wide-ranging work on pure ABA has established that it is a very active inducer of dormancy in non-dormant buds of a variety of woody plant species (El-Antably *et al.*, 1967) and it seems likely to be the natural dormancy-inducer in such plants and also probably in storage organs such as the potato tuber. It is a potent antagonist of the dormancy-breaking action of the gibberellins, as has been shown by direct experiment on bud growth (see Fig. 42).

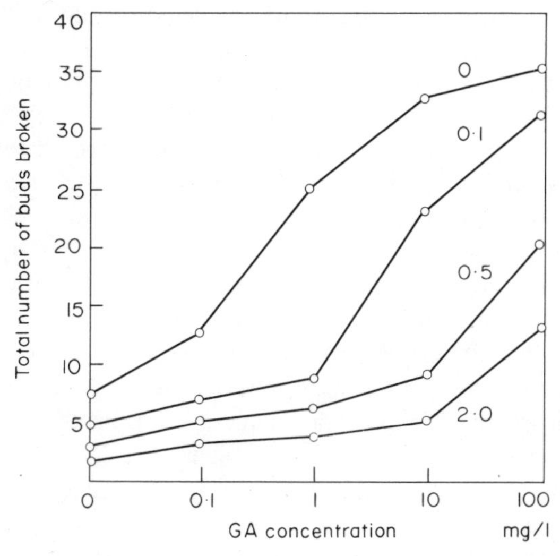

Fig. 42. Graphs showing the interactions of abscisic acid and gibberellic acid in the control of bud growth in blackcurrant stem segments. The figures against the individual graphs show the relevent concentrations of abscisic acid applied in ppm. The marked promotion of bud growth by gibberellin is progressively prevented by increasing doses of abscisic acid and the shape of the curves, showing that abscisic acid is less effective the higher the gibberellin concentration, suggests that these substances mutually compete at the growth centres. (From El Antably *et al.*, 1967.)

Many physiological studies suggest that the role of ABA is not confined to the control of dormancy in buds. As indicated above it is a factor in abscission control. It has been isolated from dormant tubers e.g. *Dioscorea batatas* (Hashimoto *et al.*, 1968) where it may be involved in the maintenance of dormancy. It has been shown to be present in the seed-coats of apple (Pieniążek & Grochowska, 1967) and possibly peach (Lipe & Crane, 1966) from which it disappears after stratification. Since it will also inhibit germination when applied to seeds (Nitsch, 1967*a*) or excised embryos (Pieniążek & Grochowska, 1967) it could eventually prove to be a key substance in the regulation of seed germination. Since it antagonizes the actions of auxins, gibberellins and cytokinins it is likely to be an important natural regulator in a wide spectrum of plant growth and development processes. (Addicott & Lyon, 1969).

Abscisic acid has been detected in a wide variety of species of vascular plant (see Addicott & Lyon, 1969). It occurs as a glucose ester, (+)-abscisyl-β-D-glucopyranoside in yellow lupin fruits (Koshimitzu, Inui *et al.*, 1968). A very close relation of abscisic acid which occurs in the seeds of *Phaseolus multiflorus* is phaseic acid (Fig. 41 CXXXVIII) (MacMillan & Pryce, 1968); so far little is known of its physiological activities although, like ABA, it will promote abscission.

The similarity of the structure of abscisic acid to that of the ubiquitous yellow pigments of plants, the xanthophylls, suggest a common metabolic origin. The work of Taylor & Smith (1967) have shown that the irradiation of xanthophylls, particularly violaxanthin (Fig. 41 CXXXIX), with blue light gives rise, presumably by photo-oxidation, to inhibitors closely resembling ABA, and they proposed that this is how the substance may be produced in xanthophyll-rich leaves when chlorophyll is disappearing at the end of the summer. This received strong support from the subsequent study of synthetic growth inhibitors closely resembling ABA. One (Fig. 41 CXL) having the ring structure of violaxanthin (Fig. 41 CXXXIX) has been shown to have physiological actions of the same order as those of ABA (Tamura & Nagao, 1969). It could well be an intermediate in the conversion of violaxanthin to abscisic acid. Subsequent work on the photo-oxidation product of violaxanthin has established its structural formula (Fig. 41 CXLI) and has also shown it to be identical with a naturally occurring inhibitor from seedings of dwarf bean (*Phaseolus vulgaris*) and wheat (*Triticum vulgare*) cultivars (Taylor & Burden, 1970*a* & *b*). This new inhibitor has been given the name *xanthoxin*. However, as has been pointed out by Addicott & Lyon (1969), it is unlikely that abscisic acid is exclusively a degradation product of ageing organs since it is also present in young immature tissue.

Little work has so far been done on structure/activity relationships in compounds closely related chemically to ABA. One study has suggested that high activity is associated with the one double bond in the ring, since its removal by dehydrogenation or the addition of a second double bond brings about considerable loss of activity. Loss of the = O group on the ring or the formation of an epoxide, such as that on the ring of violaxanthin, have little effect on activity (Sondheimer & Walton, 1970).

Chapter VII

Synthetic Inhibitors of Growth

INTRODUCTION

The growth of plants can be inhibited by a variety of chemicals operating in a variety of ways. Metabolic poisons, operating directly to block biochemical processes essential for growth, have been used as research tools but, with few exceptions, have no practical use. The synthetic auxins, if applied at concentrations above optimal, bring about growth reductions which have been explained in several ways, for example on the two-point attachment theory of auxin action in terms of a clogging of the 'growth enzymes' by excess substrate (auxin) (see p. 306) or by inducing the production of the volatile inhibitor ethylene (see p. 209–11). Even higher concentrations lead to abnormalities of growth, toxicosis and death. This phenomenon has been exploited in weed control, the ring-substituted phenoxyacetic acids, 2,4-dichlorophenoxy-, 4-chloro-2-methylphenoxy- and 2,4,5-trichlorophenoxy-acetic acids being among the most commonly used selective weed killers. Fortunately, in addition to these toxicants, there exist several categories of synthetic plant growth regulator which inhibit growth effectively at concentrations which are not toxic, and in this chapter the chemistry of these compounds and the possible basis for their modes of operation will be considered. There are two main categories, the antagonists of auxin action (antiauxins), which are structurally very closely related to the auxins, and those which I have called the growth suppressants, a mixed bag of compounds, of wide range of structure, whose modes of action are still not completely understood, although some appear to operate, at least in greater part, by suppressing the synthesis of the natural hormones, particularly the gibberellins.

ANTIAUXINS

Strictly speaking, the term antiauxin should be reserved for those compounds which compete with auxin for some specific reaction centre in the growing cell. The extent to which auxin action is suppressed will then depend on the relative amounts of auxin and antiauxin competing, according to the laws established for substrate-inhibitor interaction in competitive enzyme inhibitions. Just as the inhibition of bacterial growth by the competitive inhibitor sulphanilamide can be relieved by administration of the essential metabolite *para*-amino-benzoic acid* so also should antiauxin inhibition of extension-growth be relieved by treatment with additional auxin. Furthermore, antimetabolites act by virtue of the close similarity of their molecular structures to that of the revelant metabolite; certain portions of the molecule of both metabolite and antimetabolite must possess the same 'shapes' in order to be able to 'fit' into the reactive centre. True 'antiauxins' should, therefore, be close chemical relatives of the auxins, possessing *some, but not necessarily all*, of the characteristics of structure necessary for auxin activity. Both these requirements are fulfilled for a considerable number of the antiauxins so far studied. Indeed, under certain carefully prescribed conditions, the growth response of *Avena* coleoptile cylinders to solution mixtures of auxins and their antagonistic homologues have been shown to follow, with a very surprising degree of precision, the mathematical relationships that have been established for the competitive inhibition of simple enzyme systems (McRae & Bonner, 1953). (See Fig. 45A.) For a thoroughgoing analysis of the phenomena and their implications the reader is referred to Housley (1961).

The proper funtioning of auxin in normal growth control can be prevented, however, by means other than direct intervention at auxin-reaction centres. For instance, interference with the auxin metabolism system may cause an alteration in the rates of synthesis or destruction in the tissue and thus change concentration levels. Such an action may also show apparent competitive effects. For example, if the antagonizing compound impedes auxin production and consequently lowers auxin concentration in the organ concerned, the resulting inhibition of growth may be offset by auxin application. This kind of competitor may, or may not, be an auxin homologue. Thus 2,3,5-tri-iodobenzoic acid (TIBA) (Fig. 43 CXLII), which might be expected to be a direct auxin competitor, has been shown to lower the auxin levels of treated roots to vanishingly low values, and may exert part of its antagonistic action in this way (Audus & Thresh, 1956*a* & *b*).

*For a full discussion of 'antimetabolites' in bacterial growth and the biochemistry of this action the reader is referred to Woolley (1952).

A TIBA-promotion of auxin destruction has been directly demonstrated in pea tissue (Winter, 1968). A similar reduction in auxin levels has been claimed to result from the treatment of potato tubers with ethylene (Michener, 1942), which is certainly not an auxin homologue. The mechanism for such lowering of auxin (IAA) levels probably lies largely in the action of the enzyme (IAA-oxidase), which catalyses the degradation of this auxin in tissues. Thus phenols are substances which can act as co-factors of this enzyme and thus promote the destruction of the endogenous IAA and thereby reduce growth. This subject is dealt with in more detail in Chapter XI. However the synthetic auxins do not seem to be degraded by this enzyme and so the antiauxin propensities of the phenols would not be evident in interaction experiments with the synthetic auxins.

Another way in which the normal functioning of auxin may be impeded is by blocking its movement about the plant (see Chapter XI). This kind of action may underlie the antagonisms of TIBA (Niedergang-Kamien & Skoog, 1956; Kuse, 1953; Vardar, 1959) and also *N*-1-naphthylphthalamic acid (Fig. 43 CXLIII) (Morgan, 1964) and 9-fluorenolcarboxylic acid (Fig. 43 CXLIV) (McCready, 1968*b*). Furthermore, when interactions are being studied between growth substances applied to experimental tissue the question of uptake cannot be ignored and there is good evidence to suggest that

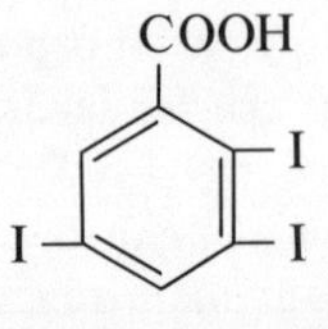

COOH, I, I, I

CXLII

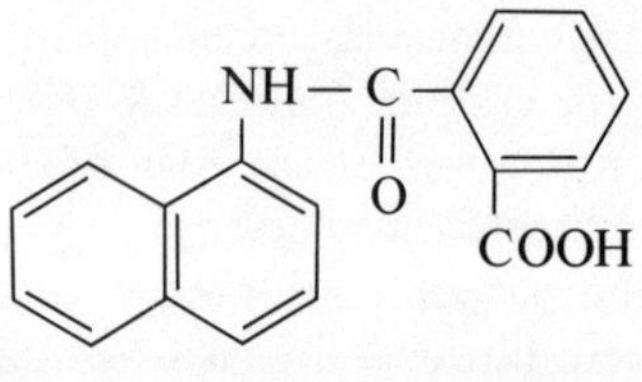

NH—C(=O), COOH

CXLIII

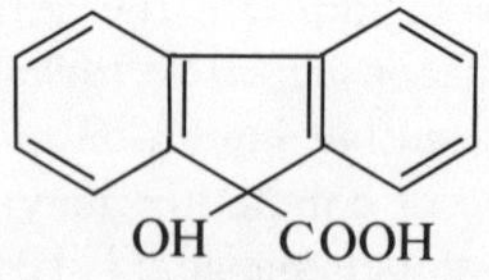

OH COOH

CXLIV

Fig. 43. Inhibitors of polar auxin transport.

CXLII	2,3,5-Tri-iodobenzoic acid (TIBA).
CXLIII	*N*-naphth-lyl-phthalamic acid (NPA).
CXLIV	9-Fluorenolcarboxylic acid. Morphactin.

TIBA may block IAA uptake, for example in segments of *Avena* coleoptiles (Sabnis & Audus, 1967). Yet another effect seems to be that TIBA will render IAA immobile in pea tissue, presumably by promoting its binding to proteins (Winter, 1968) and this has been suggested as the main cause of its blockage of IAA transport.

The precise characterization of an antiauxin is therefore very difficult, even when undeniable interactions with auxins have been established. Unfortunately, considerable confusion has tended to result from careless use of the term antiauxin for substances which produce growth responses opposite in nature to those produced by auxins. Such growth responses might have little if anything to do with auxins. The author has fully expressed his views on this subject elsewhere (Audus, 1954).

To return to the true antiauxins, i.e. auxin homologues that have little or no auxin activity and show *competitive* antagonism to auxin action, a classification was first drawn up mainly on a basis of the two-point attachment theory of auxin activity (McRae & Bonner, 1953). Four categories were recognized:

(*a*) Compounds with at least one free ortho position in the ring but lacking a terminal carboxyl group on the side-chain, e.g. 2,4-dichloranisole (Fig. 44 CXLV).

(*b*) Compounds having *both* ortho positions blocked, e.g. 2,6-dichlorophenoxyacetic acid (Fig. 44 CXLVI) and 2,4,6-trichlorophenoxyacetic acid (Fig. 44 CXLVII). A demonstration of the competitive nature of the inhibition of 2,6-D is seen in Fig. 45A.

(*c*) Compounds in which the structure and configuration of the side-chain are such that two-point contact with the receptive site is prevented. Into this category come a very large number of compounds, e.g. the iso-butyric acid homologues of auxins such as 4,chlorophenoxy-iso-butyric acid (PCIB), α-(naphth-lyl-methylthio)-propionic acid (NMSP) (Fig. 44 CXLVIII), etc. The more recently discovered phenylpropiolic acid (Fig. 44 CXLIX) may also come into this category (Åberg, 1963).

(*d*) Weak auxins, e.g. phenylacetic acid, phenoxyacetic acid (Fig. 8 XXVII), γ-phenylbutyric acid, 4-fluoro-3-nitrobenzoic acid (Fig. 44 CL), naphth-2yl-acetic acid (Åberg, 1959) and 5-(indol-2yl-methyl)-tetrazole (Hamilton *et al.*, 1960).

This classification was premature, since it was based on rather restricted data. For example, the antiauxin activity of 2,4-dichloranisole (Fig. 44 CXLV) and other similar compounds cannot be established in some tests (Audus & Shipton, 1952; Åberg, 1952). Again, as some 2,6-substituted and isobutyric acids are weak auxins, categories (*b*) and (*d*) may overlap.

Important new categories of antiauxin have since been introduced mainly as the result of the work of Fredga & Åberg in Sweden. Particularly

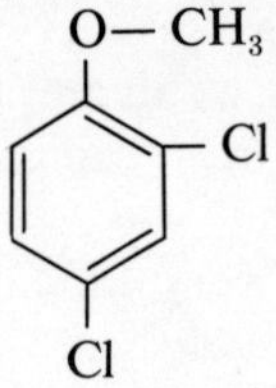

CXLV 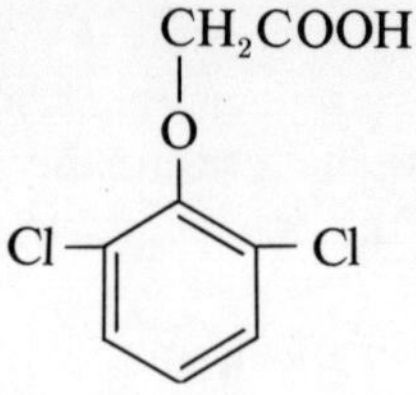 CXLVI 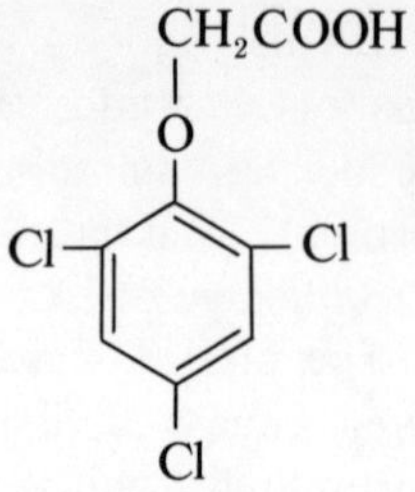 CXLVII

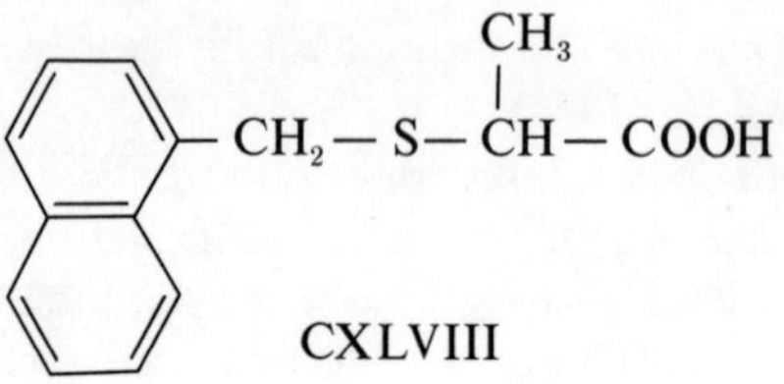

CXLVIII

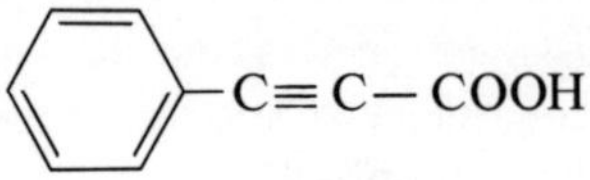

CXLIX

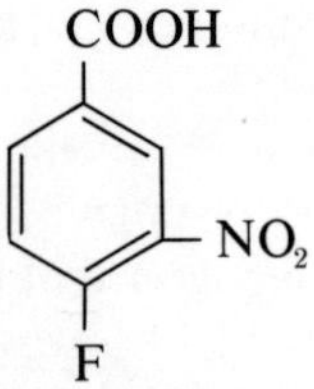

CL 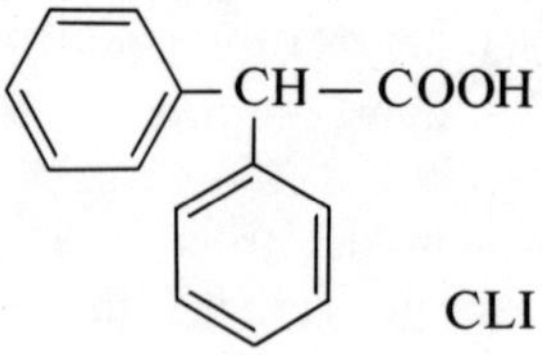 CLI

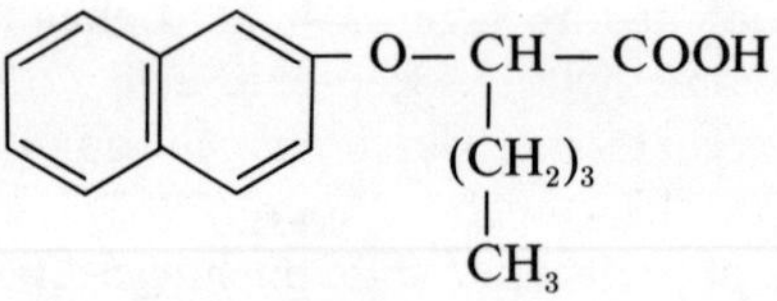

CLII

Fig. 44. Auxin antagonists.

CXLV	2,4-Dichloranisole
CXLVI	2,6-Dichlorophenoxyacetic acid
CXLVII	2,4,6-Trichlorophenoxyacetic acid
CXLVIII	α-(Naphth-lyl-methylthio)-propionic acid (NMSP)
CXLIX	Phenylpropiolic acid
CL	4-Fluoro-3-nitrobenzoic acid
CLI	Diphenylacetic acid
CLII	α-(Naphth-2yloxy)-caproic acid

interesting is the influence of optical configuration already mentioned in relation to auxin activity *per se*. While the D(+) enantiomorph of any particular pair is almost invariable an auxin the L(−) enantiomorph is a competitive antiauxin. From the very large number of compounds studied there are virtually no exceptions to this rule (see Jönsson, 1961; Fredga & Åberg, 1965). Comparisons are sometimes difficult however, since a number of D(+) enantiomorphs (e.g. of certain α-alkylphenylacetic acids) are such weak auxins that they come into category (*d*) above and exhibit 'antiauxin' properties almost as marked as the corresponding L(−) forms (Åberg, 1963). Such properties in this class of antiauxins emphasizes the importance of group-configuration round the asymmetric α-carbon atom and would, in general terms, be strong support for the three-point attachment theory. However the activity differences between pairs of enantiomorphs, although qualitatively consistent, exhibit such large quantitative variations, that a return to the less rigid theory of two-point attachment has been favoured (Fredga & Åberg, 1965). A similar pattern of behavior with *cis-trans* isomerism also conforms to such a theory. Thus in cinnamic acid (Fig. 14 LXXVII and LXXVIII) and some ring-substituted derivatives the *cis*-forms are auxins while the *trans*-forms are antiauxins (Åberg, 1961).

The view of the antiauxin picture is still obscured by ambiguities intrinsic to the biological testing methods so far employed. True antiauxin activity can be, and has been established for many compounds by rigorous analysis of data from interaction studies with IAA (or other authentic auxins). But these techniques are tedious, and simpler, indirect methods have come into favour. The most used of these is based on the growth responses of certain roots, notably flax and wheat. It has been widely assumed, although with little direct evidence, that roots are so sensitive to auxins that their own natural auxin content is supra-optimal for growth, which, in normal roots, is consequently maintained below the maximum possible with optimal auxin levels. Any reduction in the activity of this native (endogenous) auxin would consequently accelerate growth. Early observations with compounds possessing true antiauxin activity, e.g. α-(naphth-lyl-methylthio)-propionic acid (NMSP) (Fig. 44 CXLVIII) (Åberg, 1950) and 4-chlorophenoxyisobutyric acid (PCIB) (Fig. 12 LXII) (Burström, 1950), supported this suggestion and led to the establishment of root-growth acceleration as a criterion of antiauxin activity and the development of the root test for antiauxins. The subsequent application of these tests has shown that, where comparisons have been made, there is indeed a close correlation between true antiauxin and root-growth-accelerating activities (Åberg, 1959, 1963, 1965; Fredga & Åberg, 1965; Jönsson, 1961).

But caution is still required with such simple tests since certain compounds show exceptional behaviour. Thus some antiauxins have little action on roots, e.g. certain aryl- and aryloxy-acetic acids with bulky α-substituents. Compounds CLI and CLII on Fig. 44 are of this kind (Veldstra & Åberg, 1953). Other root-growth stimulants (e.g. PCIB) have definits auxin properties (Ng & Audus, 1964). Finally IAA itself, under some conditions can, at very low concentrations, accelerate the growth of roots (see Larsen, 1955*b*, 1961). This state of affairs has prompted the suggested segregation of a new class of growth substances, the 'root auxins', as distinct from 'shoot auxins' (Hansen, 1954). Root auxins are defined as substances which can stimulate extension growth of roots while they may or may not be antagonists of 'shoot auxins'.

This still leaves the situation confusing, but it may be that we are too ready to accept IAA as the natural auxin in control of the extension growth of all organs, particularly those used in the above tests. It is not impossible that the natural auxins of roots is not IAA but an even more active compound, which IAA, in the role of a weaker auxin (category (*d*) above), antagonizes, hence stimulating growth.

This sort of dual activity of the growth of coleoptiles has been used by Åberg to characterize category (*d*) antiauxins (weak auxins). Such compounds as α-(naphth-2yloxy)-*n*-butyric acid, 3-methylphenoxyacetic acid, 3-methoxyphenoxyacetic acid, 3-isopropenylphenoxyacetic acid and 2-hydroxyphenylacetic acid, all of which stimulate root growth at low and inhibit it at high concentrations, have the converse effects on *Avena* coleoptiles (Åberg, 1965, 1967, 1969) (see Fig. 45C). These low-level inhibitions of

Fig. 45. *Top.* Graphs showing examples of competitive and non-competitive inhibition of auxin action, based on the assumption that auxin is acting as a typical substrate in a 'growth-enzyme' system. The simple kinetics of a pure enzyme system can be expressed by the equation: $\frac{1}{V} = \frac{K}{V_{max}} \cdot \frac{1}{S} + \frac{1}{V_{max}}$ where V is the rate at substrate concentration S, V_{max} is the rate at saturating substrate concentrations and K is the dissociation constant of the enzyme/substrate complex. If 1/V is plotted against 1/S such a relationship gives a straight line, cutting the vertical axis at $1/V_{max}$ and with a slope of K/V_{max}. Plotting 1/(growth rate) against 1/(auxin concentration) gives such a straight line under some conditions (lowermost line in (A) for 2,4-D action on the growth of *Avena* coleoptile segments). In other cases a straight line is not obtained unless a constant (presumably representing the natural auxin already present in the test tissue) is added to the applied substrate concentration. This is the case for the lowest line in (B) for the action of IAA on wheat coleoptile segments. In the presence of a competitive inhibitor, V_{max} is not altered but the value of K is increased by an amount proportional to the concentration of the inhibitor. This is shown in (A) for the anti-auxin 2,6-dichlorophenoxyacetic acid by the increasing slope of the lines as the inhibitor concentrations are increased. In the presence of a *non-competitive* inhibitor, V_{max} is decreased and it can be shown that the slopes of the lines are so increased by increasing inhibitor concentrations that, when extrapolated back, they all intersect at a point *behind* the vertical axis (i.e. where 1/S is

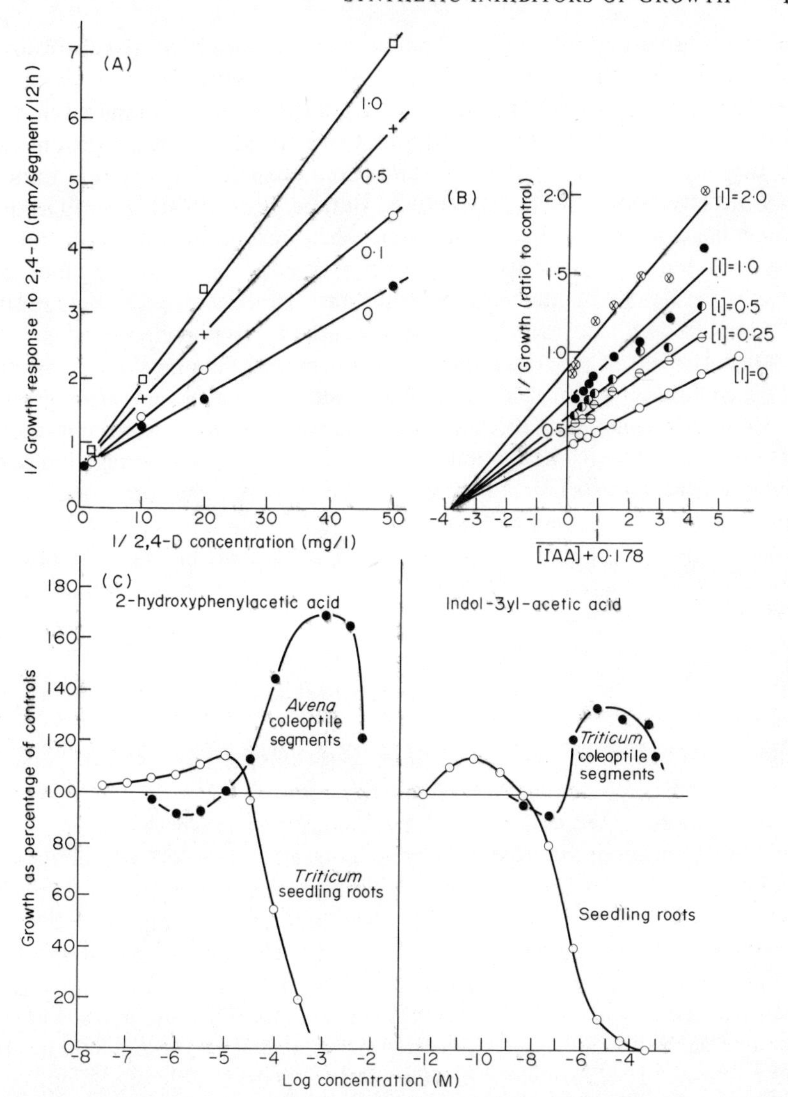

negative). This is clearly the case in (B) which shows the effects of abscisic acid on IAA action. On this basis abscisic acid would not be a true anti-auxin. In (A) the figures on the lines are the respective concentrations of 2,6-dichlorophenoxyacetic acid in mg/l. In (B) they represent abscisic acid concentrations. (A) from data of McRae & Bonner, 1952; (B) by kind permission of Prof. R. L. Wain. (From Rothwell & Wain, 1964).

Bottom. Graphs showing the concentration/activity relationships of a weak auxin, 2,hydroxyphenylacetic acid and an active auxin, indol-3yl-acetic acid (IAA) on the growth of coleoptile segments and intact roots. (Data for 2-hydroxyphenylacetic acid from Aberg (1967); data for IAA and wheat coleoptiles from Barlow (1957); data for seedling roots (average for several species and IAA from Larsen (1955*b*).

coleoptile growth are taken as evidence of antiauxin action, the implication being that the native auxin (IAA) at sub-optimal levels in the coleoptile, is here being antagonized. But again we meet the puzzling complication that IAA itself may, under some conditions have the same dual action on coleoptiles, inhibiting at low (10 μg/1) and stimulating at higher concentrations (e.g. in very young wheat coleoptiles; Barlow *et al.*, 1957). (see Fig. 45C). A similar conclusion, that in the young wheat coleoptile the native auxin is not IAA but a more active compound, immediately springs to mind; but since other hormones such as gibberellin and cytokinin may also be involved in early coleoptile growth (Wright, 1963) such judgement would be premature. However these phenomena do underline the dangers of too generalized conclusions drawn from simple tests such as the root test.

A complication which further clouds the issue is the discovery that certain substituted benzoic acid 'antiauxins', such as 3,4-di-iodo-4-hydroxybenzoic acid, can stimulate root growth up to three times the control value but only when the roots are illuminated (Wain, 1968). The basis of this unexplained response, when elucidated, may help to clear up the uncertainties concerning the operation of 'root-auxins', 'shoot-auxins' and antiauxins in root growth.

The Growth Suppressors

High growth-inhibiting concentrations of the auxins and of the anti-auxins are invariably accompanied by distortions and malformations of organs, consequent upon the disturbance of cell polarities on which the proper coordinating action of natural auxins depend (see Chapter XI). In recent years a completely new class of plant growth inhibitor has emerged with the special characteristics that they retard over-all growth processes in some organs, for example cell division in the meristems and cell elongation in stem internodes, without inducing these gross distortions of plant and cell form. I shall call these substances the growth suppressors. They include at the one extreme such substances as maleic hydrazide, which can completely inhibit some phases of cell growth and thus produce considerable modification of plant form, and at the other the so-called 'growth retardants', whose action is only to slow down growth without gross formative effects. This division may ultimately prove to be artificial and untenable although it is at present accepted for convenient categorization (see Cathey, 1964*a*).

(A) MALEIC HYDRAZIDE

The first synthetic growth suppressor to be described was maleic hydrazide (Fig. 46 CLIII). Schoene & Hoffmann, (1949) found that a solution sprayed

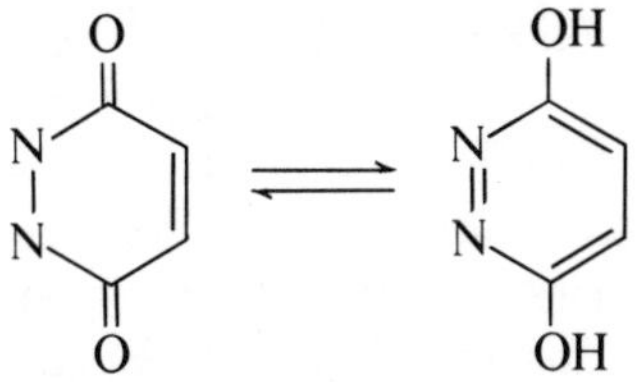

Fig. 46. CLIII Maleic hydrazide.

on tomato plants induced cessation of stem elongation and, in addition, destroyed the control which the stem apex normally exerts on the growth of lateral buds (apical dominance). In treated plants therefore the lateral buds grow out and a 'bushy' plant is the result. Because the dominating action of the apical meristem has been regarded as an auxin-mediated phenomenon, this action of maleic hydrazide seemed likely to put it into the class of antiauxins; indeed some direct support for this contention was provided by Leopold & Klein (1952) who showed that maleic hydrazide strongly inhibited auxin (IAA) action in the split pea epicotyl test but that this inhibition was completely relieved by high IAA concentrations, the main criterion of a competitive inhibition. On the other hand Johnson & Greulach (1953) could obtain no evidence for such an antiauxin action in *Avena* coleoptiles; IAA did not release inhibition in segment growth tests and there was no influence on plant curvature responses to gravity and unilateral light, which are auxin-dependant growth responses. It has been suggested that the antiauxin action of maleic hydrazide could be due to its promotion of IAA oxidation in plant tissue (Andreae & Andreae, 1953) although there is no evidence from direct assay of tissues that native auxin levels are reduced by maleic hydrazide treatment (Kulescha, 1955; Audus & Thresh, 1956*a*), and Pilet showed later that IAA oxidation in lentil roots was accelerated only in excessively high maleic hydrazide concentrations (Pilet, 1957*a*). Other observations show that in *Avena* coleoptiles maleic hydrazide inhibition is removed most effectively by small traces of heavy metals, one cobaltous ion for example being able to neutralize the inhibition by 10^3 molecules of maleic hydrazide. (Suda, 1960). Again very low concentrations of ferric sulphate and of boric acid have similar effects (Johnson & Greulach, 1953). Suda suggests that maleic hydrazide specifically blocks biochemical processes catalysed by heavy metals and that auxin action is dependent on these processes. On the other hand it will block the growth-promoting action of gibberellic acid in dwarf peas, whereas its own inhibition of growth in tall peas is relieved by gibberellic acid. This suggests that it could also be a gibberellin antagonist (Brian & Hemming, 1957). It must be remembered however that maleic hydrazide action seems to be concerned with the blocking of cell division processes, although cell extension is also affected. Auxins regulate cell extension whereas gibberellins may regulate both phases of growth and the complexity of the interactions reported above may be at a deeper level underlying all growth processes and not at any site specific to one or other hormone. Such a fundamental action is supported by the many observations showing that carbohydrate metabolism is seriously disturbed by maleic hydrazide treatment in a number of different tissues. It is usually characterized by an increase in sugars at the expense of polysaccharide reserves (Phouphas & Goris, 1952; Goris & Bouriquet, 1953; Greulach, 1953; Peterson & Naylor, 1953; Pilet & Margot, 1955). Opinions on the significance of these effects differ. One extreme suggests that they are not directly connected with the growth inhibitions (Goris & Bouriquet, 1953). Another proposed that a blockage of carbohydrate utilization may necessitate drawing on proteins as a source of respiratory fuel, and so disturb protein metabolism and cell division (Peterson & Naylor, 1953). Yet another attributes sugar accumulation in cotton to a blockage of the export from treated leaves by the induced collapse of the phloem cells (McIlrath, 1950). This, however, could not account for carbohydrate changes in tissue culture (Phouphas & Goris, 1952; Goris & Bouriquet, 1953). Another theory attributes the disturbance of carbohydrate metabolism to an inhibition of pyrimidine biosynthesis by maleic hydrazide. This then results in the observed depression of the levels of uridine and cytidine phosphates which belong to the energy-rich group of compounds involved in di- and polysaccharide biosynthesis (Pavlinova *et al.*, 1967).

The respiration of tissue is also affected by maleic hydrazide, for example it is depressed in root tips of various species (Naylor & Davis, 1951). However the exact point of action on respiratory enzymes remains obscure, although there are indications that the inactivation of enzymatic sulphydryl groups is involved (Hughes & Spragg, 1958). On the other hand it inhibits strongly certain enzymes which control amino-acid metabolism and may therefore create considerable disturbances of protein synthesis (Kim & Greulach, 1963; Suzuki, 1966). The fact that 5-fluorouracil, an inhibitor of nucleic acid synthesis, has the same effects as maleic hydrazide on amino-acid metabolism of *Chlorella* (Kim & Greulach, 1963) and that maleic hydrazide inhibition of the growth of carrot callus (Butenko & Buskakov, 1961) and young tomato plants (Povolockaja, 1961) is relieved by uracil, one of the constituent bases of ribonucleic acid, points to a disturbance of protein synthesis via some change in the genetic message.

There now seems good evidence that maleic hydrazide, when injected into mice at low doses, can induce hepatomas (Epstein & Mantel, 1968) and this also implies an interference with nucleic-acid metabolism. Its close structural resemblance to uracil underlines this, as does the demonstration that it can induce a high frequency of uracil-independant strains of yeast (*Saccharomyces cerevisiae*) (Callaghan *et al.*, 1965). The complicated interactions with auxins and gibberellins already described above may ultimately find their explanations in terms of the effects at the nucleic-acid level. For a discussion of the role of nucleic acids in growth and differentiation and its implications for hormone control see Chapter XII.

A range of other derivatives of hydrazine have been tested for growth-inhibiting activity (Huffman *et al.*, 1968). A few show activity comparable with that of maleic hydrazide. One is *N*-amino-*N*-methyl-β-alanine (Fig. 47 CLIV). Another is *N*-dimethylaminosuccinamic acid (Fig. 47 CLV) (B995) (Riddell *et al.*, 1962), which is classified as a growth retardant (see later section).

Yet another group is constituted by the hydrazinium salts of which the most active are *N,N*-dimethyl-(2-chloroethyl)-hydrazinium chloride (Fig. 47 CLVI), its bromine analogue and the iso-propyl- and allyl-dimethyl-hydrazinium bromides (Fig. 47 CLVII & CLVIII) (König, 1968). Whether these should be categorized with maleic hydrazide or with the growth retardants such as CCC (see next section) must await more information on their biochemical actions.

(B) THE GROWTH RETARDANTS

(i) Types of compound

In the same year that saw the advent of maleic hydrazide, Mitchell *et al.*, (1949) announced a compound which would, when applied at a concentration of one percent in lanolin paste, dramatically reduced the internode extension of bean plants. This was the compound 2,4-dichlorobenzyl-nicotinium chloride (2,4-DNC) (Fig. 48 CLIX). Further studies on quaternary ammonium compounds disclosed a group of very effective growth retardants, typified by its most active member 2-isopropyl-4-dimethylamino-5-methylphenyl-l-piperidine carboxylate methyl chloride (Fig. 48 CLX). This was given the code name AMO-1618, by which it is still known. The effects of these compounds were distinct from those of maleic hydrazide in that they reduced stem elongation without a complete stoppage of cell division in the apical meristems and the correlated release of lateral bud growth. A little later a further group of quaternary ammonium compounds was added to the list of the growth retardants. These were analogues of the extremely important natural organic base choline; the most active was chlorocholine chloride (christened CCC) or (2-chloroethyl)-trimethyl-ammonium chloride (Fig. 48 CLXI) (Tolbert, 1960). This substance proved

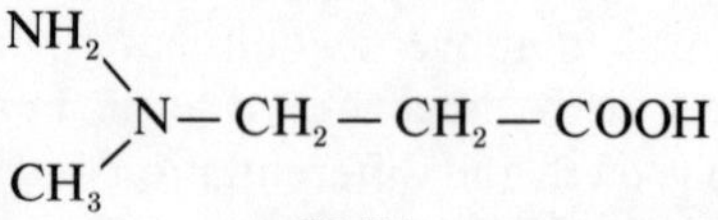

$$NH_2(CH_3)N-CH_2-CH_2-COOH$$

CLIV

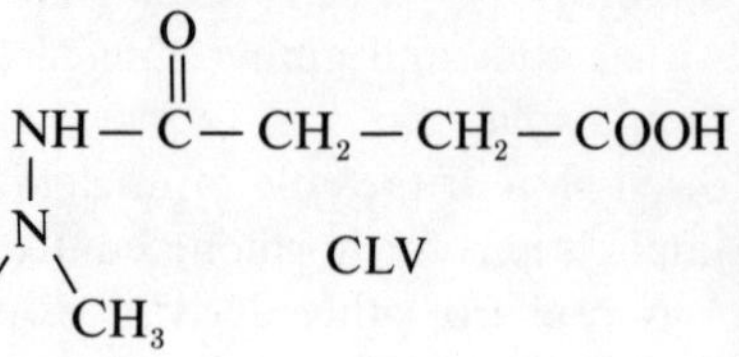

$$NH-\overset{O}{\overset{\|}{C}}-CH_2-CH_2-COOH \quad (NH \text{ bonded to } N(CH_3)_2)$$

CLV

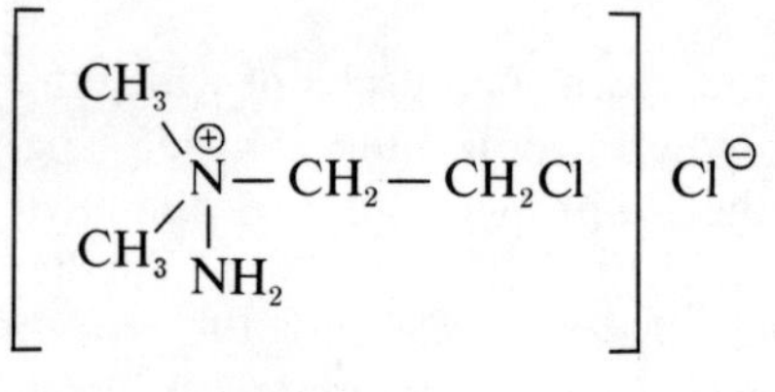

$$\left[(CH_3)_2\overset{\oplus}{N}(NH_2)-CH_2-CH_2Cl\right] Cl^{\ominus}$$

CLVI

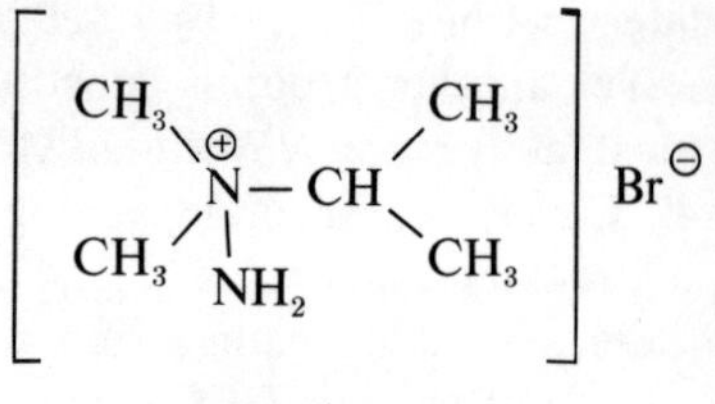

$$\left[(CH_3)_2\overset{\oplus}{N}(NH_2)-CH(CH_3)_2\right] Br^{\ominus}$$

CLVII

$$\left[(CH_3)_2\overset{\oplus}{N}(NH_2)-CH_2-CH=CH_2\right] Br^{\ominus}$$

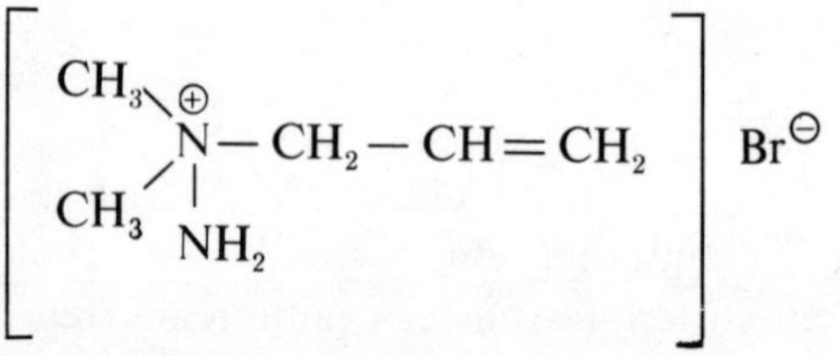

CLVIII

Fig. 47. Growth suppressants.

CLIV	*N*-Amino-*N*-methyl-β-alanine
CLV	*N*-Dimethylaminosuccinamic acid. B995
CLVI	*N,N*-Dimethyl-(2-chloroethyl)-hydrazinium chloride
CLVII	*N,N*-Dimethyl-isopropylhydrazinium bromide
CLVIII	*N,N*-Dimethylallylhydrazinium bromide

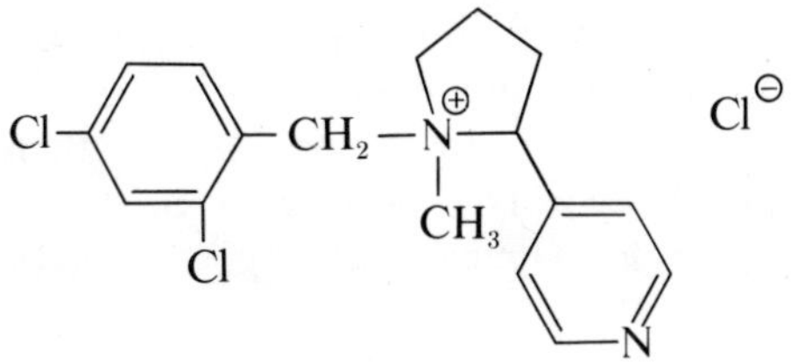

Cl– –CH_2–$N^{\oplus}$ (–CH_3) $Cl^{\ominus}$
Cl
N

CLIX

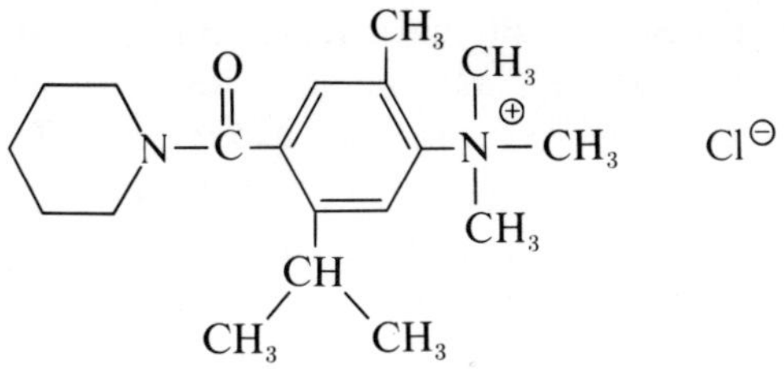

CH_3 O CH_3
N–C– –$N^{\oplus}$–CH_3 $Cl^{\ominus}$
CH_3
CH
CH_3 CH_3

CLX

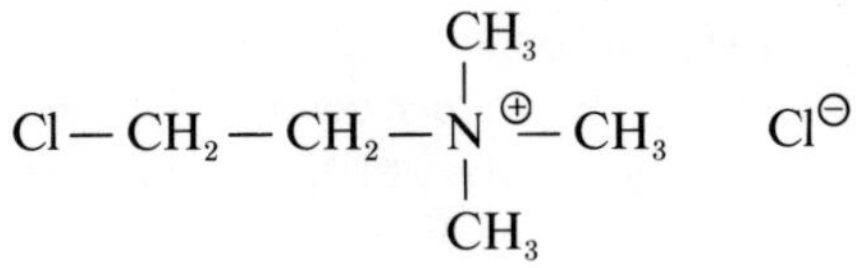

$$Cl-CH_2-CH_2-\overset{CH_3}{\underset{CH_3}{N^{\oplus}}}-CH_3 \quad Cl^{\ominus}$$

CLXI

$$Cl-C_6H_3(Cl)-CH_2-\overset{C_4H_9}{\underset{C_4H_9}{P^{\oplus}}}-C_4H_9 \quad Cl^{\ominus}$$

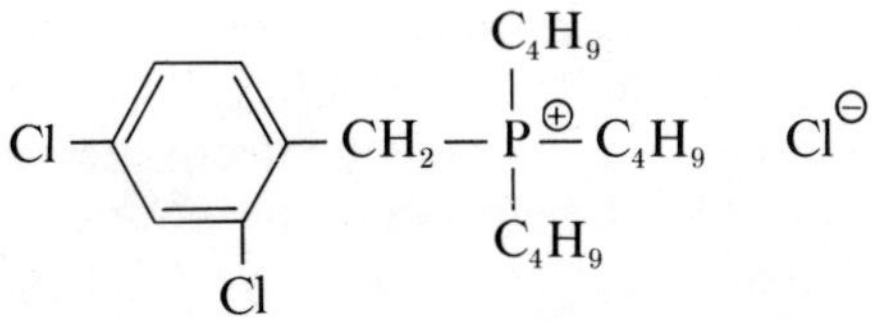

CLXII

Fig. 48. Growth retardants.
- CLIX 2,4-Dichlorobenzylnicotinium chloride. 2,4-DNC
- CLX 2-Isopropyl-4-dimethylamino-5-methylphenyl-1-piperidinecarboxylate methyl chloride. AMO-1618
- CLXI Chlorocholine chloride: (2-chloroethyl)-trimethylammonium chloride. CCC
- CLXII 2,4-Dichlorobenzyltributylphosphonium chloride. Phosphon D

to be effective on a much wider variety of plant species than the two previously described groups of quaternary ammonium compounds (see Plate 11).

Related compounds, where the quaternary nitrogen atom is replaced by phosphorus, also proved to have closely similar growth retarding properties (Preston & Link, 1958). The most active of this group was 2,4-dichlorobenzyltributylphosphonium chloride (Fig. 48 CLXII). This received the name Phosphon-D (see Plate 11).

The last compounds with these physiological actions to be discovered had a basic chemical structure quite different from that of the preceding retardants although they possessed some similarities with maleic hydrazide in that they all contained a -C-C-N-N- sequence in the molecule. They have already been mentioned in the preceding section, and include certain substituted maleamic and succinamic acids (Ridell *et al.*, 1962). Unfortunately the more active maleamic acids are unstable in water but the stable succinamic form is a very active retardant. The compound most frequently in use is *N*-dimethylaminosuccinamic acid, designated B995 (Fig. 47 CLV).

(ii) Biological activities

One of the most surprising and puzzling aspects of the action of these various growth retardants is their specificity of action and the apparent lack of any correlation between these patterns of specificity and the taxonomic position of the plant itself. On the one hand relatively unrelated species may show closely similar reactions to the same compound while on the other different cultivars of the same species may be widely divergent in their sensitivities (see Cathey, 1964*a*). The only indications of a taxonomic pattern of responsiveness is that most of the sensitive species so far demonstrated belong to the dicotyledonous division of the flowering plant kingdom; the monocotyledons have few sensitive plant species. Thus in the grass family wheat (*Triticum vulgare*) seems to be very responsive (Tolbert, 1960) while most other cereals and grasses are insensitive (see Cathey, 1964*a*). How much of this diversity can be attributed to differences in effectiveness at the growth centres and how much is due to differences in the penetration of the applied retardant and its availability at growth centres is not yet clear. Such problems of specificity have been known for over twenty years in the study of selective weed killers, where probably a variety of mechanisms operate.

Nevertheless, when an action is exerted, the same kind of quantitative change is usually evoked. Often the rate of cell division is considerably slowed down, particularly in internode regions, so that much shorter stems result. At the same time there is little or no disturbance of the pattern of cell division in the apical meristems so that, for example, leaf shape and arrangement are not disturbed. In this way a rosette-like growth habit can

(A)

(B)

Plate 11. Dwarfing effects of plant growth retardants.

(A) Tobacco treated with Phosphon D at 50 g per ft^3 of soil. *Right*. Treated plant; *Left*. Untreated control.

(B) Two wheat varieties treated with CCC. In order from left to right; Cultivar Opal (standard variety) untreated; Cultivar Opal (standard variety) sprayed with CCC solution at the 5-leaf stage; Cultivar Gaines (dwarf variety) untreated; Cultivar Gaines (dwarf variety) sprayed with CCC at 5-leaf stage.

Note that CCC treatment has reduced the stature of the tall variety to that of the dwarf and has had only a slight effect on the dwarf variety itself. (Photographs by kind permission of Dr. E. C. Humphreys, Rothamsted Experimental Station, Harpenden).

be induced, e.g. by AMO-1618 in such plants as *Chrysanthemum* (Sachs & Lang, 1961). A reduction in the total extension of cells in the main axis also contributes to the dwarfing action of these retardants. In bean plants axial cells of AMO-1618-treated plants were not much more than one quarter the length of normal axial cells (Scherff, 1952). However to illustrate the complexity of the situation it should be noted that in low concentrations these retardants may, under some conditions, actually promote the growth of some species, e.g. *Zinnia* (Cathey & Stuart, 1961), *Antirrhinum* (Halevy & Wittwer, 1965). The explanation of these exceptional responses remains obscure.

The growth in area of leaves may sometimes be restricted, with a correlated increase in thickness and in chlorophyll content, e.g. AMO-1618 action on bean leaves (Krewson *et al.*, 1959), and the action of CCC on tomato (Laborie, 1963). On the other hand there may be no effect on blade growth, e.g. CCC applied to wheat (Humphreys *et al.*, 1965) or there may even be an increased leaf area, perhaps because of reduced competition for raw materials from the stunted stems, e.g. in mustard (*Sinapis alba*) treated with CCC (Humphreys, 1963). Although the effects are not so dramatic, there are reports that the retardants also suppress root growth and development (see Cathey, 1964*a*).

Accompanying this suppression of elongation along the main axis is often an augmented expansion of the stem in the lateral direction, either by the production of wider cells or by greater cell proliferation in cambia or sub-apical meristems (Scherff, 1952; Sachs & Kofranek, 1963). This is clearly shown in the response of cereal haulms which are thereby considerably strengthened as well as shortened and this greatly alleviates 'lodging' in heavy rain and wind (see Plate 12).

As we shall be seeing later (Chapter X), very marked changes take place in the behaviour of the plant as it enters on the phase of reproduction. Subtle changes in the biochemical happenings in the growing points of the plant bring about the production of flowers instead of foliage leaves and the vegetative growth of the plant usually stops when reproductive growth begins. Such changes are controlled by hormones. One of the most interesting phenomena which has emerged from the widespread application of the growth retardants to a vast range of plants is that the chemical suppression of vegetative growth by these substances seems to release the processes leading to reproductive growth. There is an ever-increasing accumulation of observations where the growth retardants promote a variety of growth processes associated with reproduction. Thus they can bring about the initiation of flowers in such genera as *Rhododendron* and *Azalea* (Stuart, 1961) and in some herbaceous plants such as the tomato (Wittwer & Tolbert, 1960). On the other hand flowering may be suppressed, particularly in

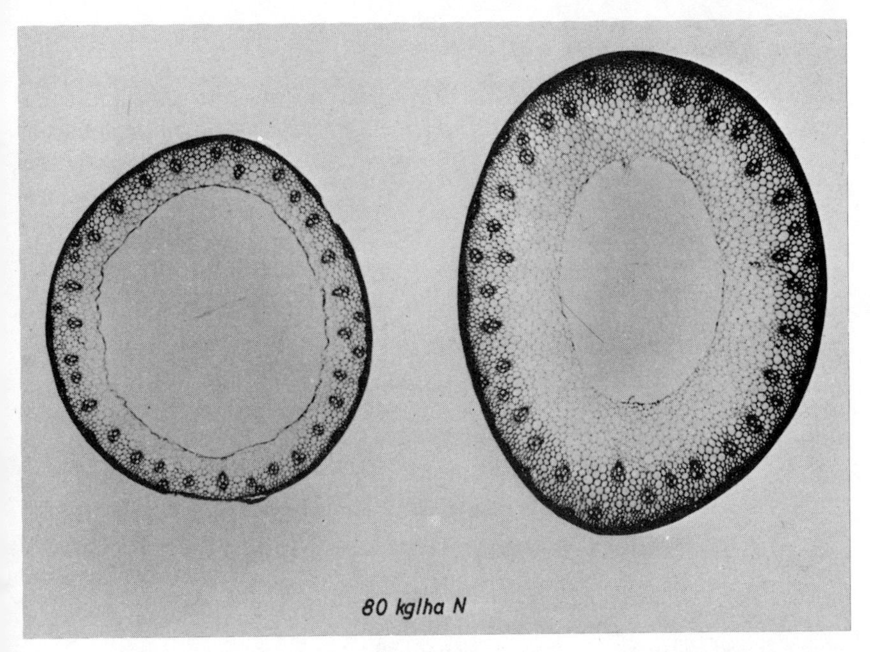

Plate 12. The effects of CCC on the stem anatomy of winter wheat. Cultivar Tassilo, supplied with nitrogen at 80 kg/ha, was sprayed with CCC at 16 kg/ha.
Left. Control untreated plant. *Right*. CCC-treated plant.
Photographs of sections of the second internode. Note the greater diameter of cross section, the greater thickness of the cortex and the larger number of vascular bundles in the treated plant. (Photographs by kind permission of Drs. H. H. Mayr and E. Presoly, Österreichische Stickstoffwerke A. G.)

those genera (e.g. *Kalanchoë*) where flowering can be promoted by gibberellic acid (Marcelle & Sironval, 1964; Zeevaart & Lang, 1963) (see Plate 13). The situation will be considered in more detail in a later chapter (Chapter X) where we deal with the hormone control of flowering.

(iii) Structural requirements for activity

As a general rule the structural requirements for an active retardant molecule are fairly closely prescribed and not much variation is permissable before activity is substantially lowered or lost altogether. In those four groups which are quaternary compounds, the quaternary structure itself is an absolute requirement for activity. Other portions of the molecule are susceptible of some degree of change with relatively minor modifications of activity.

In the quaternary ammonium compounds typified by AMO-1618 the =N-(C=O)- (carbamate) group is an obligate requirement although the

Plate 13. Experiments with *Kalanchoë daigremontianum* (short-day plant) showing the suppression of flowering by CCC after induction in short days and the reversal of this suppression by gibberellic acid.

All the plants were photoperiodically induced by 25 short days and were then grown subsequently for 43 days in long days and then photographed.

From left to right the additional treatments were; 1, Control; 2, Treatment with 4000 mg/l CCC; 3, Treatment with CCC plus 1.5 mg/l gibberellic acid; 4, Treatment with CCC plus 5.0 mg/l gibberellic acid; 5, Treatment with CCC plus 15 mg/l gibberellic acid. Note that in treatment (2) CCC has greatly reduced stem elongation and has completely prevented the formation of flowers. In the subsequent three treatments gibberellic acid at all concentrations has nullified the flower-inhibiting action of the same concentration of CCC and has almost completely restored the stem height to that of the control untreated plant. (Photographs by kind permission of Dr. J. A. D. Zeevaart and Prof. Anton Lang, Michigan State University. From Lang (1966)).

exact molecular configuration allows some latitude. Thus substitution of the six-membered ring derived from piperidine (Fig. 49 CLXIII) by a five-membered ring (pyrrolidine; Fig. 49 CLXIV) or a six-membered ring containing oxygen (morpholine; Fig. 49 CLXV) reduces activity but does not destroy it. The dimethyl-carbamate homologue (dimethylamino group; Fig. 49 CLXV) reduces activity but does not destroy it. The dimethyl-carbamate homologue (dimethylamino group; Fig. 49 CLXVI) of AMO-1618 is very active (Krewson *et al.*, 1959) although increases in the size of the aliphatic substituents progressively reduce activity. Phenyl substituents (e.g. diphenylamino; Fig. 49 CLXVII) also greatly reduce activity (Cathey, 1965). In this same group of compounds a similar small variation of the

$-N$ (piperidine ring) CLXIII

$-N$ (pyrrolidine ring) CLXIV

$-N$ (morpholine ring, O) CLXV

$-N(CH_3)_2$ CLXVI

$-N(C_6H_5)_2$ CLXVII

$Cl-C_6H_4-CH_2-\overset{\oplus}{N}(C_4H_9)_3 \quad Br^{\ominus}$

CLXVIII

$CH_2=CH-CH_2-\overset{\oplus}{S}(CH_3)_2 \quad Br^{\ominus}$

CLXIX

Fig. 49. Growth retardants (continued).

CLXIII	Piperidyl
CLXIV	Pyrrolidinyl
CLXV	Morpholinyl
CLXVI	Dimethylamino
CLXVII	Diphenylamino
CLXVIII	4-Chlorobenzyltributylammonium bromide
CLXIX	Allyl dimethylsulphonium bromide

structure of the terpene moiety was also consistent with retention of activity. Thus high activity required the substituent on the benzene ring to be either methyl, isopropyl or tertiary butyl groups, although the various isomeric arrangements gave specific patterns of activity that varied from species to species (Cathey, 1965). The compound unsubstituted on the phenyl ring has a relatively low activity.

In the phosphonium compounds susbtitution on the benzene ring also has a marked effect on activity, halogen or methyl substitution at positions 3 and 4 giving the maximum activity. The compound with the unsubstituted benzene ring has low activity which is comparable with that of compounds having other groups such as cinnamyl or allyl instead of benzyl (Cathey, 1964*b*). The introduction of an oxygen bridge into the molecule to give 2,4-dichlorophenoxymethyltributylammonium salts reduce the growth activity (Knight *et al.*, 1969). The size of the other three substituents on the quaternary phosphorus atom seems even more critical, only the three butyl groups giving high activity. Larger or smaller aliphatic groups reduce activity. (Cathey, 1964*b*).

The sulphonium analogues of Phosphon-D are without activity (Knight *et al.*, 1969) although they produce strange morphogenic effects.

Turning to CCC and its analogues, an earlier report (Tolbert, 1960) suggested that only *aliphatic* quaternary ammonium salts were active as retardants and that the trialkyl substituent had to be a methyl group. However more recently chlorobenzyl- and naphthylmethyl-tri-*n*-butylammonium salts have also been shown to be active (Knight *et al.*, 1969). Again chlorine substitution on the benzene ring produces an activity pattern similar to that for corresponding phosphon-D analogues with a substitution in position 4 being particularly effective (Fig. 49 CLXVIII). The pattern is somewhat reminiscent of that for auxin activity of the substituted phenoxyacetic acids with one exception; the 2,3,6-trichlorobenzyl- analogue was active (cf. 2,3,6-trichlorophenoxyacetic acid which is an antiauxin). A sulphonium analogue of CCC, allyl-dimethylsulphonium bromide (Fig. 49 CLXIX) is a growth retardant, and, like CCC itself, is most effective on monocotyledonous plants (Knight *et al.*, 1969).

(iv) Mode of action

In many of their actions on intact plants the effects of the growth retardants are virtually the reverse of those of the gibberellins. They suppress the elongation of internodes whereas gibberellin promotes it; they prevent leaf expansion making leaves thicker and greener whereas gibberellins have the opposite effects. Furthermore they are often mutually antagonistic both in their actions on stem growth and on flowering (see Cathey, 1964*a*). This latter is dramatically shown in Plate 13. One might therefore suppose that they should be classified as 'antigibberellins; indeed they are often so called. But an antigibberellin, in the same way as an antiauxin, would be expected to be chemically similar to the hormone it is antagonizing and to act competitively with that hormone at the site of its action in the cell. Certainly there is little chemical similarity to be seen between the retardants and the gibberellins (or indeed any other known hormone). What evidence is there for competitive action?

In intact beans the interaction of CCC and Phosphon-D with gibberellin had all the hallmarks of a direct competitive action (see Fig. 50A), in strong contrast to the apparently gibberellin-independant inhibition by maleic hydrazide, and this led Lockhart (1962) to accept a direct antagonism in the growth system. However more specific tests on isolated systems do not support such an action. For example the inhibition of the growth of *Avena* leaf segments (a system very sensitive to the promotive action of the gibberellins) by Phosphon-D or AMO-1618 was only partially reversed by gibberellin (see Fig. 50B), although there was partial reversal by IAA also (Cleland, 1965). Even more convincing was the demonstration that no types of growth retardant had any effect whatever on what is perhaps the most specific of gibberellin effects, the induction of the synthesis of the starch-hydrolyzing enzyme, α-amylase, in barley endosperm (Paleg *et al.*, 1965).

On the other hand there are data implicating the growth retardants in auxin metabolism. Cleland (1965) suggested such an action from his results on *Avena* leaf growth. Kuraishi & Muir (1963) found similar results with

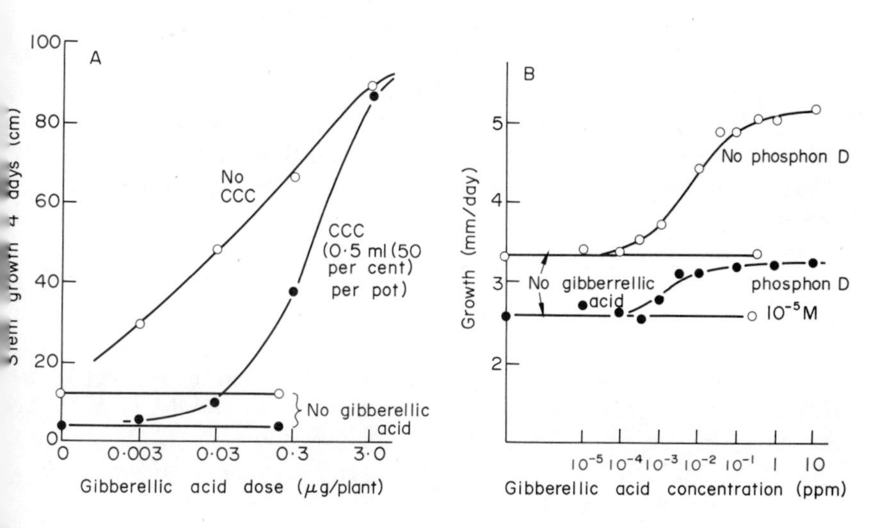

Fig. 50. Graph showing the interactions between gibberellic acid and the growth retardants in the control of extension growth.

(A) Interactions with CCC in the stem growth of Pinto bean seedlings. Gibberellic acid applied to the shoot; CCC applied to the soil. (Data from Lockhart, 1962.)

(B) Interaction with Phosphon D in the extension growth of segments of *Avena* leaf in water. Both gibberellic acid and Phosphon D applied in the aqueous growth medium. (Data of Cleland, 1965.)

Avena coleoptiles and pea stem segments, where the inhibition of growth by CCC was overcome by IAA but not by gibberellins. Furthermore in pea stems CCC reduced the content of diffusible auxins in the growing tips to one seventh normal. A similar reduction in the content of both IAA and tryptophan has been seen in wheat seedlings following CCC treatment (Norris, 1966). Such evidence poses the question as to how auxin levels could be reduced by the retardants. One probability is made evident by the work of Halevy (1963) who has shown that all five types of retardant applied to cucumber seedlings, increases their content of IAA-oxidase (see Chapter XI), the enzyme system which may play a part in controlling IAA levels in growing plants. An augmentation of IAA-oxidase activity would reduce IAA concentrations and hence IAA-dependant growth; such an inhibition could be reversed by applying IAA externally.

However the quite strikingly opposing actions of gibberellins and the quaternary retardants has encouraged much further work and it is now quite clear that one of the major actions of these growth retardants is to block the biosynthesis of the gibberellins. This was first demonstrated in the fungus *Gibberella* (*Fusarium*) *moniliforme* treated with CCC (Ninnemann *et al.*, 1964) and with AMO-1618 (Ruddat *et al.*, 1965*a*). A similar action of AMO-1618 and Phosphon-D was demonstrated in the endosperm and nucellar tissue of the wild cucumber (*Echinocystis macrocarpa*) (Dennis *et al.*, 1965), where the actual biochemical site of its operation was pin-pointed. This seems to be the ring-forming steps between the two intermediates *trans*-geranylgeraniol pyrophosphate and (–)-kaurene (see Chapter IV). The action of CCC seems to differ in that it comes later in the sequence between (–)-kaurene and gibberellin itself. In *Gibberella* however CCC seems to inhibit the formation of (–)-kaurene but not the conversion of (–)-kaurene to gibberellin (Barnes *et al.*, 1969). These blocking actions of the retardants are probably common to all higher plants since further evidence is now accumulating of the suppression of gibberellin levels in a number of species, e.g. in the seeds of *Pharbitis* after treatment of the parent plant with CCC (Zeevaart, 1966) and in the bleeding sap from the roots of balsam (*Impatiens glandulifera*) after CCC application and cutting away the shoot (Reid & Carr, 1967). An interesting feature of this last-mentioned work was that there was a changed pattern in the kinds of gibberellin produced, supporting the suggestion that CCC interferes in the synthetic pathway late in the sequence, i.e. after (–)-kaurene. Such a change in pattern would also explain some of the reports of a growth stimulation by CCC by switching of the biosynthetic pathway from production of he normal gibberellin to an abnormal (and incidentally more active) homologue.

However not all growth-retardant effects can be explained in terms of actions via the suppression of gibberellin synthesis. Thus there are examples of CCC-inhibition of plant growth which cannot be overcome by gibberellin applications, e.g. in pea and in crown-gall tissue of *Helianthus* (see Lang, 1970). Furthermore there are a number of examples of growth-stimulating effects of CCC, e.g. the promotion of stem growth in snapdragons (*Antirrhinum*) (Halvey & Wittwer, 1965), peas (Sebanek & Hink, 1966) and *Gladiolus* (Halevy & Shilo, 1970). In this last species it also increases the number of flowers per spike. Running parallel with these growth stimulations there has been observed *increases* in the gibberellin content of the tissues concerned, e.g. in peas (Carr & Reid, 1968) and in *Gladiolus* (Halevy & Shilo, 1970); in the latter case only one component of the total gibberellin was affected, namely the neutral water-soluble and not the acidic ethyl acetate-soluble fraction. Information so far available does not allow us to decide whether or not these are direct effects on gibberellin metabolism.

Then CCC has inhibitory effects on other aspects of plant metabolism, for example it interferes with respiratory processes in isolated mitochondria (Heatherbell *et al.*, 1966; De Leo *et al.*, 1968), an action which might well block energy flow into growth processes. It also disturbs choline metabolism in spinach leaves (Tanaka & Tolbert, 1966). Again all three growth retardants, Phosphon D, CCC and B995 strongly inhibit chlorophyll synthesis in the cotyledon of pumpkin (*Curcurbita pepo*) much more effectively than they inhibit growth (Knypl, 1970). The growth effects could be partially reversed by gibberellic acid (and by benzyladenine), which has no actions on the retardant inhibition of chlorophyll synthesis. This suggests that all these growth retardants have at least a dual action, one on growth and acting via gibberellin metabolism, and one on some component of the chlorophyll biosynthesis system which is unrelated to gibberellin (or cytokinin) action.

The interaction of the retardants with auxins are still puzzling but might find their explanation via the gibberellin system, since there is evidence to indicate that gibberellins may modify IAA metabolism, e.g. by suppressing the activity of IAA-oxidase. These possibilities will be further discussed when we deal with mechanisms of these hormones (Chapter XII).

The action of B995 seems to differ fundamentally from that of the quaternary retardants. It does not, for example, inhibit gibberellin biosynthesis in *Gibberella* (Ninneman *et al.*, 1964). One suggestion is that it may block the synthesis of IAA in plants, since the substance 1,1-dimethylhydrazine, which might be formed from B995 in plant cells, can markedly inhibit the oxidation of tryptamine to indol-3yl-acetaldehyde by enzymes prepared from pea tissue; in addition B995-treatment of pea tissue lowers the activity of these enzymes prepared from it (Reed *et al.*, 1965).

We must not however lightly dismiss the idea of direct gibberellin antagonism. One group of synthetic growth regulators demonstrated to have extensive effects on the growth and morphogenesis of plants are the morphactins.

We have already noted earlier in the chapter that they are extremely active inhibitors of auxin transport. One member of the group, 2-chloro-9-fluorenol-9-carboxylic acid, is also claimed to be a direct gibberellin antagonist. In pea plants treated with CCC to free them of native gibberellins, this substance antagonizes the action of applied gibberellins. Since it has no effect on gibberellin biosynthesis in *Gibberella*, the authors conclude that the morphactin must be a competitive gibberellin antagonist, i.e. a true anti-gibberellin (Ziegler *et al.*, 1966). The fact that there are certain similarities in their basic molecular structures would support this view.

Chapter VIII
Ethylene

Historical

In these days of highly sophisticated plant biochemistry, when the isolation and identification of new plant hormones demands the use of the most refined of modern chemical techniques, it may perhaps come as a surprise to the reader to learn that the earliest plant growth regulator to be recognized was a very simple organic substance, the gas ethylene. Although its effects can be dramatic and have been widely recognized for well over fifty years, belief in its role as a native regulator has never been universal; however a recent renewal of interest, based on more precise data, justifies the inclusion of a chapter, albeit short, on this still highly enigmatic regulator.

It is now over a century since it was recognized that illuminating gas (coal gas), which contains small quantities of ethylene, could cause extensive damage to plants. For example in 1864 in Berlin an escape of such gas injured road-side trees and the symptoms were described in a scientific journal by Girardin. However many years had to pass before the active constituent was recognized as ethylene. Then Neljubow in 1901 showed that the abnormal growth responses of seedling sprouts to laboratory air, i.e. stunting, stem thickening and prostrate habit, caused by an impaired response to gravity, could also be produced by ethylene. These three symptoms constitute what has come to be known as the 'triple response' to ethylene. Closely associated is an equally striking growth response known as *epinasty* (see Plate 6). This is most clearly seen in leaves, which may drop and curl downwards so that the blades assume a vertical instead of a horizontal position. This is not a wilting effect such as that which comes from the loss of leaf turgor when water supply is deficient; it is an active movement caused by the more rapid growth of the upper side of the leaf stalk (petiole) and sometimes of the main vein itself. The action of ethylene in causing this response was first shown by

Knight *et al.*, in 1910. The combination of these various responses (triple response and epinasty) can produce grossly abnormal plants, with short fat shoots and stunted roots, both often contorted into irregular spirals and showing no particular orientation with respect to gravity (Borgström, 1939) (see Plate 15).

Also in these early days studies were being made on the 'curing' of lemons. These fruit are gathered when they are still green and need to be 'cured' in a warm store before they ripen and turn yellow. This work showed that ripening could be greatly promoted by the gases produced from kerosene stoves (Sievers & True, 1912) the active component being established later as ethylene (Denny, 1924).

In the next few years ethylene was shown to exhibit other remarkable growth actions. At the end of their useful life the leaves of higher plants are shed by the separation of cells in a special 'abscission' zone, usually at the base of the petiole. It has been known for a long time that illuminating gas and tobacco smoke would induce premature shedding of leaves in house plants and in 1926 Wallace showed that this was due to ethylene which these gases contain. In 1927 Vacha & Harvey demonstrated that ethylene would break dormancy in potato tubers, in *Gladiolus* corms, in some seeds and in the buds of a number of hardwood cuttings. Then again it was accidentally discovered many years ago in the Azores that smoke would force early flowering in pineapple plantations and in 1932 that also was shown to be due to ethylene in the smoke (Rodrigues, A. G., quoted by van Overbeek, 1951). In many plants roots can be induced to form in unusual places, e.g. on stems or petioles, by continuous application of ethylene. This is a phenomenon controlled by plant hormones. In 1933 Zimmerman and Hitchcock showed that ethylene was very effective in promoting such adventitious root production in many species of plant. It can even induce branching in roots which normally never do so, e.g. the adventitious roots from the base of bulbs such as onion (Borgström, 1939) (see Plate 15). All these phenomena can be evoked by low concentrations of the gas; an extreme sensitivity is shown by the tomato plant, in which epinasty can be caused by concentrations as low as one part in 60 million parts of air (Croker *et al.*, 1935).

Is Ethylene a Natural Plant Regulator?

This question resolves itself naturally into two parts; first, is ethylene produced by plants or plant parts as a normal outcome of their metabolism and second, if this is so, does this natural production of ethylene play any part in the control of normal growth and development of plants?

The answer to the first question in undoubtedly yes! The first identification of ethylene as the volatile product of ripening apples was by Gane (1934) who trapped it in liquid bromine at –65°C and then identified it by reaction with aniline to form *N,N′*-diphenylethylenediamine. Then followed considerable studies of a range of fruits most of which appeared to produce ethylene as they ripened. However there were some apparent exceptions, of which mango and *Citrus* fruit are suitable examples (see Biale, 1960). Naturally these earlier studies were handicapped by relatively insensitive methods of identification, but recent refinements in technique, notably the adoption of gas/liquid chromatography (see pp. 45) has made it possible to show that ethylene production is a ubiquitous concommittant of ripening processes in all fruit studied. The most recent survey of these data is by Burg (1962). In this same review is set out the full range of plant organs in which ethylene production normally occurs. All vegetative parts of plants are probably capable of producing ethylene although whether they do so under normal growth conditions is still a moot point. Certainly damage of one sort or another can greatly promote ethylene evolution, for example in leaves which have been wounded or infected with virus or fungal parasites, or treated with toxic chemicals. Flowers also produce ethylene and this production is also greatly increased when they are plucked.

The answer to the second question cannot be so clear-cut, although it now seems that naturally-produced ethylene has a part to play in a number of physiological processes. The most well-founded role is that in fruit ripening, which has long been known to be 'contagious' under storage conditions; indeed it was the fact that shipped bananas ripened in 'pockets' in the storage chambers that led Ridley (1923) to demonstrate that the 'contagion' was a volatile product of the ripening fruit itself. Ethylene was first regarded as a natural ripening 'hormone' by Kidd & West (1933) but, in view of early doubts of the universality of the phenomenon, the view was not accepted and ethylene was proposed as an incidental product of biochemical ripening processes (Biale *et al.*, 1954). Furthermore there were doubts as to whether such a highly volatile product as ethylene could, under normal conditions, accumulate in tissues to levels, sufficient to produce the ripening response, such as are attained in artificial storage, where ethylene could accumulate in the ambient air. But, as we have seen, the advent of sensitive new techniques has changed the situation and it has been established for all fruits so far studied that just before the onset of ripening the internal concentration of ethylene rises to at least 0.1 ppm or higher and that this is quite sufficient to trigger off the ripening process (Burg, 1962; Burg & Burg, 1965).

The possible role of ethylene in the control of organ shedding (abscission) has long been a subject for debate, in which the pros and cons have been hotly argued. It is not intended to discuss these problems at this point; however it should be said that the main arguments against the participation of naturally-produced ethylene was, as for fruit ripening, the doubt whether sufficiently high concentrations would arise in the tissues concerned. Here again sensitive modern techniques of measurement have dispelled this doubt and have shown, at least in one species (*Phaseolus vulgaris*), a close parallelism between leaf-shedding and ethylene production under a variety of conditions (Rubinstein & Abeles, 1965). Ethylene may well be a natural regulator of abscission.

Very recently a possible role has been disclosed for ethylene in the control of shoot morphology during its passage through and emergence from the soil. During the early stages of germination the young shoot has to force its way through the soil and this exposes the delicate apex and young leaves to possible damage by friction with sharp soil particles. In many dicotyledonous plants this is minimized by the development of a plumular hook, whereby the brunt of the mechanical interactions between soil and shoot is borne by the 'neck', formed by the relatively sturdy stem itself, and the more delicate tissues of the leaves and apical growing point are dragged up in the channel made in the soil by this hook. When the hook reached ground level it 'opens' and the young leaves expand. One major factor triggering this hook opening response is light (see Plate 14 (1) and (3)), particularly that of the long wavelengths in the red end of the spectrum (660 nm) (Klein *et al.*, 1956), but it is now clear that light is not the only factor concerned with the control of hook behaviour. Thus in *Phaseolus*, a species much used for these studies, the triggering action of red light can be completely suppressed by very low concentrations of ethylene (0.1 ppm). This is shown for *Pisum* in Plate 14 (4). Then it was demonstrated that the hook would open *in the dark* if mercuric perchlorate (an ethylene absorbant) were included in the growth chamber, indicating that ethylene produced by the hook might be responsible for maintaining it in the closed position (Kang *et al.*, 1967). Ethylene is produced by the young plumule of pea (*Pisum*) apparently solely in the hook region and at a sufficient rate (6 μl kg^{-1} h^{-1}) to give a physiologically effective concentration in the issues (Goeschl *et al.*, 1967). Exposure to red light reduces the production of ethylene by the hooks of both *Phaseolus* and *Pisum* and this would then appear to be the basis of the opening action of the red light (Kang *et al.*, 1967; Goeschl *et al.*, 1967). A most interesting correlated phenomenon is the action of physical stress on pea plumule tissue; if young pea sprouts are forced to grow against a physical restraint such as a neoprene plug in a glass tube or are made to grow through closely-packed glass beads, then their production of ethylene is greatly increased above that of free-growing shoots (Goeschl *et al.*, 1966). The concentrations of ethylene thus produced (*c.* 0.14 ppm) are sufficient to induce the lateral expansion of the stems characteristc of the response to ethylene (see Plate 14 (2)). Such thickened sprouts would have an increased mech-

Plate 14. Ethylene responses in seedlings of *Pisum sativum*.

(1) Normal etiolated plants grown for 8 days in complete darkness and showing the characteristic apical hooks.

(2) Plants grown as in (1) but treated on the last two days with ethylene. The extension growth has been stopped but induced lateral expansion of the sub-apical region has given rise to swellings typical of ethylene action.

(3) Plants grown for 6 days as in (1) and then placed in light on the seventh day. This has caused a complete opening of the apical hook and the leaf blades have started to expand.

(4) Plants grown for five days as in (1) and then placed in light on the sixth day. This caused hook opening on the seventh day (as in (3)). They were then placed in ethylene in the light on the eighth day and the apical hooks reformed. (Photograph by kind permission of Dr. S. P. Burg.)

anical strength (proportional to the fourth power of the radius in a homogeneous cylinder), facilitating soil penetration. The picture which emerges is very persuasive of a key role of ethylene in these early phases of shoot development. The friction of the soil against the penetrating shoot maintains a high rate of ethylene production which in its turn keeps the plumular hook closed and promotes lateral expansion for improved penetrating power. On emergence the mechanical soil action on the hook disappears and this change, coupled with light action, causes a drop in ethylene production and consequent release of the hook-opening mechanism.

There seems little doubt therefore from observations such as these that ethylene is a natural plant growth regulator; whether it can be regarded as a hormone is still a subject for argument. If action at a distance is the necessary criterion then a case could be made out for ethylene as a hormone since it should obviously move easily in plant tissues by diffusion, although studies of its transport have not yet been attempted. However, more serious objections have been raised on the ground that it operates only through auxin, whose behaviour in tissues is supposed to be modified in such a way that the growth responses characteristic of ethylene are evoked. For example the confinement of the ethylene-induced lateral expansion of shoots to those regions where auxin concentrations are highest, suggested that ethylene action depended on the presence of auxin. In fact, in a comprehensive analysis of the information available at that time Borgström (1939) was able to explain all ethylene responses in terms of an auxin 'leak' from its normal channels of polar transport, with consequent loss of growth control by both internal and external factors (see Chapter XI for the evidence of control *via* auxin transport). The idea of an action obligately dependant on auxins persisted for well over two decades and has stifled any suggestions that ethylene could be a hormone in its own right.

But again new studies with sensitive ethylene-measuring techniques brought a disturbing suggestion that our obsession with auxins may have made us look at the whole question the wrong way round and that many auxin responses may be due to the ethylene which has been induced to form under the action of the applied auxin. There is now much experimental evidence in support of this possibility. It was first shown that the synthetic auxin 2,4-D induced the production of ethylene by cotton plants, which are particularly sensitive to this compound. The insensitive species *Sorghum* could not be induced to form ethylene (Morgan & Hall, 1962). Similar responses were later evoked in cotton by the application of the natural auxin IAA in doses sufficient to induce epinastic and other responses characteristic of ethylene action (Morgan & Hall, 1964). Then it was shown that the concentration of IAA necessary to inhibit the growth of pea and of sunflower stem segments were precisely those which produced a marked stimulation of ethylene production (Burg & Burg, 1966, 1968) (see Fig. 51). This auxin action does not seem to arise from metabolic damage or other toxic effects but from the direct induction of an enzyme concerned in ethylene production, since it is blocked by inhibitors specific for protein (and hence enzyme) synthesis (Abeles, 1966).

This new light on auxin/ethylene relations has revealed that many responses, so far regarded as auxin-mediated, may in fact be ethylene responses. The inhibition of extension growth and the promotion of lateral expansion of main axis tissue by 'supra-optimal' auxin concentrations were the first to be implicated (Burg & Burg, 1966, 1968; Warner & Leopold, 1967).

The opening of the plumular hook of seedling sprouts in red light is inhibited by auxins, apparently by their promotion of ethylene production (Kang *et al.*, 1967, Kang & Ray, 1969). Applied auxins will suppress flower-

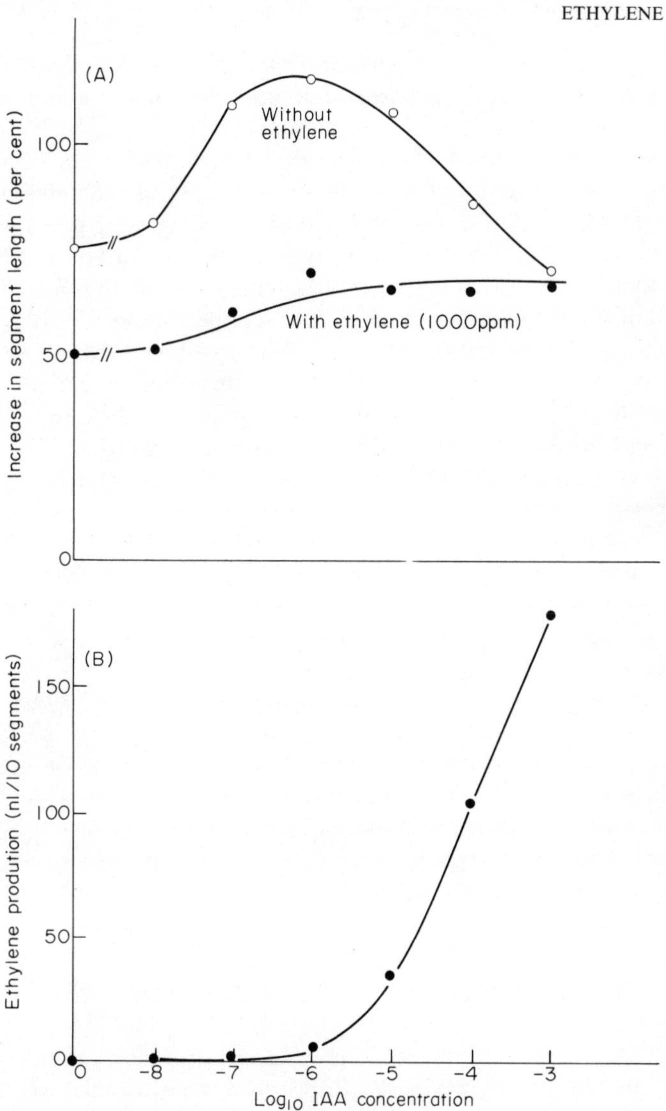

Fig. 51. Graphs showing the relationships between the concentration of applied auxin (IAA) and ethylene production in sub-apical segments excised from etiolated hypocotyls of sunflower (*Helianthus annuus*). The upper graphs (A) show that the segment response to auxin is almost completely eliminated by the application of 1000 ppm ethylene. The lower graph (B) shows that ethylene production by the segments begins at a concentration of applied auxin corresponding precisely with the point of maximum growth response and that the progressive decline in growth with increasing (supraoptimal) concentrations of auxin is exactly paralleled by increasing ethylene production (data from Burg & Burg, 1968).

ing in many plants (e.g. the cocklebur, *Xanthium pennsylvanicum*) and promote it in a few others (e.g. pineapple). In both these plants effective auxin concentrations promote ethylene production to levels which, by themselves, may produce similar flowering responses (Abeles, 1967). In cucumber, auxin applications promote the formation of female *vis-à-vis* male flowers (see Chapter X). Ethylene has the same action, suggesting that the auxin effect may be via the stimulation of ethylene production (Nitsch, 1965). In the growth of roots, which is particularly sensitive to auxins, a 50 percent inhibition of elongation being induced by concentrations of IAA or 2,4-D as low as 1 ppm, these same low concentrations cause an ethylene production sufficient by itself to account for the root growth inhibition, and it has been suggested that the normal control of root growth and its responses to gravity (geotropism) may be mediated by ethylene (Burg & Burg, 1968; Chadwick & Burg, 1970). There is also some evidence that plant growth regulators other than auxins may exert some of their actions via a stimulation of ethylene release. For example kinetin and benzyladenine treatments, which release pea-nut seeds from dormancy, bring about an enhancement of ethylene production by those seeds (Ketring & Morgan, 1971); such results suggest that the active agent here is again ethylene and that cytokinin action is therefore an indirect one.

Impressive as these data are it must be remembered that all these phenomena, with the possible exception of plumular-hook opening, are associated with applied concentrations of auxins not established as normally present in plant cells growing under natural conditions and it is difficult to decide whether or not these phenomena involving ethylene play any part in the normal growth of the plant, or, in other words, whether we are dealing here with normal hormone control.

The Effect of Ethylene on Auxin Physiology

In the last section we have seen that new observations have raised the question as to whether many auxin activities cannot be attributed to ethylene induced to form in high auxin concentrations. At the same time recent work has confirmed the far-reaching effects of ethylene on auxin physiology.

In the pioneering years of auxin studies the possibility of ethylene interference had already been recognized when van der Laan (1934) demonstrated that exposure to ethylene would suppress the amounts of diffusible auxin that could be obtained from seedlings of *Avena sativa* and *Vicia faba*. Such an effect could have arisen in three ways. Firstly ethylene could have promoted the destruction of auxin in the tissues (see Chapter XI). This has been rejected by some workers since the concent of radioactive carbon (^{14}C) from labelled IAA fed to segments of pea stems was not affected by ethylene (Burg & Burg, 1966). However if cotton plants are pretreated for 3 to 15 hours before segment excision, then the rate of decarboxylation of IAA-1-^{14}C was considerably increased at the cut surfaces, presumably by an increase in the activity of IAA-oxidase there (Beyer & Morgan, 1969). The relevance of these findings to intact tissues is not clear. Secondly there could be an inhibition of the biosynthesis of auxin; this is much more probable since studies on cell-free preparations from stems of *Coleus* and *Pisum* have shown that the enzymatic release of $^{14}CO_2$ and production of auxin from labelled tryptophan (see Chapter XI) by enzymes from ethylene-treated tissue are much less than the corresponding activities of enzymes from untreated tissues (Valdovinos, Ernest & Henry, 1967). Finally transport, which can regulate the amount of diffusible auxin collected from tissue, might be modified by ethylene. We have already seen that a possible action of this sort was the basis of early theories of ethylene action (Borgström, 1939). Modern studies have to some extent given support to such thinking. Thus, although ethylene seems to have no *direct* action on the rate of IAA movement through isolated pieces of hypocotyl (*Helianthus, Phaseolus*), coleoptile (*Zea*) or petiole (*Gossypium, Phaseolus, Coleus*) (Abeles, 1966), yet it does disturb the organization of intact plants such that auxin movement is impaired *subsequent* to ethylene treatment (Morgan & Gausman, 1966; Beyer & Morgan, 1969). This is still a puzzling situation and calls for a more careful analysis.

Nevertheless in spite of the obvious lacunae in our knowledge, there seems little doubt that IAA and ethylene can mutually influence each other's behaviour and this has led to the suggestion of their function in a kind of feedback mechanism of control (Burg & Burg, 1966). Thus IAA, in rising to levels likely to impair the growth mechanism, could induce the formation of ethylene, which, in its turn, would suppress further auxin synthesis and set a limit to the concentrations reached. How far such a phenomenon might participate in the homeostatic control of the processes of normal growth and development is still far from clear; much more careful physiological studies are needed.

The Biogenesis of Ethylene

The biochemical route whereby ethylene is normally produced in plants is still very much of a mystery. The use of glucose labelled with ^{14}C in different positions in the molecule led Burg & Burg (1965) to postulate its production from fumaric acid, which is one of the key intermediates in respiratory metabolism. Lieberman & Mapson (1962, 1964) have shown that the fatty acid, linolenic acid, can act as a substrate suitable for ethylene formation either in an enzyme system prepared from apples or in a non-enzymic system containing ascorbic acid and small amounts of copper. Later it was shown that the amino acid methionine could serve as a substrate for ethylene production by apple slices, being produced apparently by oxidation in a kind of peroxidative copper-containing enzyme system (Lieberman *et al.*, 1966). How any of these systems can be affected by auxin is still a problem for the future.

Molecular Requirements for Ethylene-like Activity

Other unsaturated olefine molecules produce effects similar to those of ethylene. Apparently only unsaturated hydrocarbons of low molecular weight can exert ethylene-like action as judged from the inhibition of pea stem growth. There must be an unsaturated bond adjacent to a terminal carbon atom. There is indirect evidence that the action of these unsaturated olefines involve complexing with a metal-containing site in the cell, possibly an enzyme intimately concerned with growth and presumably fruit-ripening processes (Burg & Burg, 1967). This implied that either growth and ripening are biochemically related or that ethylene action is non-specific and effective at many different sites in plant cells.

(2-Chloroethyl)phosphonic Acid (Ethrel, Ethephon)

In recent years (2-chloroethyl)phosphonic acid has been shown to produce the full spectrum of ethylene effects when applied to plants in aqueous solution at concentrations of the order of 100 to 5000 ppm (see Plates 6 and 15 (B)).

Plate 15. (A) The effects of ethylene on the growth of onion roots. The roots were grown continuously for fourteen days in tap water through which a very slow stream of coal gas was bubbled (a few cm^3 per hour). Three typical responses to the ethylene in the coal gas are clearly shown. Firstly the growth of the roots is stunted. Secondly there is the loss of a capacity to respond to gravity. The roots grow pointing randomly in all directions and one shows the closely correlated phenomenon of strong curling into a 'corkscrew' (bottom right). Thirdly the roots are branching, a phenomenon not observed in normal onion roots. This is best seen in the two roots towards the top right of the picture and also less clearly with other roots in the centre of the clump.

(B) The effect of ethrel on the growth of onion roots. Roots were placed in a solution of 0.05 percent ethrel when about 2 cm long and were then grown continuously in the same solution for about 5 weeks. The same stunted and distorted growth of the first-formed roots is seen as for ethylene (in (A)) together with the induction of a considerable number of lateral roots (also very unusual for onion). In the later stages these laterals have grown out when the ethrel had been entirely decomposed and no further ethylene was being produced.

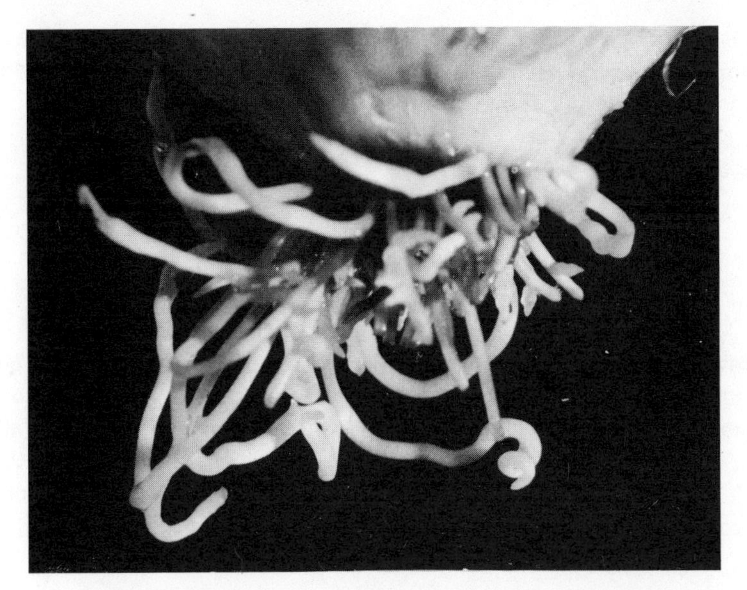

(A)

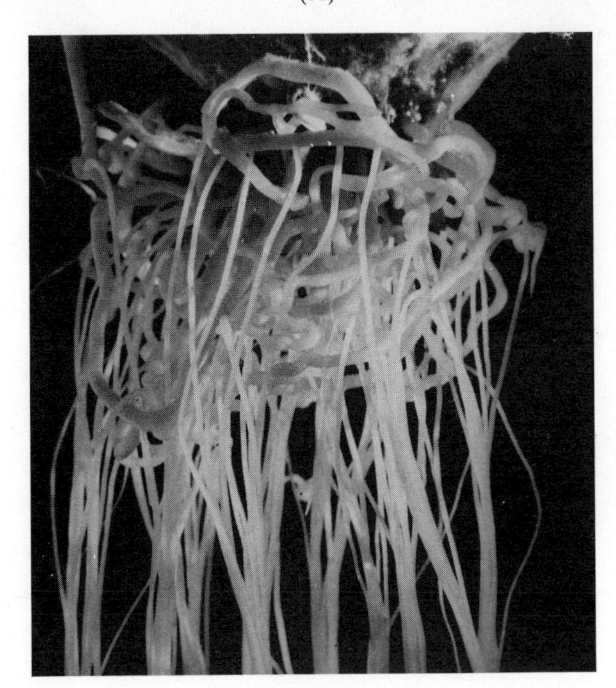

(B)

This is due to the release of ethylene by a base-catalysed elimination reaction which takes place at pH values above 4.0, i.e. in the cells of the plant which are normally more alkaline than this. The reaction is as follows:

$$\text{Cl-CH}_2\text{-CH}_2\text{-}\overset{\displaystyle \text{O}}{\overset{\|}{\underset{\underset{\displaystyle \text{O}^-}{|}}{\text{P}}}}\text{-OH} + \text{OH}^- \longrightarrow \text{CH}_2 = \text{CH}_2 + \overset{\displaystyle \text{O}}{\overset{\|}{\underset{\underset{\displaystyle \text{O}^-}{|}}{\text{P}}}}\text{-(OH)}_2 + \text{Cl}^-$$

This compound is being marketed as a practical growth-regulating compound under the name *ethrel* to effect, in crop plants, growth and other physiological responses similar to those which could be produced by ethylene gas (see Amchem Technical Service Data Sheet. Subject H-96, Ethrel, April, 1969). The practical application of this regulator is not a subject for this volume.

Chapter IX

Growth Substances in the Soil

Introduction

Soil is the normal substrate for the growth of all higher plants. On its constitution and qualities depend the yield of all vegetation, whether natural or planted by man. It is the almost exclusive source of all the mineral elements needed by the plant for its biochemical machinery and the construction of its physical body; the efficiency of roots in abstracting these essential salts from the soil is therefore a most important factor in plant growth control in natural communities and in man's planted crops. We cannot embark here on the subject of the mineral nutrition of the plant and the supply and function of the many inorganic salts necessary for normal plant development. While the majority of plant physiologists would hold that provision of a properly balanced mixture of such inorganic nutrients is all that is required by plants to give maximum growth, and while this may well be true, yet in the past many claims have been put forward for beneficial effects of organic compounds present in natural manures, and recent evidence suggests that these claims cannot be dismissed as lightly as has previously been the custom. The soil may also acquire, from many sources, specific growth substances that may exert a marked effect on the growth and development of plants rooted in it. We have already seen (Chapter VI) that many seeds are a rich source of inhibitors. Soil is therefore a repository of a vast number of organic compounds, many of which may have far-reaching actions on the growth and development of rooted plants. In the present chapter it is hoped to draw a rough picture of the complexity of this 'soil biochemistry', and to indicate what part organic growth regulators of various types and specificities may play in the growth, not only of the higher plants rooted in it but of the many kinds of micro-organism that contribute so variedly to soil fertility.

The Biochemistry of Soil

The ultimate source of all organic matter in the soil is, of course, the dead plants and animals that accumulate in it. The complex proteins and other compounds that make up the structure of the dead organism do not stay long as such, but are rapidly decomposed by some component of the soil microbial population to liberate a vast complex of simpler organic (and inorganic) products into the soil: these, in turn, may be attacked by other organisms and further broken down into simple inorganic salts, the gas carbon dioxide, and water. During this complicated sequence of events, populations of many different organisms may wax and wane as the availability of the particular substances serving them as a source of raw materials and chemical growth energy similarly rises and falls. Moreover, the micro-organisms themselves die, and their proteins enter into the cycle and serve as food for their successors.

Thus, the structure of the soil microbial population will be extremely complex and determined very largely by the nature and amount of organic food available. Many different organisms will have the same nutritional requirements and this will naturally result in intense competition for food. Here other soil factors, such as acidity and the accessibility of oxygen from the atmosphere, will exert a controlling influence and determine the success or failure of any one class of organism. As the chemical processes employed to obtain energy and raw materials from any given organic source will differ from one type of organism to another, it will be seen that these soil factors will affect the balance of the breakdown products and hence the whole subsequent sequence of soil populations: in this way the final end-products of the breakdown process may differ very widely. To quote an extreme example, in the waterlogged soil of marshes, where oxygen is virtually lacking, the gas methane is formed from the decomposition of organic carbon compounds – instead of carbon dioxide and water which are normally formed. There may be present, in the decomposing matter, substances that resist breakdown, and these may accumulate temporarily in the soil – until suitable organisms capable of attacking them have grown in sufficient numbers. The inhibitor from *Encelia farinosa* leaves is a case in point (see later, p. 236).

During the growth of soil micro-organisms, specific by-products of their particular metabolism will arise. Many soil organisms, such as fungi and actinomycetes, can synthesize large amounts of compound (antibiotics) which are specific growth inhibitors of other micro-organisms. In Chapter II it was noted that IAA was similarly produced by certain moulds. Gibberellins and similar growth-promoting compounds such as helminthosporol (see Chapter IV) are produced by fungi which may be soil inhabitants. It is possible that these by-products may also be formed in soil and will diffuse out from the cells into the soil solution, there exerting a regulating effect on the growth of roots or soil micro-organisms (Brian, 1949).

Micro-organisms are not the only living things that can contribute simple, soluble organic molecules to the soil solution. A large variety of such compounds can diffuse out from the roots of higher plants. Amino acids, the building bricks of proteins, have been demonstrated by several independent experimenters to pass out into the soil in this way (Katznelson *et al.*, 1954, 1955; Linskens & Knapp, 1955; Rovira, 1956; Rovira & McDougal, 1967). It has long been a practice to grow leguminous plants such as clover side by side with non-leguminous crops, for the latter grow much better as a result of the association. In the nodules which are caused to form on the roots of these leguminous plants after invasion by specific associated bacteria (Rhizobia), atmospheric nitrogen is converted into organic nitrogen compounds. Here there seems little doubt that the benefit derived by the associated non-leguminous crop is from the augmented nitrogen supply provided by amino acids diffusing from the leguminous nodules (Nicol, 1934). Modern studies using radioactive amino acids have shown that they can be removed intact from the soil by plant roots and utilized in growth, e.g. arginine by sugar-cane (Nickell & Kortschak, 1964). Pea roots have also been shown to release nucleotides and flavanones* into the external solution (Lundegårdh & Stenlid, 1944). The vitamins, biotin and thiamin, can be released by flax roots (West, 1939). Sugars are also excreted. A recent comprehensive survey of the literature (Rovira & McDougall, 1967) reveals that a high proportion of the low molecular weight intermediates of plant metabolism (e.g. 10 sugars, 21 amino acids, 11 organic acids, 10 vitamins) have been shown to exude from plant roots. Even macromolecules such as the enzymes invertase (Knudson, 1920), amylase (Knudson & Smith, 1919) and phosphatase (Rogers *et al.*, 1942) seem to escape in the same way.

There is no doubt, therefore, that the soil solution, in addition to inorganic mineral salts, may contain a very considerable variety of simple organic molecules arising in the metabolism of the living organisms which inhabit it, and it is very likely that many of these compounds may play a vital role in the intense competition that goes on there. Experience shows, however, that organic compounds, no matter how toxic, will eventually disappear from the soil, and this disappearance seems to depend on the activity of soil micro-organisms that are insensitive to their toxic actions. What evidence there is suggests that, in response to the appearance of any particular organic com-

*Nucleotides are complex organic molecules containing phosphorus. They are the units from which the nucleoproteins of the cell nucleus are very largely constructed. The flavanones are yellow plant-pigments very closely related to the flavones.

pound in the soil, a new population of a specific micro-organism, capable of attacking it, will arise from the vast and infinitely varied pool of the normal population, will proliferate to a peak, and will then decline to normal levels when the source of its energy, the particular organic molecule, is used up. Thus an appreciable accumulation of these simple organic molecules, be they hormones, vitamins, amino acids, sugars or specific growth inhibitors, will not be usual in ordinary unsterilized soils. In such soils the most that can be expected is the maintenance of a low level of concentration, the outcome of the establishment of a balance between production and breakdown, although these low concentration levels may still be high enough to have a decided effect on plant growth. Only in soils where conditions (e.g. waterlogging) do not favour the growth of the particular micro-organism involved in the decomposition, may concentrations reach a high level.

Growth Substances and Soil Micro-Organisms

(A) INTERACTIONS BETWEEN MICRO-ORGANISMS

From what has been so briefly outlined above, it should have become apparent that the various populations of soil micro-organisms are in a condition of very delicate equilibrium, maintained by a variety of factors. The most obvious of these is their relative success in the competition for the available nutrients. In addition the availability of specific growth factors may discriminate between one class and another. Thus a high proportion (between a sixth and fifth) of soil bacteria seem not to be able to synthesize their own supply of vitamins such as thiamin and biotin. About 7 percent require vitamins B_{12}*. Smaller numbers of species require other well-known vitamins such as nicotinic acid, riboflavin, pantothenic acid, folic acid etc. (Lochhead, 1958). It seems possible that these organisms may receive these factors from other soil organisms which do synthesize them. Another factor of such a kind is the 'terregens factor'. This is produced by the soil bacterium *Arthrobacter pascens*, and is essential for the growth of another soil organism. *Arthrobacter terregens* (Lochhead & Burton, 1953). It seems to be a peptide and is active at a concentration of 1 part in 10^9 (Burton *et al.*, 1954).

In such ways the growth of one species of micro-organism might assist the growth of another and not antagonize it. This equilibrium is also very much under the influence of conditions such as soil acidity and degree of aeration, factors which the soil organisms themselves may partially control. It may

*Vitamin B_{12} is the anti-pernicious anaemia factor present in liver and recently found to be synthesized by numerous soil micro-organisms. It is a complex organic compound containing the metal cobalt.

also be affected by the nature of the debris from the plants growing on the surface, as active bacterial growth inhibitors have been demonstrated in leaves and roots of many flowering plants. (See Winter, 1952.)

The second factor which may determine microbial equilibrium in the soil is their susceptibility to or tolerance of inhibiting metabolic by-products of other organisms. Such substances are called *antibiotics* and the phenomenon itself *antibiosis*. This phenomenon, although familiar to botanists for a long time, came into great prominence during the Second World War, when it was shown that a substance, produced by a strain of the mould *Penicillium notatum*, completely inhibited the growth of a large number of bacteria causing disease in man. The fact that this compound, later christened penicillin, was harmless to man himself, enabled it to be injected into infected individuals, with the startling curative results that the general public have long since come to accept as commonplace. Thus, a vast new branch of biological science sprang up almost overnight, and, in the intervening years, a multitude of moulds and other micro-organisms have been studied and their chemical products tested for inhibiting action on pathogenic bacteria. The most important of these new drugs are penicillin from *Penicillium notatum* and streptomycin from the soil micro-organism *Streptomyces griseus*.* (See Plate 16.) The chemical structures of some of these compounds have been worked out, and some have even been synthesized, e.g. penicillin and chloromycetin. What is striking about them is that they are all of diverse chemical structure,† and it is not surprising, therefore, that some are quite specific in their action on different classes of organism.

Now these antibiotic inhibitors are in a somewhat different class from those described in Chapter VI; like the hormones those inhibitors were active in the higher plants in which they were produced. Here the inhibitors are by-products of metabolism to which the generating organism is often very largely or sometimes completely indifferent; such indifference is not shared by other plants with a different biochemical constitution. In these latter sensitive organisms the by-product, owing to its moelcular shape, clogs up certain essential enzyme systems, and so prevents the normal growth processes from going forward. The clogging action arises because the by-product has certain structural characteristics closely resembling or identical with those of normal essential metabolic reactants (metabolites). They can therefore compete successfully with these metabolites for enzyme surfaces and are therefore placed in the general category of *anti-metabolites*. Such

**Streptomyces griseus* belongs to that group of organisms known as Actinomycetes. These are very simple organisms that are closely related to certain groups of bacteria. Their natural habitat is the soil.

†The interested reader will find the subject treated in detail in Waksman (1948), Brian (1951) and Gottlieb & Shaw (1967).

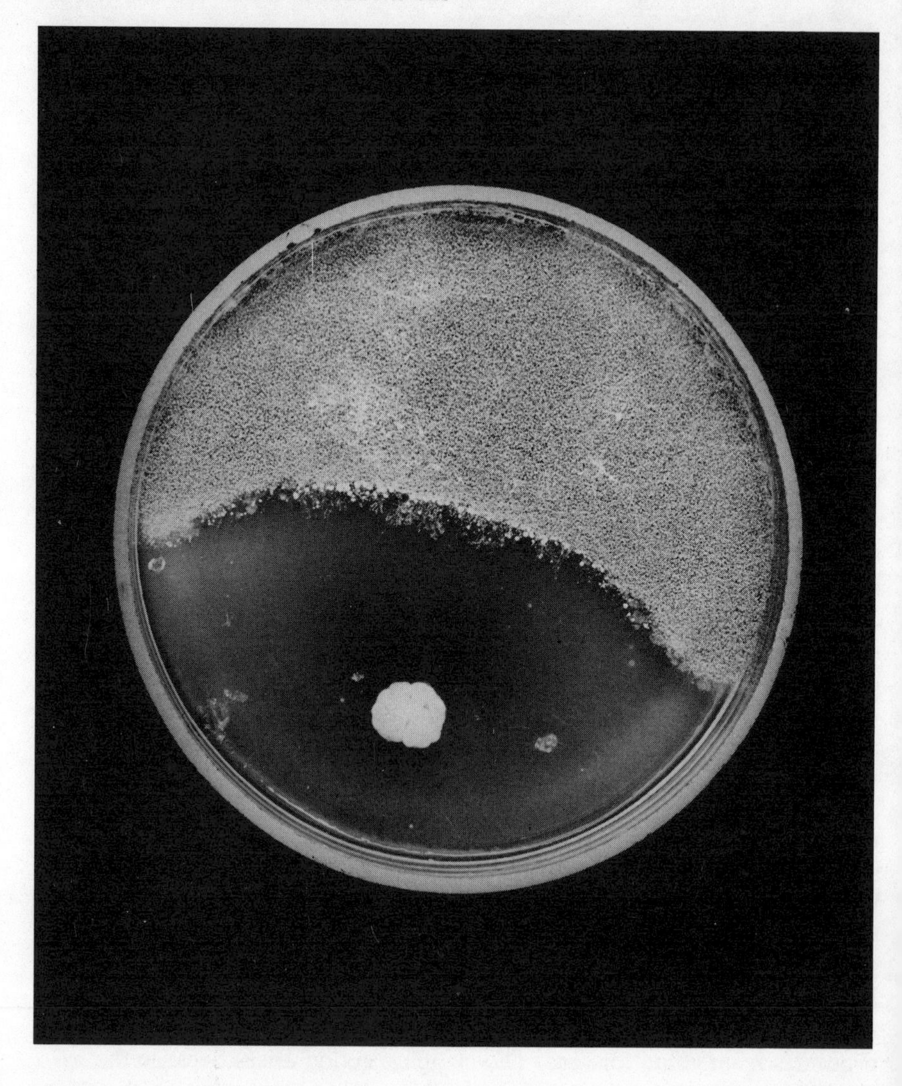

Plate 16. The phenomenon of antagonism in the growth of micro-organisms. The photograph shows a Petri dish of nutrient agar on which two micro-organisms are growing. The large crescent-shaped colony is of the rapidly-growing onion-disease fungus *Botrytis allii*. The smaller colony is of the slower-growing soil organism *Streptomyces* sp. In the absence of the latter the plant pathogen would have covered the dish, but all growth has been prevented here in an area surrounding the *Streptomyces*. This is due to the diffusion of a fungistatic substance from the cells of the latter into the agar, forming a wide circular area of growth inhibition around the colony. (Photograph by kind permission of Prof. P. W. Brian.)

enzyme clogging may happen in the organism producing the antibiotic but not until much higher concentrations are reached. The reason for the insensitivity of the generating organism to its own metabolic by-product is by no means clear and probably varies from organism to organism. It is possible, for example that the particular sensitive enzyme system is not essential for its own growth, or that it possesses an alternative non-sensitive enzyme system that is not inhibited by the antibiotic. Whatever the underlying causes may be, the final effect is clearly apparent. The organism, which by some happy evolutionary chance has acquired the power of synthesizing and, at the same time, growing in the presence of such growth inhibitors, will gain a great advantage over its competitors in the acquisition of nutrients. In the soil, for example, in the absence of antibiotic production, the efficiency with which it can utilize the limited nutrient supply will determine the success of any one species of organism. If, on the other hand, it suppresses the growth of any of its competitors by means of a chemical inhibitor, it gains an overwhelming advantage in the struggle for survival. The significance of such chemical factors in the continuous competition of soil organisms has been somewhat overlooked in the past.

There is no doubt that a wide range of antibiotics are excreted by organisms into the soil. This can be easily demonstrated in sterile soils if supplementary organic matter is supplied to promote the growth of the secreting micro-organisms (Grossbard, 1952). In non-sterile soil, because the antibiotics are rapidly decomposed by other organisms, they cannot normally be demonstrated except when organic supplementation is extremely high so that antibiotic production is boosted to such an extent that it exceeds destruction and some accumulation can take place (see Brian (1957) for details).

Much scepticism has therefore been expressed about the possibility of *effective* antibiotic production in soils. It seems possible however that antibiotics may occur locally in quite high concentrations at specific spots – e.g. around the excreting organism. These concentrations may be high enough to suppress the growth of fungi in the neighbourhood, but not high enough to allow detection by present methods of bulk extraction and assay (Jefferys & Brian, 1954). Such a state of affairs has been indicated in sterile soil inoculated with various species of actinomycetes. Although no antibiotic could be extracted, yet the growth of a test fungus, *Helminthosporium sativum*, on agar films inserted into the soil, showed growth distortions characteristic of antibiotic action (Stevenson, 1954, 1956). This seems clear proof of localized antagonism in otherwise sterile soil.

The production of fungistatic* substances has been demonstrated in

*A fungistatic substance is one that prevents the growth of fungi (i.e. an antifungal antibiotic).

natural soil. Activated charcoal, buried in cellophane bags in Wareham soil, adsorbed these compounds, which could be demonstrated on subsequent extraction of the charcoal (Jefferys & Hemming, 1953). In sterilized Wareham soil inoculated with *Trichoderma viride*, gliotoxin† production has been demonstrated by paper partition chromatographic methods. Composts greatly stimulate its production even in non-sterile soil (Wright, 1954*a*). The gliotoxin may even accumulate in composting material, as it has been found possible to extract it from straw buried for some time in a soil inoculated with an appropriate fungus (Wright, 1954*b*). Similar accumulation of gliotoxin takes place in the seed coats of white mustard and peas in soils infected with *Trichoderma viride* (Wright, 1956).

There is also indirect evidence that local concentrations of antibiotics may control the growth of parasitic organisms of roots. The root is such a rich source of nutrients which diffuse into the adjacent soil (see later section). In this soil film antibiotic-producing organisms proliferate on these nutrients and may chemically prevent the growth of parasitic organisms on the roots themselves (Brian, 1957). Interestingly enough plant parasites seem to be particularly sensitive to antibiotics (Garrett, 1950). Indeed one day we may be able to effect the control of plant diseases by the cultivation of suitable antibiotic-producing organisms in the soil (Brian, 1954).

Many other factors will, however, disturb the balance of pathogenic and non-pathogenic organisms in the soil, and thereby determine the incidence of disease in plants. In East Anglia the root rot of pines caused by *Fomes annosus* spreads much more rapidly on alkaline than on acid soils (Rishbeth, 1950). This results from the fact that acid conditions greatly stimulate the growth of *Trichoderma viride*, which antagonizes the growth of the pathogenic *Fomes* and greatly reduces its spread and the correlated disease incidence. There are now many examples in the literature of antagonistic action of saprophytic fungi such as *Trichoderma viride* on plant pathogenic fungi in the soil, resulting in a great reduction in disease incidence. The most important have been reviewed by Brian (1949), Garrett (1950), and Wood & Tveit (1955). These considerations underlie the practice of seed 'bacterization', which has received considerable attention in Russia. By incubation treatment of moist seed, the natural microbial flora of the seed coat is increased. The antagonistic action of these non-pathogenic organisms on the pathogens in the soil is expected thereby to reduce attack when the seed is sown. Again, in certain soils heavily infected with potato-scab disease, 'take-all' disease of wheat, or root rot of cotton, the incorporation of green manures has been used with considerable success to disturb the

† Gliotoxin is an antibiotic produced by *Trichoderma viride* and *Aspergillus fumigatus*, both soil fungi. Gliotoxin is also fungistatic.

microbiological equilibrium in the soil and encourage the growth of harmless antagonistic organisms. Much consideration has been given to the possibility of controlling soil-borne diseases by the inoculation of soil with suitable harmless antagonistic micro-organisms. A critical review (Wood & Tveit, 1955) states that, so far, no great measure of practical success has been achieved. The field of study shows great promise, however, in that an elucidation of the ways in which antagonistic organisms may shift biological equilibria in the soil in favourable directions, may point the way to efficient chemical or cultural methods for doing the same thing.

A marked effect of this type has been demonstrated in soils in an afforestation area in Wareham Heath in Dorset, England, where pine seedlings usually fail to become established. This was investigated over a period of years by Rayner and by Neilson-Jones (see Rayner & Neilson-Jones, 1946), who found that not only did roots and shoots become stunted, but that the normal association of the roots with their specific mycorrhizal fungi (see later section) was greatly disturbed. It was suggested that this poor development of the plant and its associated fungus was caused by the presence of a toxic compound produced by a soil micro-organism, probably a species of *Penicillium* (see Brian, 1949). Sterilization of the soil by heat or by disinfectants killed this fungus, and subsequent inoculation of the soil with a suitable mycorrhizal fungus allowed the vigorous growth of pine seedlings. This same effect could be obtained, without sterilizing, by treating the soil with certain composts which presumably acted by so altering the microbiological equilibrium in the soil that the growth of the toxin-producing Penicillia was suppressed.

However there is very little direct evidence that antibiotic production is a significant factor in determining the levels of those micro-organisms necessary for high soil fertility. In culture in the laboratory most antibiotics tried have inhibited the growth of such important soil organisms as the nitrogen-fixing and nitrifying bacteria at very low concentrations. But in soil studies concentrations may have to be increased thousands of times before significant effects can be obtained. In the case of some antibiotics that are basic in nature (e.g. streptomycin) this may be the result of absorbtion on to soil colloids so that concentrations of free and effective antibiotic may be greatly reduced; not such high soil concentrations of the acidic antibiotics, e.g. actinomycin, are needed to suppress bacterial growth (see survey by Pramer, 1958).

Antibiotic action is not, moreover, confined to the interactions between micro-organisms. A large number of species of lichens contain specific antibiotics, the significance of which in the life of these plants is quite obscure. (See review by Bustinza, 1954.) Higher plants may also have in their cells substances which will prevent the growth of certain classes of micro-organism. For example, garlic (*Allium sativum*) produces allicin to which the formula (Fig. 52 CLXX) has been given, and which is inhibiting to a very wide range of bacteria. Many families of higher plants contain species which possess similar antibiotic substances, particularly in their seeds and fruits. Lists of such species have been compiled by Ferenczy (1956) and Nickell (1959). The latter contains well over a thousand species from 157 families and shows that the antibiotics they contain can be effective against a wide range of micro-organisms, i.e. bacteria, fungi and even viruses.

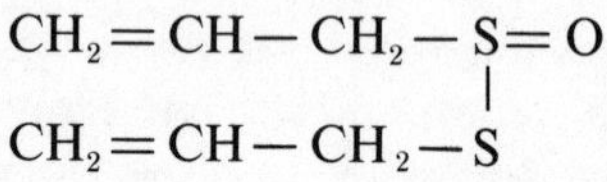

$$\begin{array}{l} CH_2{=}CH{-}CH_2{-}S{=}O \\ \qquad\qquad\qquad\quad | \\ CH_2{=}CH{-}CH_2{-}S \end{array}$$

CLXX

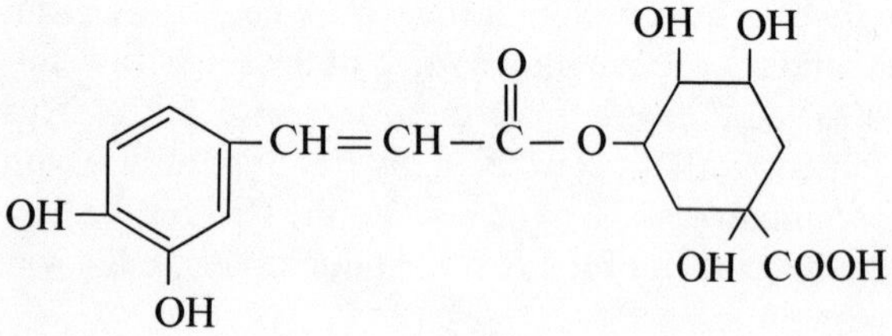

OH OH
CH=CH–C(=O)–O–
OH
OH
OH COOH

CLXXI

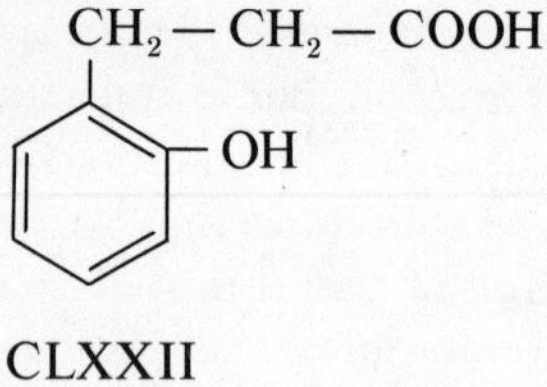

CH_2-CH_2-COOH
OH

CLXXII

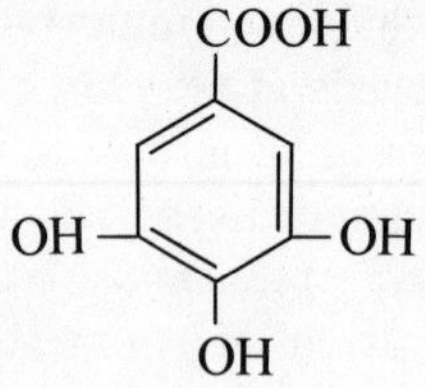

COOH
OH OH
OH

CLXXIII

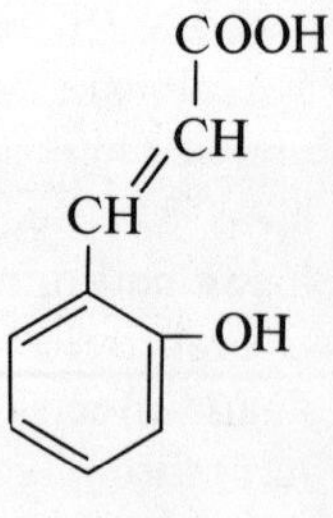

COOH
CH
CH
OH

CLXXIV

CH=CH–CH=CH–COOH

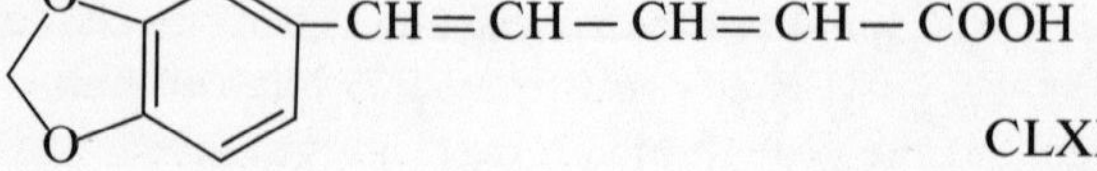

CLXXV

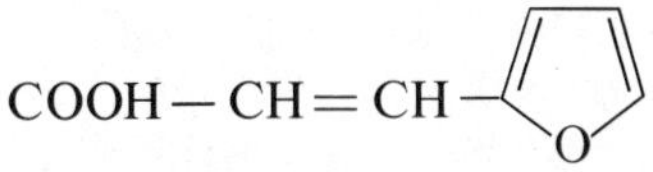

$COOH-CH=CH-$ (furan ring, O)

CLXXVI

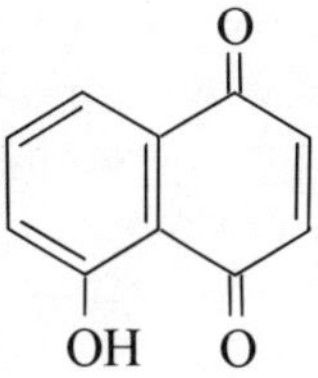

O, OH, O

CLXXVII

CHO, OH

CLXXVIII

Fig. 52. Inhibitors in root exudates.

CLXX	Allicin
CLXXI	Chlorogenic acid.
CLXXII	Melilotic acid
CLXXIII	Gallic acid
CLXXIV	*o*-Coumaric acid
CLXXV	Piperic acid
CLXXVI	2-Furanacrylic acid
CLXXVII	α-Hydroxynaphthoquinone. Juglone.
CLXXVIII	*p*-Hydroxybenzaldehyde

(B) INTERACTIONS BETWEEN ROOTS AND MICRO-ORGANISMS

As we have already suggested plant roots can influence the growth of micro-organisms in their vicinity and it would be surprising if micro-organisms could not in their turn directly affect the growth of roots and through them that of the plant as a whole. In this section we will consider these two aspects of plant/micro-organism interaction in more detail.

(i) Root metabolites and the growth of soil micro-organisms

The fact is well established that the roots of plants will affect and be affected by this microbiological equilibrium in the soil. Quantitative and qualitative studies on the microscopic flora of soils have demonstrated that the distribution of this microbial population is by no means uniform. In particular, there is a manifold increase in numbers as we pass into the immediate vicinity of the roots, and the greatest numbers are found actually on the surface of the root itself. Furthermore, it has been shown that the predominant organisms in this thin, highly populated soil-sheath surrounding the root (the so-called *rhizosphere*), are those which have relatively complex food requirements, especially for certain amino acids and vitamins such as thiamin. Presumably this accounts for their presence in the neighbourhood of the roots, which may supply these requirements in their exudates. On the other hand, there is also an increase in the rhizosphere in the populations of the less exacting organisms, and so obviously other influences cannot be ruled out. Antibiotics may also play a part in determining the exact constitution of the microbial flora of the rhizosphere, as some plant roots seem to excrete antibiotics into the soil, thereby inhibiting the activity of nitrifying bacteria (Stiven, 1952).

The underlying causes of the whole disturbed equilibrium in the soil population in the neighbourhood of the root is still far from being solved.* The extreme case of this augmented microbial population on and near root surfaces is seen in the widely occurring *mycorrhizas*. These structures arise, in a very great variety of plants, as small branched lateral rootlets that have become covered with a thick layer of interwoven threads of a fungus, which is usually specific to the particular plant affected. In some instances the fungus may even penetrate the tissue of the host root, where it apparently behaves as a weak parasite. This close association of fungus and plant root has been extensively studied for many years, especially by foresters, as it is in forest trees that this phenomenon is most clearly seen. The exact significance of this association is still obscure. Foresters claim that the higher plant definitely gains, in that more vigorous growth is shown when mycorrhizas are present, and it has been severally suggested that the plant receives from its fungal associate nutrients and growth factors including growth hormones and vitamins. It is probable that the fungus also derives some similar benefit from the host root.

The effect of root exudates on the soil micro-flora may result in both advantages and disadvantages to the plants concerned. There is an increasing volume of evidence to support the view that the resistance or susceptibility to invasion by pathogenic soil organisms of the roots of certain crop plants may be directly correlated with their effects on the micro-flora of the soil immediately surrounding them. Thus, in two varieties of flax (Bison, which is resistant, and Novelty, which is susceptible to the soil-borne *Fusarium* wilt disease), it has been shown that root exudations from the susceptible variety Novelty stimulate the growth of the responsible fungus, whereas exudations from the resistant variety, Bison, do not. On the other hand, exudates from

*The reader is referred to the following authors for an exhaustive survey: Katznelson *et al.*, 1948; L. J. Harley, 1948, 1969, Garrett, 1960 (see Bibliography).

Bison do stimulate the growth of the soil fungus *Trichoderma viride*, which, as we have seen previously, is an active producer of fungistatic compounds (Timonin, 1941). It is very probable that these exudates play a fundamental part in determining the resistance of flax plants to wilt disease.

Antibiotic substances present in the coats of seeds may account for a natural resistance to infection by soil pathogens. For example the sweet pea seed contains a substance which is active in suppressing the growth of the pathogenic bacterium *Corynebacterium fascians* and the differing ability of several strains of this organism to infect sweet pea seeds is inversely correlated with their sensitivity to the seed-coat inhibitor (Jacobs & Dadd, 1959).

Bacterial growth inhibitors produced by higher plants may also play an indirect though important part in determining the ecological succession in natural communities. Rice *et al.*, (1960) studied the natural succession of plants invading abandoned fields in Oklahoma. They discovered that the order of succession of the main invaders, i.e. *Aristida oligantha, Andropogon scoparius* and *Panicum virgatum*, was the same as their order of increasing requirement for nitrogen and phosphorus. Later it was shown (Rice, 1964, 1965*a*, 1968) that many of the invading species excreted bacterial inhibitors from their roots and these suppressed the growth of free-living nitrogen-fixing bacteria (*Azotobacter* sp.) and also the nodule organisms (*Rhizobium* sp.). A similar pattern of inhibitors was observed for the growth of a number of nitrogen-fixing blue-green algae present in the soil populations in these abandoned fields (Parks & Rice, 1969). The compounds seem to be phenolic in nature, i.e. mainly chlorogenic, caffeic, ferulic, tannic and gallic acids and their glucose esters (Rice, 1965*b*, 1967; Blum & Rice, 1969). The nitrifying organisms, *Nitrosomonas* and *Nitrobacter*, which convert ammonia to nitrate, were also inhibited. It seems not improbable that these pioneering species, with their low nitrogen requirements may, by preventing further soil nitrification, acquire for themselves a distinct competitive advantage over more nitrogen-demanding species. In the high-veldt savanna grasslands of Rhodesia two abundant species of the grass genus *Hyparrhenia* secrete a toxic substance which suppresses the growth of nitrifying bacteria (Boughey *et al.*, 1964). This explains the well-known poverty of such savanna soils in available nitrogen and could again be of competitive advantage to the grass species.

(ii) Antibiotics from micro-organisms and plant growth.

We have seen earlier (p. 219) that antibiotics belong to that class of biologically active substances called anti-metabolites. Since their antagonistic actions are usually exerted in biochemical processes which are common to a very wide range of organisms it is not surprising that they can also inhibit enzyme activity and hence growth in cells and tissues of higher plants. Indeed certain antibiotics have become well-established chemical tools in the study of growth mechanisms in higher plant tissue. For example actinomycin D (from the soil micro-organisms *Streptomyces antibioticum* and *S. chrysomallus*) is a specific anti-metabolite for the synthesis of ribonucleic acid (RNA) on the deoxyribonucleic acid (DNA) template of the nucleus (see Chapter XII). Again puromycin (from *Streptomyces alboniger*) is a specific inhibitor of protein synthesis on the RNA template of the protein-synthesising organelles of the cell (ribosomes). Both these antibiotics, and many others, are now widely used to sort out the roles of protein and nucleic acid synthesis in the growth and development of plants (for further details see Chapter XII; for a review of the biochemical reactions of antibiotics see Gottlieb & Shaw, 1967).

We have known for a considerable time that a large number of antibiotics can be taken up by the roots and distributed throughout the aerial parts of plants. Fortunately many of them are innocuous to higher plants and there is a considerable possibility that such substances may ultimately find use in a kind of chemotherapy of plants to afford protection against pathogenic micro-organisms susceptible to them. On the other hand some antibiotics are distinctly toxic to higher plants (Brian, 1957) and this raises the question whether soil micro-organisms can affect plant growth via their antibiotic production. These possibilities have received little direct experimental attention. However a few observations suggest that their importance cannot be ignored. Thus root growth in such plants as cucumber and barley can be greatly inhibited by low concentrations of the antibiotics polymyxin, oxytetracycline, oligomycin, L-cycloserine, azaserine and duramycin (Norman, 1955, 1959*a* & *b*, 1960) and this is coupled with a comparable reduction in uptake of the essential mineral nutrient potassium (Norman, 1959*a* & *b*). It seems that these growth inhibitions may result from interference with nuclear division (Wilson & Bowen, 1951), i.e. via protein and nucleic acid metabolism. Streptomycin, well known for its medicinal use, has a specific destructive action on chloroplasts, bleaching the chlorophyll which is essential for photosynthesis (see Provasoli *et al.*, 1951). However there is no evidence that such deleterious actions are ever exerted by antibiotics produced by soil micro-organisms under natural conditions.

It is a curious thing that antibiotics under certain conditions will markedly *stimulate* the growth of higher organisms. The effects of feeding farm animals with residues from antibiotic production processes are now well known. It was at first thought that higher plants were always inhibited, but the use of low concentrations (less than 50 ppm) of highly purified samples of some compounds have revealed similar growth stimulations.

The compounds are apparently working on a system which directly affects the growth process of the plants and not via the antagonism of harmful micro-organisms in the environment, as these results were observed on duckweed (*Lemna minor*) growing under *sterile* conditions (Nickell & Finlay, 1954). The physiological reasons for these stimulations remain obscure. It has been suggested, among other things, that the detoxication of staling products of metabolism of the higher plant may be the cause. Effects on the metabolism of natural growth substances are also possible (Nickell, 1955). These stimulatory effects are first preceded by an inhibition and there is now evidence that they are correlated with a progressive decomposition of some of the antibiotics in the tissues of the plant. These may, at least partially, account for the observed stimulations which may have little to do with antibiotic action *per se* (Nickell & Gordon, 1960).

Organic Substances Promoting Growth in Higher Plants

The question of the relative merits of organic farmyard manure and artificial fertilizers has long been a bone of contention between farmers and scientists alike. Those who favour the use of farmyard manure would have us believe that, quite apart from its beneficial effect on the physical properties of the soil, it provides specific organic growth substances essential for the best growth of the plant. On the other hand, their opponents, who are in the majority, hold that optimal growth of a plant can take place if inorganic salts only are provided in a balanced mixture of optimal concentrations, and they point for proof to the success of the 'soilless culture' of market garden produce, or 'hydroponics', as it is unnecessarily called.

The adherents to the former view are not without experimental support. Nearly fifty years ago, Bottomley, one of the earliest proponents of the theory, showed that 'bacterized peat', prepared by subjecting raw peat to the action of certain unspecified aerobic 'humating' bacteria, gave watery extracts which, he claimed, markedly stimulated the growth of wheat and other crop plants (Bottomley, 1914): this, he suggested, was due to the presence of specific organic growth-promoting substances to which he gave the name 'auximones'; later, he showed them to be present in other organic sources, such as well rotted stable manure, leaf mould, yeasts and germinating seeds. Subsequently (Bottomley, 1917), experiments were extended to duckweed (*Lemna minor*), the small floating green fronds of which are characteristic of stagnant fresh-water ponds. From the results it was concluded that 'auximones' were essential for the sustained growth and reproduction of this plant, and some evidence was obtained that the effective compounds might be derived from the breakdown of nucleic acids (Bottomley, 1920). Although much of the stimulating effects noted by Bottomley may have been evoked by the provision in his extracts of small traces of inorganic nutrients (e.g. easily assimilated iron salts, etc.) otherwise present in sub-optimal concentrations in his untreated culture medium (Bonner, 1946), there still remained evidence of additional effects of organic substances with their own specific actions (Bassalik, 1934). Although it is now clear that *Lemna* can grow for long periods on a purely inorganic culture medium, yet a number of subsequent workers (see Chesters & Street, 1948) have demonstrated the marked growth-stimulating effects of small traces of organic matter, e.g. sterilized extracts of fresh horse-dung, on the growth of *Lemna* and the closely allied genus *Spirodela*.

Although the 'auximone' interpretation of the stimulation effects observed by Bottomley on higher plants has never been accepted by the majority of plant physiologists, yet there has been, in the interval, a steady trickle of experimental results, suggesting that under certain conditions organic factors in manures may produce significant stimulations of plant growth. Some 'enthusiasts' have even claimed that organic manures produce crops that have a higher nutritive value to man than have those receiving artificial fertilizers only, this being the result of their higher vitamin content. Unfortunately, many of these 'positive' results can be explained by the stimulating effect of small traces of mineral nutrients in the organic manures and the extracts used. On the other hand, as has been pointed out by Chesters & Street (1948), in all water culture experiments so far carried out, where plants have been grown supposedly in the complete absence of any external supply of organic compounds, no adequate precautions have, in actual fact, been taken to exclude such sources as dust and micro-organisms from the air. Recent experiments have shown that very appreciable quantities of nitrogen and phosphorus compounds are present in submicroscopic dust particles (natural aerosols) even in the cleanest country air, and such material tends to be returned to the soil in rain (Ingham, 1950).

Chesters & Street carried out some of the most careful experiments in sand cultures in which lettuce and radish were grown from seed in solutions containing optimal amounts of all the known mineral nutrients, including trace elements. The effects of adding aqueous extracts of leaf-mould, yeast, and the products of hydrolysis of casein, were compared with controls in which extra nitrogen, supplied as inorganic nitrate, was added in amounts equivalent to that in the organic extracts (Chesters & Street, 1948; Street, 1950). The addition of the leaf-mould extract significantly increased the growth of these plants. Although these extracts contained small traces (about one part in 10^{10}) of IAA, this was not responsible for the effects noted, and the absence of response to the casein hydrolysate and yeast suggested that neither vitamins, nucleic acids, nor amino acids were involved in the stimulation. It is, however, possible that these effects may not be the direct result of the action of a specific organic compound present in the extract, but may be indirectly due to some interaction with the microflora of the sand culture. This is suggested by other results (Swaby, 1942), in which organic materials were shown to have no effect on plant growth in steriles and culture, but exhibited such effects only under conditions favouring the development of micro-organisms. In view of the stimulating effect that small amounts of antibiotics have on plant growth (see Nickell, 1955), it is not impossible that traces of such antibiotic by-products, from the micro-organisms stimulated to grow by these organic materials, may be the actual growth promoting factors in such cultures.

Considerable attention has been paid by plant physiologists in Poland to the beneficial effects of *humus* on plant growth. Humus is a term applied to a very important component of soils which comprises a very heterogeneous complex of products (mainly colloidal) of the decomposition of ligno-cellulose, the material of the plant cell wall mainly responsible for its rigidity and strength. This is not the place to embark on a description of this extremely complicated group of soil organics; details should be sought elsewhere (e.g. Mortensen & Himes, 1964; Hurst & Burges, 1967). Suffice it to say that they are polymers of aromatic degradation products, mainly phenolic in nature, and containing nitrogenous components derived from proteins. The Polish workers have shown that aqueous extracts of raw humus can greatly promote the growth of tomato plants if applied to the roots in a culture medium which is not fully aerated (Gumiński & Gumińska, 1953). Similar action was also obtained from extracts of the leaves of beech and oak and with the oxidation products of gallic acid and tannins and this led them to suppose that the action was due to the promotion of a heightened oxidation potential at the root surface, not to a specific biochemical effect of the compounds on the growth processes. Later the effects of these humic materials were related to their capacity to adsorb certain cations, particularly iron, since they were greater under unfavourable mineral nutrient conditions (Guminski & Sulej, 1967). On the other hand those components of humus which are of relatively small molecular volume can be taken up by plant roots and, if supplied in sufficient quantity, can bring about significant increases in the dry weights of test plants (e.g. rye seedlings; Flaig *et al.*, 1960)). The mechanism of this action remains obscure but it is interesting that small rates of application of the phenolic herbicide 2-methyl-4, 6-dinitrophenol, which disturbs energy-transforming mechanisms in cells, can cause a similar increase in crop yield (Bruinsma, 1962). Other examples of humus promotion of growth are in Flaig (1965).

But phenolic acids in soils can be extremely toxic to plants. This subject is dealt with in the next section.

There is evidence that both auxins and vitamins are present in measurable quantities in many different types of soils and organic manures. The vitamins, as we have seen, arise mainly from the metabolism of soil micro-organisms, and the auxins presumably result from the decomposition of dead plants, or may even be secreted from plant roots (Vančura & Hovádik, 1965). Naturally the highest concentrations of those compounds occur in the surface layers of the soil, where organic debris has accumulated and the activity of associated micro-organisms is highest: there even seems to be a positive correlation between the concentrations of these growth substances in the soil and the latter's fertility. Details of authors and results are given in a review by Schmidt (1951). There is, however, very little evidence that these growth substances can, in any direct way, stimulate the growth of crop plants. The concentration of auxins (not yet chemically identified) in soils is of the order of 10^{-7} M or less in terms of IAA equivalents (Whitehead, 1963) and is not one that would be expected to stimulate plant growth. With the vitamins, very little work has been done. A few claims have been put forward that an external supply of thiamin, ascorbic acid, etc., increased root and shoot growth in a number of plant species when given in solution to the roots, but they have remained unconfirmed by subsequent investigators. Indeed, some of the early claims of positive effects have now been withdrawn. (See Chesters & Street, 1948; Schmidt, 1951.)

Both on experimental evidence and from theoretical considerations, it seems unlikely, therefore, that vitamins and auxins present in the soil produce any stimulation of the vigour of plants growing in that soil. Any normal plant in receipt of optimal nutrient supply is able to fulfil, by synthesis, its full requirements of these particular growth factors, and any additional external supply is unlikely to exert anything but the harmful effects of supra-optimal concentrations. Such effects may account, at least partially, for the 'sickness' of over-manured soils, where growth factors may be present in high concentrations together with specific organic inhibitors.

Growth Substances and the Interaction between Plants in Natural Habitats—Allelopathy

The question now arises as to whether chemical antagonisms, similar to those demonstrated for micro-organisms, are exerted between species of higher plant growing together with their roots intertwining in close proximity in the soil. Is it possible that plants may excrete from their roots, inhibitors that retard the growth of other susceptible plants growing with them? The answer is that there has indeed accumulated over the last century a considerable amount of evidence of marked antagonisms between crop plants and their associated weeds. The decline in vigour and productivity of crops grown continuously on the same soil, in spite of adequate cultivation and fertilization of that soil, and the depressing effect of some crops (e.g. maize, rye and buckwheat) on certain succeeding crops (e.g. fruit trees, onions, tomatoes), implies the accumulation in the soil of a growth-inhibiting factor from the roots of the offending plants. Furthermore there is direct evidence that unnatural substances such as herbicides, when applied to the aerial part of one plant, may be excreted later by its roots and absorbed by the roots of a neighbouring plant. This has been shown to occur in bean plants treated with α-methoxyphenylacetic acid (Preston *et al.*, 1954). This phenomenon of 'the influence of one plant upon another under natural conditions and exerted by chemical means other than nutritional ones' (Evenari, 1961) has been called *allelopathy**.

As early as 1832 the famous botanist, A. de Candolle, pointed out the apparent specific inhibition of flax by spurge (*Euphorbia* sp.) and of oats by thistles (*Cnicus* sp.), and postulated the production of specific toxic substances by these weeds. His ideas were not accepted at the time, however, and for the remainder of the century competition for essential inorganic nutrients was held to underlie all such antagonistic effects. Interest in the possibility of excreted toxic substances was aroused in America at the beginning of the present century by the loss of fertility in certain soils as the result of continued cropping of one plant species for a series of generations. Workers in the U.S. Department of Agriculture took such soils, made water extracts of them, and showed that these extracts prevented the growth of wheat seedlings (Livingstone, 1907; Schreiner & Reed, 1907). Unfortunately, this work produced no evidence that the inhibitors had their origins in the plants grown previously in those soils. The part played by exhaustion of nutrients in the loss of soil fertility was also ignored and, as a result, the implications of this work were not seriously considered at the time.

Nevertheless observations of the toxic effects of plant products in the soil have continued to be made. For example the walnut (*Juglans* sp.) has for well over half a century been known to inhibit the development of certain herbs growing within range of its root system (see review in Wood, 1960). Again young trees are very difficult to establish in ground that has previously been a peach orchard, and this fits in with the observation that old peach roots contain a substance inhibitory to the growth of young peach trees (Proebsting & Gilmore, 1940). Soils from old *Citrus* orchards are quite

*This aspect of plant growth inhibitors has been comprehensively reviewed by Loehwing (1937), Bonner (1950*a*), Grümmer (1955), Woods (1960), Garb (1961) and Evenari (1961).

inhibitory to the growth of young *Citrus* seedlings, 50 to 75 percent more growth being made on similar soils from outside the orchard area (Martin, 1950). The presence of harmful soil organisms and the deterioration of soil structure were experimentally ruled out as contributory causes. Only treatment of the soil with 2 percent sulphuric acid or 2 percent caustic potash, followed by saturation with calcium, was found to restore these orchard soils to the full fertility of the normal soils. Since tomato seedlings were not affected in these soils, the most reasonable explanation of these results is that toxicity of these soils is caused by the accumulation of a growth inhibitor specific for *Citrus* itself and excreted by its roots. Presumably this inhibitor is normally adsorbed into the soil colloids so that it is not washed out by rain, and it can be removed or destroyed only by treatment with relatively strong acids or alkalies.

Just before the Second World War the commercial implications of such antagonisms in rubber and coffee estates in the Far East led to a comprehensive series of experiments with the different plant species used for 'cover crops'. In the tropics, where jungle has been cleared and young plantations established, the removal of natural vegetation cover may have very serious effects on the structure of the soil so exposed. The very violent rainfall will, in the absence of suitable precautions, result in serious erosion of the soil, and, in any event, a rapid removal of humus components accumulated slowly in the natural jungle. To combat these effects it is customary for the planter to grow, on the soil between and under his young tree seedlings, certain vigorous herbaceous plants that form a continuous cover to the bare soil. These so-called 'cover crops' bind the soil with their roots, thus preventing erosion and allowing the accumulation of humus. The use of leguminous plants, e.g. *Centrosema pubescens*, which bring about the enrichment of the soil in nitrogen compounds gives an added advantage. The deleterious effect of such cover crops on the growth of young trees had been known since the introduction of the practice. The effect was at first regarded as due to competition for nutrients, but a series of widespread experiments in young coffee plantations in Java indicated that toxic root exudates were the basis of the competition (van der Veen, 1935). For example drainage water from soil in which *Salvia occidentalis* was growing was shown to have a marked toxic effect on the growth of coffee seedlings when applied to their roots. These results are very similar to those obtained nearly twenty years earlier on the toxicity of grass to apple trees in orchards (Pickering, 1917). Fortunately all cover crops are not similarly toxic, and rubber reacts differently from coffee. Some of the worst offenders, in addition to the *Salvia*, are *Paspalum conjugatum*, *Cynodon dactylon*, *Passiflora foetida* and *Amaranthus spinosus*. The use of 'smother crops' (e.g. barley) to suppress the growth of undesirable weeds may have allelopathy as its basis rather than competition for nutrients as originally supposed (Overland, 1966).

Numerous observations have indicated the excretion of toxic compounds from various grass roots. For example rape (*Brassica napus* var. *oleifera* and *B. rapa* var. *oleifera*) germinated and develop very poorly in fields where couch grass (*Agropyron repens*) was growing well (Osvald, 1947*b*). Watery or dilute ammonia extracts of dried powdered roots and rhizomes* showed that these organs contained a toxic substance which prevented the germination of rape seed and allowed it to be attacked and destroyed by moulds. This suggested that the well known success of grasses in competition with other species in natural grassland communities might reside in a similar toxin production, and led to the study of soil extracts from such grassland and from agricultural land. The soil from a particular meadow, dominated by an almost pure stand of red fescue grass (*Festuca rubra*), was extracted with alcohol and with dilute ammonia: these extracts stopped the rape seeds from germinating, whereas similar extracts from cultivated soils had no such ill effect (Osvald, 1949). The production of the inhibitor is greatly promoted in waterlogged, and hence oxygen-deficient soils (Grümmer, 1964). This could be attributed to changes in the pattern of decomposition of the dead or moribund rhizome induced by oxygen lack (Welbank, 1963). Other species such as *Agrostis tenuis* and *Lolium multiflorum* can also produce toxins under the same conditions. In barley (*Hordeum vulgare*) however more inhibitor is produced by living than by dead roots, suggesting an active secretion of the inhibitor by living tissue (Overland, 1966).

Some evidence has also been obtained that certain crop plants may be mutually inhibitory to each other's growth when grown in close association in the same plot. Spinach and radish are such a pair, and it has been suggested that the saponins of the spinach, and the mustard oils of the radish, may be the inhibitory compounds involved (Schuphan, 1948). Italian rye-grass and red clover are another pair (Mann & Barnes, 1953). It is suggested that flax can, as it were, 'foul its own nest', possibly by the excretion of toxic substances, which depress the development of subsequent plantings of that species in the same soil (Becquerel & Rousseau, 1941). That this may also be true of wheat roots has been indicated by experiments in which much better growth was obtained with continuously renewed culture solution than in non-flowing solutions in which exudates were allowed to accumulate (Eliasson, 1959).

It was in the early days of the second world war that experiments in America on the production of rubber by the guayule (*Parthenium argen-*

* Rhizome is the botanical term for the long underground stems or runners by which this grass spreads so effectively and rapidly.

tatum)* resulted in the isolation and identification of a toxicant in root exudates. In plantations of this species, mutual antagonisms were obvious; individuals in the centre of a block of plants showed much feebler growth than those at the edges. This was not caused by competition for water or nutrients but presumably by toxic root exudates, as roots of adjacent plants did not intermingle but grew out into areas unoccupied by other guayule roots. Pot experiments confirmed this antagonism and finally 1.6 g of a highly toxic crystalline material were obtained by soaking 20,000 roots of normal living plants in distilled water. This material was subsequently identified as *trans*-cinnamic acid (Fig. 14 LXXVII), a normal constituent of the guayule plant. These observations show that the toxicity of the soil water under normal conditions was due to an exudation from living roots and not to a release following the death of the old roots.

There can be little doubt from this accumulated evidence that roots and underground stems of a very wide range of plant species produce compounds that are active inhibitors of seed germination and root growth: it seems highly probable that these compounds are normally excreted into the soil, where they may play a rôle in the selective suppression of the growth of competing seedlings. There is, however, another source of growth inhibitor in the soil, namely, the normal decomposition of plant organs when the plant dies (see reports on *Agropyron* rhizome above). Such inhibitors may also play an important part in determining the nature of the vegetation the soil supports. This has been clearly shown recently in America for the desert shrub, *Encelia farinosa*, which, in contrast to most other desert shrubs, does not shelter a small colony of annual plants under its branches. This is an effect of a chemical inhibitor, passing out from its leaves, into the surrounding soil after they have been shed and covered by the sandy soil under the bush. In the south-western U.S.A. the degeneration of grassland areas where creosote-bush (*Larrea tridentata*) is present, may be due to toxic substances leached out from the leaves and fruits (Knipe & Herbel, 1966). In another species, *Artemisia absinthium*, the wormwood, inhibitors present in the leaves may be washed out and severely stunt plants of certain species growing in its shelter (Bode, 1940). In southern California, water condensed from fog and dripping from the leaves of *Eucalyptus globulus*, is almost completely suppressive of the growth of herbacious annual plants under the trees (del Moral & Muller, 1969). Beech leaf litter is the source of an inhibitor of the germination of pine, fir, wheat, oats and cress (Winter & Bublitz, 1953). The straw of various cereals also contains inhibitors of seed germination and seedling growth. Wheat, rye, barley and oats all produce

* Guayule is a Central American plant of the family Compositae, which has been used for rubber production for many years. (For a survey of rubber-forming plants, see Bonner & Galston, 1947.)

inhibitors which are effective on all four species (Winter & Schönbeck, 1953). Other crop plants such as broad bean and broccoli decompose to give highly toxic products (Patrick *et al.*, 1963).

How long these organic substances from living roots or decomposing plant organs can remain in the soil is at present unknown. It is possible that they may disappear quickly under the action of micro-organisms, except perhaps under certain unfavourable conditions (e.g. poor aeration in waterlogged or heavy clay soils). When active production is going on, however, concentrations may be maintained at equilibrium levels high enough to exercise a considerable inhibiting influence on surrounding plants, even under conditions favouring rapid decomposition. Thus, the old dead roots of brome grass (*Bromus inermis*) liberate into the soil a toxic substance that has a markedly deleterious effect on the growth of seedlings of the same species, and this has been held to account very largely for the gradual loss of vigour of pure stands of this grass growing in the same place for a number of years (Benedict, 1941). The same phenomenon is presumably operating in some species which grow in 'colonies', spreading out from a centre of origin while the older central parts die as a result, it has been suggested, of the accumulation of accumulated toxins, e.g. in certain genera of the Compositae, *Helianthus*, *Antennaria*, *Aster*, and *Erigeron* (Cooper & Stoesz, 1931; Curtis & Cottam, 1950). Again in the *garigue* of western Provence, a region of France dominated by rosemary and heathers, there are virtually no annuals in the natural communities of plants. This has been shown to be due to the production of inhibitors by the dominant perennials (Deleuil, 1950–1). A similar paucity of annuals in a *Brachypodium pinnatum–Bromus erectus* grassland in northern France may be due to toxic materials coming from the perennial inhabitants. The majority of important plants of this community were found to produce some inhibitors, but the most active were three species of hawkweed (*Hieracium*), cat's ear (*Hypochaeris radicata*), and goldenrod (*Solidago virgaurea*) (Guyot *et al.*, 1951; Guyot, 1954). Strangely enough the dominant grasses showed only weak toxin production. In abandoned fields in central Oklahoma *Sorghum halepense* (Johnson grass) may be an early pioneer, when it often dominates the natural return to prairie for long periods. This is due to the production of various inhibitors by roots and rhizomes which actively prevent the establishment of other invading species (Abdul-Wahab & Rice, 1967). Another striking example of chemical competition between plants in natural vegetation is seen in the coastal regions of southern California where vast areas are colonized almost exclusively by the two species *Salvia leucophylla* and *Artemisia californica.* These areas are adjacent to natural uncultivated grassland and between these two plant communities exists a wide zone of almost bare ground on which few plants will become established and grow to maturity (see Plate 17). It has been clearly demonstrated (Muller, 1966) that this phenomenon is due to the production of volatile toxic substances from these two plants, which inhibit the germination and growth of the grassland species. Similar volatile inhibitors are produced by *Eucalyptus globulus* and interestingly enough these are ineffective on *Eucalyptus* itself (Baker, 1966).

Even lichens can influence chemically the growth and development of associated grasses, as has been shown for *Peltigera canina* on the sand dunes of Glamorgan (Pyatt, 1967).

One very interesting ecological situation involving inhibitors has been shown in the California chaparral when the shoots of *Adenostoma fasciculatum* often forms pure stands and herb populations are entirely suppressed by leachates from its leaves. This chaparral is characterised by periodic fires which play an important part in determining the nature of the vegetation. After a fire, not only do the shrubs regenerate but for a time there is a considerable flush of growth of herbacious plants, released from inhibition by the destruction of the allelopathic inhibitors from the shurbs. Eventually of course inhibition is reimposed when the shurbs have grown. These herbacious plants have apparently evolved under the selection pressure of these conditions in such a way that their seeds can be maintained dormant for very long periods by the inhibitor and can then germinate promptly as a result of fire action, which not only destroys the inhibitor but also directly promotes seed germination (McPherson & Muller, 1969).

For further surveys of the role of allelopathic substances in the interactions and successions of plants in natural communities the reader is referred to Muller (1968, 1970).

In some members of the grass family, root excretions may have very important biological effects. In the tropics of the Old World, many grasses are parasitized by a genus of plants closely related to the familiar yellow-rattle (*Rhinanthus*) of temperate pastures. This genus, which is called *Striga*, sends small suckers into the roots of the host grass and so considerably supplements its own food supply. One striking thing is that the seeds of this semi-parasite will germinate only in the presence of the living roots of a suitable plant, usually the grass host; this has been shown to be due to the presence, in the diffusate from the host's roots, of a specific germination stimulant (Brown & Edwards, 1944). Its chemical nature is not yet certain; it has been tentatively identified as a pentose sugar (Brown, Johnson, Robinson & Todd, 1949; Brown, Robinson & Johnson, 1949). Some species of *Striga* are troublesome pests. Thus *Striga lutea* (witchweed) causes serious damage to cereal crops in South Africa. It is extremely difficult to eradicate by ordinary cultural practices, as the seeds are very plentiful (about half a million per plant) and may remain dormant in the soil for up to twenty years, ready to be

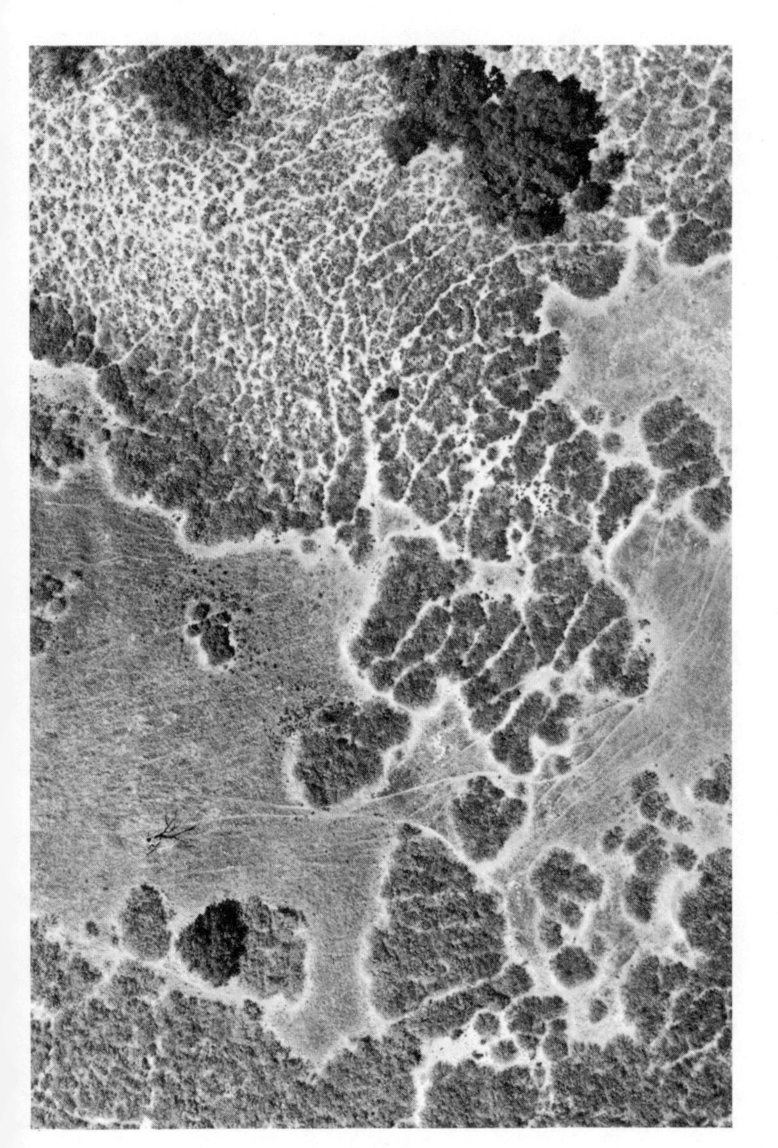

Plate 17. Aerial photograph of intermixed *Salvia leucophylla* and *Artemisia californica* shrubs invading annual grassland in the chaparral of the Santa Ynez Valley, California. For a distance of two metres or so in front of the advancing edge of the shrub zone the ground is virtually devoid of all vegetation as a result of the production of volatile and toxic inhibitors by the shrubs themselves. (Photograph by kind permission of Professor Cornelius B. Muller, University of California, Santa Barbara.)

stimulated into germination by a suitable root exudate. It is fortunate that some plant species, not parasitized by it, produce root exudates that trigger off its germination. Eradication can then be effected by growing such crops, e.g. sunflower, cowpea, groundnut, etc., and then spraying the growing plants of *Striga* with herbicides, which would otherwise have no action on the dormant seeds.

A similar germination stimulant for the genus *Orobanche* has been identified as an unsaturated lactone (Brown *et al.*, 1952).

Inhibitors Involved in Allelopathy

(A) ROOT EXUDATES

It is now widely recognized that a very great variety of chemicals are excreted by the roots of plants. Many, such as sugars and amino acids, serve as nutrients to the rich bacterial flora which characterizes the thin film of soil surrounding the roots (rhizosphere). Substances suppressing the growth of micro-organisms may also be excreted. Much less is known however about the nature of the toxic root exudates responsible for allelopathic interactions of higher plants although a few have been identified.

Scopoletin (Fig. 40 CXXVI) is given off by the roots of various species of plant (Eberhardts Martin, 1957) but in view of its rapid microbial destruction its action may be fugitive. Many organic acids have also been implicated; they include oxalic, salicylic (Fig. 40 CXXVII), *trans*-cinnamic (Fig. 14 LXXVII) (from guayule, see p. 235), chlorogenic (Fig. 52 CLXXI), melilotic (Fig. 52 CLXXII), ferulic (Fig. 40 CXXI), gallic (Fig. 52 CLXXIII), *o*-coumaric (Fig. 52 CLXXIV), phenylpropiolic (Fig. 44 CXLIX), piperic (Fig. 52 CLXXV) and 2-furanacrylic (Fig. 52 CLXXVI) acids (for references see Woods, 1960; Stevenson, 1967). In Johnson grass (Abdul-Wahab & Rice, 1967) the inhibitor seems to be *p*-hydroxybenzaldehyde (Fig. 52 CLXXVIII), presumably arising from the cyanogenic glucoside dhurrin, present in the rhizome. It is interesting that apple root tissue will liberate the glucoside phlorizin into soil and this is broken down there ultimately to yield *p*-hydroxybenzoic acid which may account for the apple re-plant problem in old orchards (Börner, 1959) and a similar problem in peach, *Citrus*, etc. It is intriguing that many of the acid inhibitors contain unsaturated aliphatic groups. How long they persist in the soil is unknown.

The toxic principle from walnut roots (see p. 233) is undoubtedly juglone, α-hydroxynaphthoquinone (Fig. 52 CLXXVII) (Davis, 1928); the pure material is extremely toxic to the growth of seedlings of alfalfa and tomato.

(B) LEAF LEACHATES AND DECOMPOSITION PRODUCTS

Toxic materials may be washed out from leaves into the soil. For example a toxic chemical may come from the leaves of *Artemisia absinthium* and inhibit the growth of seedlings in the neighbourhood. This material is probably the glucoside absinthin (a camphor glucoside) (Bode, 1940). The compound leached from the leaves of the desert shrub *Encelia farinosa* (see p. 236) is 3-acetyl-6-methoxybenzaldehyde (Fig. 53 CLXXIX) (Gray & Bonner, 1948), although this may disappear rapidly under the action of micro-organisms (Muller, 1953). From *Eucalyptus globosus* the inhibiting compounds are mainly chlorogenic, *o*-coumarylquinic and genistic (Fig. 53 CLXXX) acids (del Moral & Muller, 1969).

CHO
CH_3O
$CO{\cdot}CH_3$
CLXXIX

OH
COOH
OH
CLXXX

COOH
OH
OH
CLXXXII

$$CH_3-(CH_2)_7-CHOH-CHOH-(CH_2)_7-COOH.$$

CLXXXI

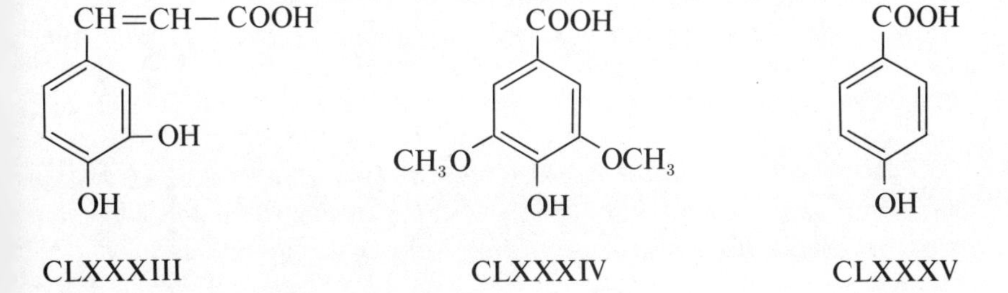

Fig. 53. Inhibitors in leaf leachates etc.
CLXXIX 3-Acetyl-6-methoxybenzaldehyde
CLXXX Genistic acid
CLXXXI Dihydroxystearic acid
CLXXXII Protocatechuic acid
CLXXXIII Caffeic acid
CLXXXIV Syringic acid
CLXXXV *p*-Hydroxybenzoic acid

Leaf and strawberry litter may also be the source of toxic materials that prevent seedling establishment. The long-chain fatty acid dihydroxystearic acid (Fig. 53 CLXXXI) has long been regarded as the major component responsible for the low fertility of worn-out soils (Schreiner & Lathrop, 1911). Cereal straw produces such aromatic acids as protocatechuic (Fig. 53 CLXXXII), caffeic (Fig. 53 CLXXXIII), chlorogenic, *p*-coumaric, *p*-hydroxybenzoic (Fig. 53 CLXXXV), ferulic, syringic (Fig. 53 CLXXXIV) and vanillic (Fig. 40 CXXXII) acids (Köves & Varga, 1958; Guenzi & McCalla, 1961; Wang *et al*., 1967). There has been considerable debate as to whether these compounds persist in the soil long enough to have any significant allelopathic effects (see discussion in Evenari, 1961), although under some conditions such persistance is observed and may bring about concentrations sufficient to inhibit growth in test plants, e.g. in sugar-cane soils (Wang *et al*., 1967). It is possible that quinones formed during the decomposition of conifer needles could be responsible for the paucity of species in many conifer forests (Bautz, 1953). Under conditions of poor aeration, for example in poorly drained rice fields of the Far East, the decomposition of organic matter by micro-organisms results in the accumulation of aliphatic acids. In these soils the growth of young rice roots may be severly restricted and root rot may ultimately occur. These growth inhibitions are most probable caused, at least in greater part, by these aliphatic acids, of which acetic, butyric and formic are the most plentiful and butyric the most toxic (for a detailed summary and references see Stevenson, 1967). Under even more reductive conditions aliphatic alcohols (methanol, ethanol, *n*-propanol and *n*-butanol) may be produced. Of these *n*-butanol is very growth-inhibitory and may also be a factor contributing to the infertility of these waterlogged soils (Wang & Chuang, 1966).

Further details of phytotoxic substances produced in soil by the decomposition of plant residues can be found in a review by Patrick (1971).

(C) VOLATILE TOXICANTS

One of the most widespread of volatile organic products of plants is the gas ethylene which has a wide spectrum of physiological actions in plants. These have already been described in Chapter VIII. It seems very unlikely that in nature ethylene production could ever arise to such high levels that allelopathic actions might be evoked.

However other more complex volatile substances may be involved in allelopathy. We have already seen such a case in *Salvia leucophylla* and *Artemisia californica* of southern California. The volatile emissions of *S. leucophylla* are terpenes and have recently been separated by gas chromatography and their phytotoxicity tested on cucumber seedlings (Muller,

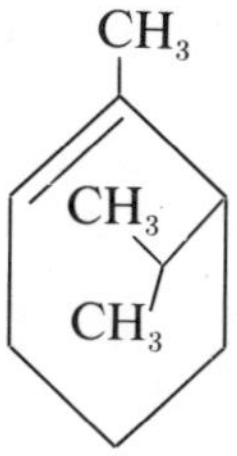

CLXXXVI 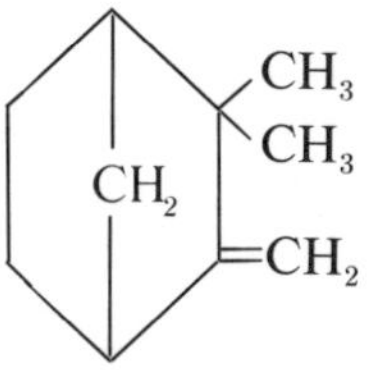 CLXXXVII 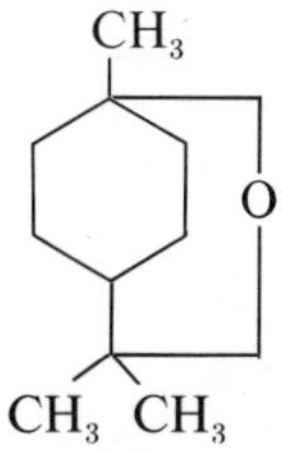 CLXXXVIII

Fig. 54. Volatile inhibitors.
CLXXXVI α-Pinene
CLXXXVII Camphene
CLXXXVIII Cineole

1965, 1966; Muller & del Moral, 1966). The most plentiful compounds were α-pinene (Fig. 54 CLXXXVI), camphene (Fig. 54 CLXXXVII) and cineole (Fig. 54 CLXXXVIII). One of the most interesting aspects of these studies was the demonstration that these substances can be absorbed from the atmosphere by the dry soil during the period of maximum production and can be retained there during the summer drought period. When the winter rains arrive they inhibit the germination and growth of seeds which have been dispersed into the affected area. The growth inhibition is closely paralleled by a reduction of oxygen uptake by mitochondria prepared from plants and this may be responsible for the poor growth of affected seedlings. (Muller *et al.*, 1969). Seedlings are not killed directly by these vapours but probably succumb from water deficiency in these arid regions because of the much reduced root growth.

(D) 'SICK SOIL' TOXICANTS

There is still much uncertainty as to whether or not soil sickness (see p. 234) is due mainly to the accumulation of toxic organic substances from previous plant inhabitants. However there is evidence that in certain cases such substances may play an important role. For example in old peach orchards studies by Patrick (1955) indicate the glucoside amygdalin from the old roots is attacked by micro-organisms to release toxic products into the soil. The hydrolysis of amygdalin yields glucose, hydrocyanic acid and benzaldehyde. In old apple orchards another glucoside phlorizin is present in the bark of apple roots and can be liberated into the soil. The glucoside itself and its aglycone (phloretin) can inhibit the growth of both roots and certain bacteria (Börner, 1959). Again these compounds may not persist for long enough periods to impede seedling establishment directly. However, long-continued leak of small quantities of these compounds into the soil of old orchards may so unbalance the microbial flora that 'sickness' results. The subject has been reviewed by Börner (1960) and McCalla & Haskins (1964).

Chapter X

Hormones Controlling Reproduction

THE FLOWERING PLANT

Introduction

One of the most refreshing and heartening sights after the cold dreariness of winter is the appearance of the spring flowers in our gardens and hedgerows. What a glory are the russets and golds of the autumnal chrysanthemums that grace our flower beds long after the earlier summer blooms have withered and fallen. This never-failing rhythm of reproduction, each species producing flowers and seeds in its proper season is as commonplace to us as our very breathing. But the miraculous way in which plants succeed in keeping to the seasonal time-table, in spite of wide fluctuations in prevailing weather from year to year, has been, until relatively recent times, one of the greatest enigmas of plant physiology. Why should the crocus be one of the first of our spring flowers, while the meadow-saffron (*Colchicum*), with a remarkably similar flower structure produces its flowers in the autumn? Why should annual plants be able to develop mature flowers and fruit in the first season of their growth, whereas biennials fail to do so until their second summer? Why should spring varieties of cereals in temperate regions be able to bear seed in the same season after spring sowing, whereas winter varieties need to spend the winter in the ground before ears can be formed? Why does such a predominating number of tropical plants flower all the year round with no perceptible rhythm in their fecundity? These are the sorts of puzzling questions that we are constantly meeting in connection with the apparently inexorable cycles of sexual reproduction in plants; but gradually over the last few decades, we have been getting at the answers.

As a mass of data accumulates concerning plant behaviour in a vast variety of plants, it has become clear that it is the regular alternation of certain environmental seasonal factors which imposes on the plant this developmental rhythm of reproduction. Environmental stimuli, the nature of which we shall consider later, set in train a series of plant responses which determine whether the appropriate buds shall produce flowers or vegetative branches. Stimuli and reactions are both complex, and there is much that is highly speculative in the explanations of our experimental results, but an imposing amount of evidence is piling up to suggest that the vital co-ordinating factors between stimulus and ultimate flowering reactions are hormonic in nature. The known hormones already described are certainly implicated and there is direct evidence for the existence of other substances specifically concerned with the control of flowering. Although these specific hormones have not yet been satisfactorily isolated in crystalline form, and therefore occupy the same position of doubt as did the auxins before the advent of Went's *Avena* assay, yet their interest and implications for the future are so great that this chapter on them has been written.

THE NATURE OF THE FLOWERING PROCESS

Perhaps we should pause to consider briefly what happens when a higher plant is 'switched' from a purely vegetative condition to one in which flowers are produced. I have deliberately chosen to use the word 'switch' because this described exactly the dramatic change-over which takes place in the plant's behaviour. As we have seen in Chapter I, leaves and branches develop from vegetative buds, and are the products of the activity of the growth centres they contain, the small conically-shaped terminal meristems. The highly organised patterns of cell division in these meristems determine the final form and arrangement of the leaves on the main stem (e.g. spirals, whorls, etc.). During the vegetative phase of development this pattern continues virtually unchanging and is thus responsible for the characteristic shape and structure of the vegetating plant.

When environmental change and the hereditary make-up of the plant interact to operate the 'flowering switch', a remarkable change takes place in a very short time in certain predetermined meristems. The pattern of cell division therein changes radically with the result that the shape of the meristems and the nature and arrangement of the organs they produce are dramatically affected. Instead of a series of very similar green photosynthesizing leaves, a succession of very different structures follow, coloured perianth segments, pollen-bearing stamens and finally ovule-containing carpels. Unlike the vegetative stem, the flowering axis elongates little if at all, and the arrangement of its floral parts (petals, sepals, stamens and carpels) is very different from that of the green leaves on the elongated vegetative shoot. As a rule, once this 'switch' is thrown, the events thus set in train cannot be reversed and, since the meristem is 'used up' in the generation of the flower, all subsequent growth ceases at that site.

If this happens in all terminal meristems the plant is doomed and will die when all its seeds are mature. This is what happens in ephemeral plants and in true annuals, where the complete switch of all meristems to the flowering state takes place in the first, and therefore only, season. The annual weeds Shepherd's Purse (*Capsella bursa-pastoris*) and mayweed (*Anthemis cotula*) are examples of this type of plant. Biennial plants live for two seasons, one of the factors operating the flowering switch being the low temperatures of the first winter. They too use up all their meristems and die at the end of the second season.

Extreme examples of such catastrophic behaviour are the 'century plants' (*Agave americana*) of tropical America, and the talipot palm (*Corypha umbraculifera*) of Ceylon and Indomalaysia. These plants live for a great number of years (up to a hundred in the case of the century plant) in a purely vegetative state, building up a vast vegetative body and accumulating large quantities of reserve food. Then suddenly, for reasons not yet understood, the switch is operated and a massive inflorescence is produced. As soon as the seeds mature the whole plant structure collapses and dies.

Fortunately a large proportion of plants have evolved in such a way that only a proportion of their terminal meristems are switchable by the effective environmental or internal factors. Such plants are perennials and are potentially immortal; indeed the big trees (Sequoias) of North America live and reproduce for many thousands of years and would never seem to die purely of old age and maturity.

The Environmental Control of Flowering

(A) RESPONSE TO DAY-LENGTH

The idea of a flower-inducing substance is by no means new. It figured prominently in the theoretical discussions of the nineteenth-century pioneer, Sachs (1865), long before the word hormone had been coined. But Sachs's theories were not very seriously considered by those immediately following him, since, at the close of that century, and in the first decade or so of the present one, the rapid advance of our knowledge in the purely nutritive aspects of plant physiology dominated contemporary ideas on growth control. Such an approach to the problem culminated, in 1918, in a theory in which the initiation of flowering was attributed to the attainment of a certain balance between the carbon and nitrogen nutrition of the plant (Kraus & Kraybill, 1918). The theory was based on experiments carried out on the nutrition of tomatoes. An abundance of nitrogenous manures greatly lowered the fruitfulness of these plants and, when coupled with conditions favouring active manufacture of carbon compounds in the leaves, gave rise to lush vegetative growth. When the supply of nitrogen was lowered, while maintaining a high level of carbon nutrition, there resulted plentiful fruiting and reduced vegetative growth, while low levels of both carbon and nitrogen nutrition, as would be expected, greatly reduced both vegetative and reproductive growth. But later, as more data accumulated, it became increasingly obvious that many species of plant would flower over a very wide range of carbon/nitrogen ratios, by which measure this balance was distinguished. The theory was obviously of very limited application.

It is to the French botanist Tournois (1911) that we owe the idea that day-length has a part to play in flowering behaviour, when he showed that Japanese hop, grown under glass in the short days of winter, showed precocious flowering. But the full significance of day-length as a major controlling factor was not realised until the publication of the works of two botanists in America, W. W. Garner and H. A. Allard who, in 1920 were experimenting on tobacco breeding near the city of Washington D.C. In contrast to the usual behaviour of tobacco, a new variety, Maryland Mammoth, refused to flower in the open in the Washington growing season, and did so only in the greenhouse in the autumn and winter. Crossings could not therefore be carried out between this new variety and other varieties. These two botanists tried unsuccessfully every horticultural trick that they could think of to make the plant flower in the summer. Finally the last possible variable, length of day, was investigated. By artificial shortening of the day-length during the summer the new variety was brought into regular and prolific flowering. They called this new phenomenon *photoperiodism.*

This startling discovery naturally led to an outburst of research which has steadily increased in volume in the intervening period, and has established that the flowering of a great proportion of plants is controlled by the length of day. Three classes of plants have been recognized on a basis of their photoperiodic behaviour. Firstly, there are those which resemble the Maryland Mammoth tobacco, in that they will come into the flowering state only when the day-length is decreased to a certain level. These plants have been called short-day plants and include many of our spring and autumn flowers such as *Chrysanthemum, Salvia, Cosmos*, goldenrod (*Solidago*), etc. The second class comprises long-day plants which will flower only if a certain minimum day-length is attained or exceeded. This embraces all the summer-flowering plants of the temperate regions of the earth, such as most of the temperate grasses, the spring varieties of cereals, beet, radish, lettuce, potato, etc. The remaining plants fall into the last or day-neutral class that can flower in any day-length. Examples of these plants are tomato, buck-wheat, and holly.

The classification of plants is in no way paralleled by the natural classification into families, etc., and there is no telling what the behaviour of any one species will be from a knowledge of its nearest relatives. Even varieties of the same species, as we have seen in the tobacco, can show widely differing responses to day-length. What is, however, of considerable importance to the plant geographer is that the world distribution of plant species may be largely determined by their photoperiodic responses. In the tropics, where the day-length varies little from twelve hours throughout the year, only short-day or day-neutral plants will be able to reproduce sexually and survive. Similarly, long-day plants will be restricted to the regions nearer the poles, where long summer days allow flowering: such plants in the tropics will remain non-flowering indefinitely, as the visitor to the tropics learns to his chagrin if he is misguided enough to try and grow temperate summer flowers there.

These very remarkable reactions of plants have for some time been used by horticulturalists in flower culture. We have seen their potential use in facilitating the crossing of varieties with widely spaced flowering times. Low intensity supplementary illumination during the winter nights is being used in greenhouses to maintain Begonias (short-day plants) in a vigorous vegetative condition, so that cuttings can be taken and grown continuously throughout the winter. Short periods of artificial illumination at midnight in the autumn are being used in *Chrysanthemum* cultivation to prevent flower development in the long autumn nights, and thereby spread the supply of blooms over a much longer period.

(B) RESPONSE TO TEMPERATURE

Day-length, however, is not the only climatic factor that may affect flowering. The temperature during one or more of the developmental phases of the growth of the individual plant may determine the future flowering behaviour. Very good examples of this are the spring and winter varieties of temperate region cereals such as wheat and rye. Spring varieties of these plants, if sown in the spring, will come into ear in the summer of the same year, while winter varieties will not. These latter need to be sown in the autumn and so spend the winter in the soil before ears will mature. If sown in the spring, they will not ear until the following season, i.e. they behave as biennials.

This phenomenon has been known for a very long time, and over a hundred years ago an American agriculturalist described an empirical temperature treatment for winter grain that would make it behave as spring grain

(Klippart, 1858). This consisted of allowing the wheat to germinate slightly in the autumn or winter, while keeping it from vegetating by a low temperature until it could be sown in the spring. The fact that it is the low temperatures of the first winter which induces biennials to flower in their second season was understood at the beginning of the century, since the famous german plant physiologist Klebs has shown by 1906 that many biennial plants (e.g. sugar-beet, *Verbascum*, *Digitalis purpurea*) could be kept vegetative over the winter in a heated greenhouse. In 1910 Gässner, his compatriot, first postulated a specific 'cold requirement' for flowering in such plants. But the idea received only passing attention until ten years after the first World War when the Russian agronomist Lysenko (1928) adopted it to work out a practical technique for the Russian farmer and called it *jarovizacija*, the English translation of which is vernalization. His technique for inducing in winter wheat the flowering properties of spring wheat was virtually the same as that of Klippart. Thus the cereal was heaped on the concrete floor of the barn and soaked in water sufficient to cause the grain to germinate, and when roots just began to form, the grain was cooled by opening the barn door. Subsequently the grain was periodically turned to ensure good aeration and prevent heating, and the water content kept up to the optimum value indicated above. After about three months the grain could be dried and stored at a higher temperature without losing its acquired spring-flowering characteristics. This technique was used in the U.S.S.R. to ensure early crops in regions where growing periods were curtailed by early frosts or droughts, or even to obtain two crops per season on the same land.

Needless to say, the phenomenon of vernalization has received much attention since its rediscovery, and other crops of agricultural importance have been shown to have their flowering behaviour modified by temperature treatment in the early growing stages. Mustard grown in India will flower much earlier after cold treatment during germination. In the tropics on the other hand, the same effect is brought about in rice by high-temperature treatment. A more complicated state of affairs exists in the temperature requirements for flower development in bulbous plants, as shown by systematic research on tulips and hyacinths in Holland (see Hartsema 1961). One brief example will illustrate the complexity of the flower-forming process. The tulip at the time of lifting is ready to initiate the flower in the centre of the thick fleshy scale leaves of the bulb. This initiation takes place most rapidly at 20° C. and is complete in about 2 to 3 weeks. For the development of this rudimentary flower to the state when it is ready to come above ground, temperatures of 8 to 9° C. are optimal, and this period lasts for 13 to 14 weeks. When elongation of the flower stalk actually begins, the optimum temperature slowly rises, reaching a maximum at about 23° C. at full bloom.

(C) THE COMPLEXITY OF ENVIRONMENTAL CONTROL

The picture so far presented of the environmental control of reproduction is grossly oversimplified. In most plants a *sequence* of environmental stimuli is required. Thus a photoperiodic requirement is often closely coupled with a temperature requirement. For example many plants which have an absolute requirement for low temperature exposure also have a subsequent requirement for long days, e.g. *Hyoscyamus niger*, a plant much used for vernalisation studies. It is an interesting fact that very few cold-requiring species (i.e. one or two *Chrysanthemum* spp.) need subsequent short days for flowering (see Plate 18). This is possibly connected with the fact that plants of temperate regions have evolved under a rhythm of lengthening spring days following the short cold days of winter.

Plate 18. Interaction of day-length and low temperature vernalization in *Chrysanthemum rubellum*. From left to right: unvernalized plants grown in short days; unvernalized plants grown in long days; vernalized plants grown in long days; vernalized plants grown in short days. Only the last plant has flowered, showing that short days will not bring about flowering in this species unless flowering meristems have been induced by low temperatures. (Photograph by kind permission of Prof. W. W. Schwabe, Wye College, London University.)

A further variation comes with the behaviour of some of the temperate cereals where short days may substitute for low temperatures. This has emerged from the work of Purvis & Gregory (1937) on rye (see also Purvis, 1961) which has brought to light the most complicated interactions between day-length and low temperature vernalization, in the initiation and maturation of the ears. These investigations have involved the painstaking dissection of the growing points of many thousands of plants in every stage of development, and have demonstrated unequivocally that the various discrete phases of ear development (e.g. initiation of young flowers, maturation and emergence from sheath) may each be separately determined by temperature or light treatment.

It was shown that the vernalization of winter rye by low temperatures was not an essential step for ear development, since such plants, grown through the winter in a warm greenhouse, matured normal ears in the following summer. This suggested that the internal changes brought about in the partly germinated *seed* under the cold winter soil could be induced also by the short days of winter in the *growing plant*. Unvernalized winter rye sown in spring will, however, eventually initiate flowers, though the ears which are produced late in the summer never emerge, but die within their sheaths. This failure to emerge is caused by the shortening days of late summer which stop 'shooting' or ear stalk elongation, which is promoted by long days. In spring rye and low-temperature-vernalized winter rye, in which the ears are initiated early in the season, the long summer days bring about emergence. A period of short days, however, will speed up the initiation of ears in non-vernalized winter varieties, and if long days are subsequently given, such varieties may be brought to ear in the first season.

An even tighter linking of day-length and temperature factors is seen in those plants whose responses to photoperiod may be radically changed by different temperature treatments. The strawberry is an excellent example of this. At temperatures above 14°C., this species behaves as a short-day plant and will not form flowers in day-lengths longer than 12 hours, the optimum being an 8-hour day. At temperatures below 14°C., the behaviour is intermediate, and flowering takes place even in continuous illumination (Went, 1954).

From what has been said above about hyacinth and rye it should be clear that flowering is a multi-stage process, each step of which may need a specific environmental stimulus for its completion. Very often different photoperiods are necessary for the different phases. We have seen already that short days followed by long days promote the formation of mature ears in unvernalized rye. A similar situation is seen in some grasses such as *Poa pratensis* and *Dactylis glomerata*. Other broad-leafed plants such as the devil's bit scabious (*Scabiosa succisa*) and clover (*Trifolium repens*) require

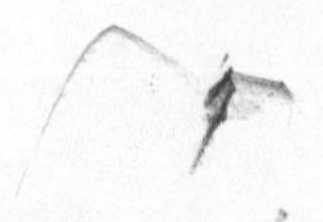

short days followed by long days; they are called short-long-days plants. Others, such as species of the succulent genus *Bryophyllum* respond to long days followed by short days and are called long-short-days plants; indeed the permutations and combinations of different shades of requirement are very great and, taking both temperature and day-length into account, Salisbury (1963) has distinguished forty-eight different response types. For a full exposition of these phenomena the reader is referred to Lang (1961), Salisbury (1963) and Evans (1969).

The Mechanism of Operation of the Stimulus

(A) PHOTOPERIODISM

One of the most significant of the characteristics of photoperiodic response is that it is of an 'all-or-none' type. Thus the 'quantity' of stimulus, that is the number of cycles (light period plus dark period), has to exceed a certain threshold before flowers are initiated. But once this has been achieved, the 'switch' is operated and flowers will be formed irrespective of the conditions to which the plants are subsequently exposed, including a photoperiod otherwise unfavourable for flowering. The plant in this condition is said to be 'induced' and the process of switching is called *photoperiodic induction.* The number of cycles necessary to operate the switch varies from species to species; in one of the classical experimental short-day plants, *Xanthium pennsylvanicum*, some varieties need only one short day. At the other extreme we have such plants as *Echeveria harmsii*, which requires 20 short days followed by 10 long ones to give even a minimal response. (Rünger, 1962). However minimal induction of this kind may switch only a few meristems and continuing exposure to inductive photoperiods results in an increasing production and more rapid maturation of the flowers. Coupled with this quantitative aspect of the response is the duration of the induced state. Thus some plants, e.g. the short-day *Xanthium* and the long-day *Nigella damascena*, once induced, never revert to the vegetative state. Others, e.g. short-day soybean (*Glycine soja*) need continuous induction to maintain flowering. As we shall see later, the biochemical changes which constitute the initial phases of the induction process take place in the leaves which are thus the sites of perception of the photoperiodic stimulus.

Since the stimulus is a light stimulus it is clear that a perceptor pigment must be involved in the process. A knowledge of its constitution and biochemical properties should be of great value in elucidating the mechanisms of photoperiodic induction and flowering control. Our effective access to this field of study was first made possible by the discovery that the critical part of the photoinductive cycle was the dark period and not the light period (Hamner & Bonner, 1938). Thus if the long dark period of the photoinductive cycle for a short-day plant is interrupted about half-way through by a brief period of illumination, then flower induction is prevented. Similar treatment of long-day plants kept vegetative by a short-day cycle will induce them to flower (see Fig. 55A). Clearly these short-day plants ought to be called 'long-night plants' and the long-day plants should be 'short-night

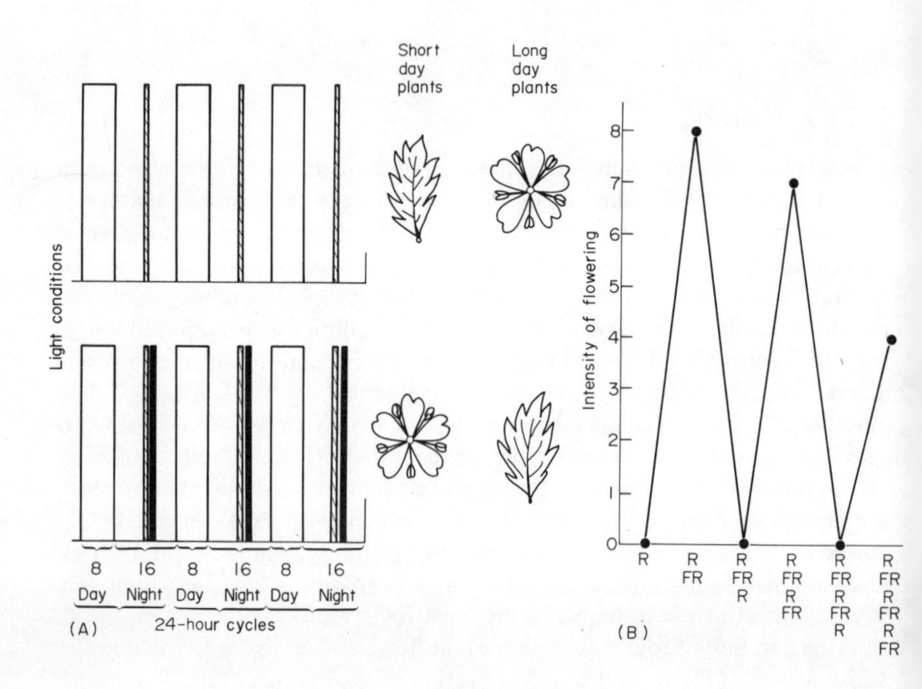

Fig. 55. Diagrams and graphs illustrating the effects on the flowering of plants grown in a short-day regime (8 hour day; 16 hour night) of interrupting the dark period at its mid-point by flashes of red or far-red light. (A) Experimental procedure and quantitative responses. The white rectangles represent exposure to normal daylight. The narrow cross-hatched and black rectangles represent flashes of red and far-red light respectively. On the right the responses show that red flashes keep short-day plants vegetative and cause long-day plants to flower. A far-red flash following a red flash reverses this situation. (B) Graph showing the quantitative responses in *Chrysanthemum* to successive alternating flashes of red and far-red light in a short-day regime. The treatments are shown vertically in descending sequence under the horizontal axis. (R = red flash; FR = far-red flash). The flowering response is determined by the last (bottom) flash of the series, i.e. red stops flowering (*Chrysanthemum* is a short-day plant) while far red promotes it. (Data from Borthwick, 1959).

plants', since it is the length of the uninterrupted dark period which controls the induction process. This ability to control the switch by brief light exposures allowed a team of workers in the U.S. Department of Agriculture Plant Industry Station at Beltsville, Maryland to study, for the first time with a precise quantitative technique, the effect of the wavelength of the light used on the effectiveness of the 'light-break'; in other words it enabled them to draw an action spectrum and hence obtain valuable information on the nature of the pigment concerned (Parker *et al.*, 1946). The action spectrum was found to be the same for the effect on both short- and long-day plants and was characterized by having one relatively narrow peak at the red end of the spectrum (maximum effectiveness at 660 nm). This indicated that a blue-green pigment was involved. Perhaps the most exciting discovery however was that the effects of irradiation with this red light could be completely annulled by immediate subsequent irradiation with light of a slightly longer wavelength (730 nm), i.e. in a region now called the far-red (see Fig. 55A). Furthermore the far-red light effect was itself easily reversed by a second exposure to red light; in fact the whole process of reversal could be repeated over and over again, the plant ultimately reacting to the wavelength of the last irradiation almost as if the preceding light exposures had not been given (see Fig. 55B). Further application of this technique has enabled the 'action spectra' of the two light actions to be worked out to show the relationship of the wavelength of the light to its effectiveness in exercising these two actions on flowering (Figs. 56A & B). As regards their spectral characteristics the actions of light on flowering are the same as those on many other plant growth processes, either triggered or controlled by light, such as the germination of light-requiring seeds, the control of leaf and stem growth, the production of the red pigment in the skin of apples and tomato etc. Descriptions of these widespread phenomena involving red and far-red light stimuli are obviously beyond the scope of this book and interested readers are referred to reviews by Borthwick & Hendricks (1961) and Hendricks & Borthwick (1964).

This easy and mutual reversibility of the effects of light of one wavelength by light of another suggested that the pigment concerned with light absorption ought to exist in two forms, one absorbing red light at 660 nm and the other far-red light at 730 nm. Furthermore the act of light absorption at one wavelength would change the chemical structure of the molecule to one absorbing at the other wavelength; i.e. red irradiation would convert the molecule to the far-red-absorbing form (symbolized now by P_{730}) while far-red light would convert P_{730} back to the red light-absorbing form (called P_{660}). This pigment system of twin forms was called *phytochrome* and its probable function in the photoperiodic control of flowering is illustrated in the following diagram:

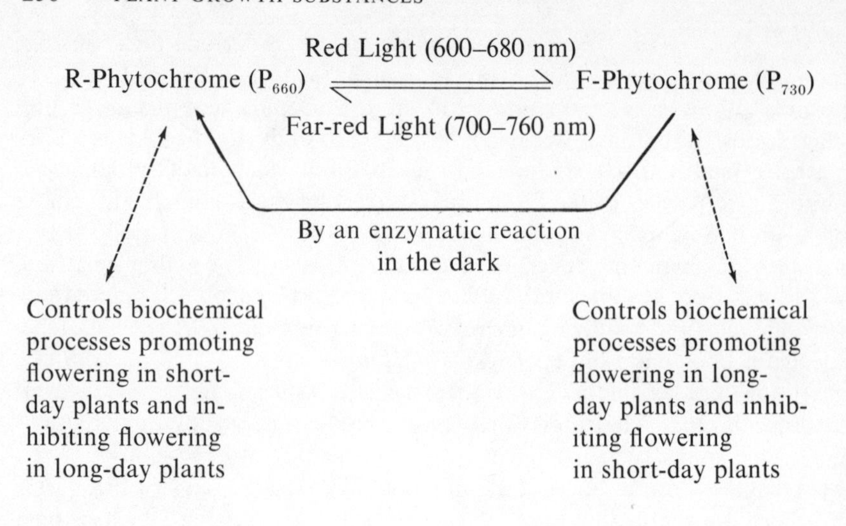

Naturally such a theory demanded experimental proof but success was long in coming owing to the extremely small quantities of the pigment involved in such a highly light-sensitive process. The pigment was first demonstrated in plant tissues by the use of a highly sophisticated instrument (a dual-wavelength difference photometer), and shown to have the interconversion properties described above (see Fig. 55) (Butler *et al.*, 1959). Further long and painstaking work ultimately led to the isolation of phytochrome in reasonably pure state in solution, when it proved to be a blue-green protein with a molecular weight of the order of 100,000 (Siegelman & Firer, 1964). We now know a very considerable amount about this extremely important protein and its phytochemical properties and much effort is now being put into a study of its behaviour in plants in relation to the physiological processes it undoubtedly controls. But again such topics are outside the scope of this book and further information should be sought in the specialized reviews of Butler *et al.* (1964), Siegelman & Butler (1965) and Hillman (1967).

The question naturally arises as to how the plant is able to respond to photoperiod, which inescapably involves the measurement of time, and thus the possession of some kind of internal biochemically-operated clock; as a corollary there is the question of the part played by phytochrome in this clock mechanism.

There is now overwhelming evidence that many, perhaps all, living organisms, exhibit regular rhythms in various aspects of their physiological activities. The most easily observed are the up-and-down movements of the

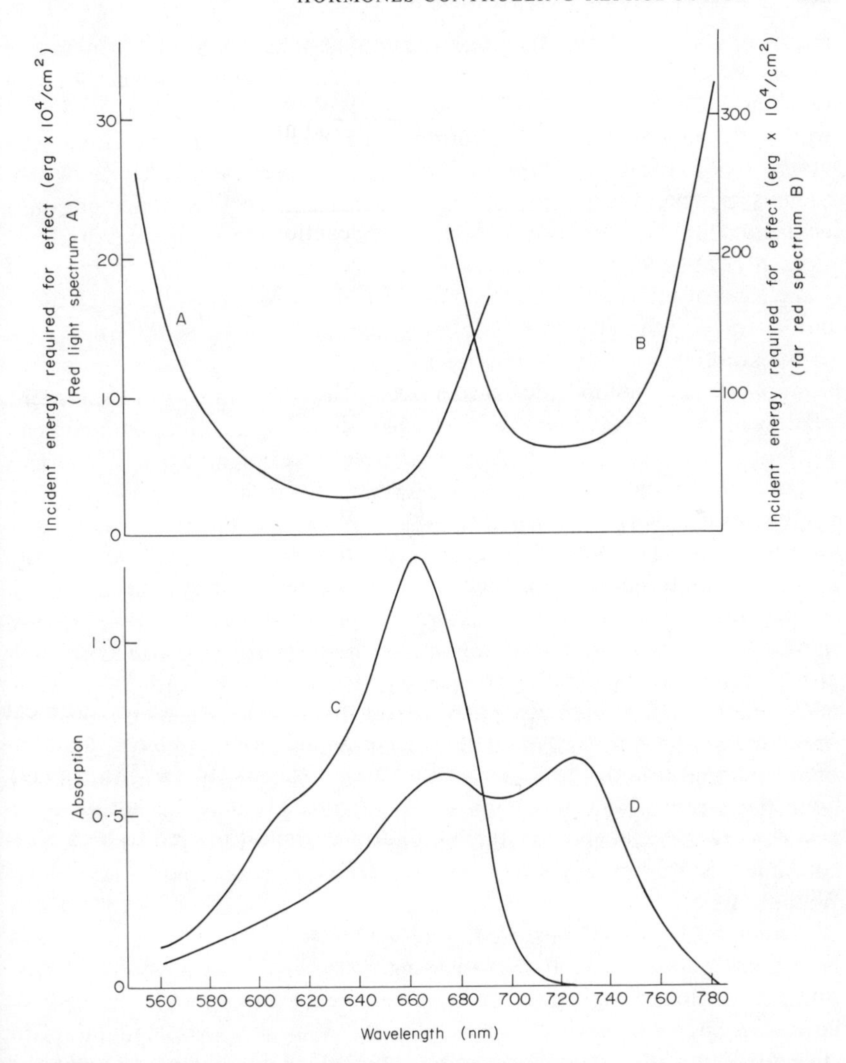

Fig. 56. (A) and (B). Action spectra for the control of flowering in cocklebur (*Xanthium*) by red and far-red light flashes. (A) Energy in flashes of red light in the long dark period necessary to induce half of the experimental plants to flower. (B) Energy in flashes of far-red light, given after a saturating dose of red light, necessary to inhibit the flowering of half the experimental plants (the saturating red flash alone causes full flowering). Maximum sensitivity of course corresponds with the *trough* of these curves (optimum wavelength for maximum sensitivity). (Data from Salisbury, 1963.)

(C) and (D) Absorption spectra of phytochrome. (C) After prolonged far-red illumination (i.e. red-absorbing from Pr); (D) After prolonged red illumination (i.e. far-red absorbing form Pfr). (Data from Siegelman & Butler, 1965.)

leaves of such plants as *Phaseolus*, the common bean, and the opening and closing of certain flowers. The realisation that organisms are sensitive to rhythms in their physical environment and that their behaviour can be controlled by such rhythms (i.e. photoperiodic control of flowering) has promoted studies which have revealed many subtle oscillations in the activity of organisms which can continue for considerable periods under constant environmental conditions. These have been called 'endogenous rhythms' since they appear to be generated inside the organism by the operation of some kind of internal oscillator system of biochemical change. Such oscillations include the growth rates of certain plant organs (e.g. the coleoptiles of oat seedlings), the light emission (luminescence) of the marine organism *Gonyaulax*, the metamorphosis of insects, the CO_2-emission of succulent plants such as *Bryophyllum* and even human physiological functions and their attributes such as body temperature, water secretion and pulse rate. The remarkable characteristic of all these oscillations is that they have a period (repetition interval) of approximately 24 hours and for this reason have been called circadian rhythms (Latin *circa* = about, *dies* = a day). Although they seem to be self-maintaining they are not completely independent of the environment since they need an external impulse to set them in motion. This impulse could be a period of illumination (the most usual), or a temperature change etc. Just as a pendulum, after a lateral push, will continue to swing with a period determined by its length, so will the internal pendulum of the organism continue to oscillate under unvarying conditions once the impulse of the external stimulus has started it off. Unravelling the mechanism underlying these remarkable biochemical oscillations presents one of the most fascinating of challenges to the biologist. The interested reader should continue his studies elsewhere, starting with the recent lucid survey by Wilkins (1968).

Clearly such endogenous oscillators, or 'biological clocks' as they have been called, constitute a time-measuring mechanism which would operate for the detection of day-length and thus mediate the photoperiodic control of flowering. Such a theory was proposed by Bünning who postulated the existance in plants of a circadian rhythm controlling flowering responses to photoperiod. In each 24-hour cycle he distinguished two periods, a photophile (light-loving) and skotophile (dark-loving) period, unfortunately not yet definable in any exact biochemical terms. For the flowering process to be set in train the external photoperiod had to be 'geared in' to this circadian rhythm; light exposure during the photophile phase would promote flowering and suppress it during the skotophile. (For full details of the theory see Bünning, 1960, 1967; Lang, 1965).

The difficulties involved in the interpretation of many flowering phenomena by this theory (see Salisbury, 1963; Lang, 1965) and the advent of phytochrome gave rise to a simpler hypothesis based on the observations that the P_{730} form of the pigment would (see diagram p. 256) revert spontaneously to the P_{660} form in the dark. At the end of the light period all (or most) of the pigment would be in the P_{730} form and during the subsequent dark period the rate of its spontaneous disappearance would constitute the timing process measuring the length of the dark period, just as the loss of sand from the upper bulb of an hour-glass has for centuries been used to measure time. But, as for the circadian rhythm theory of Bünning, this simple system cannot explain many of the experimental observations and we are still far from understanding the detailed mechanisms of the detection of and reaction to photoperiod.

Whatever it may be, this timing process constitutes the first stage in overall flower induction. Once the requisite number of inductive cycles have been measured off, the biochemical switch is turned and the next stage would seem to be the synthesis of a specific flowering hormone. The direct experimental evidence for this will be considered in detail shortly.

(B) VERNALIZATION

There are very many different types of response to vernalization treatment, for the details of which the reader should consult the review by Chouard (1960). But despite this a few basic characteristics suggest the nature of the response mechanism.

Firstly it would seem to be a process involving the use of metabolic energy, possibly for the synthesis of complex from simple molecules. Thus the exclusion of oxygen from the system under study will inhibit the process. For example exposing winter cereal grain to cold in an atmosphere of nitrogen (Gregory & Purvis, 1938) or submerged in water where oxygen access is severely restricted (Filippenko, 1940) will very greatly delay vernalization. Similarly certain inhibitors of the enzymes catalyzing terminal oxidation in respiration will also reduce the effectiveness of cold treatments (Krekule, 1961).

Secondly if cold treatments are immediately followed by exposure to high temperatures the vernalized state is destroyed; this is known as devernalization and plants so treated will not flower. If the devernalization treatment is delayed for a period of 3 to 5 days then it is ineffective since by then the vernalized state seems to have become established. Such behaviour suggests that low temperatures promote the synthesis in plants of some substance necessary for flowering and that this substance is caused to disappear at high temperatures. This idea is set out below in a reaction sequence suggested by Purvis and Gregory (1952).

$$A \underset{b}{\overset{a}{\rightleftarrows}} A' \xrightarrow{c} B$$

A is a substance acting as a precursor which is converted by pathway *a* to *A'*. At the same time this could revert to *A* by pathway *b* or be transformed to *B* by pathway *c*. *B* is the substance immediately responsible for starting the flowering processes (see later discussion on hormones). The vernalization and devernalization effects can be explained if the effects of temperature on the conversion pathways *a* and *b* are different. At low temperatures *a* is supposed to be much faster than *b* so that substance *A'* accumulates and is then converted slowly to *B*. As the temperature rises both processes are accelerated but *b* is promoted much more than *a* so that at a certain temperature *b* overtakes *a* and then becomes faster so that there is a net reversion of *A'* to *A* and consequently the production of *B* will be slowed and eventually stopped. This is of course not proven by any experimental observations but would seem to be most plausible, since a similar system seems to underlie the accumulation of sugar (sweetening) of potato tubers kept at low temperatures just above freezing. The stabilization of the vernalized state after 3 to 5 days would presumably be the outcome of the accumulation over that period of sufficient of substance *B* to allow the flowering processes to proceed, the subsequent loss of *A'* at high temperatures now being immaterial.

Thus all this evidence points to the promotion of the synthesis of a flower-controlling substance or substances. The theoretical cold-induced hormone has been called *vernalin* by Melchers (1939). The direct evidence for hormones of this kind will now be considered.

Evidence for Specific Flowering Hormones

(A) Evidence from Photoperiodism

The first fact to head the evidence for flowering hormones is that the leaves are the organs of the plant which perceive the photoperiodic stimulus. The first demonstration of this was by Knott in 1934 who established that it was necessary to hold only the leaves of a spinach plant in long days for it to flower. Shortly afterwards a more elaborate series of experiments were performed by the Russian plant physiologist Chailakhyan (1936) on *Chrysanthemum*. A number of similar plants with a terminal bushy habit were taken, and all the leaves were removed from the upper halves. These plants were now divided into four sets (see Fig. 57). The first set was left to grow in the long summer days, with the result that no flowers were formed at the apex of the upper leafless branches. In the second set, the lower halves with the intact leaves were covered each afternoon with a light-tight box, so that this part of the plant received short days, and the upper leafless part received long days. All these plants produced flowers at the apices of the leafless branches. In the third set, the tops were provided with boxes and given short days, and the leafy lower parts received normal long days, but no flowering resulted. The fourth set, in which the whole plant had short days, flowered as did the second set (see Fig. 57). Obviously, the stimulus of short days was received only by the leafy basal part of the plant and this stimulus was transmitted, presumably by means of a hormone, to the apical meristem of the upper defoliated part where flowers were induced to form. The upper part, being robbed of its leaves, could not receive the stimulus. Chailakhyan christened this theoretical hormone 'florigen'.

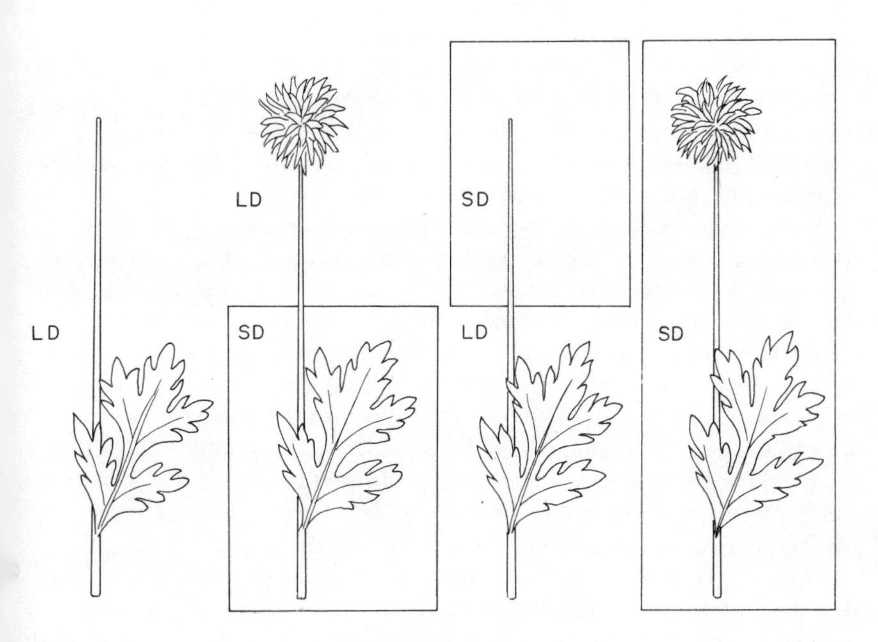

Fig. 57. Diagram of Chailakhyan's experiment with *Chrysanthemum* to demonstrate the transmission of a flowering hormone (florigen) from the receptors of the photoperiodic stimulus in the leaves to the apical meristem which is then induced to form an inflorescence. LD = long-day conditions; SD = short-day conditions.

Since then the leaves have been unequivocally established as the sites of photoperiodic perception in a number of other plants by similar techniques (see Lang, 1965, p. 1398). One most dramatic demonstration was with the little labiate *Perilla frutescens* with particularly hardy leaves which can be detached from the plant and will remain healthy for many days if kept properly moist. If such detached leaves are kept under inductive conditions (short days) and then grafted on non-induced plants in long days those plants will flower (Lona, 1949; Zeevaart, 1957). Properly induced leaves can be remarkably effective, one square centimetre of blade being sufficient in some instances to produce a substantial response (Zeevaart, 1958).

At the present time a very large number of grafting experiments have established that without doubt an influence (presumably a chemical substance) can be trasmitted by grafting from induced to non-induced parts of plants. Not only can this take place, as described above, between individuals of the same species and different response type (see Plate 19), or between different species of the same genus and response type (e.g. *Xanthium pennsylvanicum* and *X. americanum*), it can also occur between different genera of the same or even different response type. For example it can be transmitted in long days from annual henbane (*Hyoscyamus niger*) (long-day plant) to Maryland Mammoth tobacco (short-day plant) by grafting an induced portion of the former on to the latter (Melchers & Lang, 1941). Many other similar examples from more recent work are described by Lang (1965, pp. 1403–4). This is very strong evidence that the flowering hormones of long- and short-day plants are identical and that they are completely non-specific. A most interesting observation in this respect is that the broomrape (*Orobanche minor*), which is a root parasite of red clover, will flower only when its host does (Holdsworth & Nutman, 1947), suggesting that the flowering hormones are identical in these two widely separated families of flowering plants. However, for some unknown reason, transmission of this kind is not always possible as other species of broomrape (*O. ramosa* and *O. speciosa*) can flower on *Coleus* when this host is in the purely vegetative condition, and, furthermore, no flowering stimulus is thereby transmitted to it (Kribben, 1952). Again, flowering in dodders (*Cuscuta* spp.) will not cause their vegetative hosts to flower (Naylor, 1953; Harada, 1962), the flowering of the parasite being quite independent of that of the host.

Other characteristics of this transmission phenomenon point to the moving 'influence' being a chemical substance (hormone). For example if leaves are removed from an intact plant immediately after exposure to an inductive treatment then no flowers will be initiated. This implies that the influence is unlikely to be any kind of electrical message, like that in animal nerve fibres, since this should be propagated rapidly from cell to cell of the plant. If however induced leaves are left on the plant for a few hours after

Plate 19. The transmission of the flowering stimulus through a graft union. The scion is a short-day variety of tobacco, 'Maryland Mammoth' and the stock is a long-day variety. The composite plant is growing in long days. Flowers were removed from the stock as they were formed but the hormone induced by long days in the leaves has been transmitted to the short-day scion and induced it to flower. (Photograph by kind permission of Prof. Anton Lang, Michigan State University).

induction then flowering will take place as if the leaves had been left in position all the time. This argues strongly in favour of the movement of a hormone which would be subject to the same laws of transport as other soluble substances moving about the plant; it would need a considerable time to move out from the leaf and along the stem to the growing points where the flowers are formed. In fact leaf removal at various intervals after an inductive treatment has been used as a method of determining the speed of florigen movement and the factors controlling it (Salisbury, 1955).

There have been many attempts to get information on the nature of the hormone by studying its transmission through a graft union. For example, if it is a simple molecule one might expect it to diffuse through a watery medium in the same way as auxin diffuses in agar or gelatin (see Chapter II). Some workers have explored the possibility of such diffusive transmission through a water gap. This was done by fixing a leaf, which had received the correct day-length for flowering, to a debladed leaf stalk on a plant in an unfavourable day-length. The two cut ends, separated by about a millimetre, were kept surrounded by water. In *Chrysanthemum* the passage of a stimulus through this gap has been claimed (Moshkov, 1936), but these findings have very largely lost their significance as we now know that, although in this species the development of flowers from flower buds may be brought about by short days, yet the *initiation* of these flower buds is independent of day-length, although it is greatly hastened by low temperature vernalization of the vegetative plants. More recently, experiments on soya bean, similar to those of Moshkov, met with no success (Galston, 1949*b*). It is also likely that grafting experiments result in the actual organic union of tissue to a greater or lesser degree, and would therefore afford no positive proof of diffusive transport of a simple chemical substance. In one experiment on the American plant cocklebur (*Xanthium pennsylvanicum*), lens paper* was inserted between scion and stock to prevent this union, and the usual transmission of flowering influence was demonstrated (Hamner & Bonner, 1938). In the absence of union, such a transmission could have taken place only by diffusion of a soluble hormone through the water which was soaked up into the lens paper, although it has been claimed that tissue union may be established through such paper (Withrow & Withrow, 1943).

Confirmation of this suggestion comes from measurements of the time taken by the flowering 'message' to get across after a graft has been made; it corresponds exactly to that necessary for the tissue union to take place (Zeevaart, 1958). Furthermore extension of these studies on the movement of the sugar sucrose, labelled with radioactive carbon (^{14}C) for easy detection, showed that it too traversed the union at the same time. All this indicates that florigen travels only in those tissues responsible for the transport of sugars and other metabolites, i.e. in the phloem, and that for this an organic tissue continuity across the graft is necessary. There are other convincing experiments, which cannot be described in detail here, which demonstrate that florigen travels along the same channels as and together with the products of photosynthesis from the leaves. Thus if the intervening portion of stem between meristem and leaves is cooled, then no stimulus will pass through, in the same way as the movement of sugars etc. in the phloem is blocked by cold. The stimulus will also not pass through a portion of stem on which there are mature leaves subjected to a day-length unsuitable for flowering. It was originally thought that such leaves actually produced a substance which destroyed the hormone or prevented its movement. However it now seems fairly certain, from the very careful experiments of Chailakhyan and Butenko (1957) in which the route taken by photosynthetic products from such leaves was followed by allowing the leaves to

*Lens paper is very fine soft absorbent tissue paper made of pure cellulose and containing no weighting or dressing materials. It is used for cleaning the surfaces of very delicate lens systems such as are used in the compound microscope.

assimilate radioactive carbon dioxide, that the inhibition of flowering was caused by the deflection of florigen away from the meristems by an opposing stream of assimilates coming from the intervening leaves in the wrong photoperiod (and therefore themselves producing no florigen). As excellent survey of this work will be found in Lang's review (1965, p. 1411).

(B) EVIDENCE FROM VERNALIZATION

Evidence that vernalization processes also cause the formation of a transmissible chemical inducer came from grafting experiments similar to those previously described for photoperiodic induction. Such transmission takes place not only from vernalized to non-vernalized plants of the same species but also from non-cold-requiring to cold-requiring species and even between different species and genera. A few examples of many (see Lang, 1965) will serve to illustrate the situation. Thus grafts of biennial vernalized *Hyoscyamus niger* and *Lunaria biennis* can induce flowering in non-vernalized plants of the same species (Melchers, 1937; Wellensiek, 1961). Summer annual (i.e. non-cold-requiring) white mustard (*Sinapis alba*) flowering in long days will cause flowering in the non-vernalized biennial *Brassica oleracea* if they are grafted together (see Lang, 1965). Usually the effective transmission is promoted by the removal of the flowering buds from the 'donor' and of the mature leaves from the 'receptor' plant. The reasons for these effects are presumably the same as those for similar behaviour in photoperiodism experiments, i.e. the obligate movement of the transmitted hormone in the assimilate stream from actively photosynthesizing leaves.

The most interesting implication of such results is that the transmitted hormone is completely non-specific and therefore effective in all plants. Furthermore it seems to be the same substance whether it is produced by the process of photoperiodic induction or by vernalization, since non-vernalized cold-requiring plants can be induced to flower by grafting with photoperiodically induced non-cold-requiring plants of different species as well as by vernalized cold-requiring plants of the same or of other species.

Considerations of the Role of Known Hormones in Flowering Phenomena

It is naturally much easier experimentally to adopt what has been called the pharmacological approach and to study the effects of applying known plant growth substances on flowering behaviour, than it is to chase an elusive, chemically uncharacterized florigen and try to relate its behaviour to flowering phenomena. Consequently there is now a vast amount of data accumulated from the former type of experiment, much of it confusing and of little help to our understanding of the problems in hand. Nevertheless some attempts will be made below to sort out the main characteristics of the applied hormone responses. More direct approaches have also been made by following changes in endogenous growth regulators at various stages during the initiation and development of flowers. These will be outlined in a subsequent section. But before we embark on such an analysis it is important to recognize that the final emergence of a mature flower is the result of a complex sequence of morphological (and causative biochemical) processes, which can be categorized as follows in the order in which they occur: flower initiation (the turning of the 'switch'), flower organization involving the genesis of the floral parts and flower differentiation which marks the final stage in the maturation of those parts. Whereas the hypothetical florigen is presumably concerned only with the first of these phases, the known growth substances may affect all three.

(A) THE AUXINS AND ANTIAUXINS: EFFECTS OF APPLICATION

(i) On ripeness to flower

It is a general phenomenon that most plants must undergo a certain amount of vegetative development before they can be induced to flower. They must pass through what has been called the 'juvenile phase' of their development and reach 'ripeness-to-flower' before photoperiodic induction or vernalization treatments are effective. As an indication of the possible complexities of hormone relationships in the over-all flowering process it should be noted that applied auxins can affect the precise timing of this 'ripeness-to-flower' stage. Thus in Brussels sprouts, which is a cold-requiring biennial, treatment with auxin (NAA) brings forward by about two weeks the attainment of this developmental stage at which plants begin to respond to vernalization treatments (de Zeeuw & Leopold, 1955). These hormone treatments were applied to developing plants but similar effects have been claimed to arise from seed treatments (called 'seed hormonization' by Cholodny, 1936). The nature of these effects apparently varies from species to species; in some the flowering of plants grown from such treated seed is brought forward, in others it is retarded; many seeds are unresponsive (for literature survey see Kruyt, 1954). It is likely that these auxin effects result from non-specific actions on the general vigour of vegetative growth, on which the speed of attainment of 'ripeness-to-flower' may to some extent depend.

(ii) On photoperiodic responses

Much more frequently the question of the role of auxin in flowering has been considered in rather closer relationship to photoperiodic requirements. By a suitable choice of conditions, plants have been brought into the flowering, or kept in the non-flowering state, and attempts have been made to deflect the course of development so determined, by the application of auxins and related substances. The results have been somewhat conflicting, but some kind of pattern has emerged. Plants have usually to be brought up to the flowering threshold by a suitable photoperiod – so that they can be just 'eased' by the applied compound, either into the flowering condition or out of it.

Starting with short-day plants, it has been found that flowering can be inhibited under short-day conditions by the application of auxin sprays (see Plate 20). Thus *Xanthium*, subjected to four cycles of short days (8 hours), will subsequently flower in long days; but if NAA or IAA solutions (50 to 500 ppm) are applied, flowering is considerably delayed (Bonner & Thurlow, 1949). Other short-day plants, e.g. *Kalanchoë* and *Chenopodium*, behave similarly, and it would appear that auxin is effective only when applied to the leaves and during the dark period of the cycle (see Bonner & Liverman, 1953.) Since the degree of inhibition depends on both the concentration of IAA applied and on the duration of the light period, it has been suggested that IAA interacts with some flower-promoting product of the

Plate 20. The suppression of flowering in *Cucurbita* by auxin treatment. (A) Control, untreated. (B) Treated by the application of 500 ppm NAA solution. (Photograph by kind permission of Prof. J. Heslop-Harrison.)

light period, thereby making it ineffective for subsequent processes of induction (Hammer & Nanda, 1956). If these auxins are applied after the completion of the induction in *Kalanchoë* then there is no effect on the subsequent flowering response (van Senden, 1951). This, and other kinds of indirect information (see Lang, 1961), has led to the suggestion that in short-day plants auxin inhibition of flowering is the result of an interference with the synthesis of florigen.

This inhibition of induction processes has been confirmed using much simpler experimental conditions. Thus it is possible to grow sterile internode segments of some plants in culture, e.g. *Plumbago indica*, where they will produce vegetative buds under non-inductive and flowering buds under inductive conditions. By applying growth substances to the agar medium one can observe their effects on flowering under much more precisely prescribed conditions (Nitsch & Nitsch, 1967) (see Fig. 59 & Plate 24). In such work auxins (and also tryptophan and isatin which can act as auxin precursors) prevent induction in short days in *Plumbago indica* (Nitsch, 1968) (see Fig. 59B).

The problem has also been tackled by using substances supposed to counteract auxin action, i.e. the so-called 'antiauxins' or auxin antagonists. The auxin antagonists chosen for these experiments on flowering have been 2,3,5-tri-iodobenzoic acid (TIBA) (Fig. 43 CXLII), 2,4-dichloranisole (Fig. 44 CXLV), and ethylene (as the chlorhydrin). All these compounds have some claim to be antiauxins (see Chapter VII). The first report of a flowering effect of such compounds was in 1942 (Zimmerman & Hitchcock), when it was shown that, after applications of solutions of TIBA (4 to 200 ppm) to the soil around the roots, or as a spray to aerial parts of the tomato plant, clusters of flowers grew from axillary buds where shoots normally appear, and that terminal buds were similarly replaced. These observations have since been verified by a number of independent investigators, and flower buds have actually been induced to form in the very earliest post-germination stages when plants were only an inch or so high and had produced about three leaves (deWaarde & Roodenburg, 1948). Thus treatment with this anti-auxin appears not only to augment flower production but also to hasten the onset of 'ripeness-to-flower' (see auxin effects on p. 266). But tomato is insensitive to day-length, and it has been suggested that the above responses are merely general stimulations of flower development contingent upon modifications of vegetative growth by the growth regulator (Lang, 1961). Whether an antagonism of endogenous auxin is involved or not is still not established.

But to return to short-day plants, *Xanthium*, kept just in the vegetative condition by supplementing short days with very low light intensities, can be brought in flower by the application of TIBA and 2,4-dichloranisole

(Bonner, 1949*a*). Furthermore, plants which have already been brought into the flowering state can be made to flower more prolifically by TIBA, e.g. soybean (Galston, 1947). Ethylene chlorhydrin would seem to be a much more active antagonist as, when applied to *Xanthium* leaves, it can induce flowering even under long-day conditions (Khudairi & Hamner, 1954). However negative results have been obtained with other species such as soybean and *Kalanchoë* (see Lang, 1961, p. 921) and TIBA effects, when they occur, may have little direct relevence to the hormone control of flowering.

When we turn to long-day plants, the auxin—antiauxin effects are even less clear-cut. In a number of species that have been studied, the application of auxins to plants under long-day (flowering) conditions results in partial inhibitions or delays in flowering, but many either do not respond at all, or do so only as one aspect of a general suppression of overall growth. In one series of experiments with fourteen species of long-day plants, only six showed any specific inhibitions of flowering (von Denffer & Grundler, 1950).

However there are records of the *stimulation* of flowering response under long-day conditions. For example, small quantities of NAA solution (from 0.01 to 400 ppm) were applied in small phials to cut ends of leaves of Wintex barley which completely absorbed them. In a suitable long day, low concentrations of the auxin (0.01 to 1 ppm) augmented the total number of flowers formed, but high concentrations greatly reduced the number. The vegetative growth of the plant, moreover, showed very closely correlated growth responses, indicating that the action of the auxin was a general one on the whole growth of the plant and not merely specific to flower production (Leopold & Thimann, 1949). Dissection of ears in the early stages revealed much longer spikes in the treated plants. The auxin effect in this Wintex barley can be interpreted as a general effect on growth vigour, resulting not only in an augmented vegetative growth but also in closely correlated effects on the growth of flower primordia already initiated by the flowering hormone.

Experiments have been done with long-day plants in short days. In one experiment, *Hyoscyamus niger* was grown in an 8-hour day and brought almost up to the flowering threshold by being given short exposures to very low supplementary light in the dark periods. These plants could be forced over the threshold and brought into flower by dilute IAA sprays (Lang & Livermann, 1954; Liverman & Lang, 1956). Furthermore, the lower the supplementary light intensity (i.e. the more removed the plant was from the flowering threshold), the higher was the auxin concentration necessary to induce flowering. From such experiments it might be thought, since the response of these long-day plants to auxin seem to be opposite to that of short-day plants (i.e. a promotion *vis-à-vis* an inhibition), that photoperiodic induction ultimately operates through auxin levels. Unfortunately this is not upheld by extended experiments on *Hyoscyamus niger* (Lang, 1961) in which similar flowering promotion was obtained with the anti-auxins TIBA and 2,4-dichlorophenoxyisobutyric acid. Accelerations of flowering in the paw-paw (*Carica papaya*) have similarly been caused by applications both of auxins (IAA, NAA and particularly benzthiazole-2-oxyacetic acid) and the antiauxin TIBA (Dedolph, 1962). Such experiments are very difficult to interpret in terms of an obligate action of auxin in the flowering response proper. It is possible that they might be explained in terms of induced ethylene production by near-toxic concentrations of all these substances (see Chapter VIII p. 208).

(iii) In vernalization

The possible rôle of auxins in vernalization phenomena has not been much considered. The available data are mostly unsatisfactory in that they do not allow us to distinguish between effects on flower initiation, akin to vernalization effects, and effects on flower development, i.e. post-vernalization effects. The soaking of non-vernalized seeds seems to produce small inhibitions. Thus in radish both TIBA and IAA treatments delay flowering (Kojima *et al.*, 1957). Similar effects were observed in a wide range of plants (e.g. Wintex barley, soybean, oats, maize, Alaska pea, etc.) in dilute NAA solutions (0.0001 to 1 ppm) and subsequent germination at room temperatures (18°C) had a slight retarding effect on the subsequent flowering of the plants (Leopold & Guernsey, 1953*a* & *b*). If auxin soaks are combined with low-temperature treatment various responses have been reported. Thus low-temperature treatment of pea seeds caused a delay in flowering but this can be overcome by a previous soaking in auxin (NAA); such interactions led to the suggestion that auxin is also necessary for low-temperature vernalization processes (Leopold & Guernsey, 1954). On the other hand, in *Brassica campestris*, the effect of pre-vernalization treatment of seeds has been shown to depend on the substance employed. IAA, IBA and NAA all shorten the vegetative period following vernalization, whereas 2,4-D and the 'antiauxin' TIBA both have the opposite effect (Chakravati & Pillai, 1955). Similarly in *Lupinus albus* NAA accelerates flowering whereas IAA has no effect (Sechet, 1953). In other cases marked inhibitions of flowering have been observed, for example after treatment of winter wheat with IAA, NAA and TIBA (Chouard & Poignant, 1951).

Treatment of non-vernalized cold-requiring plants may hasten slightly the formation of flowers, e.g. in *Arabidopsis thalliana* (Sarkar, 1958) and *Sinapis alba* (Tang & Loo, 1940) but this is unlikely to be directly related to the vernalization processes themselves. Theories of a direct auxin action in vernalization processes must therefore be accepted with considerable reserve.

(B) ENDOGENOUS AUXIN CHANGES ASSOCIATED WITH FLOWERING PROCESSES

When we turn to consider direct evidence of auxin levels obtained by extraction and assay, the picture becomes more confused. Surveying the position up to 1952 Bonner & Liverman (1953), pointed out that the lack of reproducible results can be attributed to the presence, in the extracts, of substances interfering with auxin action in the assay techniques employed. What evidence then existed suggested that, in short-day plants (e.g. *Perilla* and *Kalanchoë*), individuals grown under long-day conditions contained more auxin than those grown under short (flower-inducing) days. Since then, further studies in *Xanthium* (Cooke, 1954) have demonstrated a rise in auxin content during short-day induction. As the flower initials form, however, auxin levels fall, whether the plant is in long or short days. Subsequently, flower development seems independent of auxin levels, which rise in long days and fall in short days. In soy bean and Maryland Mammoth tobacco, a chemical indicator (*p*-dimethylaminobenzaldehyde) used in a colorimetric estimation of indole compounds on paper chromatograms of extracts, has shown that short-day induction brings about a hundred fold increase in free (i.e. ethanol-extractable at $-10°$C.) auxin in the tissues (Vlitos & Meudt, 1954). However, as pointed out by Lang (1961) the changes were observed long after the photoinduction had been completed and are therefore associated more with flower development than with flower initiation.

Observations on long-day plants are few but here again long days seem to be associated with higher auxin contents suggesting that auxin levels are general responses to lighting conditions and are unrelated to the flowering behaviour (Lang, 1961).

As regards vernalization, only one observation calls for comment. This concerns the decorative plant *Streptocarpus wenlandii* which, throughout its life, possesses one persistent foliage leaf, which requires cold treatment as a pre-requisite for flowering. During temperature induction the auxin content of this leaf (probably IAA and IAN) decreases in the initial stages and then rises (Hess, 1958), the opposite of what happens in short-day induction in *Xanthium*. In all these varied observations there is no convincing evidence for a causal connection between changes in auxin content and flower initiation.

We may well summarize the situation by following Lang (1961) who distinguishes four categories of auxin behaviour. Firstly there is that in which flowering is promoted by a general enhancement of the growth vigour of the plant, such as may arise by seed treatments (hormonization). Then there are inhibitory effects due to excessive non-physiological auxin doses. Thirdly there are indirect effects where the promotion of vegetative vigour may, by some correlative and antagonistic action, suppress reproductive growth.

The last type of effect is where auxin appears to act directly in the induction processes. Here however critical studies are still needed and in their absence the role of auxin in induction processes still remains an open question.

(C) GIBBERELLINS: EFFECTS OF APPLICATIONS

In the early work on the bakanae disease of rice, which led to the discovery of the gibberellins (*see* p. 104), it was noticed that infected plants which reached maturity flowered earlier than normal plants. This observation has led to many investigations on the possible rôle of gibberellin-like factors in the hormonal control of flowering and the relationship of these substances to photoperiodic and vernalization phenomena. There is now no doubt that gibberellin treatment can very markedly modify the flowering behaviour of many different species. Indeed, the first time vegetative plants were induced to flower by substances obtained from higher plant tissue was when the endosperm of *Echinocystis macrocarpa*, which is rich in gibberellins was applied to the centre of the leaf rosette of vegetative biennial *Hyoscyamus niger* and *Samolus parviflorus.* Both species 'bolted' and flowered profusely in response to the treatment (Lang *et al.*, 1957) (*see* Plate 21.)

(i) In long-day plants

It appears that a great number of long-day-requiring plants will flower under non-inductive conditions if treated with gibberellins. Plate 21 illustrates the response in *Samolus parviflorus*. Lang (1965) gives a list of twenty-three different species from eleven families which react in this way. In some other species where daylength has quantitative effects and a few flowers are formed in short days, e.g. *Centaurea cyanus*, *Brassica juncea*, *Petunia hybrida* (Wittwer & Bukovac, 1957*a* & *b*) gibberellin application augments the flower production. In a few other species however there was either no response (*Sedum telephium*, *Anagallis arvensis*, *Urtica* sp.) (Chouard, 1957) or only a promotion of the elongation of the flowering axis (*Centaurea calcitrappa*, *Lactuca scariola*, *Aethusa cynapium*) (Lona, 1956; Lona & Bocchi, 1956). Lang has suggested that the lack of response in some of these species may have arisen because insufficient amounts of gibberellin had reached the sites of action or because the gibberellin applied (A_3) had been the 'wrong' one for the species concerned. This last suggestion was based on the fact that the various gibberellins (see Chapter IV) are not equally effective in flower induction and furthermore the order of effectiveness varies from species to species. Thus Michniewicz & Lang (1962), have shown in the long-day plant *Silene armeria* that the activity of gibberellin A_7 far exceeds that of any of the eight others tested, even including gibberellic acid. In another long-day plant *Crepis parviflora* gibberellins A_7 and A_4 had equally high activities not greatly ex-

Plate 21. The effect of gibberellin treatment on flowering in the long-day plant *Samolus parviflorus* under short-day non-inducing conditions.

From left to right; Control untreated plant; Plants treated with 1, 2, 5, 10 and 20 μg respectively of gibberellic acid daily. This treatment produced the same flowering response as growth in continuous long days. (Photograph by kind permission of Prof. Anton Lang, Michigan State University.)

ceeding those of A_1 and A_3. The stem-elongating action of these gibberellins follows yet another pattern (see Chapter IV). Still different orders of activity were shown by the cold-requiring plant *Centaurium minus* and *Myosotis alpestris* (Michniewicz & Lang, 1962) and by a number of different races of *Arabidopsis thalliana* (Napp-Zinn, 1963). This implies that gibberellic acid is not the gibberellin active in flower induction but needs to be converted to such an active compound if it is to have any effect. Ability to do this may well vary from species to species.

(ii) In short-day plants

In the converse situation of obligate short-day plants grown under long days gibberellin has no flower-initiating action. Examples are *Kalanchoë blossfeldiana* (Harder & Bünsow, 1956), Biloxi soybean and *Xanthium* (Lang, 1957), Maryland Mammoth tobacco and *Perilla* spp. (Chailakhyan, 1957).

If gibberellin is applied to short-day plants in short days there may be an enhancement or an inhibition of flower production. Only one exception to this general situation has so far been reported; *Impatiens balsamina*, a short-day plant, can be induced to flower by gibberellic acid application in long days at certain times of the year (Nanda *et al.*, 1970), a differential response presumably conditioned by the temperature situation preceding the treatment.

There are a few cases of gibberellin inhibition of flowering reported in plants of both photoperiodic types and one wonders how much of this is an indirect effect arising from the promotion of a vegetative growth antagonistic in some way to flower formation, as was suggested by Lang (1961) for some auxin actions. Such a negative correlation of effects on flowering and vegetative growth respectively have been shown in the strawberry (Guttridge & Thompson, 1964). On the other hand in tissue culture (see Chapter V) of internode segments of the short-day plant *Plumbago indica*, flower buds can be induced by short days and such induction is inhibited by three gibberellins (A_1, A_3, A_7) (Nitsch, 1967) under conditions where competitive effects from vegetative growth could scarcely arise.

(iii) In cold-requiring plants

In many biennial plants that require exposures to low temperatures before they will flower (e.g. biennial *Hyoscyamus*, *Digitalis purpurea*, *Centaurium minus*, *Lactuca sativa*, *Daucus carota*) gibberellin treatment can replace the period of exposure to cold and will induce flowering in addition to the characteristic stem elongation (see Lang, 1965) (see Plate 22). These effects are usually obtained only under long-day conditions; under short days the flowering stalk elongates only and does not produce flowers (Carr, *et al.*, 1957; Lang, 1957; McComb, 1967). In other similar plants however the situation is not so clear-cut, since in beet and several other biennials stem elongation (bolting) is promoted but no flowering is induced unless plants are given regular applications of gibberellins and in addition exposed to low temperatures approaching those normally required for flower induction (Bukovac & Wittwer, 1957). Unvernalized winter rye and several other cold-requiring cereals and grasses are similarly insensitive and elongate without flowering (see Lang, 1965).

It is interesting to note that helminthosporol has a flower-promoting action in lettuce and radish similar to but much smaller then that of gibberellic acid (Suge & Rappaport, 1968).

(D) GIBBERELLINS: ENDOGENOUS CHANGES ASSOCIATED WITH FLOWERING

There is no unequivocal evidence relating changes in the levels of specifically

Plate 22. The effect of treatment of carrot (*Daucus carota*) with gibberellic acid showing how it can substitute completely for the cold requirement otherwise necessary for flowering of this biennial plant. All the plants were grown in long days. *Left*. Control plants treated with water containing a wetting agent only. It shows the rosette of leaves of the purely vegetative plant. *Centre*. Plants treated daily with 10 μg of gibberellic acid applied as a spray to the growing region. The solution was in water plus wetting agent. *Right*. Plants which has received eight week cold treatment and showing normal second-season flowering. Note that the gibberellin-treated and the cold-treated plants are almost identical. (Photograph by kind permission of Professor Anton Lang, Michigan State University. From Lang (1957).)

identifiable gibberellins to flowering behaviour in plants. However there are several independent investigations which show that substances with the physiological properties of gibberellins change in concentration during flower induction and development.

The situation is clearest in long-day plants where inductive conditions cause an increase in the activity of such compounds. In *Hyoscyamus niger* one compound present in short days greatly increases in amount and another different gibberellin is induced to form in long days (Lang, 1960). Similar increase in gibberellin levels occur when spinach is transferred from short to long days (Wareing *et al.*, 1968). The most extensive observations have been made by Harada & Nitsch (1959, 1964) and Harada (1962) on *Rudbeckia speciosa* and *Nicotiana sylvestris*. Here, in the early stages of the inductive treatment, auxin-like substances accumulate and then decline but at the point when stem elongation and flower initiation commences a new compound, 'substance E', makes its appearance; this coincidence implies a possible causal relationship. Substance E, which has been extracted in milligramme quantities from hollyhock apices (*Althaea rosea*) (Harada, 1962), possesses some but not all of the properties of the gibberellins. It will also induce flowering in *Rudbeckia* under non-inductive day-lengths. Although its chemical structure has not yet been elucidated, tests so far suggest that it has neither an indole nor a gibberellin ring structure and is not an acid. Similar changes in gibberellin-like compounds have been demonstrated in the closely-related species *Rudbeckia bicolor* (Pont Lezica, 1965).

The hypothesis that long-day induction brings about a rise in the level of gibberellins to values requisite for flowering is supported by experiments with the growth retardants (see Chapter VII). Thus Zeevaart & Lang (1963) have shown that CCC, which blocks gibberellin biosynthesis, inhibits flower formation in induced plants of *Kalanchoë daigremontianum* (long-day plant) (see Plate 13). Normal flowering is restored if gibberellic acid is applied at the same time. Similarly with *Samolus parviflorus* both AMO-1618 and CCC suppress the flowering of long-day induced plants and this suppression can be relieved by increasing the number of inductive cycles or by the application of gibberellic acid (Baldev & Lang, 1965).

Qualitative changes in the gibberellin complement may well be involved in induction. In red clover (*Trifolium pratense*) the production of diffusible gibberellins was much greater in short day (non-inductive) conditions than in long days. However transfer from non-inductive to inductive conditions caused a change in the proportions of three different gibberellins making up the total complement (Stoddart & Lang, 1967*b*) (see Fig. 58). The authors suggest that these three gibberellins are linked in a biosynthetic sequence and that light may be required for at least one of the steps. In this way therefore day-length may be adjusting the level of a specific flowering gibberellin via its synthesis from other gibberellins. This impression is strengthened by studies on the long-short-day plant *Kalanchoë daigremontianum* (Zeevaart, 1969*b*). This plant contains two major gibberellins (called I and II since they were not characterized). In continuous short days the levels of both I and II remain very low; in continuous long days II increases considerably. In long days followed by short days (flower-inducing) there was a further increase in II. However II is not produced from gibberellic acid since that gibberellin, when applied, although inducing flowering, is not itself metabolized over a 45 day period and furthermore did not alter II levels.

In short-day plants (*Chrysanthemum*, *Perilla*, *Xanthium* and Maryland Mammoth tobacco) similar studies (Harada & Nitsch, 1959; Harada, 1962) have demonstrated no similarly clear-cut correspondence between the appearance of substance E and the induction of flowering.

In plants requiring vernalization, cold treatment seems also to induce changes in gibberellin content which may be causally linked to flower

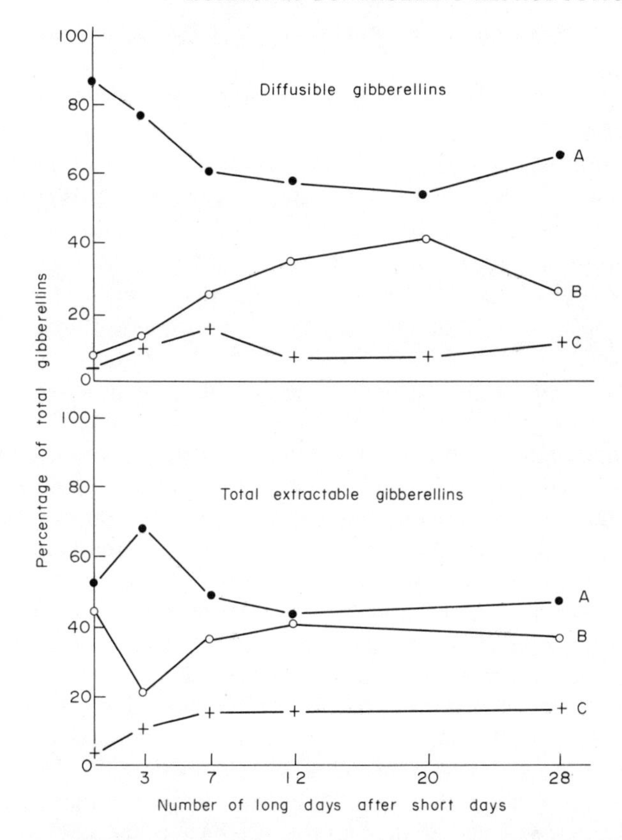

Fig. 58. Graphs showing the time-course of changes in the levels of three different gibberellins (A, B and C) in the leaves of red clover during exposure to long-day flower-inducing conditions. The top graphs show the relative proportions of the three components in the diffusible gibberellin fraction (collected in agar blocks from the cut ends of the petioles) and the lower graphs the proportions of the same components in acetone extracts of pre-frozen leaves. A, B and C not positively identified but have properties in common with GA_1, GA_5 and GA_9 respectively. Gibberellin levels determined after chromatography by the α-amylase, the lettuce hypocotyl and the dwarf maize bioassays (Data from Stoddart & Lang, 1969.)

induction. In the Japanese chrysanthemum 'Shuokan', which requires a period of exposure to low temperatures before it will flower, substance E begins to accumulate after the minimum chilling time has been passed (Harada & Nitsch, 1959); furthermore substance E applied to non-vernalised plants will cause them to flower (Harada, 1962). In winter wheat, vernalization brings about an increase in gibberellin-like substances in the grain and a vernalization inhibitor, sodium azide, prevents this accumulation

(Kentzer, 1967). In the honesty plant (*Lunaria annua*), gibberellic acid will not substitute for the cold requirement and there is little change in two major gibberellins of leaves and shoots on vernalization. However when vernalized plants were transferred to normal temperatures and flowers began to form there appeared a third gibberellin (Zeevaart, 1967), which may therefore be specifically involved in the flowering process. In the Wedgewood iris flowers are initiated in the bulb during periods of storage at relatively low temperatures (13°C). Flowering can be promoted in pieces of stem excised from these bulbs by treatment either with extracts of the scale leaves of vernalized bulbs or with gibberellins (Robrigues Pereira, 1961). Analyses of scale leaves and of the buds of these bulbs before and after vernalization indicate that cold treatment does not induce the formation of gibberellins, which are always present in the scale leaves, but activates their transport into the buds, thereby initiating flowers (Rodrigues Pereira, 1964). On the other hand cold-treated tulip bulbs seem to contain more extractable gibberellins (Aung & Hertogh, 1967), suggesting that the gibberellin economy (and hence flowering) may be controlled by low temperatures differently in the bulbs of different plants.

All this raises the question as to the true relationship between the gibberellins and, on the one hand the photoinduced 'florigen' of Chailakhyan and, on the other, the cold-induced 'vernalin' of Melchers. From grafting experiments it seems clear that one florigen may be common to both long- and short-day plants. Since gibberellins are inactive on short-day plants they cannot themselves constitute florigen. Chailakhyan (1958, 1961) has tried to resolve this difficulty by postulating that florigen is really a system of two hormones, one a gibberellin and the other a hypothetical 'anthesin' (see Plate 23). Both are needed in adequate quantities to induce flower formation. In long-day plants anthesins are always adequate under any day-length but gibberellins are at sub-threshold levels in short days and have to be boosted by synthesis in long days before an effective florigen system is constituted. In short day plants gibberellins are adequate under any day-length but it is the anthesins which are lacking in long days and their synthesis needs to be promoted to effective levels by short-day treatment. Unfortunately this theory is unacceptable since the grafting of non-induced short-day to non-induced long-day plants should give an effective combination of anthesins and gibberellins and both types of plant should flower; this does not happen in practice. *Kalanchoë daigremontianum* is a species requiring long days followed by short days in order to flower. Gibberellin treatment will induce flowering in continuous short days but not in continuous long days. However if a non-induced portion of a plant is grafted on a stock which has been induced by gibberellin in short days then the non-induced scion will flower *in long days* (Zeevaart & Lang, 1962). This can only mean that some flowering

(A)

(B)

Plate 23. The induction of flowering by plant extracts. Experiments showing the effects of growth in long days in promoting the synthesis of gibberellin-like substances in tobacco varieties, extracts of which will induce the flowering of *Rudbeckia* under non-inductive short-day conditions.

(A) *Left.* Plant of *Nicotiana sylvestris* growing under inductive long-day conditions. *Centre.* Plant of *Rudbeckia bicolor* growing under non-inductive short days but treated with an extract of the *Nicotiana sylvestris. Right.* Control untreated *Rudbeckia* plant growing under short days.

(B) *Left.* Plant of Maryland Mammoth tobacco plant growing under non-inductive long-day conditions. *Centre.* Plant of *Rudbeckia bicolor* growing under non-inductive short days but treated with an extract of the Maryland Mammoth tobacco plant. *Right.* Control untreated *Rudbeckia* plant growing under short days.

In both the long-day and the short-day varieties of tobacco, long days have promoted the formation of gibberellin-like substances which will induce flowering in *Rudbeckia* under short-day conditions. Although this gibberellin seems to promote flowering in the long-day variety of tobacco it does not do so in the short-day variety, because another essential flowering factor is probably missing. Chailakhyan has called this missing factor *anthesin* (see text). (Photographs by kind permission of Professor M. Chailakhyan, Timiriazev Institute of Plant Physiology, Moscow, U.S.S.R.)

hormone, *other than gibberellin*, is being transmitted from the gibberellin-induced stock to the scion. Lang (1965) concluded that gibberellin could be a factor causing, under appropriate conditions, the synthesis of florigen. This would take place in the leaves since a leaf, from a plant grown in short days, when treated with gibberellin and grafted on a receptor stock in long days, will induce flowering in that stock (Zeevaart, 1969*a*), suggesting that the long-day phase of induction, which must precede a 15-day phase of short days in *K. daigremontianum*, involves the build-up of high levels of gibberellins in the leaves. In cold-requiring plants, as we have seen, the effects of gibberellins vary, sometimes inducing bolting and flowering and sometimes inducing bolting only. This would suggest that vernalin in these latter plants is not gibberellin but another hormone independantly induced by cold treatment, and that gibberellin is concerned only with the phenomenon of bolting, a conclusion reached by Suge & Rappaport (1968) from gibberellin analyses in vernalized cereals. On the other hand Chailakhyan (1968) regards vernalin as a precursor of gibberellin; he suggests that under long days it is converted to gibberellin which then induces flowering but under short days no conversion takes place and the plant remains vegetative. Thus the long-day plant *Rudbeckia bicolor* will flower under short days if treated with extracts of vernalized winter wheat grown in long days but not by extracts of similar wheat grown in short days. Apparently the gibberellin necessary for flowering had been produced in the vernalized wheat only under long days.

(E) OTHER GROWTH-REGULATING SUBSTANCES

(i) Cytokinins and allied substances

Over a decade a few isolated studies have been made of the influence of applied cytokinins on flowering responses. Unlike the situation with the gibberellins, cytokinin applications seem to promote flowering in short-day plants. In *Perilla* spraying of plants with kinetin solutions under inductive conditions greatly accelerates flowering (Lona & Bocchi, 1957). In *Pharbitis* seedlings the stimulating effect was observed only when applications were made to the cotyledons and not to the apical bud (Ogawa, 1961). Furthermore its promotive effect was more marked for applications either just before or just after the commencement of the inductive dark period (Ogawa, 1961; Nakayama *et al.*, 1962). This suggests a kind of kinetin 'potentiation' of the induction process in the cotyledons or leaves; however, isolated terminal buds of *Perilla* grown on nutrient media can be induced to produce flower primordia by kinetin even under long days (Chailakhyan & Butenko, 1959). Similarly flowers have been induced to form on stem segments of the

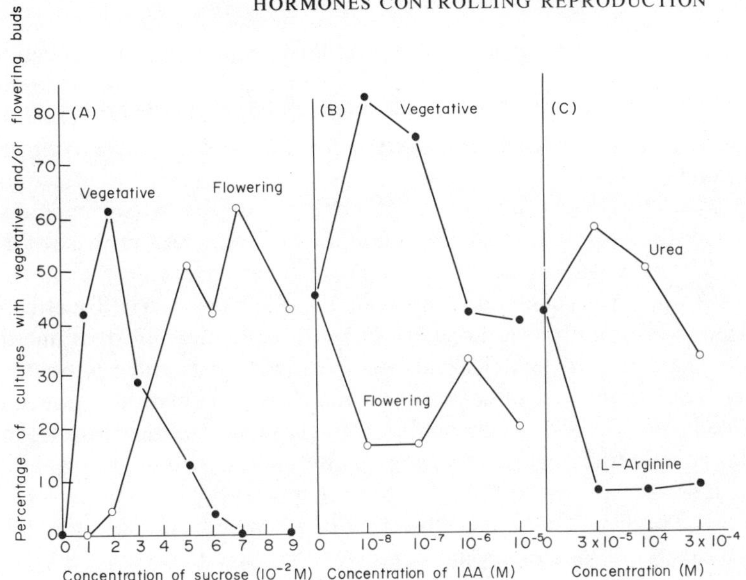

Fig. 59. Graphs showing the effects of various metabolites in the culture medium on the production of vegetative and flowering buds on excised internode segments of *Plumbago indica* grown in sterile culture under short-day (inducing) conditions.

(A) Effects of the concentration of sucrose in the medium on the percentage of vegetative and flowering buds. Medium contains in addition 1 mg/1 of kinetin and 10^{-4}M adenine (Data after C. Nitsch, 1967).

(B) Effect of a range of concentrations of IAA on the percentage of vegetative and flowering buds. Medium also contains benzyladenine (10^{-6}M), adenine (10^{-4}M) and 2 percent sucrose.

(C) Effects of L-arginine and urea on the percentage of flower buds. Medium also contained kinetin (1 mg/1), adenine (10^{-4}M) and sucrose (1.5 percent). (Data from Nitsch & Nitsch, 1967).

Bud counts were made after about one month of culture.

obligate short-day plant *Plumbago indica*, growing in culture in long days, by a mixture of a cytokinin and the base adenine (Nitsch, 1967, 1968) (see Fig. 59). This suggests that cytokinins have a much more direct action on induction processes in the apex and that these are allowed to proceed in isolated apices in long days because of the absence of leaves from which a positive inhibitory influence would emanate under non-inductive conditions (Chailakhyan, 1968).

Although in the long-day plant *Rudbeckia* kinetin application greatly suppressed flowering both in whole plants and in buds in culture (Chailakhyan & Khlopenkova, 1961) it induces flowering of the long-day plants *Arabidopsis thalliana* and *Callendula officinalis* in short days (Michniewicz &

Kamienska, 1965; Baszynski, 1967). Similar flower initiation was claimed for the cold-requiring plants *Cichorium intybus* and Petkus winter rye grown under warm non-inductive conditions (Bruinsma & Patil, 1963; Michniewicz & Kamienska, 1964) which suggested a possible action via the control of gibberellin levels.

Nevertheless these effects are not confined to the cytokinins but are shared by the closely-related bases adenine, guanine and uracil, and also the derived nucleosides adenosine and guanosine (Marushige & Marushige, 1962; see other references in Lang, 1965; Chailakhyan, 1968). There is also an extensive literature (see Lang, 1965, pp. 1463) which clearly shows that certain analogues of nucleic acid bases, which act as antimetabolites in nucleic acid syntheses, effectively prevent flower formation. Clearly the flowering switch involves a switch in gene expression on the chromasomes and this involves changes in the type of ribonucleic acid synthesized there (for a fuller discussion of this topic see Chapter XII). Cytokinins and the related bases very likely affect flowering by somehow influencing this synthesis directly; no-one as yet has suggested that cytokinins are florigens.

(ii) Vitamins

Vitamin C and nicotinic acid promote flowering quite markedly under inducing conditions (e.g. in *Rudbeckia* and *Perilla*) but do not cause flowers to form under non-inducing conditions, which suggests that they are not directly concerned with the induction processes (see Chailakhyan, 1961). On the other hand Michniewicz and Kamienska, (1964, 1965) claim that Vitamin E (as α-tocopherol acetate) can induce flowering in the long-day plant *Arabidopsis thalliana* under short-day conditions and in the cold-requiring *Cichorium intybus* under warm conditions where an indirect effect via gibberellin levels was suggested. Vitamin E can also replace the chilling requirement for flowering in winter rye (Bruinsma & Patil, 1963).

(iii) Sterols

In 1937 Chouard reported that the application of the female sex hormone oestradiol to plants of *Callistephus sinensis* greatly speeded up the onset of flowering in short days. Later Sironval (1957) made extracts of flowering plants of strawberry and found a particular fraction of it would promote flower formation in the same plant; this fraction contained Vitamin E, shown in the previous section to be active in other plants, and sterols, a chemical group to which the animal sex hormones belong. Although these were not cases of flower induction they aroused interest in the sterols as possible flowering hormones. Czygan (1962) succeeded in stimulating flowering in the duck-weed (*Lemna minor*) with animal ovary extract (containing the steroid oestrogens). Then Bonner *et al.*, in 1963 treated induced plants of *Xanthium* and *Pharbitis* with synthetic compounds which inhibit the biosynthesis of cholesterol (a very important sterol) and showed that they were powerful inhibitors of flowering. One compound, tris-(2-diethylaminoethyl)-phosphate trihydrochloride (SK and F 7997-A_3), strongly inhibited the incorporation of the precursor mevalonic acid into sterols in the leaf, suggesting that florigen may be a sterol. However since the recognized sterols, β-sitosterol and stigmasterol, are manufactured in the leaves of both vegetative and flowering *Xanthium* plants, it is unlikely that they are flowering hormones and this situation has led Lang (1965) to suggest that a minor sterol component, as yet undetected, might be florigen.

Bonner *et al.* (1963) claimed that the greatest effect of this sterol inhibitor was obtained with applications on the leaves just before the inductive long night and was ineffective on buds or leaves after induction, suggesting that it was indeed sterol (flowering hormone) biosynthesis during long-night induction that was inhibited. On the other hand when this compound is injected near the shoot apex of *Lolium temulentum* plants *after* long-day induction flowering is similarly prevented (Evans, 1964), which indicates that the induction process in the apex itself is also disturbed. We do not of course know what other biochemical processes may be influenced by these inhibitors.

(iv) Abscisic acid

Strangely enough the dormancy-controlling inhibitor abscisic acid seems to be able to induce flowering in certain short-day species, e.g. *Pharbitis nil*, *Chenopodium rubrum*, *Ribes nigrum* (El-Antably & Wareing, 1966) and in *Plumbago indica* stem segments growing *in vitro* Nitsch, 1967, 1968). Conversely it has been shown to inhibit flowering in the long-day plant *Lolium temulentum* (Evans, 1966) and *Spinacea oleracea* (El-Antably *et al.*, 1967). In the latter plant it drastically reduces gibberellin levels and thus its inhibiting effect may be indirect. The reduction in gibberellin levels may be the result of the suppression of the synthesis of RNA and hence of the enzymes of the gibberellin biosynthetic pathway (Wareing *et al.*, 1968). This question as to whether all these effects are direct or indirect via interactions with gibberellins or auxins will have to wait our better understanding of the growth-inhibiting actions of abscisic acid.

(v) Phenolic compounds

The addition of certain phenolic acids, e.g. *p*-coumaric acid and chlorogenic acid (Fig. 52 CLXXI) to the culture medium on which fragments of the root of chicory (*Cichorium intybus*) are growing, induces the formation of flower buds (Paulet & Nitsch, 1964). Again in the course of photoperiodic induction in the short-day plant Maryland Mammoth tobacco, the concentration of phenolic acids such as chlorogenic acid and two of its derivatives increases in the leaves up to the time when induction becomes irreversible (10 days). Then it suddenly decreases until flower primordia begin to differentiate when it rises again (Zucker *et al.*, 1965). The significance of these phenomena remain obscure although one is tempted to see a possible indirect connection with IAA, since phenolic compounds have long been known to modify the activity of IAA oxidase, and might therefore be acting on flowering by way of changes in endogenous auxin levels (see Chapter XI).

(vi) Nucleotides

Studies of diffusates from vernalized grain of winter rye have shown that they contain a substance which greatly promotes the onset of flowering in test plants of *Poa annuua* and winter wheat when applied to the grains (Tomita, 1962, 1963, 1964). The substance seems to be uridylic acid, purified preparations of which evoke the same responses.

(vii) Sugars

The new technique of studying flower induction in tissue culture (see above) has raised once again the question of carbon and nitrogen in flowering control. Thus flower buds will not form on cultures of *Plumbago indica* unless the concentration of disaccharide sugars (sucrose, cellobiose and maltose) is above a certain minimum; otherwise only vegetative buds result (see Fig. 59A). This is a specific chemical and not a general osmotic effect since isotonic mannitol solutions will not substitute for these sugars (Nitsch & Nitsch, 1965). However, increasing the supply of nitrate nitrogen also increases flowering in the presence of these high sugar concentrations (Nitsch, 1967) thus burying once and for all the early C/N ratio theory of Kraus & Kraybill (see pp. 247–8). In fact the effects of various nitrogen sources differ widely; some amino acids and amides suppress flower formation while urea favours it (Nitsch, 1967) (see Fig. 58).

(viii) Animal hormones.

There is some evidence suggesting the linkage of the animal nerve-ending hormone, acetylcholine, with phytochrome-mediated phenomena in mung bean (*Phaseolus aureus*) roots. Thus red light inhibits the formation of secondary roots and also induces root tips to adhere to a negatively-charged glass surface. Far-red light counteracts both these effects. Acetylcholine is present in all organs of *Phaseolus aureus*, particularly the growing points and an efflux of this hormone from secondary root tips can be demonstrated; red light augments this efflux and far-red light reduces it. Furthermore the application of acetylcholine to plants in the dark will mimic the effects of red light in reducing secondary-root formation and in inducing the adherence of root tips to glass. Both these simulations of red-light action by acetylcholine are prevented by acetylcholine esterase which hydrolyses the acetycholine to choline and acetic acid. Eserine, which inhibits acetylcholine esterase activity prevents the far-red light action in releasing the adherence of root tips to glass (Jaffe, 1970). From such evidence Jaffe concludes that acetylcholine acts as a local hormone in plants, as it does in animal nerve endings by regulating ion fluxes across membranes and that phytochrome action is mediated via the acetylcholine status of the plant cell. If this is so and it has any relevence to flowering, then acetylcholine could be immagined as in some way influencing the release of florigen from the cell of the leaf, or even its synthesis there. Obviously these highly speculative ideas need experimental checking in relation to flowering phenomena.

(viii) Ethylene

As we have seen short-day plants are unresponsive to treatment with gibberellins; indeed they are particularly refractory to chemical induction. One exception is the pineapple (*Ananas sativa*) which can be induced to flower by NAA applications (Clark & Kerns, 1942), an effect which was later attributed to the promotion of the production of ethylene, the supposed true inducing agent. This received strong support by the demonstration that (2-chloroethyl)-phosphonic acid (see Chap. VIII) also had the same flower-inducing properties (Cooke & Randall, 1968). This has been further extended to the short-day plant *Plumbago indica* which can be induced to flower in long days by application of this acid to entire plants or to pieces of stem generating buds in sterile culture (Nitsch & Nitsch, 1969) (see Plate 24).

Attempts to Extract a Specific Flowering Hormone

Ever since the concept of a specific florigen was formulated attempts have been made to extract it from induced plants without, as yet, any wholly convincing success. As early as 1936 a claim was made that such an extract had been prepared from the stigma of the crocus (Ulrich, 1939), but subsequent attempts at confirmation failed (Melchers & Lang, 1941). Kerosene-oil extract of flowering cocklebur (*Xanthium echinatum*) yielded a water-soluble material of a crystalline nature, apparently a calcium soap, which was claimed to have induced flowering in vegetative cocklebur plants (Roberts, 1951). The effects were however, very slight, and the material is very unlikely to be the natural 'flowering hormone'. Extracts of leaves of flowering strawberries have been shown to contain a substance or substances which promote the formation of flowers in vegetative test plants (Sironval, 1957) and we have already seen that activity seems to be associated with an unsaponifiable fraction, containing small quantities of vitamin E and a larger

Plate 24. The induction of flowering in *Plumbago indica* by ethylene.

(A) Experiments with intact plants, grown under non-inductive 16-hour days. *Left*. Control untreated plants. This plant is purely vegetative. *Right*. Plant sprayed on two successive occasions, three days apart, with 240 mg/l ethrel. The plant is flowering.

(B) The induction of a flower on a segment of internode grown in sterile culture in long non-inductive days by 0.1 percent ethylene in air. (Photographs by kind permission of Mme. Colette Nitsch and Dr. J. P. Nitsch, Laboratoire du Phytotron, Gif-sur-Yvette, France. (B) from C Nitsch (1968).)

quantity of a sterol. Similar flower-inducing extracts possibly containing sterols have been obtained from *Chrysanthemum* (Biswas *et al.*, 1966). We have also discussed (pp. 275–7) the extensive work of Harada & Nitsch who have extracted substance E from induced long-day plants which will induce those plants to flower in short days. The fact that this substance, although not a gibberellin, has a number of physiological properties characteristic of the gibberellins, suggests that it may be acting as such and may not, for reasons set out earlier in this chapter, be regarded as a true florigen.

This well-established response of long-day plants to gibberellins makes them unsuitable as test plants for a non-specific florigen and much more reliance should be placed on studies with short-day plants. In this respect attention has recently been concentrated on *Xanthium* (Lincoln *et al.*, 1961; Lincoln *et al.*, 1966; Carr, 1967); young leaves of induced plants have been freeze-dried and then extracted with absolute methanol. The extract contains a substance which induces the early stages of flowering in *Xanthium* plants in long days if applied to the backs of the leaves. The really interesting observation is that the effect of this extract is greatly increased by gibberellic acid which, by itself of course, has no action, under non-inductive conditions (Carr, 1967). Gibberellic acid also greatly augments the flowering response due to short-day induction and this has led Carr to suggest that gibberellins have a specific role to play also in the flowering of *Xanthium* where they act in co-operation with the active principle demonstrated in the extract. This situation is highly reminiscent of Chailakhyan's theory (see p. 278) that florigen is really a complex of gibberellin and another hormone, anthesin.

However the situation remains confused since we have to reconcile the apparently easy diffusive transmissibility of the substances in such extracts with the observations that transmission from induced plants can take place only through the cell-to-cell contacts of a graft union (Withrow & Withrow, 1943). These latter characteristics of 'florigen' are strikingly similar to the properties of plant viruses. Again the flowering state, like many plant viruses, does not persist through the seed. These similarities have led to the suggestion that 'florigen' may *not* be the kind of hormone we are acquainted with in auxin, i.e. a small diffusible molecule, but that it may be of a complex macromolecular nature (Bonner & Liverman, 1953). It would not be transmitted by diffusion but, like the virus particle, would induce the formation of identical molecules in neighbouring cells and thus spread through the plant.

Carr (1967) has suggested that plant physiologists have been too ready to accept the concept of a single non-specific florigen, whereas there is a fair amount of evidence against it. In view of the complexity of the hormone system controlling the production of sexual organs in lower plants (see p. 293), he is of the opinion that a similar complex substances ought to be controlling the much more complicated processes of flower production. He suggests that there may be at least two florigens; one, which is water soluble and transmitted by diffusion is the immediate product of the photoinduction of leaves and the active principle of the *Xanthium* extracts described above; the other, the true flowering hormone, would be that which is transmitted only by cell-to-cell contact and would be the factor responsible for conferring the flowering state from an induced to a non-induced individual (see p. 262). The relationship between these two florigens is not clear and further progress must wait for a closer chemical characterization of these elusive substances.

The Hormone Control of Sex Expression in Flowering Plants

(a) Introduction

The majority of flowering plants possess fully developed male and female organs in each flower, which are therefore said to be hermaphrodite or monoclinous, but in some groups the flowers may be unisexual and therefore to two kinds, one with only male and the other with only female organs; such flowers are diclinous. In these plants, both male and female flowers may develop on the same individual parent (monoecious condition) or they may be confined to separate male and female plants (dioecious condition). The distribution of sexes between the flowers in the former instance and between individual plants in the latter is determined by hereditary laws, which means that the nature of the floral organs produced after flower induction and their distribution about the plant are under the control of the nuclear genes. The regular sequence of flower parts (calyx, corolla, stamens and carpels) produced from the flowering meristem is a consequence of the programmed reading of closely related information sequences in these genes in the meristem cells. A possible basis of this programmed reading is outlined by Heslop-Harrison (1963). Normally this programme is immutable but in a considerable number of plant species, external factors, such as the state of the plant's nutrition, etc., may greatly influence sexual expression in these plants.

The effects of soil nutrients on sexual expression were among the first to be studied from a range of diclinous species. Over the years there has come consistent evidence that moist soils with a high content of nitrogenous nutrient favour femaleness relative to maleness. e.g. they induce an increase in the proportion of female flowers in the monoecious cucumber; dry soils with a low nitrogen content promote maleness. Temperature also has a marked effect; in some monoecious species such as *Cucurbita pepo* low temperatures favour an increase in the proportion of female flowers. But perhaps the most interesting environmental influence, in the present hormone context, is that of day-length; in fact in the first observations ever made on the effect of day-length on flowering behaviour it was shown in the dioecious species hemp (*Cannabis sativa*) and Japanese hop (*Humulus japonicus*) that short days would markedly alter sex expression. Female plants remained strictly female and produced only carpellate flowers but normally male plants exhibited a considerable degree of sex inversion of the male flowers in which stamens showed potential or complete transformation to carpels (Tournois, 1912, 1914). Since then there has accumulated much data establishing that in many diclinous short-day plants continuous short days induce a definite shift towards femaleness. For a comprehensive survey of these environmental effects the reader is referred to Heslop-Harrison (1957). These effects clearly imply that hormone changes, similar to those which may regulate the onset of flowering, may also be concerned in sex expression.

(B) THE EFFECTS OF KNOWN HORMONES

One of the first observations on the effects of applied growth substances on sexual expression in plants was made with mammalian sex hormones (Löve & Löve, 1945). Thus, the application of the mammalian female hormones oestrone* and oestradiol* in lanolin paste to the leaf axils of male plants of *Melandrium rubrum*, caused a partial suppression of the male organs (stamens) in the flowers which subsequently arose in those axils, and promoted the formation of the female organ (pistil). The male hormone, testosterone.† had the opposite effect, suppressing the expression of femaleness in flowers on female plants, and promoting the formation of stamens. Unfortunately, these observations have yet to be confirmed, and a subsequent careful repetition of the work on two British races of the same species of plant failed to obtain any response except some slight stimulation of growth vigour (Heslop-Harrison, 1948). There seems little doubt, however, that the Swedish results are real.

The mammalian female sex hormones are widely distributed in plants, and their male counterpart also occurs; we have already seen that specific inhibitors of sterol synthesis strongly inhibit flowering while the external application of sterols may promote it. Clearly then the possible involvement of sterols in the reproductive development of plants needs further close study.

Auxin is a more recent arrival on the scene of sex determination in flowers. Attention was first drawn by Laibach (1952) to the fact that in a very large number of species that have male and female flowers on the same plant, the transition from vegetative to flowering state is usually marked by the preponderance of female flowers. This is followed later by a wave of production of male flowers. Now it seems possible from the previous discussion (p. 271) that the transition from vegetative to flowering state is marked by a fall in the auxin content of the plant. If this fall persists into the period of flowering, then female flower production is associated with a higher auxin level than male flower production. This suggestion receives experimental support from experiments on cucumber (*Cucumis sativus*) (Laibach & Kribben, 1950) and pumpkin (*Cucurbita pepo*) (Laibach & Kribben, 1951), where applications of the auxins IAA and NAA caused a considerable increase in female flowers at the expense of male flowers. In hemp (*Cannabis sativa*), dilute auxin sprays can induce the formation of female flowers on male plants at sites which would normally be occupied by male flowers (Heslop-Harrison, 1956).

*Oestrone and oestradiol are hormones elaborated by the ovary in mammals and are of fundamental importance in the control of the uterine cycle.

†Testosterone is the male mammalian hormone elaborated in the testes and is responsible for the control of male sexual characteristics.

Treatment of plants (e.g. *Cannabis sativa*, *Mercurialis ambigua*) with low concentrations of the gas carbon monoxide, which inhibits the enzymes probably responsible for the destruction of auxins in plants, also results in a considerable increase in female flowers at the expense of male flowers (Heslop-Harrison & Heslop-Harrison, 1957*a* & *b*). This response may be a reflection of raised auxin levels brought about by the reduced enzymatic oxidation of IAA.

In the last decade these results have been largely confirmed and extended to other species such as *Hyoscyamus niger, Silene pendula*, and *Zea mays*, in which latter moneocious plant female flowers were induced to form in the male inflorescences (tassel) (Heslop-Harrison, 1961). On the other hand some plants are in this aspect unresponsive to auxins; for example, the balance of sex in hermaphrodite flowers of *Solanum nigrum* cannot be altered by NAA (Kiermayer, 1961) and neither can sex expression in the dioecious species *Mercurialis annua* and *Melandrium rubrum* (Heslop-Harrison, 1963). The effects of the presumed antiauxin TIBA (see Chapter VII) on flower development are varied and puzzling (see review by Heslop-Harrison, 1964) and in view of the uncertainty surrounding the basic mode of action of this substance, are of no help with the problem of the role of auxins in sex expression. However the positive effects of auxins are substantial and consistent enough to be accepted as physiologically significant.

This promotion of femaleness by heightened auxin levels in the plant is puzzling in view of the low auxin concentrations which appear, from the previous discussion, to be necessary during the *induction* of the flowering state. To resolve this enigma, Heslop-Harrison (1957) has suggested that although auxin levels may have to fall before flowers can be *initiated*, yet a much higher auxin level is required for flower *development*. He also proposes that the auxin concentrations necessary for female organ development may be much higher than those for male organ development.

In comparison with the relatively clear-cut action of the gibberellins on flowering, their effects on sex expression are a little puzzling. For example in the hermaphrodite flowers of *Hyoscyamus niger* and *Bryophyllum proliferum* and in monoecious *Zea mays* its effects resembled those of auxin in suppressing maleness and promoting femaleness. On the other hand it has the opposite effect in cucumber (for reference details see Heslop-Harrison, 1964) and in pumpkin (*Cucurbita pepo*) (Splittstoesser, 1970). Similarly it will induce anther and pollen development in male-sterile tomato and barley mutants (Phatak *et al.*, 1966; Kasembe, 1967). Even more complications are seen in *Luffa acutangula* where both gibberellic acid and the antigibberellin CCC promote femaleness and suppress maleness (Bose & Nitsch, 1970) (see Fig. 60). The action of ethylene, which we have seen earlier very often mimics the action of auxin at high concentrations, is also varied; it can

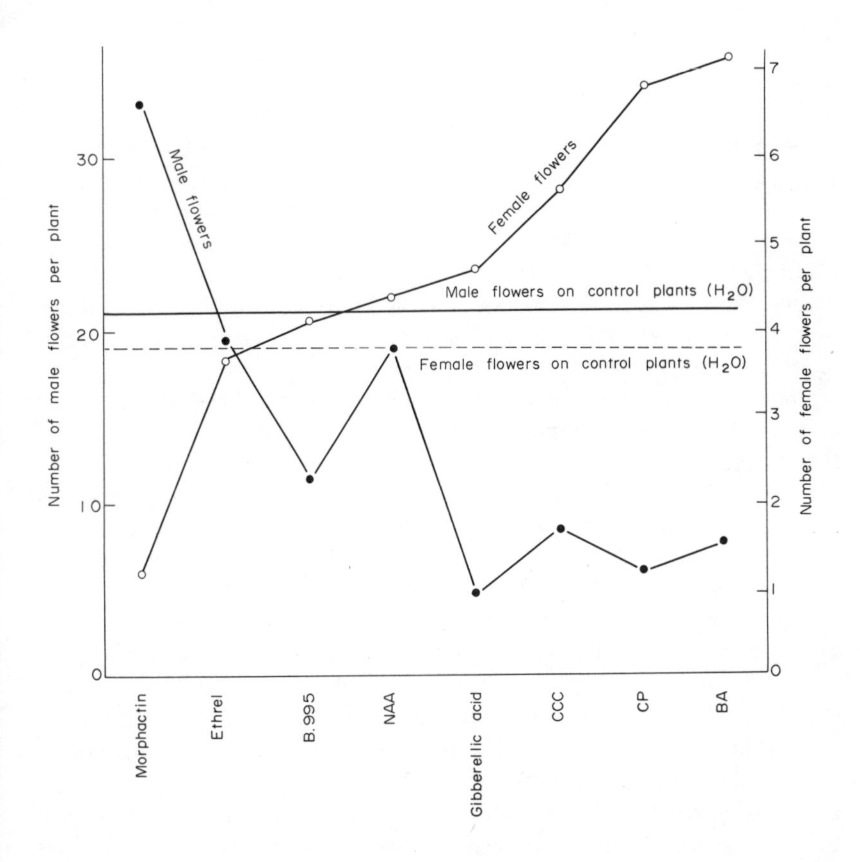

Fig. 60. Graphs showing the relationships between the number of male and female flowers per plant on the first ten nodes of *Luffa acutangula* plants grown from seed treated with a range of plant growth substances. The substances have been ranged along the horizontal axis in order of increasing promotive action on 'femaleness.' It will be seen that there is a distinct inverse correlation between the numbers of male and female flowers, a decrease in five male flowers per plant corresponding roughly with an increase in one female flower per plant (Data from Bose & Nitsch, 1970.)
Morphactin = 10 mg/1 methyl ester of 2-chlorofluorenol-9-carboxylic acid
Ethrel = 10 mg/1 (2-chloroethyl)-phosphonic acid
B995 = 100 mg/1 *N*-dimethylaminosuccinamic acid
NAA = 100 mg/1 naphth-lyl-acetic acid
Gibberellic acid = 100 mg/1
CCC = 1 g/'1 2-chloroethyltrimethylammonium chloride
CP = 100 mg/1 *N*-3-chlorophenyl-*N'*-phenylurea
BA = 10 mg/1 Benzyladenine

promote femaleness in *Cucumis sativus*, *C. melo* and *Cucurbita* pepo (Rudich *et al.*, 1969; Splittstoesser, 1970), but has no action on *Luffa acutangula* (Bose & Nitsch, 1970). A dramatically opposite effect is shown by morphactin (i.e. the methyl ester of 2-chlorofluorenol-9-carboxylic acid), which belongs to the group of growth regulators which have a marked blocking action on the polar transport of auxins (see p. 324); in *Luffa acutangula* seed treatment with this compound markedly promotes the subsequent production of male flowers while female flowers are correspondingly suppressed (Bose & Nitsch, 1970) (see Fig. 60).

Shifts towards femaleness also characterize the response to cytokinins. In a dioecious form of *Vitis vinifera* a synthetic cytokinin, 6-(benzylamino)-9-(2-tetrahydropyranyl)-9H-purine, applied to male plants during flower development induced the formation of female organs in otherwise male flowers (Negi & Olmo, 1966). Kinetin application to developing inflorescences of *Bryophyllum crenatum* increased the number of ovules produced per flower (Catarino, 1964). In *Luffa acutangula* seed treatment with benzyladenine and with *N*-chlorophenyl-*N′*-phenylurea both greatly increase the number of female flowers subsequently produced on the plant and suppress the male flowers (Bose & Nitsch, 1970) (see Fig. 60).

A new approach to the problem in recent years has employed the culture of flower-bud primordia on sterile nutrient solutions. In cucumber IAA promoted the formation of ovaries in potentially male flower primordia and gibberellic acid counteracted this effect. These two hormones did not influence in any way the normal development of potentially female or hermaphrodite primordia (Galun *et al.*, 1963), In *Aquilegia formosa* all three groups of hormones (represented by IAA, gibberellic acid and kinetin) seem to be necessary in the culture medium before flower primordia will develop. However the only clear-cut specific effect so far demonstrated seems to be that of IAA, which has to be present for carpel growth to take place. TIBA prevents the action of IAA (Tepfer *et al.*, 1963, 1966). These results confirm those from whole plants and establish that auxins have a direct determinative role to play in sex expression but still leaves as an open problem the roles of other hormones.

(C) THE RELATIONSHIP OF SEX EXPRESSION TO ENDOGENOUS HORMONE LEVELS

If native hormones do regulate sex expression then a relationship should be demonstrable between the levels of endogenous hormones in the flowering apex or plant as a whole and the type of floral organ being produced. However data so far obtained have not helped to settle the issue. In cucumber, hermaphrodite plants contained more auxin than andro-monoecious plants (possessing male as well as hermaphrodite flowers) (Galun *et al.*, 1965) and in

Salix caprea female catkins always appear to have more auxin than male catkins (Heslop-Harrison, 1964). These results are in conformity with those of experiments involving auxin applications and suggest that femaleness is promoted by this hormone. On the other hand in hemp plants during the early stages of flowering, male plants contain more auxin than female (Conrad, 1962) which runs counter to that hypothesis.

With regard to the gibberellins one investigation on cucumber indicates that plants with a high ratio of male to female flowers have a higher content of gibberellins A_1 and A_3 than plants with a correspondingly low ratio, suggesting that gibberellins promote maleness (Atsmon *et al.*, 1968). This contrasts with the implications of bud-culture studies where gibberellin was shown to stimulate all floral organs *except* stamens in *Aquilegia* (Tepfer, 1965).

Clearly then hormones are involved in sex expression in higher plants but no clear pattern of control has yet emerged. Since hormones do not operate independantly and may interact in a complicated manner in their operation, it is clear that the study of hormones in isolation may give grossly misleading results. The sterile culture of isolated floral parts seems to offer the best prospects for the solution of the problem.

HORMONES AND SEXUALITY IN SIMPLE PLANTS

From what has been said before in this chapter, one might be led to regard reproduction in the flowering plants as being ultimately controlled by a relatively simple hormone system. Indeed, much evidence implies that the ultimate hormone causing flower-initiation is the same, whether it is produced by vernalization or by long or short photoperiods. When we turn from these highly organized flowering plants, with their great complexity of structure and organs, to the simply organized plants such as colourless moulds (fungi) we find evidence of a very complex system of hormones controlling their reproductive processes. This state of affairs makes one wonder whether we are not grossly over-simplifying the whole hormone set-up in the higher plants.

The rapid growth and ease of culture of such simple organisms, under the carefully controlled environmental conditions of the laboratory, have enabled us to learn considerably more about the chemical regulation of their reproduction than about that of the highly organized flowering plants. Needless to say, many species and conditions have been investigated, but in a book such as this it will be possible only to outline some of the more important findings to illustrate this hormonal complexity.

The Fungi

Let us begin with the moulds. These are, for the most part, colourless plants devoid of the green colouring matter chlorophyll, and therefore incapable of using the radiant energy of the sun for their growth processes. The body of the mould is composed of microscopic branched threads called hyphae, which ramify and spread over the organic medium serving it as food. Sexual reproduction, which occurs ultimately in the life of the majority of known moulds, takes place usually by the fusion of the contents of two separate parts of the hyphae, which are often differentiated into distinct male and female organs. It is the differentiation of these organs and their behaviour during the processes leading up to actual sexual fusion that are apparently controlled by hormone complexes.

There is now much well-established evidence that nutrition, particularly the provision of certain specific chemical growth factors, exerts a marked influence on the sexual reproduction of moulds. Perhaps the best known work is that on *Phycomyces blakesleeanus*, which is a close relative of the pin mould commonly found growing on old moist bread. This fungus requires vitamin B_1 (thiamin, aneurin) to grow at all, since it is incapable of making this essential biochemical catalyst by itself. (See review by Schopfer, 1943.) Sexual reproduction is brought about by the close juxtaposition of two hyphae, the subsequent fusion of their contents, and the formation, at the point of fusion, of a large thick-walled spore, the so-called *zygospore*. For this sexual fusion to take place, at least two organic substances other than thiamin must be provided in minute traces. One of these factors is hypoxanthine (6-ketopurine) (Robbins & Kavanagh, 1942). Some yeasts, a group of microscopic unicellular fungi, which normally multiply solely by simple cell division,* show sexuality by the fusion of two cells. In one such yeast, *Zygosaccharomyces* sp., sexual fusion is induced by vitamin B_2 (riboflavin) and glutaric acid acting together (Nickerson & Thimann, 1943). There are indications that natural sexual fusion, which occurs most frequently in old cultures, may be encouraged by the accumulation of these compounds produced by the ageing cells themselves.

Some of the parasitic disease-causing fungi are very difficult to persuade to reproduce sexually in culture in the absence of the host plant; for such fungi sexual reproduction can sometimes be induced by adding plant extracts to the growth media. Recent studies have shown that one group of active constituents may be the sterols, which will promote the formation of male and female organs in the blight genus *Phytophthora* and the genus

*Actually, of the special type known as 'budding'.

Pythium which causes damping-off of seedlings (Hendrix, 1964; Leal *et al.*, 1964; Harnish *et al.*, 1964; Sietsma & Haskins, 1967).

Endogenous sexuality-inducing substances are essentially hormones, as we usually define them, and are the most interesting from our present point of view. The first demonstration of the production of such substances by fungi came from studies on *Mucor mucedo*, a white pin mould. This mould exists in two distinct strains that are structurally indistinguishable under the microscope, a phenomenon known as heterothallism. Purely for the sake of convenience, these strains have been designated (+) and (–) respectively. The only difference between them is in sexual reproduction, which takes place, as in *Phycomyces blakesleeanus*, by the fusion of two hyphal tips, which have to be from different strains. Tips from identical strains will not fuse. The participation of a diffusible hormone in such a process is indicated by the fact that, considerably before the (+) and (–) hyphae come into contact, the approaching tips begin to swell; the so-called zygophores are induced. A mutual stimulating influence must therefore have crossed the intervening space between one hyphal tip and the other, i.e. either over the surface of or through the nutrient material on which the mould was growing; such a stimulating influence is most likely to be chemical. This was established experimentally when an agar block with (+) hyphae was placed upon (–) hyphae, also growing on agar, with a film of collodion between them. Although the hyphal tips could not pass the membrane and so fuse, yet they swelled up in the usual preparatory manner, which could have been brought about only by the diffusion of chemical substances from one strain to the other through the permeable membrane (Burgeff, 1924). It seemed likely that at least two different substances were involved, one produced by the (+) strain and evoking a reaction in the (–) strains and *vice versa*.

Modern development of this study has indicated that at least six separate hormones may be involved in the mating process of this heterothallic *Mucor mucedo*. It seems that one strain produces a hormone (called a progamone) which stimulates the formation of another hormone (gamone) in the opposite strain. This in turn induces the swelling of the hyphal tips (zygophore formation) in the first strain. Since a similar sequence operates for the other strain a total of four distinct hormones may be involved (Plempel, 1963). There are indications that these hormones may be yellow carotenoid pigments (van den Ende, 1967). Then each zygophore secretes another specific hormone (thus two more in all) which induces the zygophore of the opposite strain to grow towards it, thus ensuring mutual contact and fusion and the ultimate production of the sexual reproductive spores (zygospores). There is strong evidence that these two zygotrophic hormones are volatile substances since they can cross air gaps (Banbury, 1955).

The type of sexual reproduction so far discussed has been relatively simple. In some other fungi we find a much more complicated sequence, with the differentiation of male and female organs of markedly different structure (see Fig. 61C). Such male and female organs may be borne on the same plant body, i.e. on the same hypha, which may be simple or much-branched. On the other hand, as in *Mucor mucedo*, there may be two distinct strains, one bearing female organs only and the other male only: such a plant is said to be *heterothallic*. This situation also occurs in some species of mould which grow in fresh water on plant debris such as twigs, berries, leaves, etc. It is in one of these genera of water moulds,* *Achlya* spp., that a most amazing complex of sexual hormones has recently been demonstrated. Here the reproductive sequence begins with the production by the female plant of two hormones (A and A_2) which diffuse to the male plant and there induce the formation of male sexual hyphae, which are short lateral branches of the main hyphae. The total number of such branches is determined by a variety of external physical factors such as temperature, salt supply, etc., and also by the relative concentrations of a chemical activator (hormone A_1), and a chemical inhibitor (hormone A_2) from the male plant.

Soon after they have been formed, the male sexual hyphae in *Achlya* secrete another hormone (hormone B) which diffuses to the female plant and causes the female organ to be formed. These short, swollen female branches were then thought to produce a third hormone (hormone C), which also diffuses out into the medium and causes the male hyphae to grow along the concentration gradient towards them. It now seems fairly clear is that hormone C is identical with hormone A (Barksdale, 1963). When the male branch finally reaches the female branch, the male organ is induced to form, under the joint action of contact and hormone A, as a short club-shaped terminal cell containing dense protoplasm and several nuclei. This stage of the process is illustrated in Fig. 61C. As this male cell develops, it secretes a fourth hormone (hormone D) which in its turn brings about the maturation of the female organ and its contents. This ripe female organ consists of a large spherical cell at the end of a short branch and contains a number of spherical masses of protoplasm, each with its own nucleus (the eggs). Fertilization takes place by the growth of a fine tube from the male into the female organ, the passage of male nuclei into that organ, and the ultimate fusion of male and female nuclei. Such a fusion is followed by the formation of a thick-walled spore from each fertilized egg, which, on germination, gives rise to new plants. This oversimplified description has obviously omitted many important points of detail, and the interest reader should refer to the discoverer's own account in his reviews, where reference

*Water moulds are usually to be found growing on plant debris, e.g. fleshy fruits of wild rose, hawthorn, etc., submerged in freshwater streams and ponds.

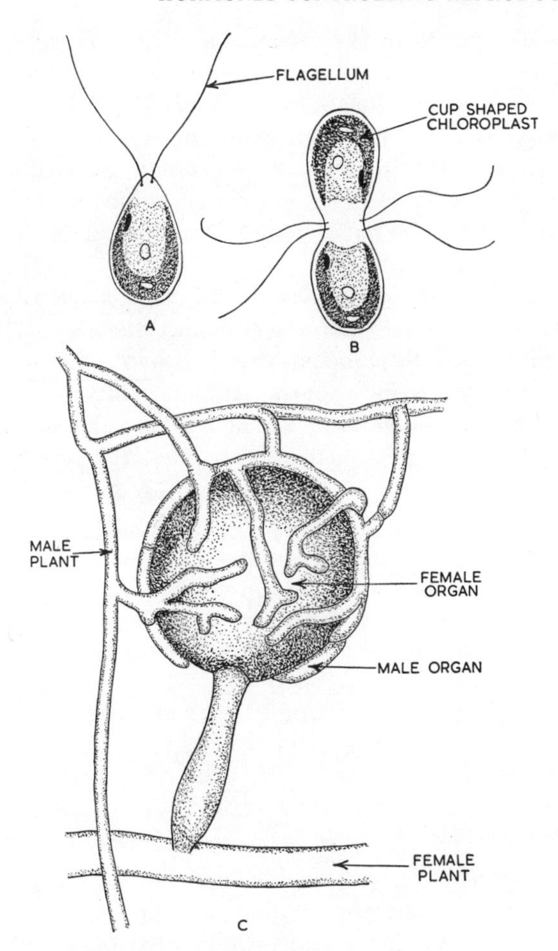

Fig. 61. (A) *Chlamydomonas* plant. (B) Sexual fusion of two *Chlamydomonas* gametes (× 1,500). (C) Envelopment of a young immature female organ of the water mould *Achlya ambisexualis* by numerous male hyphae.

to the original papers will be found (Raper, 1951, 1952, 1957) (see also Barksdale, 1967).

So far these various hormones have not been identified with any known chemical substance, owing primarily to the incredibly small amounts of the compounds involved. For example, in one experiment nearly 1,450 litres (320 gallons) of culture fluid, in which female plants had been growing, were concentrated until 2 mg of highly active substance was obtained: when this was dissolved in 10^{13} (ten billion) times its own weight of water, it could still induce the formation of male branches on male plants. The occurrence of these hormones in such extremely low concentrations makes the work of direct chemical identification extremely difficult.

However some progress has been made towards its identification. The extraction of about 6,000 litres of culture medium followed by chromatographic purification yielded 20 mg of pure crystalline hormone. Its exact formula is not yet known but it appears to be a steroid (McMorris & Barksdale, 1967; Barksdale, 1967).

Similarly intensive investigations have been made on a hormone secreted by the female organs of another water mould, *Allomyces*. This hormone causes the motile male cells (gametes) to swin towards the female organ and has been given the colourfully appropriate name sirenin (Machlis, 1958). After ten years of painstaking study sirenin has been characterized and has formula CLXXXIX (Fig. 62) (Machlis *et al.*, 1968). It is active at a very high dilution of 10^{-10}M.

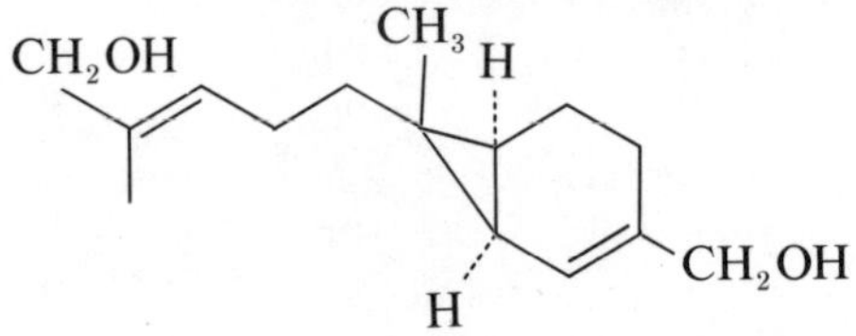

CLXXXIX

Fig. 62. CLXXXIX Sirenin

Other similar hormone systems have been demonstrated but not characterized chemically in other groups of fungi, e.g. the Ascomycetes, exemplified by the genera *Ascobolus*, *Saccharomyces* (yeasts), *Glomerella*, etc. and in certain cellular slime moulds e.g. *Dictyostelium discoideum*. A full discussion of the situation in these plants is out of place in a book of this kind, and the interested reader is referred to the comprehensive reviews of Raper (1957) and Machlis (1966).

The Algae

The fungi are not the only lower plants in which sexual processes are apparently controlled by diffusible hormones. In the algae, a group which includes the microscopic plants constituting the green scums of stagnant ponds and slowly flowing streams, evidence has recently been obtained of a similar complex of hormones controlling sexual reproduction. The genus of organism most intensively studied is a microscopic unicellular form of fresh-water alga called *Chlamydomonas*. The individuals of this plant genus are egg-shaped structures about 0.025 mm (25 μm) in diameter. Each contains a green protoplasmic body (chloroplast) for the synthesis of food in the light, and two thin whip-like protoplasmic structures (flagella) attached to the more pointed end of the cell. The rhythmic movement of these flagella propel the cell through the water in which it lives. If these motile organisms are transferred to the surface of a solid medium, e.g. agar-agar, the flagella are lost and the cells become embedded in a jelly-like matrix secreted by the cells.

Reproduction of *Chlamydomonas* may be sexual or asexual. When asexual reproduction takes place, individual cells simply divide into four or eight identical daughter cells which are exact miniature replicas of their mother. Under some conditions, however, such daughter cells fuse in pairs, this fusion constituting the sexual act: this is illustrated in Fig. 61 B. Even ordinary vegetative cells can fuse in some species without prior division to form sexual cells (called *gametes*). The earliest extensive investigations into the chemical control of motility and sexual fusion of these algae were carried out in Germany on a species, *Chlamydomonas eugametos*, which is characterized by producing two distinct strains of cell – a male and a female strain. Such a phenomenon has already been noted in the fungal genera, *Mucor* and *Achlya*. Moewus (1950), in an elaborate series of experiments, claimed to have demonstrated the existance of a complex of interconvertable carotenoid pigments which controlled the induction of motility in non-motile cells, the activation of the conjugation (fusion) between motile cells (gametes) and even the sex of the gametes themselves. However much uneasiness has always been felt concerning many details of his elaborate theory, the most disturbing of which was the many unsuccessful attempts to repeat Moewus' work and to isolate these chemical evocators of sexual fusion in several species in this genus (see Lewin, 1954) and also of associated sex-determining substances (see G. M. Smith, 1951). The theory was finally demolished by Raper (1957) in a critical survey of the established facts.

However it is now well established that both sexes produce specific copulatory substances; these substances are not carotenoids but are high molecular weight proteins associated with sugars (glycoproteins) (Hartmann, 1955). When male and female gametes are mixed they aggregate by the agglutination of their flagella. Subsequently they form a protoplasmic bridge between their pointed ends and fuse. This agglutination is due to the sex-specific glycoproteins which constitute the surface of the flagella (Wiese & Jones, 1963). The male flagellum surface has a high affinity for the female flagellum surface but flagella of the same sex are not mutually attractive. These glycoproteins can be obtained from cell-free filtrates of gamete suspensions and can therefore be detached from the flagellum surface. These agglutination substances have been called gamones and have been isolated and purified (Föster *et al.*, 1956). The solution from a male gamete can cause the mutual aggregation of female gametes and *vice versa* demonstrating that a particular gamone can become attached to the flagella of the opposite sex, there acting as a glue sticking similar flagella together. The subsequent fusion of male and female gametes seems also to involve a mutual surface attraction of a different kind. Whether gamones are true hormones, i.e. are liberated from flagella in nature and play a part in natural copulation, is still a moot point.

Sexual hormones are also found in more highly organized algae. Thus in the brown seaweed *Ectocarpus siliculosus*, male and female gametes are produced from morphologically very distinct reproductive organs (sporangia). Sexual reproduction is by the fusion of motile male and female gametes. It has recently been shown that the female gamete produces a highly volatile substance of small molecular weight and characteristic odour which is an attractant for the male gametes (Müller, 1968). In the green filamentous alga *Oedogonium*, a native of ponds and freshwater streams, a situation almost as complex as that of the fungus *Achlya* seems to exist. In this genus the female cells take the form of large eggs (ova) produced singly in certain cells, while the male cells are motile and are formed, usually in pairs, in other cells of the filament. In some species these male cells are liberated from the filament cells which produce them, swim towards the egg-containing cells, become attached thereto and develop into tiny filaments, the 'dwarf males'. These dwarf males produce the male gametes which, on liberation, swim through a pore in the egg cell and fuse with the eggs (ova). This complex sequence of happenings may be controlled by hormones. Thus the mature oogonium (egg-containing cell) releases an attractant substance for the male cells, which, having become anchored, in their turn provide a chemical stimulus for cell division which produces the mature egg. Finally the egg itself may produce another attractant for the motile male gametes, causing them to swim to and through the pore in the oogonium and fuse with the egg cell (Rawitscher-Kunkel & Machlis, 1962).

When we consider the complexity of the hormone systems concerned with reproduction in these simply organized plants, we come to suspect that the state of affairs in the flowering plants may be even more complicated, and that simple theories of control involving the level and balance of auxins, gibberellins and cytokinins are gross oversimplifications. If the development, in proper order, of such simple structures as the male and female organs of *Achlya* requires a series of hormones, how much more complicated a chemical organization might be necessary for the proper development of a flower?

The Ferns

The sexual reproduction of the ferns takes place on a small green, usually heart-shaped structure (prothallus) produced by the germination of the fern spore. The prothalli are completely independant of the large spore-bearing fern plant and produce the distinct male and female organs (antheridia and archegonia respectively). Male sperm liberated from the tiny globular antheridia swim to and pass down the neck of the flask-like archegonium and fuse with the egg at its base. From this fertilized egg the new spore-bearing plant develops.

In 1950 Döpp demonstrated that the formation of antheridia on young prothalli of *Pteridium aquilinum* (bracken fern) and *Dryopteris filix-mas* (male fern) was induced by a substance which could be extracted from mature prothalli. It was given the name *antheridogen.* Concentration and purification of all biochemical reactions, the ultimate irreducible components of complex unsaturated carboxylic acid, containing no phosphorus or peptide bonds and needing a free acid group for activity. However there may be a whole family of antheridogens since two other active materials, each specific for the species, have been found in *Aneimia phyllitidis* (Näf, 1959) and *Lygodium japonicum* (Näf, 1960) respectively.

It was not long before gibberellins were shown to hasten and augment antheridia formation (von Witsch & Rintelen 1962; Schraudolf, 1962). A limited but real activity is also shown by helminthosporol (see Chapter IV) (Schraudolf, 1967). However, chromatographic studies of the antheridogen of *Aneimia phyllitidis* showed that it could not be identified with any of the gibberellins A_1 to A_9 (Voeller, 1964) although it has some of the properties of gibberellic acid (A_3) (i.e. amylase induction in barley endosperm). (Näf, 1968). Nevertheless its relative activities in different fern species shows that it is much more species-selective than gibberellin A_3 (Näf, 1968). Until these antheridogens have been isolated in amounts sufficient for chemical identification we shall not know whether or not they are gibberellins or whether the known gibberellins act, in the words of Schraudolf (1966), as 'pick-locks' or 'skeleton keys', merely imitating in a non-specific fashion, the actions of the native antheridogens. Details of the physiology of the antheridogens can be found in two recent reviews by Näf (1962) and Bopp (1968).

Chapter XI

The Control of Hormone Levels at the Sites of their Actions

INTRODUCTION

The problems of growth control in organs and organisms are perhaps the most challenging in all science. The reason for this is not far to seek; it lies not simply in the almost inconceivable complexity of organization of living cells and tissues but mainly in the fact that a growing system is a changing system of which the mosaic of components, identified, suspected or unknown, wax and wane in their degree of participation in the over-all growth progress as cells come into being, grow, mature and age. The rate and direction of all biochemical reactions, the ultimate irreducible components of growth, are determined in the last analysis by the concentrations and activities of the molecules participating. The chemist, studying such component reactions, can closely define and control his system; he can specify the nature and concentration of his reactants and, knowing what they are, can follow their changing patterns with precision. The biologist, dealing with organized living tissue, is in a very different position. He can only partially define the system he is studying since there may be many important components unknown to thim; he can only partially control the experimental environment of his system since much of it is internal, i.e. in the cell itself, and therefore subject to complex self-adjustment to change so characteristic of all life processes.

Nowhere is this situation brought out more clearly than in the field of the hormone control of growth in plant tissues. Here the major controlling factors of the key growth processes are still not clearly delineated and consequently a rigorous analysis, using the approach of the chemist with his relatively simple systems, is still a distant dream. Nevertheless, no matter how complex the complete integrated system may be, the plant physiologist is often able to study one or more of its components in reasonable isolation, once that component has been recognized and conditions established for its separate study. He can then hopefully apply the methods of the chemist and the physicist. In the two remaining chapters of this volume we shall be looking at the picture which has emerged from the attempt of plant physiologists and biochemists to dissect the growth system of the plant into its components, particularly as it concerns the participation of plant growth-regulating substances. The present chapter will be concerned with the various mechanisms which determine the concentrations of growth regulators at the sites of their growth-regulating actions in the growing cell. The last will survey the possible biochemical mechanisms whereby they exert this regulation.

The Relationship between Hormone Concentrations and Growth Rates

Ever since the pioneering studies of Went (see Chapter I), it has been the tacit assumption of plant physiologists that the hormone control of plant growth is operated via its concentration in the responding cell. There is now a vast amount of data, mostly indirect, which supports this contention and it would be a rash physiologist who would doubt its general validity. The demonstration that auxin was produced in the tip of the coleoptile, that it was transported downwards to the extending cells below and that its supply regulated growth, led Went to make his dictum, 'Without growth-substance there is no growth', the implication being that the concentration of auxin, determined by the rate of its production in the tip and the efficiency of its transport to the extending cell controlled the rapidity of cell expansion. The characteristics of his famous *Avena* coleoptile curvature assay, indeed of all subsequent hormone assays, are in line with this contention and, as we have seen, the response of the cell, e.g. the rate of its extension or the rate of its senescence (see the assay of cytokinins and gibberellins) bears a precise relationship (i.e. linear or logarithmic) to the applied concentration of the growth regulator.

In spite of this impressive indirect evidence it is a sobering fact that, even under conditions where cells exhibit their quantitative responses to applied growth regulators, there is no *direct* evidence that the normal response of the cell is controlled via the modulation of *natural* growth-hormone levels. Observations have been made from time to time of the amounts of hormone which will diffuse out of or can be extracted from growing cells in an attempt to correlate these levels with the growth rates of the corresponding cells. Sometimes rough positive correlations have been observed (see Fig. 63) but

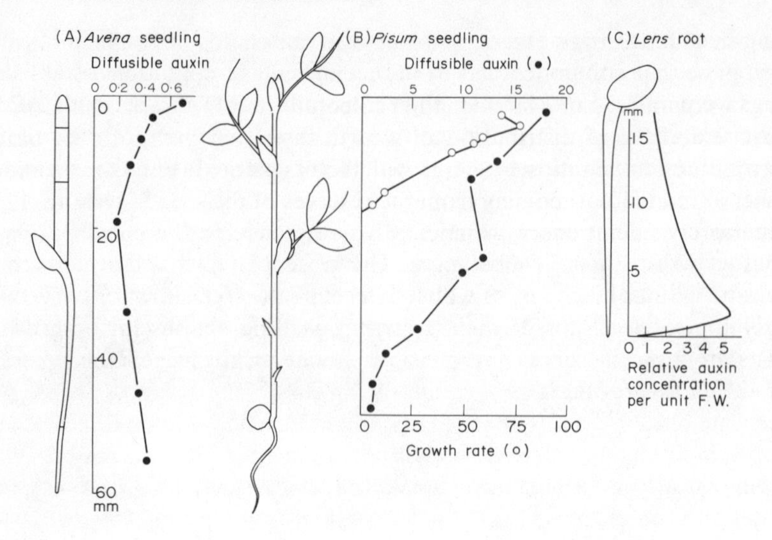

Fig. 63. Diagrammatic and graphical representation of the relationships between growth rates and endogenous auxin concentrations in seedling organs.
(A) Seedling of *Avena* (Data from Thimann, 1934)
(B) Stem of *Pisum sativum* seedling (Data from Scott & Briggs, 1960)
(C) Radicle of *Lens culinaris* (Data from Pilet & Siegenthaler, 1959).

this is not always so, for example in the third internode of *Zebrina pendula* growing under a variety of mineral nutrient conditions (Al-Omery, 1968). Not infrequently little hormone can be detected in the most rapidly-growing cells. The explanation of these discrepancies must of course lie in our present ignorance of the growth systems involved, of the identity of the sites where the regulators are operating and where their concentrations are relevant. At the moment we can study only their concentrations in the tissue as a whole and have not yet discovered the means of determining concentrations of endogenous hormones in any particular part of the cell. We cannot even be sure that the hormones we study are in their active molecular form when we have extracted them and that some transformation is not necessary before they can operate at their relevant active centres.

A major complication is of course the fact that hormones are not the only internal factors which affect growth. Since growth involves the manufacture of cell components it is obvious that a limitation of the supply of raw materials for their synthesis can also restrict and hence control growth. In other words, in the growth production-line of the cell there can be as many bottle-necks as there are essential components for growth and the supply of that component which is the most scarce, be it hormone or raw material for cell construction, will determine the overall growth rate. This of course has long been recognized (Went, 1935) and in the early days of hormone physiology several simple analyses of such a multiple control of growth in seedlings were made. For example in the coleoptile itself Thimann explained the observed curve of distribution of growth rate from the tip to the base in terms of the distribution of two growth factors, an auxin from the tip and the other a 'food factor' coming from the reserves of the seed (Thimann, 1934). Similar considerations were applied to other seedlings with other growth characteristics (Went & Thimann, 1937). As is now realized, these were gross oversimplifications, particularly in their neglect of the growing cell as a dynamic, constantly changing system. For example it is now realized that in its progression from inception to maturity the cell may have a succession of different growth requirements, being under the control, during each separate phase, of a different set of hormones. For example the wheat coleoptile in the early stages of its growth is sensitive only to the controlling action of applied gibberellins and cytokinins and auxin is without effect. Later the organ loses sensitivity to the first two hormones and shows its greatest response to applied auxin (Wright, 1963).

In the case of the auxins many attempts have been made to study the precise quantitative relationships between hormone concentrations and growth rate with a view to elucidating the mechanism of that control, a subject we shall deal with in the next chapter. The technique has involved the incubation of growing tissue (e.g. coleoptile segments as in the assays described in Chapter II) in hormone solutions of different concentrations and to construct graphs relating growth rates to these external concentrations, the assumption being that the hormone will rapidly penetrate into the tissue and that in a short space of time concentrations at the growth centres will become equal to the external concentrations. The concentration/growth-rate curves thus obtained are usually of a particular form; if the auxin concentration is plotted on a logarithmic scale the growth response curve has the shape of a virtually symmetrical inverted bell, as is well illustrated by the results of Foster *et al*., (1952) for *Avena* coleoptile segments (see Fig. 64A). In the left-hand half, where rising concentrations increase growth, the auxin levels are sub-optimal, while on the other side, where growth decreases with concentration, auxin levels are supra-optimal. Both types of control, suppression or promotion of growth by rising hormone concentrations, may operate, depending on the organ concerned and its stage of development.

Mathematical models, based on hypothetical biochemical mechanisms, have been severally devised to simulate these observed growth curves. Thus Bonner & Foster (1955*b*) have shown that they can be closely fitted by the kind of rectangular hyperbola which describes the relationship between the rate of an enzyme action and the concentration of its substrate, the implica-

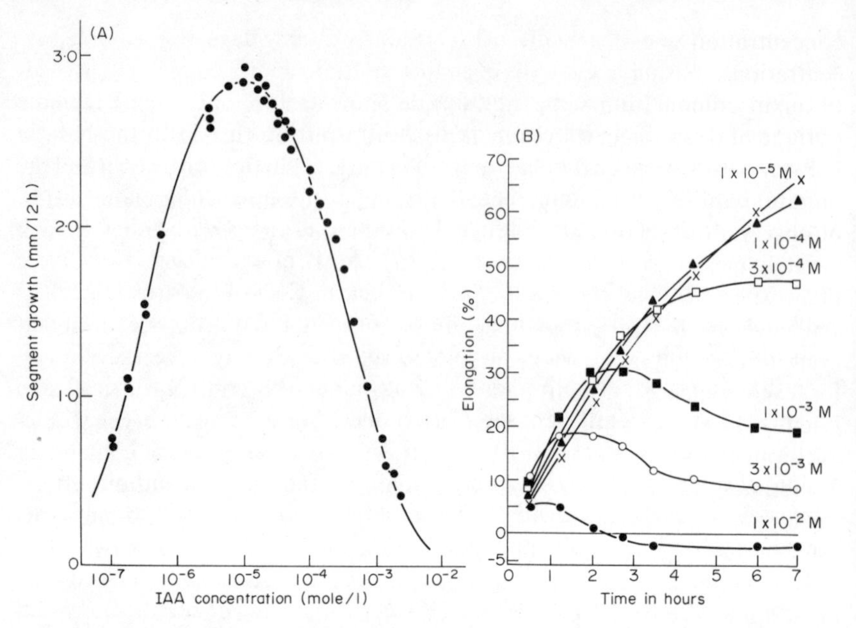

Fig. 64. Graphs showing the relationships between auxin concentrations and growth kinetics in coleoptile segments of *Avena*.

(A) Segment growth in the first 12 hours after excision plotted against IAA concentration on a logarithmic scale. The smooth curve drawn through the observed points is one fitted from theoretical formulae based on the application of enzyme kinetics to the two-point attachment theory of auxin action (see p. 98). The decline in growth at supra-optimal auxin concentrations is accounted for there by the competitive blocking of 'growth-enzyme' sites by ineffective one-point attachments of super-saturating auxin molecules. (Graph from Foster *et al.*, 1952.) (For full discussion of this theoretical treatment of auxin growth kinetics see Housley, 1961.)

(B) Graph showing the time-course of changes of segment length in a range of concentrations of IAA (marked on respective graphs). Note that, with the exception of 10^{-2} M, a highly toxic level of IAA, the initial growth rate of segments in the first half hour increases with increase in concentration and there is no evidence of a supra-optimal concentration. With time however the growth rate falls off and the rate of fall also increases with concentration; at the higher levels segments ultimately shrink, due presumably to a progressive damage to protoplasmic membranes, bringing with it loss of solutes and water (Graph from Marinos, 1957).

tion being that control is exerted on a 'growth enzyme' for which auxin acts as the substrate. But the adsorption of a solute on the surface of a colloid yields a similar kind of relationship with concentration, and a model of this kind, based on the occupation by auxin of active sites in the cell has been used by Linser & Kaindl (1951) and Kaindl (1954, 1955) to explain their experimental curves. Enzyme kinetic curves have been used by Foster *et al.*, (1952) to explain the bell-shaped curve of growth response against auxin concentration and hence the inhibition of growth by supra-optimal concentrations, basing their arguments on the two-point attachment theory of auxin combination with the enzyme sites (see Fig. 18). For a balance critique of this subject the reader is referred to the review by Housley (1961).

Even with simple enzyme systems the study of kinetics (effects of factors on rate) requires, as far as possible, unchanging conditions during the period of observation, so that constant rates of reaction can be precisely and accurately measured. Under the conditions used by Bonner & Foster (1955) in their experiments on *Avena* coleoptile segments, growth was apparently constant over periods of 18 hours or so, whatever concentrations of IAA were used, thus enabling a very accurate determination of average growth rates for each concentration. But no other workers have been so successful. Usually the growth rate declines more or less continuously from the time of excision of the segments and their initial exposure to the auxin solution. Sometimes there is the rise and fall in growth rate which is characteristic of the typical S-shaped size/time curve for organs and organisms. Most significant perhaps is the fact that, during the first hour or so after immersion in auxin solutions, growth rate increased with concentrations, even up to the highest levels studied and there is no obvious occurrence of a supra-optimal depression. But the higher the concentration of auxin, the sooner does the progressive decline in growth rate set in so that the final total growth after 18 hours or so shows the bell-shaped relationship with concentration, of Bonner & Foster's studies (Bennet-Clark & Kefford, 1954; Marinos, 1957) (see Fig. 64B).

This demonstrates clearly that under these conditions the growth system of the cell is in a state of rapid change, making it impossible to derive any meaningful estimate of a growth rate to relate to applied auxin concentration. Furthermore the tacit assumption behind these curve-fitting operations is that the auxin penetrates the tissue rapidly and that internal concentrations rise quickly and uniformly throughout the segment to steady values inducing correspondingly steady rates of growth. That this is not so is clearly shown by the extensive studies of auxin uptake by Blackman and his colleagues, who have used many different auxins and many different tissues. Full details would be out of place here but the most appropriate example to illustrate the situation can be taken from recent work with IAA labelled with ^{14}C and stem segments of pea and cotton (Kenny *et al.*, 1969). The uptake of radioactive IAA was shown to be almost entirely via the cut ends of the segment and during the whole of the subsequent period, when growth measurements were made, it remained concentrated in the cells near the cut surface. What meaning then has external-concentration/growth-rate relationships in such a complex system? Except therefore in the purely empirical procedures of biological assay, there seems little point at this stage in attempting sophisticated kinetic analyses of hormone-concentration/growth-rate data; in our present ignorance of the complexity of the growing tissue segment their interpretation is fraught with danger. Nevertheless at the stage when a particular hormone is clearly controlling the growth of the cell, it seems inescapable that this control is mainly operated by a modulation of the hormone concentration at the growth centres. Other factors, such for example as the level of co-factors or that blanket-term for ignorance, the 'sensitivity' of the growth centre, may come into the picture, but hormone concentration is most likely to be predominant. With this as a major premise we will now consider the various methods whereby this concentration may be adjusted in the growing tissue.

Control by Biosynthesis

(A) AUXINS

(i) The site of synthesis

It has long since been one of the tenets of auxin lore that the main centres of auxin synthesis are the apical meristems of plants and the young primordial organs which differentiate from them. Although there is little direct evidence that this is so, yet the general picture of the control of cell growth by auxin provided from the neighbouring meristematic tissue has received overwhelming, albeit indirect support from many sources.

The first type of evidence comes from auxin distribution studies. Thus it is reasonable to suppose that the centres of production should be the regions of highest concentration and that auxin levels should decline as one moves away from those centres to the regions into which the auxin is flowing. (There are many loopholes in this argument and these will become apparent in later sections). Many such distribution studies have been made. In the classical material of auxin studies, the *Avena* seedling, Thimann (1934) was the first to show such concentration gradients with a maximum concentration at the coleoptile apex and a minimum concentration at its base. The concentration then rose again to another maximum towards the root apex (see Fig. 63A). Similar apical/basal gradients have been shown in the stems of the broad bean (*Vicia faba*) (Thimann & Skoog, 1934) and garden pea (*Pisum sativum*) (Scott & Briggs, 1960) (see Fig. 63B). A more detailed study of auxin gradients in roots has been made by Pilet and his colleagues in the lentil (*Lens culinaris*) radicle (Pilet & Siegenthaler, 1959). Again the highest concentration is found in the meristem and just below it (see Fig. 63C), but there are very small amounts in the extreme tip, i.e. in the root cap. Other young actively meristematic organs have high auxin concentrations;

they include immature leaves (Avery, 1935; Avery *et al.*, 1937*a* & *b*), young fern croziers (Steeves & Briggs, 1960), flower and inflorescence primordia (Nitsch, 1953; Kaldewey, 1957; Söding, 1952), etc.

It is generally assumed that the synthesis of auxin (i.e. IAA) in these embryonic cells is from the amino acid tryptophan as precursor. In Chapter II we described the pioneering work of Skoog which indicated that in the *Avena* seedling, tryptophan released from the food reserve proteins of the grain, moved to the coleoptile tip where it was converted to auxin. This kind of situation has, by extrapolation, now been assumed for other organs. Tryptophan, arriving with all the other complex soluble raw material for growth and metabolism in the meristematic cells, would have two fates; one would be its incorporation once more into the structural and metabolic proteins (enzymes) of the growing cell; the other would be its degradative conversion to IAA. This is perhaps the real significance of the marked correlation between meristematic activity and auxin production. This extrapolation receives support from the fact that regions of high auxin concentration (and thus presumably active auxin synthesis) are usually rich in tryptophan and are well provided with the enzymes involved in its conversion to IAA (for documentation see Gordon, 1961).

However a note of caution must be sounded about the acceptance of concentration gradients as indications of sites of hormone production. As we shall see in a later section, differences in the ability of various tissues to destroy auxin may seriously complicate the picture. Uncertainties of this kind are most sharply met in the root where marked apical/basal gradients exist. Also in the root of *Lens* there is a remarkable inverse correlation of auxin concentration with the levels of an enzyme capable of destroying it (Pilet & Galston, 1955); this enzyme may be the major determinant of IAA levels (but see later section). Although isolated roots grown in sterile culture may produce auxins (Thurman & Street, 1960) and weak growth responses can be invoked by the application of tryptophan and tryptamine (Audus & Quastel, 1947), there is no convincing evidence to date that the root tip can convert tryptophan to IAA. Lastly recent experiments suggest that IAA moves through root tissues mainly in a direction from base to apex (Yeomans & Audus, 1964; Scott & Wilkins, 1968) (see next section) and this implies that the auxin in the root is synthesized in the shoot and flows thence into the root tip where it accumulates. This received indirect support from comparisons of the levels of auxin-synthesizing enzymes in different parts of the pea plant (Moore, 1969). Sterile cell-free extracts when incubated with tryptophan labelled with ^{14}C yielded radioactive IAA which was identified and assayed on thin layer chromatograms. The highest enzyme activity was shown in terminal buds and young leaves; the least active were the root tips with a conversion rate about 12 percent of that of the shoot tip.

(ii) Factors affecting synthesis

The question now arises as to whether any factors which can control growth, do so via a modification of auxin synthesis. Three types of factor immediately come to mind; they are the factors of the physical environment, nutrient factors and organic chemical factors such as other plant regulators.

Of the factors of the physical environment, temperature and radiation are perhaps the most important since they can both have dramatic effects on plant growth. However the temperature at which a plant grows seems not to have any effects on the levels of auxin in tissues and hence, by implication, on the synthesis of auxin (for details see Hillman & Galston, 1961).

The very short wavelengths of electromagnetic radiation, i.e. X-rays and ultra-violet radiation both have markedly stunting effects on plant growth. There seems little doubt that this reduced growth is the direct result of a lowered auxin content of the irradiated plants. This was first shown for X-rays by Skoog (1935) using *Pisum*, *Vicia* and *Avena* plants. Later Gordon (1956) confirmed this for mung bean seedlings and showed a concomitant increase in indol-3yl-acetaldehyde in the tissue, indicating that the X-rays had inactivated the enzyme responsible for the oxidation of this last intermediate in the biosynthetic pathway from tryptophan to IAA. This was later established in cell-free enzyme extracts. Similar reduction in auxin content of plants result from ultra-violet radiation (see Hillman & Galston, 1961) but there is no evidence here that a suppression of auxin synthesis is involved (see section on inactivation).

Visible radiation also reduced the growth in length of plant stems since darkness produces the tall spindly growth known as etiolation. The relationship of auxin synthesis to these effects is still very obscure. Thus many of the early observations showed that high light intensity increased the auxin which could be collected by diffusion from the apices of such plants as radish (van Overbeek, 1933), lupin (Navez, 1933) and tobacco (Avery *et al.*, 1937*b*). These findings were interpreted in terms of increased auxin synthesis in response to augmented supply of sugars from photosynthesis in the light. Such interpretations were disputed by von Guttenberg & Zetsche, (1956), who demonstrated that the increased outward diffusion of auxin in the light was correlated with and probably caused by an increased capacity of the stem tissue of tobacco, tomato and sunflower to transport the auxins (see next section); furthermore the total auxin that could be extracted by solvents was reduced by light. They suggested that it was the auxin transport which was 'powered' by the sugar 'fuel' supplied by photosynthesis. Other workers have also found more extractable auxin in darkened plants, e.g. in the tomato and maize (Gustafson, 1946), which suggests that light does not enhance auxin synthesis.

The mineral nutrient supply to the plant is an important factor controlling growth. Nitrogen and phosphorus are two major elements concerned and there are a number of observations (see Hillman & Galston, 1961) that a shortage of either of them reduces the auxin content of experimental plants such as sunflower, tobacco, radish, kohlrabi etc. The balance of nutrient ions can also have marked effects; for example in *Zebrina pendula* high calcium application coupled with good nitrogen supply causes a great increase in auxin content (Al-Omary, 1968). These effects however are likely to be only one aspect of a complex syndrome of nutrient imbalance and not the result of specific changes in the system synthesizing auxin. Zinc deficiency however seems to be more specifically associated with lower auxin levels in affected plants (Skoog, 1940; Tsui, 1948). These lowered auxin levels were correlated with a lowered content of tryptophan although the ability of leaf disks to convert tryptophan to auxin was not impaired in zinc-deficient plants (Tsui, 1948). This suggests that zinc deficiency reduces tryptophan synthesis and hence auxin production from it. Direct evidence on this is conflicting. Thus in the fungus *Neurospora* an enzyme, tryptophan synthetase, will catalyse the production of tryptophan from indole and the amino acid serine. Zinc deficiency in this organism lowers tryptophan synthetase activity (Nason, 1950). Application of zinc to barley grain after irradiation with X-rays has been shown to restore the capacity of the developing plants to synthesize tryptophan (and hence auxin), that capacity having been impaired by grain irradiation (Kutáĉek *et al.*, 1966). On the other hand in zinc-deficient tomato plants the activity of the same enzyme seems to be enhanced (Mudd & Zalik, 1958). The true role of zinc in auxin synthesis is still obscure.

There has been much speculation, a few desultory experiments but no systematic study of the possible chemical control of growth via the modulation of auxin synthesis. We have already discussed (Chapter VIII) the possibility that ethylene may invoke some of its characteristic growth effects via a disturbance of auxin metabolism. The toxicity of some of the synthetic auxins (e.g. 2,4-D) and hence their actions as weedkillers, has on occasion been attributed (but with little sound supporting evidence) to a dislocation of the natural auxin system (see critique by Audus, 1961). Anti-auxins, which characteristically suppress auxin action may probably do so by suppressing auxin synthesis, since treated plant organs may have greatly lowered IAA contents, e.g. TIBA on pea roots (Audus & Thresh, 1956*a* & *b*) and bean shoots (Rodionova, 1962). The most recent candidates for such a rôle are the naturally occurring hormones, the gibberellins, which undoubtedly augment the auxin content of many tissues and may do so by stimulating synthesis. This suggestion will be more fully dissected in the chapter which follows and which deals with hormone mechanisms.

In Chapter VII we saw that of the group of synthetic growth regulators known as the growth retardants two, i.e. CCC and AMO-1618, probably exerted their action via the synthesis of gibberellins. This would naturally imply that the rate of gibberellin synthesis is normally the limiting factor for growth in these plants, a suggestion that is supported by the reversal of retardant-inhibition by applied gibberellin. This is illustrated in Fig. 50 for CCC and Phosphon D. However it seems clear from these figures that the retardants may have other inhibitory actions, possibly on other limiting factors because growth retardation is still observed at concentrations of applied gibberellic acid which promote growth well above that set by normal endogenous gibberellin synthesis (i.e. normal growth rates). Could these other limiting factors be associated with auxin synthesis? The possibility is emphasized by the example of β-hydroxyethylhydrazine which has been shown to inhibit the oxidation of tryptamine to indol-3yl-acetaldehyde by extracts of pea seedlings (Reed, 1965); hence auxins levels and growth would be reduced by this growth retardant.

(B) GIBBERELLINS

(i) Site of synthesis

As we have already noted for the auxins, the site of synthesis for the gibberellins seems to be those regions where active cell division is going on. Thus much of the early work of extraction from the tissues of higher plants indicated that the highest concentrations were to be found in actively growing tissue such as immature seeds and young seedlings. Experiments involving organ excision implicated young leaves as the main source of gibberellin. Thus Lockhart (1957) removed the apices and young leaves of pea seedlings and found that the growth of the adjacent stem region was suppressed; subsequent gibberellin applications allowed growth to continue thus pointing to the young leaves and apex as the sources of the growth-controlling gibberellin. Later direct quantitative assays have shown that young leaves can contain higher concentrations of gibberellins than more mature parts of the plant, for example on the flowering haulms of barley (Nicholls, 1962). Red light will induce the production of gibberellins in etiolated barley leaf segments and this is prevented by the antibiotic chloramphenicoal. Since this inhibitor will prevent protein synthesis in chloroplasts it has been suggested that gibberellin synthesis may also occur there (Reid & Clements, 1968). A comparison of the amounts of gibberellin that can be extracted from stem apices by solvents with those that can be collected by protracted diffusion (i.e. 20 hours) shows that in sunflower seedlings the latter far exceeded the former; the excess could have been produced only by continuing synthesis during the 20 hour period. Similar results were obtained from root tips and actively growing regions of the stem itself (Jones & Phillips, 1966). Extracts of root tissue of rice, *Pharbitis nil* and sweet potato (*Ipomoea batatas*) and of the root exudates of the latter yielded relatively large amounts of gibberellin-like substances. However applications of gibberellic acid to mature leaves of *Ipomoea* resulted in little movement of the hormone into the roots (Murakami, 1968) which suggests that the main site of gibberellin synthesis is the roots and that shoot gibberellins are derived from that source.

However, continuing production of an assayable gibberellin does not necessarily mean steady biosynthesis, since gibberellins can apparently exist in bound forms, from which they may be liberated by proteolytic enzyme hydrolysis (McComb, 1961). The continued production of active gibberellins by roots does seem to be due to primary synthesis however. Thus roots which have had the attached shoots removed will, if given an adequate water supply, continue to exude sap from the cut stump for considerable periods. This 'exudate' contains active gibberellins (Phillips & Jones, 1964), which can continue to be produced for period up to four days (Sitton *et al.*, 1967). Incubation of the apices of these (sunflower) roots with radio-carbon-labelled mevalonate yielded labelled (–)-kauren-19-ol, an intermediate in gibberellin biosynthesis (see Chapter IV), indicating that at least some of the essential enzymes are present in roots. The presence of a gibberellin in tomato roots grown for 5 years in isolated culture is very strong evidence that it was synthesized there (Butcher, 1963). Similarly the demonstration of the ability of cell-free extracts of immature pea seeds (Anderson & Moore, 1967) and the endosperm of *Echinocystis* (Upper & West, 1967; Dennis & West, 1967) to catalyse some of the steps of gibberellin biosynthesis from mevalonate also establishes these organs as sites of synthesis. A rather special case is seen in the germinating cereals, e.g. barley grain, where gibberellin seems to be synthesized in the scutellum (Radley, 1969), but only in the presence of the embryonic axis.

(ii) Factors affecting synthesis

There is even less *direct* information about the effects of the environment of the plant on the rate of its biosynthesis of gibberellins than there is for the auxins. Conversely there is a fair amount of *indirect* evidence that a number of important aspects of growth and development may be controlled by environmental factors via their influence on gibberellin biosynthesis.

The two major factors are again light and temperature. There are two aspects of the former which are relevent here, the intensity and the duration (which in terms of the natural environment means the day-length). Light intensity has a dramatic effect on stem elongation and we have already seen that it has not been possible to relate this causally to auxin levels. Is the mechanism of light action operated through the gibberellins? Since light can suppress stem growth one might expect the gibberellin content of the stem to be lower in the light. Indeed there are observations that the growth of stems in the light can be made to equal that of dark-grown stems by application of gibberellic acid which has little effect on the stems in the dark (see survey of literature in Phinney & West, 1961; Gorter, 1961). However the light/gibberellin interactions are often complex, and light may still exhibit some inhibition at high intensities even in the presence of presumably saturating gibberellin concentrations (see Fig. 65). In some cases this gibberellic acid antagonism of light-inhibition does not take place, e.g. in *Sinapis alba* (Lockhart, 1959). Again studies of total gibberellin levels in plants growing in the light and in the dark have revealed no consistent differences to support these suggestions. Thus Kende & Lang (1964) could find no differences in extracts of light- and dark-grown pea seedlings. On the other hand the exposure of segments of etiolated barley leaves to red light put up their content of gibberellin-like substances (i.e. induced gibberellin synthesis?) (Reid & Clements, 1968). The evidence obtained by the use of inhibitors of chloroplast development, which also inhibits gibberellin production(?) in leaves, suggests that gibberellin synthesis takes place in the chloroplasts.

But it must be recalled that several gibberellins, some of which may not be, as it were, 'operational', may contribute to the total extractable gibberellins, and Jones (1967) has shown that only gibberellin A_1 can be obtained by diffusion from the apices of tall and dwarf pea seedlings, whereas gibberellin A_5 also occurs in extracts. He could find no quantitative differences in the gibberellins obtained by diffusion from light- and dark-grown apices. However qualitative changes in the components of the total gibberellin complement there may well be, as has been shown in the leaves of red clover by Stoddart and Lang (1967*a* & *b*) and these could be reflections of differential synthesis mediated by light. On the other hand Loveys and Wareing (1971) have shown that in etiolated (dark-grown) leaves of wheat most of the gibberellin-like substances are present in the bound form and that exposure to red light brings about an apparent release of free gibberellins from this bound state. It is clear that effects of light on gibberellin equilibria in plant organs can arise from actions on more than one cell system.

The most clear-cut changes of this kind have been recorded in relation to the day-length under which plants are grown. Many developmental phenomena, in particular the flowering of higher plants, are controlled by this environmental factor (see Chapter X) and there is now overwhelming evidence that the levels of gibberellin may have an important rôle in determining this behaviour and may be directly regulated by day-length. No further details will be given at this point since the whole subject has been discussed in Chapter X).

Clearly a situation is emerging in which specific gibberellins or forms of

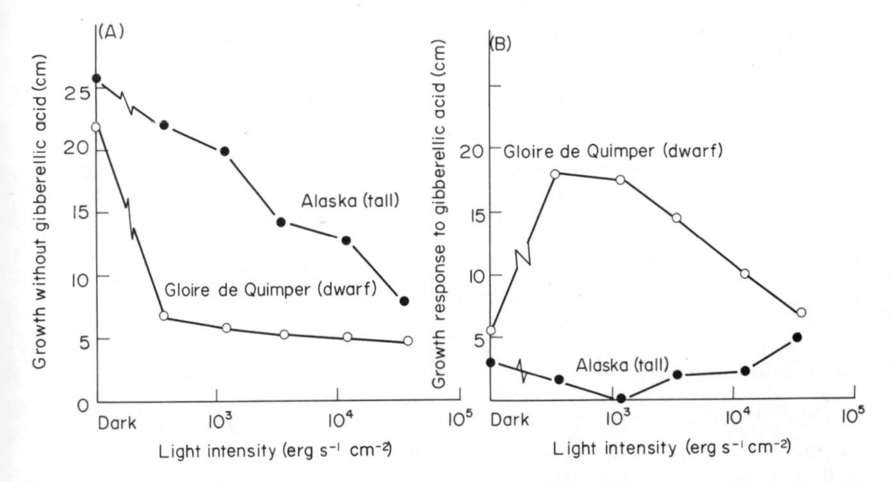

Fig. 65. Graphs showing the interactions of gibberellic acid and light intensity in the control of a dwarf (Gloire de Quimper) and a tall (Alaska) variety of *Pisum sativum.*

(A) The effect of light intensity on growth in the control plants untreated with gibberellic acid. Note that there is very little difference in the heights of the two varieties when completely etiolated in the dark but that the dwarf habit becomes evident in Gloire de Quimper only in the light, which has a much greater inhibitory effect on growth in that variety than in the tall variety Alaska.

(B) The effect of the light intensity on the *response* (i.e. increase in growth) due to gibberellic acid treatment. Note that gibberellic acid has very little effect on either variety in the dark and for the tall variety Alaska that remains so for all light intensities. However the dwarf variety is greatly stimulated at low light intensities; so much so that it becomes almost as tall as the tall variety. (Graphs from Gorter, 1961.)

gibberellin seem to be in control of different stages and aspects of the growth and development of different plant organs. Interconversions of one into another may well be under the control of environmental factors such as light and temperature. The gross synthesis of gibberellins may be relatively unimportant and studies of overall gibberellin levels may mislead rather than elucidate.

(C) THE CYTOKININS

The origins and biosynthetic pathways of the cytokinins are virtually unknown. The isoprenoid side-chain of such molecules as zeatin is presumably derived from mevalonic acid, as are the gibberellins. Cytokinins have been demonstrated in the bleeding sap from root stumps, from which the shoot has been excised, and this indicates their synthesis in roots. This observation and their occurrence in immature fruits and seeds suggest that the site of their synthesis may be in the actual meristematic cells whose division they promote. The mechanism of control of their synthesis is a complete mystery at the moment.

Control by Transport

(A) INTRODUCTION

Action at a site remote from their point of origin is fundamental to the original concept of hormones. It is not surprising therefore that hormone transport from point to point in the plant and the part that it plays in growth regulations and inter-organ correlations has been a major subject for study and speculation ever since the isolation of auxins fifty years ago. We have already seen in Chapter II that the strictly one-way (polar) movement of auxin underlay the growth curvatures produced by unilateral applications of that hormone in the classical coleoptile curvature test. From that time onwards the modulation of hormone flow by external or internal factors has been the undisputed theme underlying such growth phenomena as the tropisms (growth curvatures in relation to external unilateral stimuli such as light and gravity), apical dominance (the suppression of lateral bud growth by the apical bud), etc. Although in recent years it has become evident that some hormones may act in the cells in which they are manufactured, yet in most growth phenomena, control of growth seems to be intimately associated with the control of hormone movement.

Of course the transport of soluble materials is a phenomenon which dominates the physiology of the plant as a whole. The mechanical rigidity of such organs as leaves and the adequate supply of mineral nutrients depend on the efficient flow of water from the soil in the woody (xylem) tissues of the plant. Similarly the supply of photosynthetic products from the green leaves to the non-green portions of plants (roots, etc.) requires a very efficient transport in the specialized conducting systems known as phloem sieve tubes. In this kind of movement the keynote is speedy and efficient *bulk* transport, since new materials for growth are the substances moved. In the case of hormones bulk transport is not involved since, as compared with the requirement of raw materials for growth, the effective quantities of hormone are very much smaller and the nature of their actions very different. Consequently the nature of the flow and its control seems to be very different; in this section the characteristics of this flow will be outlined.

(B) THE AUXINS

(1) Transport in undifferentiated tissue

Characteristics. The oat coleoptile, on which the pioneering work on auxin physiology was done, is a relatively simple undifferentiated organ. It is a tube of oval cross-section and placed centrally in the tissue along the major diameter are two small longitudinal conducting strands (vascular bundles) which run the length of the organ. In the early work with the *Avena* coleoptile test it was soon discovered that the curvatures caused by the unilateral placement of the agar block containing auxin was always strictly away from the side of application, irrespective of whether the block was over one of these conducting strands or whether it was between them. It was thus clear that auxin movement took place in the undifferentiated parenchyma cells and was not confined to these specialized conducting strands. Furthermore the mere fact that a growth curvature was induced implied little lateral movement of the auxin (i.e. from the side of application to the other), and indicated a largely polar movement down the longitudinal axis of these parenchyma cells, from the distal (apical) towards the proximal (basal) regions of the coleoptile.

The first direct experiments to check this supposition were performed by Went (1928) who used small segments of coleoptile cut from just below the tip. Samples of these were placed standing on small blocks of agar with either their proximal or their distal ends downwards. On the other ends were applied other agar blocks containing auxin collected by diffusion out of coleoptile tips. The segments were then stood for a period in moist air and then all the agar blocks were assayed for auxin concentration. The loss of auxin from blocks initially containing it (donor blocks) gave a measure of movement into the segment whereas the gain of auxin by the blocks of initially pure agar (receiver blocks) gave a measure of auxin transport through the segments. As will be seen from Fig. 66 A, transport appeared to take place only from the distal to the proximal end of the segments, i.e. in a base-seeking or *basipetal* direction; there was no apparent apex-seeking or *acropetal* movement. Furthermore the direction of movement was completely independent of the concentration gradient, even taking place against it, i.e. from donor blocks having a lower concentration than receiver blocks (van der Weij, 1934) (see Fig. 66 B). Thus the transport of auxin in the coleoptile was established as strictly polar and studies of other aerial axes such a epicotyls and hypocotyls suggested that the same one-way basipetal flow of auxin occurred universally in such organs. However as more tissues were studied from different organs of different plants it became clear that there was not always strict basipetal polarity of movement, a greater or lesser degree of acropetal movement also taking place. For example in

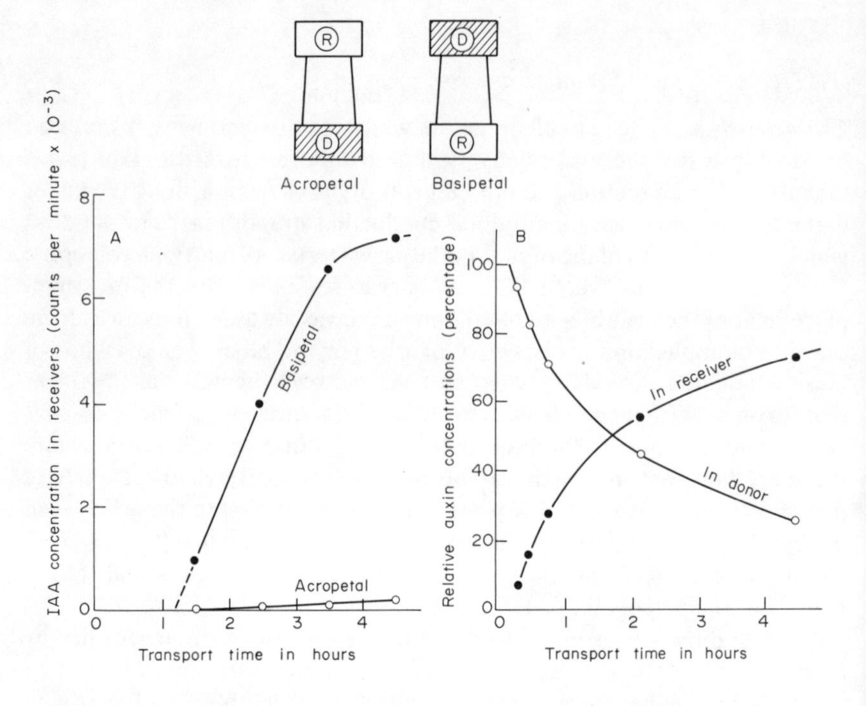

Fig. 66. The polar nature of auxin transport. The two small diagrams at the top of the figure illustrate the fundamental principles of experiments on auxin transport. Segments of the tissue to be studied are placed between two blocks of agar, one, the donor (D) containing a suitable concentration of the auxin and the other the receiver (R) to collect the auxin transported. The segments are drawn tapered to show orientation, the narrower end being the distal (apical) and the other the proximal (basal) end. The left diagram shows measurement of acropetal movement and the right hand measurement of basipetal movement.

(A) Graph showing the time course of accumulation of ^{14}C-labelled IAA in acropetal and basipetal receivers applied to 5.4 mm segments excised from the petiole of *Phaseolus vulgaris.* The slope of these graphs gives a measure of the intensity of transport while from the intercept on the time axis the velocity of transport can be determined (data from McCready, 1968*b*).

(B) Graphs showing the time course of concentration change of auxin in donor and receiver blocks applied to *Avena* coleoptile segments. It will be seen that as the concentration drops in the donor so there is a corresponding and almost symmetrical rise in the concentration in the receiver as transport takes place. Note that after 1.75 hours transport continues *against* a concentration gradient, i.e. concentrations in the receivers are higher than in the donors. Natural auxin (Data of van der Weij, taken from Leopold, 1963.)

Coleus auxin transport seems to be strictly basipetal in young vegetative stems but an acropetal movement is established and increases progressively as the plants mature and produce flowers (Leopold & Guernsey, 1953*d*). In these maturing *Coleus* stems the ratio of acropetal to basipetal movement can reach 3:1 and experiments on stem segments, with various tissues cut out, suggested that the vascular strands accounted for much of the auxin movement (Jacobs, 1961). Such a situation would mean that in differentiated organs we have to reckon with two possible channels of auxin transport, one in the parenchyma, and possible strictly polar, and the other in the normal food transport system, the vascular strands.

The advent of auxin (IAA), labelled with radioactive carbon (^{14}C) enabled much more precise and accurate observations to be made on auxin movement in a great range of tissues. The basis of the technique is substantially that of Went, except that quantitative measurements are made by electronic equipment counting particle emission from the radioactive auxin present in the agar blocks rather than by the biological assay of Went. Of course checks have to be made on the radioactive substance(s) emerging from the plant material to ensure that it is chemically identical to the auxin applied at the other end of the segment. Such experiments have shown unequivocally that in aerial organs so far studied auxin movement is not *absolutely* polarized; there is usually some acropetal movement, however slight. The degree of polarity, i.e. the ratio of auxin moving basipetally to that moving acropetally under the same conditions of application, varies widely with conditions even in the same tissue but in actively extending tissues it is usually of the order of 100/1. A most elegant way in which the polarity of movement has recently been demonstrated is by a pulse-application technique. In this, auxin labelled with ^{14}C is applied in an agar block to one end of a 20 mm segment of maize coleoptile. After a short period of time, sufficient for an appreciable 'pulse' of auxin to have moved into the end of the segment (i.e. 15 to 30 minutes) the agar block is removed and replaced by one containing unlabelled auxin. The segments are later cut into ten equal portions and the ^{14}C-content of each separately determined. By using large numbers of segments and sampling at regular intervals an auxin 'pulse' has been demonstrated to pass down the segment from apex to base (Goldsmith, 1967*b*) (see Fig. 67). This could be due only to a strongly polar transport of the labelled auxin. The pulse broadens in its passage down the organ, a phenomenon which could be explained in terms of diffusion spread.

It was originally thought that synthetic auxins, although having the same growth-regulating actions, could be distinguished from the natural auxin IAA in that their movement in growing parenchyma tissue was not polar. In 1961, using unlabelled compounds, Leopold & Lam showed that, like IAA, the closely related γ-(indol-3yl)-*n*-butyric acid (IBA) and also naphth-1yl-

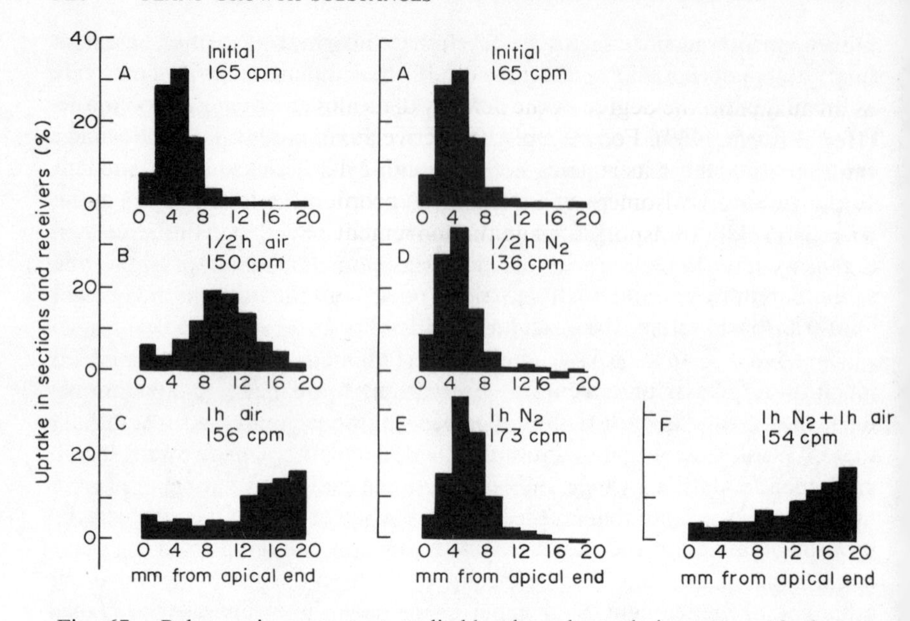

Fig. 67. Polar auxin movement studied by the pulse technique. Agar blocks containing ^{14}C-IAA were applied to apical ends of segments of *Zea* coleoptiles, 20 mm long, for 15 minutes. After this period of uptake these blocks were replaced for a further period of 15 minutes by similar blocks containing unlabelled IAA, to allow the 'pulse' of absorbed and labelled IAA to move a little distance into the segment. Then the blocks were removed and a sample of segments taken, subdivided and the distribution of radioactive IAA down the segment determined (Graph A, duplicated for convenience of comparison). Other samples were left in air and sampled after 0.5 (Graph B) and 1.0 (Graph C) hours. Still others were left in nitrogen and also sampled after 0.5 (Graph D) and 1.0 (Graph E) hours. Finally other samples were left in nitrogen for one hour and then transfered to air for a further hour before sampling (Graph F). It will be seen that in air a pulse of IAA moves down the segment, spreading a little, presumably by diffusion, and reaching the basal end in an hour. In nitrogen there is no measurable movement of the pulse but after transfer to air there is complete recovery and the pulse moves to the base of the segment in one hour, i.e. at the same speed as in segments not previously exposed to nitrogen (Graphs modified from Goldsmith, 1967*a*).

acetic acid (NAA) showed similar though not so marked polarities of movement through young sunflower stems. Subsequently use of ^{14}C-labelled auxins in this problem revealed equally strong polar movement of 2, 4-D and 2,4,5-T in the petioles of *Phaseolus vulgaris* (McCready, 1963, 1968 *a* & *b*). Under comparable conditions (25°C) the polarities (ratios of basipetal to acropetal movement) of the two compounds were approximately 17–25, for 2,4-D and 20-53 for 2,4,5-T as compared with 45-68 for IAA (McCready, 1968*a*). It seems reasonably safe to conclude that basipetal polarity of movement in young aerial axes is a common characteristic of all auxins, in fact a reasonably close correlation exists between the activity of a molecule as an auxin and the degree of the polarity of its movement in *Zea* coleoptiles (Hertel *et al.*, 1969). For example, the active auxin naphth-lyl-acetic acid is transported whereas its inactive isomer naphth-2yl-acetic acid is not. Similarly in the two stereo-isomers of α-(indol-3yl)-propionic acid the D(+) active form is rapidly transported while the movement of the L(+) relatively inactive form is much more sluggish. The same applied to active 2,4,5-T and inactive 2,4,6-T (Jacobs, 1968).

The pulse application technique described above suggested that auxins move down the main axis in a wave at a constant speed. This agrees with the conclusions drawn from all experiments using Went's classical technique, where accumulations of auxin in receiver blocks follow more or less regular patterns irrespective of the nature of the auxin. Such a time-course is shown for IAA in Fig. 66 A. It shows a considerable lag after the application of the auxin at the apical end before it starts to collect in the receiver at the basal end. The lag presumably represents the time taken for the advancing front of the auxin wave to pass down the length of the segment. Thereafter the auxin accumulates at a constant rate. Van der Weij (1932), who pioneered the quantitative measurement of the parameters of auxin movement, suggested that the time taken for the first appearance of auxin (i.e. the intercept of the accumulation curve on the time axis) could be used to measure the velocity of auxin movement in the tissue. If l is the length of the segment and t is the time at this intercept the velocity would simply be l/t. Such determinations have shown that native auxin moves basipetally through aerial tissues at speeds much greater than can be explained by simple diffusion. Van der Weij (1932) first showed this for *Avena* coleoptiles with measured rates of 10–15 mm h^{-1}. More precise and critical experiments have confirmed these measurements, i.e. 9–12 mm h^{-1} (Went & White, 1939) and 11 mm h^{-1} (Newman, 1965). Rates of the same order have been shown for IAA in other tissues, i.e. 10–12 mm h^{-1} in the hypocotyl of *Helianthus annuus* (von Guttenberg & Zetsche, 1956), 10 mm $^{-1}$ in the gynophore of *Arachis hypogaea* (Jacobs, 1951), 6 mm h^{-1} in *Carica* petiole (Yin, 1941) and *Phaseolus* petiole (McCready & Jacobs, 1963), 6.5 mm h^{-1} in the epicotyl of *Lens culinaris* (Pilet, 1965*a* & *b*), 5 mm h^{-1} in apple stems (Gregory & Hancock, 1955) and 14 mm h^{-1} in *Zea* coleoptiles (Leopold, 1963). The synthetic auxins travel rather more slowly, for example in *Helianthus* hypocotyls naphth-1yl-acetic acid moves at a rate of 6.7 mm h^{-1} and γ-(indol-3yl)-butyric acid at 3.2 mm h^{-1} (Leopold & Lam, 1961); in *Phaseolus* petioles the rate for 2,4-D is only 0.6–1.0 mm h^{-1} (McCready & Jacobs, 1963) and for 2,4,5-T also about 1 mm h^{-1}. The fact that pure diffusion of these substances would give average rates of the order of 0.05 mm h^{-1} as measured by the technique used, strongly implies that some kind of physiological acceleration is involved in the polar transport of auxins.

Another parameter of auxin movement is the rate at which it accumulates in the receiver block of agar, i.e. the *slope* of the concentration/time curve in Fig. 66. This is known as the transport *intensity*, a term coined originally by van der Weij (1932) and still in use. Apart from small early inflexions, this accumulation is virtually linear with time over the usual measurement period, which means a virtually steady unvarying flow in the transport system.

Effects of factors on transport. The temperature of the tissue has a marked effect on its auxin transport characteristics. Van der Weij, from his measurements on *Avena* coleoptiles, claimed that it has no effect on the velocity of natural auxin movement but that the intensity increased roughly three-fold for every rise of 10°C to a maximum at 35–40°C. This most unusual characteristic was difficult to explain with any rational model of a transport system. Went (1928) had suggested that the circulating cytoplasm of the cell might act as a 'carrier' and this would account for the constant mass transport and the acceleration of movement well above that for purely diffusive processes. This received indirect support from observations (Bottelier, 1934) that this protoplasmic circulation in *Avena* coleoptile cells was not affected by temperature over the range 17–35°C. This would have explained the lack of effect of temperature on the *velocity* of auxin movement; the high temperature coefficient of movement *intensity* could be explained by an increase in the *loading* of this carrier with increased temperatures. Polarity would have to be due to a loading mechanism at the distal end of the cell and an off loading mechanism at the proximal end.

However a later, completely rigorous and objective analysis of van der Weij's data (Gregory & Hancock, 1955) showed that the velocity also was affected by temperature and subsequent experiments of greater precision have demonstrated an optimal relationship in apple twigs (Gregory & Hancock, 1955) and in *Fritillaria meleagris* stems (Kaldeway, 1965) similar to that of other physiological responses, but with a rather low temperature coefficient, up to the optimimum, of the order of 1·5. In Fig. 68 the effect of temperature on both velocity and intensity in *Phaseolus* petioles (A) and the relationship between temperature and velocity in stems of *Fritillaria meleagris* (B) are illustrated.

The second important environmental factor which could influence auxin transport is light. Naturally with the conventional measurement techniques described above, complications may arise from light effect on auxin production or on auxin destruction and immobilization in the experimental tissue; their importance has still to be unravelled. Nevertheless transport itself does seem to be altered by all-round illumination. Guttenberg & Zetsche

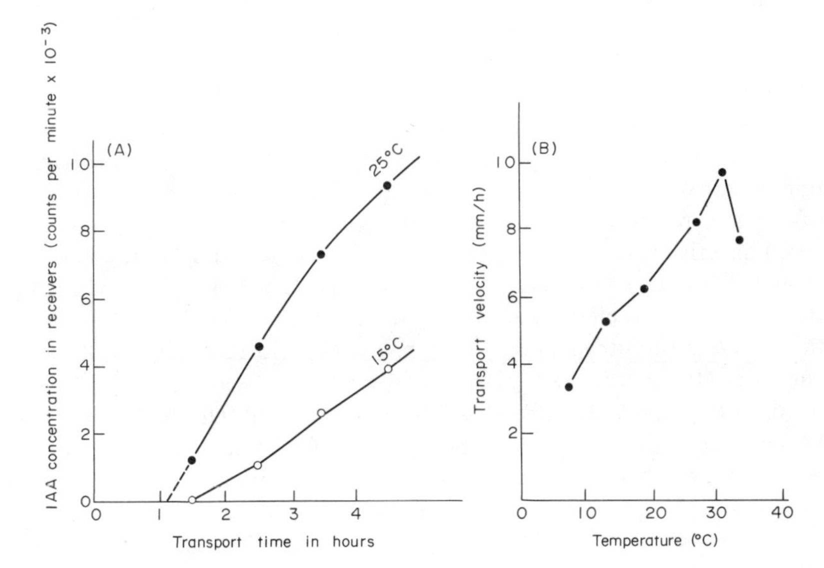

Fig. 68. Graphs showing the effects of temperature on the intensity and velocity of the polar transport of auxin.

(A) Time course of accumulation of radioactive IAA in receivers applied to the base of segments petiole of *Phaseolus vulgaris* at two temperatures (technique illustrated in Fig. 65). The temperature coefficient for the transport intensity (ratio of the slopes of the accumulation curves) over the range 15 to 25°C is approximately 2·3. (Data from McCready, 1968.)

(B) Transport velocity and temperature in the stem of *Fritillaria meleagris*. The temperature coefficient of velocity over the same range of temperature (15–25°C) is approximately 1·4. (Data from Kaldewey, 1965).

(1956) showed that exposure of *Helianthus* hypocotyl tissue to periods of darkness greatly reduced their capacity to transport auxin and attributed this to a depletion of carbohydrate substrate necessary to drive the transport mechanism. On the other hand short periods of exposure to light cause depression of the basipetal export of auxin from *Avena* coleoptiles (Thornton & Thimann, 1967). Similar inhibition by continuous light in coleoptiles has been shown by Naqvi & Gordon (1967) and Meyer & Pohl (1956). The latter explained these effects in terms of the photo-conversion of IAA to indol-3yl-aldehyde, which then acts as a transport-blocking agent. In other tissues the complete absence of effects has been claimed, e.g. in segments of light-grown Alaska pea stems (Thimann & Wardlaw, 1963) although uptake from apical donors was greatly promoted, an effect which may explain the anomalous results of Guttenberg & Zetsche (1956). Results such as these illustrate the difficulties of interpreting results of studies of auxin movement and the complexities of the factors which control it.

The relatively high rate of auxin movement, its marked polarity, often up a concentration gradient, implies that it is not a purely physical process but one that requires the expenditure of work; the energy for this must come from metabolism. Consequently it is not surprising that situations which restrict or prevent this energy provision will also inhibit the basipetal transport of auxin. Thus transport in apple shoots depends on an adequate supply of oxygen, presumably for respiratory metabolism (Gregory & Hancock, 1955) and the absence of oxygen prevents it in maize and *Avena* coleoptiles (Hertel & Leopold, 1963; Goldsmith, 1966; Wilkins & Martin, 1967). Potassium cyanide, which inhibits the cytochrome system in respiratory oxidations, and 2,4-dinitrophenol, which prevents the utilization of respiratory energy, both depress it (duBuy & Olsen, 1940); sodium azide, another respiratory poison, also inhibits it (McCready, 1968*a*). The exact mechanism by which energy is applied to power the polar transport still evades us, although the participation of particular enzymes (sulphydryl enzymes) have been indicated from the inhibitor studies of Niedergang-Kamien & Leopold, (1957). However the fact that oxygen concentrations have to be lowered to about one part in 100 of nitrogen if transport is to be reduced to a half, suggests that the oxidative system powering the transport has a very high affinity for oxygen and points to cytochrome oxidase as the responsible enzyme (Goldsmith, 1968).

The effects of other molecules. Substances other than respiratory inhibitors may have far-reaching effects on auxin transport. The most extensively studied is the suspected anti-auxin 2,3,5-tri-iodobenzoic acid (TIBA). We have already seen in Chapter VII that its physiological actions are far-ranging and its growth effects numerous; its best established action is in the virtually complete blockage of basipetal auxin transport in a variety of aerial organs, e.g. in the petiole of *Ipomoea batatas* (Kuse, 1953), in hypocotyls of *Phaseolus* (Zwar & Rijven, 1956), in segments of tobacco stem (Niedergang-Kamien & Skoog, 1956; Keitt & Baker, 1966) in sunflower stems (Niedergang-Kamien & Leopold, 1957) and in *Phaseolus* petioles (McCready, 1968) (see Fig. 69). Indirect experiments on many other tissues have also indicated similar blockage there; in fact TIBA is now accepted as a tool to study the role of auxin transport in various auxin-controlled growth phenomena such as apical dominance. The exact mechanism of its action remains elusive although it seems unlikely to be simply one of blocking metabolic energy flow into the transport system. Following reports that TIBA can function as a weak auxin, studies have been made comparing its transport-blocking action with those of chemically related molecules and of known weak auxins (Niedergang-Kamien & Leopold, 1959; Zwar & Rijven, 1956; Keitt & Baker, 1966). They have shown that the transport-inhibiting properties of these molecules are roughly an inverse function of their auxin activities. Consequently it has been suggested that TIBA and other related inhibitors of IAA movement are strong competitors for the 'carrier' involved in the polar transport of the natural auxins; they combine with it and then effectively block the access and subsequent transport of IAA. So far no attempts have been made to check this by direct experiment.

Another weak auxin, which has an exceptionally strong inhibiting action even surpassing that of TIBA, is naphth-lyl-phthalamic acid (NPA) (Fig. 43 CXLIII) (Morgan & Söding, 1958; Morgan, 1964, Keitt & Baker, 1966, McCready, 1968). Recent newcomers which are almost as effective as NPA are the flurenolcarboxylic acids (Fig. 43 CXLIV) (McCready, 1968) (see Fig. 69), which, because of their marked effects on some aspects of plant development have been given the 'descriptive' name of *morphactins*. There is no evidence to suggest that their mode of action is any different from those of TIBA or NPA.

The natural metabolite ethylene (see Chapter VIII), which has come back into prominence as a possible natural 'hormone', was many years ago proposed as a modifier of auxin transport, disturbing the polarity of its movement and in this way producing indirectly some of the well-known growth responses to ethylene (van der Laan, 1934; Borgström, 1939). It is only recently however that direct studies have been made to check these suggestions. Morgan & Gausman (1966) have claimed that the movement of IAA labelled with radioactive carbon is inhibited by ethylene in *Gossypium* and *Vinca* plants but such responses could not be confirmed by Abeles (1966) in *Gossypium* and several other genera. More recently the application of the 'pulse' technique for the study of the polar movement of ^{14}C-labelled IAA in stem segments of *Gossypium* has shown with much greater precision that pretreatment of the segments with ethylene somewhat reduced the *velocity* of the movement of the pulse, which is broadened and flattened in the process (Beyer & Morgan, 1969). This indicates that the effect of ethylene may be indirect and the outcome of a promoted 'immobilization' of the auxin in the tissue.

Other established plant hormones may also modify the transport of auxins in parenchymatous tissue. There are two reports that gibberellin applications increase the basipetal flow of IAA in plant tissues. Jacobs & Case (1965) demonstrated a greatly augmented flow down intact shoots of *Pisum sativum* and Pilet (1965*a*) observed that in isolated segments of *Lens culinaris* stems, pretreatment with gibberellic acid increased both the velocity (from 5·7 to 9·8 mm h^{-1}) and the intensity of basipetal transport of ^{14}C-labelled IAA. The situation with the cytokinins is not so clear-cut. First reports showed that treatment of segments of *Phaseolus* petiole with kinetin 13 and 20 hours after excision, increased their ability to transport auxin *vis-à-vis* the controls (McCready *et al.*, 1965), a result which could be attributed to a retardation or reversal of the deterioration of the transport system in such excised tissue. In segments of *Lens culinaris* stems, shorter (2 hour) treatments with kinetin inhibited transport (Pilet, 1965*b*). However immediate application of kinetin in the donor blocks augmented basipetal transport of IAA in *Helianthus* hypocotyl (Leike, 1967) and of 2, 4, 5-T in *Phaseolus* petiole segments (Osborne *et al.*, 1968). The implication of these findings is discussed later.

The growth inhibitor abscisic acid (see Chapter VI) is reported to reduce the 'intensity' of auxin movement in *Avena* coleoptiles but not its velocity (Kaldewey, Weis *et al.*, 1969); this may partly account for its growth-inhibitory activities.

The mechanism of polar transport. We have seen previously that, arising from the experiments of van der Weij, the transport of an auxin carrier in the cytoplasm of the cell by the phenomenon of cyclosis was proposed to account for the rapidity of movement; some differential loading and unloading mechanism at the distal and proximal ends of the cell respectively would explain the polarity. Indirect support for this hypothesis came from observations (Clark, 1938) that bile salt, sodium glycocholate, a substance which is very surface-active and which might be expected to disturb cytoplasmic streaming (cyclosis), could block auxin transport at concentrations of 10 ppm. However, although van Overbeek (1956) was able to confirm the transport-blocking effect, he was unable to observe any effects on cyclosis; furthermore saponin, which stops cyclosis, had no effect on auxin transport. Very recent experiments on *Helianthus* hypocotyl could not confirm Clark's claims of an auxin transport blockage (Vardar & Denizci, 1962). It is not surprising therefore that the participation of cyclosis in auxin movement has not figured prominently in recent theories.

One striking phenomenon which has recently emerged is that the inhibition of basipetal transport is usually accompanied by a correlated increase in acropetal transport (McCready, 1968*a* & *b*), whose speed may approach that of basipetal transport (although this has still to be established beyond doubt); then acropetal movement would also be an energy-requiring process and could not be explained by a simple back-diffusion. Such a situation has led Leopold & dela Fuente (1968) to suggest that the high velocity of movement and its polarity are expressions of 'different physiological functions', which means presumably that they are controlled by different mechanisms.

Another interesting characteristic of the polar movement of auxins is that it is closely associated with actively-growing tissue, the effectiveness and polarity declining as the cells elongate and mature (Leopold & Lam,

1962; Jacobs, 1961; Leopold & dela Fuente, 1968). This means that there is a strong correlation between the number of cells per mm along the main axis and the amount of auxin transported basipetally by those cells (Jacobs, 1961), or to look at it another way, with the number of transverse cell membranes per mm crossed during this axial movement. This implies (Audus, 1967) that these transverse membranes are in some way involved in polar auxin transport. Such concepts are the basis of two theories to account for the polarity of auxin movement. One is based on the observation that when auxin transport is inhibited by interfering chemicals or by lack of oxygen it is the export from the proximal end which is affected more than uptake by the distal end of the segment (Hertel & Leopold, 1963). Consequently the polar movement of auxin was envisaged as arising from a differential *secretion* (output) at the two ends of the cell, the metabolic 'pump' actuating this secretion being powered by energy from respiration. With the aid of a mathematical model Leopold & Hall (1966) were able to demonstrate that only a slightly greater secretion by the proximal end of the cell (i.e. of the order of 1 percent) would, when integrated over the chains of cells present in the usual experimental segment, result in virtually complete one-way movement.

Another view of the 'pump' is taken by Osborne (1968) who adopts as her vantage point the observations (Osborne *et al.*, 1968) that kinetin promotes the transport of auxin and may do so by preventing the deterioration of the transport system through cellular senescence after excision. This suggestion is based on the well known action of cytokinins in promoting the net synthesis of protein and hence preventing senescence (see Chapter XII). Her suggestions involve a promoted *intake* of auxin at the distal end of the cell and associated with specific enzymes (transpermeases) synthesized on the endoplasmic reticular membranes associated with the plasmalemma at that end. Hence actively growing cells possessing a high rate of enzyme synthesis would consequently have an active auxin-importing enzyme system at the distal end of the cell.

Auxin transport in roots. Studies of the transport of auxins in roots, as compared with those in aerial organs, have been meagre. Indeed the whole question of the role of auxin in the control of root growth has always been an enigmatic one and is still not resolved. Roots are extremely sensitive to applied auxins; their growth is somewhat stimulated by concentrations as low as one part in 10^{11} of water and at levels which promote coleoptile growth the extension of roots is greatly depressed and may be stopped entirely. In fact the responses of roots to different concentrations of applied auxins probably follows the same kind of inverted bell-shaped 'optimum' relationship as shown for the *Avena* coleoptile by Foster *et al.* (1952) (see

p. 306); roots would differ from shoots only in the concentration for optimal growth which would be about 10^{-5} to 10^{-6} that for shoots. This fact, coupled with the early observations that auxin was produced by the root tip (see previous section) led to a theory of root growth control which has been widely accepted for many years. This theory, due originally to Boysen-Jensen (1936) suggested that this control was operated by an *inhibition* of growth, since the auxin moving in the extending zone would always be at concentrations *above* that optimal for growth (i.e. supra-optimal concentrations). This was a convenient theory for explaining the fact that the tip is responsible for the reactions of the root to gravity; auxin, moving from the tip of a horizontal root, for reasons we cannot discuss at this point, accumulates in the cells on the lowermost side of the extension zone, there producing an increased inhibition of root growth and hence a downward curvature, i.e. a positive geotropic response.

Such a theory presupposed that auxin movement was polar and was also basipetal in roots, i.e. from the tip towards the base, and therefore, in relation to the main axis of the plant as a whole, in a direction opposite to that in shoots. Direct evidence on this point was always conflicting, some workers claiming that they could demonstrate this polarity while others showed that auxins could undoubtedly move towards the root tip if applied to the shoot or to the root base. Even theories of two opposing streams of auxin, one acropetal and one basipetal, were advanced. The dilemma of the first ten years of root studies is fully set out in Went & Thimann's book (1937).

The advent of auxins containing radioactive carbon did little at first to clear up the situation. Thus first claims were for a predominantly basipetal movement in *Zea* and *Vicia* roots (Hertel & Leopold, 1963) but other work could detect very little movement in *Vicia faba* roots and what movement there was seemed to be preferentially in an acropetal direction (Yeomans & Audus, 1964). However the most recent observations from experimental involving labelled IAA of high specific activity (33 mCi/m mol) have established clearly that the movement of this auxin in segments cut from the extending zone of roots of *Lens*, *Phaseolus* (Kirk & Jacobs, 1968) and *Zea* (Scott & Wilkins, 1968) is strictly polar and in a direction base to apex (see Fig. 70). Furthermore the rate, as determined by the methods already described for shoot segments, is 2·2 mm h^{-1} for *Lens* and 4 to 6 mm h^{-1} for *Zea*, which is a high rate, of the same order as that for coleoptiles. This, coupled with subsequent observations (Wilkins & Scott, 1968) that the acropetal flux is virtually eliminated by low temperatures (0°C) and lack of oxygen (Fig. 70) suggests that, like polar movement in shoots, it is the outcome of a one-way molecular pumping-mechanism powered by energy provided by respiratory metabolism. An interesting feature of the effect of lack

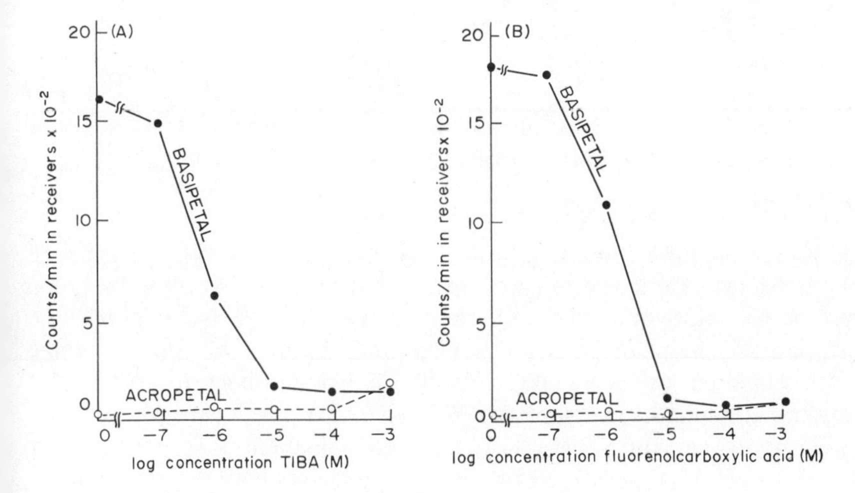

Fig. 69. The effects of two specific inhibitors of auxin transport on the intensity of basipetal and acropetal transport in petioles segments of *Phaseolus vulgaris* (A) TIBA, (B) 9-fluorenolcarboxylic acid (Graphs from McCready, 1968*b*.)

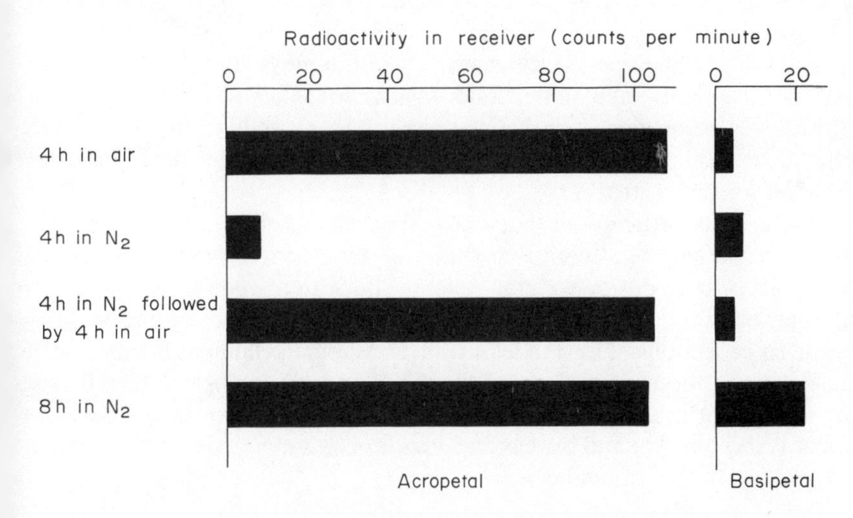

Fig. 70. Diagram showing the effects of periods of anaerobiosis (exposure to pure nitrogen) on the acropetal and basipetal transport of radioactive IAA in root segments of *Zea mays* (technique as illustrated in Fig. 66). Note the very strong acropetal polarity in air and the reversible blockage of acropetal transport in nitrogen. Note also the complete recovery of acropetal transport during prolonged periods in nitrogen (data of Wilkins & Scott, 1968).

of oxygen is that it disappears after four or so hours and acropetal transport is again established, implying that roots have a considerable ability to adapt to anaerobic conditions (Fig. 70). Sodium fluoride completely abolishes this anaerobic acropetal movement, which must therefore depend on the flow of metabolic energy (Wilkins & Scott, 1968).

(ii) Transport in food-conducting channels

Although the most intriguing problems of the auxin control of growth via its transport characteristics are centred in undifferentiated parenchyma, yet the possibility of movement in the regular specialized channels for solute transport in the plant must not be neglected. Rapid movement of organic nutrients about the plant takes place in the phloem tissue of the vascular system, although its mechanism is still far from clear. It seems logical that natural auxin might travel there also, although there is very little direct evidence on this point. However several years of study of the movement of such synthetic auxins as 2,4-D, extensively used as weed-killers, have shown that it takes place in the phloem and is subject the same laws as that of nutrient movement, i.e. under a variety of conditions it moves in the same direction as and possibly associated with sugars. The rate of movement is thus much greater than that in parenchyma tissue and is of the order of 100 cm h^{-1} as compared with 0.1 cm h^{-1} in *Phaseolus* petioles. It seems that the movement of this synthetic auxin could be directly linked to that of the sugar, since, when applied to leaves, it will not move therefrom into other parts of the plant except under conditions when sugar is also being exported. It has even been suggested that the auxin has to combine chemically with the sugar to form an ester before it will move, with the sugar, into the phloem transport stream.

There is also little doubt that when auxins are applied in *high* concentrations to one part of a plant that they will 'leak' into the water-conduction channels (xylem) of the vascular system. They can then be carried passively in solution in this water stream to all parts of the plant, depending on transpiration conditions. This mode of transport in the xylem has however little relevance to the movement of native (endogenous) auxins; the part played by movement in the phloem in the normal control of growth and development is still obscure and its relative importance *vis-à-vis* movement in parenchyma cannot be estimated.

(C) THE GIBBERELLINS

Our knowledge of the movement of gibberellins in plants comes almost entirely from studies involving external applications; movement has been detected and measured either by observations of growth responses in parts of the plant remote from the point of application or by using gibberellic acid containing ^{14}C and directly measuring the radioactivity of appropriate portions of tissue or of block of agar collecting the gibberellin from cut surfaces of segments.

Scanty though it is, the evidence suggests that the movement of gibberellins is not polarized in the same way as that of auxin. It moves freely towards the apices of various plants when applied to mature leaves (McComb, 1964; Chin & Lockhart, 1965) and potato tubers (Lazer *et al.*, 1961). It also moves basipetally in stem segments cut from *Helianthus annuus* seedlings (Kentzer & Libbert, 1961), where, as for IAA, its movement is blocked by TIBA (Libbert & Gerdes, 1964); it can also move acropetally in pea stem segments (Galston & Warburg 1959).

Thus there is no evidence of a movement polarity although the inhibition by TIBA suggests that the flow involves a physiological mechanism and is not a simple diffusion. This is borne out by measurement of the rate of movement, which is even greater than that of IAA. In maize internodes a rate of 10 mm h^{-1} was estimated (Neely & Phinney, 1957); in elongating pea stems a rate of 50 mm h^{-1} was measured for ^{14}C-gibberellic acid (McComb, 1964) while for the same tissue a value of 120–360 cm h^{-1} has been estimated by Galston & Warburg (1959). These rates are approaching those which are characteristic of sugar movement in the phloem and indeed this tissue is regarded as perhaps the major channel for the movement of applied gibberellins.

There is little evidence relating to the movement of endogenous (native) gibberellins in plants. The demonstration of their presence in the bleeding sap of de-topped roots e.g. in tomato (Reid *et al.*, 1969) and *Helianthus* (Phillips & Jones, 1964) and in the xylem sap of trees in the spring (Reid & Burrows, 1968) point strongly to their movement in the xylem sap and possibly in the phloem too.

(D) CYTOKININS

The situation with the cytokinins is no clearer. The many observations that these regulators, when applied to leaves, retard their senescence only at the point of application, led to the general conclusion that they moved little if at all about the plant (Skoog *et al.*, 1965). Such ideas were supported by observations on the lateral buds of pea seedlings where application of kinetin stimulated growth only when applied directly to the buds and not to the neighbouring tissues (Sachs & Thimann, 1964). Nevertheless there are now several instances on record of undoubted movement of cytokinins in plants, most of them suggesting a certain polarity comparable to that of IAA movement. Thus radioactive benzyladenine, injected into apple petioles, moved for considerable distances *down* the stem but not up it (Chvojka *et al.*, 1961), while in the same plant kinetin was shown to induce bud growth in regions below but not above the point of application (Pieniążek, 1964). These observations imply a strong basipetal transport. More direct experiments involving the classical van der Weij segment techniques has verified this for benzyladenine-^{14}C in *Phaseolus* petioles (Osborne & Black, 1964; Osborne & McCready, 1965; Lagerstedt & Langston, 1966) where polarity is very marked. In cocklebur petiole segments the degree of polarity is less marked whereas in cotton petioles there is no detectable movement at all in any direction (Lagerstedt & Langston, 1966). There is movement but no polarity in petioles of *Coleus* (Veen & Jacobs, 1969). In segments of *Lens* epicotyl there is a slight movement which is the same in both directions (Pilet *et al.*, 1967) although when applied to the outside of intact epicotyls benzyladenine moves predominantly upwards. This latter movement was attributed by the authors to an effect of the apical bud; one possible explanation would be that the cytokinins were travelling in the phloem in the general stream of nutrients. Similar movement in phloem might account for the strictly basipetal flow in apple shoots. But in petiole segments movement seems most likely in the parenchyma, where the polar properties vary from species to species. In this tissue there is a certain further parallelism with auxin movement since the polarity seems to be increased in the presence of IAA or 2, 4, 5-T (Osborne *et al.*, 1968) (cf. effect of kinetin on auxin movement, p. 325). Whether this means that similar systems are involved in the transport of both hormones in parenchymatous tissue is a question for the future.

There is now considerable evidence that cytokinins move up the shoot from the root (where biosynthesis presumably takes place) in the xylem sap (Kulaeva, 1962; Kende, 1965; Reid & Burrows, 1968; Burrows & Carr, 1969) and this movement is presumably passive in the transpiration stream. It probably represents, as for gibberellins, a very important channel of transport and is of great significance for the influence of the root on the growth of the shoot.

(E) ABSCISIC ACID.

A little direct information is to hand on the transport of ABA in plant tissue. In stem segments of *Coleus* it has been shown to move at a rate of 24–36 mm/hour and, in young tissue, shows some basipetal polarity (Dörffling and Böttger, 1968). More precise measurements of the movement of (2–^{14}C)-ABA in excised segments of cotton cotyledon stalks gave a velocity of 22.4 mm/hour with no trace of polarity in its movement (Ingersoll and Smith, 1971). However this movement was almost completely inhibited by low temperatures, restricted oxygen supply and the oxidative-phosphorylation uncoupling-agent 2,4-dinitrophenol. Both the high velocity, comparable with that of IAA, and its dependance on respiratory-energy supply suggest that it is an active process similar to that for the auxins in parenchymatous tissue (see p. 324).

Control by Inactivation

The last possibility of the control of hormone levels in plant tissues is by inactivation, in which, by some mechanism or other, the molecule is destroyed or its active chemical groups so altered or masked that it can no longer function in growth control. Chemical destruction, which itself can be controlled by natural or externally-imposed factors, is perhaps the major process in operation in plants; at least it is the one which has received major consideration in the last 30 years. The linking or 'binding' of the hormone via its growth active centres is another possibility which has not been so fully explored. In this section we will consider what has been discovered of these processes and a little of the part they may play in the control of growth.

(A) AUXINS

In the study of auxin degradation, particularly as it concerns the natural control of growth, attention has been confined almost exclusively to IAA. Thus natural auxin is a rather labile compound and can be decomposed by a number of physical and chemical agents. For example ionizing and ultraviolet radiation and even visible radiation in the presence of a suitable photosensitizing pigment, will degrade it. Similarly it can be easily oxidized by peroxides or by oxygen in the presence of suitable catalytic (redox) systems (see review by Gaslton & Hillman, 1961). However all these conditions are relatively extreme and it seems likely that they are only of occasional relevance to the situation in living tissue, where destruction seems always to be under enzymic control and involves oxidation.

(i) The enzymic oxidation of IAA

The very early work on extraction had shown that tissues ground up with water were very poor sources of auxins and implied that some inactivation had gone on during the grinding process. It was Thimann (1934) who first demonstrated that tissue (leaf) extracts could inactivate natural auxin and from comparative studies with different species he suggested that the inactivation was associated with the oxidizing enzymes peroxidase and catechol oxidase. Larsen (1936, 1940) first showed that a thermolabile enzyme was involved in the sap of bean seedlings and this led to the early naming of the enzyme as IAA-oxidase (Goldacre, 1949). In view of subsequent studies however it seems unlikely that a unique mechanism of oxidation by a unique enzyme system is in operation in all plants and so 'IAA-oxidase' must be regarded at the moment only as a blanket term to cover all possible enzyme systems oxidizing IAA.

Since these early studies the presence of IAA-oxidase has been demonstrated in a great diversity of plants. The most recent survey (Pilet & Gaspar, 1968) shows such sytems to be present in nine genera of monocotyledons, twenty-six genera of dicotyledons, four genera of gymnosperms, three genera of pteridophytes, eleven genera of fungi, six genera of algae and twelve genera of bacteria. There are reports of tissues lacking an IAA-oxidase system but Pilet & Gaspar suggest that it may have been present but remained undetected in the unpurified extracts studied because of the presence of an excess of natural oxidase inhibitors.

The actual nature of the enzyme or enzymes responsible is still a matter for dispute. The earlier attempts to identify it led Tang & Bonner (1947) to suggest that it was an iron-containing haemoprotein since it could be inhibited photoreversibly by carbon monoxide. On the other hand these properties were not confirmed by Wagenknecht & Burris (1950) who, on the basis of its inhibition by copper-complexing reagents, suggested that it was a copper protein. All subsequent work however has confirmed the views of Tang & Bonner and there now seems little doubt that a haemoprotein of the nature of a peroxidase is the active component of IAA-oxidase systems. An important landmark in these studies was the discovery by Galston *et al.* (1953) that highly purified horseradish peroxidase was very effective in catalysing the oxidation of IAA. Later, the fractionation of crude enzyme preparations, by ammonium sulphate precipitation and electrophoresis, demonstrated that the proteins associated with IAA-oxidase activity also had peroxidative properties (Perlis & Galston, 1955; Stutz, 1957). But there still remains evidence that a peroxidase is not the only enzyme that may be involved. On the basis of the activation of the crude enzyme preparation by visible light and its relationship to wavelength, Galston & Baker (1951) suggested that a photosensitive flavoprotein enzyme might also be involved. Although Kenten (1955) could not demonstrate the presence of flavin-containing compounds in his very active IAA-oxidase preparations from wax-pod bean roots and horseradish, yet reports of the multi-component nature of the enzyme system still came in. Thus the peroxidase of pea tissue (the most used source of IAA-oxidase) is heterogeneous and composed of a family of closely related proteins (isoenzymes) (Macnicol & Reinert, 1963; Siegel & Galston, 1967*a*); however not all these peroxidase proteins are IAA-oxidases and the peroxidase activities of these which are, do not closely correlate with their respective IAA-oxidase activities (Endo, 1968). Similarly in tobacco roots IAA-oxidase can be separated into two fractions, only one of which has peroxidase activity (Sequeira & Mineo, 1966). Furthermore if the haem group is removed from the protein of horseradish peroxidase, its peroxidative activity is lost while its ability to oxidize IAA is maintained, provided certain cofactors (the manganese ion and a monophenol – see later) are supplied (Siegel & Galston, 1967*b*). This is an interesting situation implying that one enzyme protein can perform two separate catalytic functions, depending on the substrate and the cofactors available.

There are other enzymes which catalyse the oxidation of IAA. Catalase, the enzyme normally responsible for the decomposition of H_2O_2, will oxidize IAA in the presence of high concentrations of Mn^{2+} ions and a monophenol (Waygood *et al.*, 1956; Avella *et al.*, 1966). However the lack of any consistent correlations between the IAA-oxidase and catalase activity of plant tissue under various conditions has led Pilet & Gaspar (1968) to deny catalase involvement in auxin destruction in normal plant tissue. Certain phenol oxidases are capable of destroying IAA, for example tyrosinase isolated from the fronds of the fern *Osmunda* and from higher mushroom-type fungi (Basidiomycetes) (Briggs & Ray, 1956) and laccase from a variety of fungi (see Pilet & Gaspar, 1968). The activity of these fungal enzymes may have little relevence to the oxidation of IAA in normal plant tissue, where the role of phenol oxidase in its destruction is still debatable.

The presence of chemical cofactors can greatly modify the activity of IAA-oxidase, opening up possibilities of growth control through the regulation of IAA-oxidase in the tissues. The demonstration that the stimulation of IAA-oxidase activity by the synthetic auxin 2,4-D was due to a small trace of 2,4-dichlorophenol as an impurity (Goldacre *et al.*, 1953) led to the realization that a wide range of mono-phenols will enhance the reaction. These include phenol itself together with some of its halogen and nitro derivatives, cresols, phenolic acids such as the hydroxybenzoic and the coumaric acids, hydroxycoumarins, naphthols, *p*-hydroxy derivatives of phenylacetic acids, benzaldehyde and benzyl alcohol and even the amino acid tyrosine (for full bibliographic details see Pilet & Gaspar, 1968). Since some of these mono-phenols are normal metabolites of plants, one wonders whether they might not serve as natural IAA-oxidase co-factors in growing tissues. A number of such natural co-factors have been extracted from plant tissue and some have been identified with known mono-phenols.

In general the presence of manganese ions augments the IAA-oxidase activity although under some conditions it produces an inhibition. The nature of these discrepancies is not completely understood.

In addition to a range of standard inhibitors used as tools to unravel enzyme mechanisms, a considerable number of phenolic substances can also block IAA-oxidase action. In contrast to the phenolic co-factors, which are mainly mono-phenols, these inhibitors are mostly phenols with more than one hydroxy group. A few of the more striking examples are catechol, hydroquinone, pyrogallol, phloroglucinol, dihydroxyphenylalanine, gallic acid (Fig. 52 CLXXIII) and chlorogenic acid (Fig. 52 CLXXI). However it is only fair to record that there are exceptions to this general situation, some monophenolic compounds acting as inhibitors and one or two diphenolics acting as stimulants (for documentation and references see Pilet & Gaspar, 1968). For example the possession of a methoxy group by a mono-phenol, e.g. in eugenol, ferulic acid (Fig. 40 CXXXI), vanillic acid (Fig. 40 CXXXII) and syringic acid (Fig. 53 CLXXXIV) makes it an IAA-oxidase inhibitor (Parish, 1969*b*). It is reasonable to imagine therefore that natural phenols may play an important role in regulating growth by the modulation of IAA levels via an effect on the activity of IAA-oxidase. Certainly there is very strong presumptive evidence that this could be so since in experiments involving applications to segments of coleoptiles, mono-phenolic co-factors suppress growth and the poly-phenolic inhibitors stimulate it. Similarly the growth action of IAA may be antagonized by mono-phenols and poly-phenols may produce synergisms, i.e. an amplification of the growth stimulation by IAA (Nitsch & Nitsch, 1962*b*; Tomaszewski, 1964). Since mono-phenolic and poly-phenolic compounds are easily and naturally interconverted in plant tissue, it is easy to see that such interconversions have been proposed as a further mechanism for adjusting IAA levels via the control of IAA-oxidase activity.

Investigations into the effects of light on extension growth and the IAA-relationships of irradiated tissue has lent some support to these suggested involvements of phenolic co-factors and inhibitors of IAA-oxidase in growth control. Thus Galston & Baker (1953) first showed that segments cut from the epicotyl of pea seedlings grown in complete darkness were much more sensitive to IAA (i.e. the applied IAA concentration for maximum growth response was lower) than were similar segments previously exposed to red light several hours before harvest. Red light also inhibits the growth of pea epicotyl tissue both in intact seedlings and in excised segments (Russell & Galston, 1969). It also lowers the level of IAA in *Avena* coleoptiles (Blaauw-Jansen, 1959; Briggs, 1963) and in *Phaseolus vulgaris* stems (Fletcher & Zalik, 1964). These phenomena could all be explained in terms of a red-light-stimulated destruction of IAA in the tissue. Red light has also been shown greatly to affect the metabolism of flavonoid pigments, which are compounds possessing phenolic moieties (Mumford *et al.*, 1961; Furuya *et al.*, 1962). The two most important of these are kaempferol (Fig. 71 CXC) and quercitin (Fig. 71 CXCI) in the form of their triglucosides or glucoside-coumaric acid esters. There is little doubt that flavanoid compounds can be

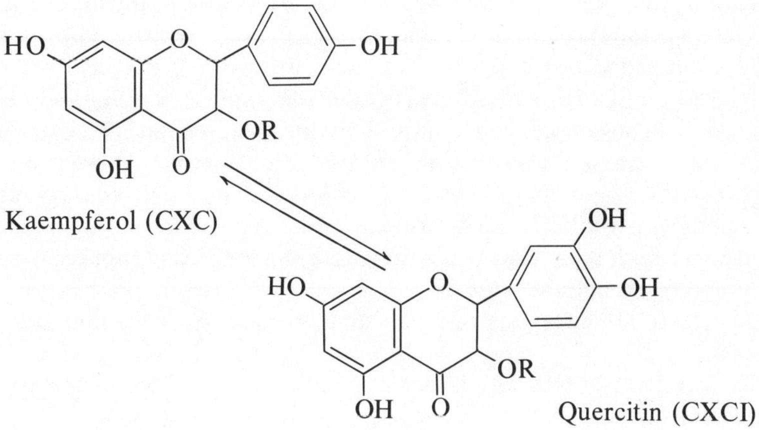

Kaempferol (CXC)

Quercitin (CXCI)

Fig. 71. CXC Kaempferol CXCI Quercitin

very active as modifiers of IAA-oxidase action, stimulating or depressing it depending on constitution (Stenlid, 1963). An intriguing fact is that these two flavonoid compounds in pea tissue are closely related biochemically, the difference being that quercitin has a diphenolic and kaempferol a monophenolic moiety. Since quercitin markedly reduced IAA-oxidase activity and kaempferol is likely to stimulate it (Stenlid, 1963), we see here a possible basis for the inhibition of growth by red light. Thus a promotion of kaempferol synthesis or a suppression of quercitin synthesis or both by an action of red light would augment IAA-oxidase activity, reduce the level of IAA and hence inhibit growth. Although red light undoubtedly disturbs the balance of these two compounds in plant tissues, the precise nature of the shifts is not yet sufficiently clear. Thus the action of red light on pea tissue has been claimed to be an increase in the content of quercitin glucoside without affecting the levels of the kaempferol glucoside (Bottomley *et al.*, 1966); later studies showed similar treatment caused a considerable increase in the kaempferol triglucoside content (Russell & Galston, 1969). The interesting feature of this last-mentioned work is that gibberellic acid, which completely blocks the red-light inhibtion of the growth of intact etiolated pea epicotyls, also blocks this red-light-stimulated synthesis of the kaempferol glucoside. However the fact that the effect of red light on growth can be detected before any significant increase takes place in the kaempferol content of the tissues (Russell & Galston, 1967) suggests that a direct action of kaempferol on

IAA-oxidase activity is not the basis of the red light response; simpler phenolic intermediates of flavanoid synthesis may still be involved (Russell & Galston, 1969) such for example as *p*-coumaric acid whose synthesis is induced in gherkin seedlings by blue light (Engelsma & Meijer, 1965).

However a more direct action of light on the *synthesis* or *activation* of IAA-oxidase is a considerable possibility. Thus in wheat seedlings light has been shown to increase the peroxidase content of coleoptiles, whose growth is suppressed by light, and to increase it in leaves, whose growth is promoted by light (Parish, 1969*a*). Light also causes changes in the IAA-oxidase inhibitors of both these organs but they were slight and were not regarded as sufficient to explain the large growth changes involved. In this plant at least IAA-oxidase inhibitors may not play an important role in light-growth responses.

The mechanism whereby IAA oxidation is catalysed by peroxidase has been subjected to much study and is still imperfectly understood. The main problems have centred in the facts that oxygen is essential for the breakdown which proceeds according to the following equation.

$$\text{IAA} + O_2 \xrightarrow[\text{(Co-factors)}]{\text{Peroxidase}} \text{Products} + CO_2$$

and the addition of H_2O_2 is not essential. The first scheme proposed that the enzyme was acting as a true peroxidase, oxidizing IAA directly with H_2O_2 as oxygen donor, the H_2O_2 being generated internally by a second enzyme component of the nature of a flavoprotein (Galston & Baker, 1951). However it was later suggested that flavoprotein was not an essential component of the IAA-oxidase system (Kenten, 1955) and that the only enzyme necessary was peroxidase. Consequently the problem crystallized into eludidating the mechanisms of the direct oxidation of IAA by a peroxidase enzyme. The earlier schemes were drawn up against the strong promoting action of manganese ions and the monophenolic cofactors and the inhibitory effects of polyphenols.

It was first proposed (Kenten, 1955) that IAA was directly oxidized by the manganic ion and that the manganous ion so formed was reoxidized by the peroxidase with the monophenolic co-factors acting as intermediate oxido-reduction systems. The following scheme (taken from Galston & Hillman, 1961) indicates such a system.

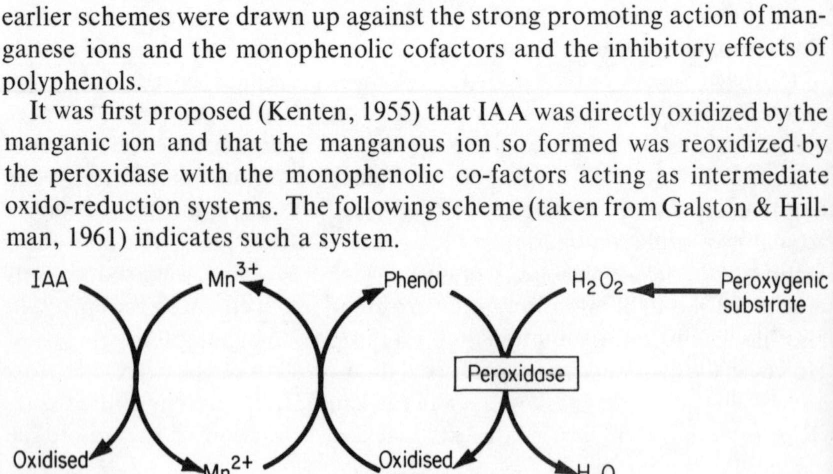

IAA → Oxidised products; Mn^{3+} ⇌ Mn^{2+}; Phenol ⇌ Oxidised Phenol; Peroxidase; H_2O_2 → H_2O; Peroxygenic substrate → H_2O_2

This scheme of course still leaves open the problem of the source of the hydrogen acceptor H_2O_2, presumed to be produced from some peroxigenic substrate. A modification of this scheme was put forward by Maclachlan & Waygood (1956); in this the initial step of the oxidative decarboxylation of IAA by the manganic ion led to the formation of an intermediate free radicle, indole peroxide. This then acted as oxygen donor for the peroxidation of the phenol co-factor, the oxidized form of which then re-oxidized the manganous ions formed in the initial step.

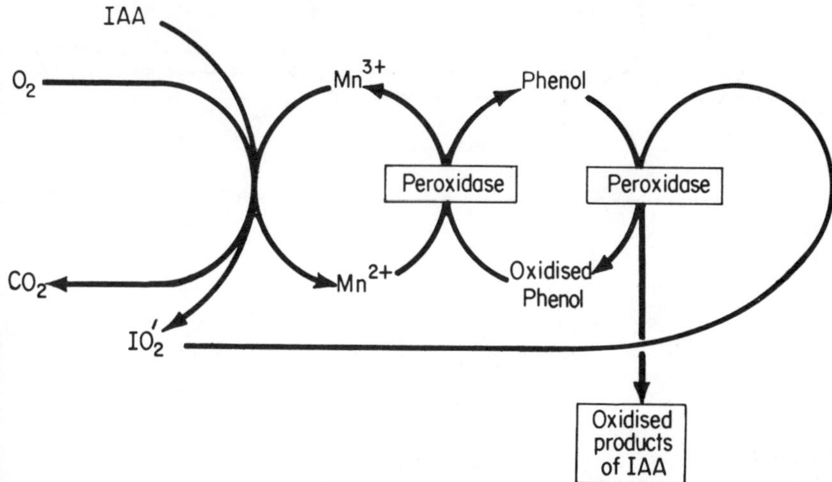

In these schemes the IAA does not react with the enzyme in the initial stages of the reaction. Polyphenolic inhibitors, being good substrates for peroxidase, might therefore be expected to inhibit because they compete with Mn^{2+} and phenol for the peroxidase enzyme. Maclaughlan and Waygood however thought that they interacted directly with the Mn^{3+} ions and the phenolic co-factors, thus blocking the action.

However studies on the effect of IAA on pyrogallol peroxidation (Ray, 1960) led him to suggest that IAA interacts directly with the peroxidase enzyme and later Bastin (1964) produced kinetic evidence of the competition of diphenolic peroxidase substrates such as catechol and guiacol with IAA for the catalytic centres of horseradish peroxidase. All this suggests that peroxidase is *immediately* involved in the oxidation of IAA and the observation (Ray, 1960) that carbon monoxide inhibits the process and that this inhibition is completely reversed by light, suggests the participation of a ferrous haemoprotein in the oxidation reaction. Furthermore the application of highly sophisticated spectrophotometric techniques have indicated the formation of enzyme-IAA complexes during the reaction (Fox *et al.*, 1965). We have already seen that Maclachlan & Waygood postulated the formation of free radicles during the oxidation of IAA by Mn^{3+} and evidence for their participation in the IAA-oxidase reaction has now come from the use of the free-radicle inhibitor $Na_2S_2O_5$ which completely blocks the oxidation (Fox *et al.*, 1965). In the same reaction system the additions of catalase, which catalyses the decomposition of H_2O_2, has no effect on the oxidation, showing that the intermediate production of H_2O_2 as a hydrogen acceptor was not a necessary part of the reaction. From such evidence IAA oxidation by peroxidase might therefore be simply represented as having the mechanism outlined below;

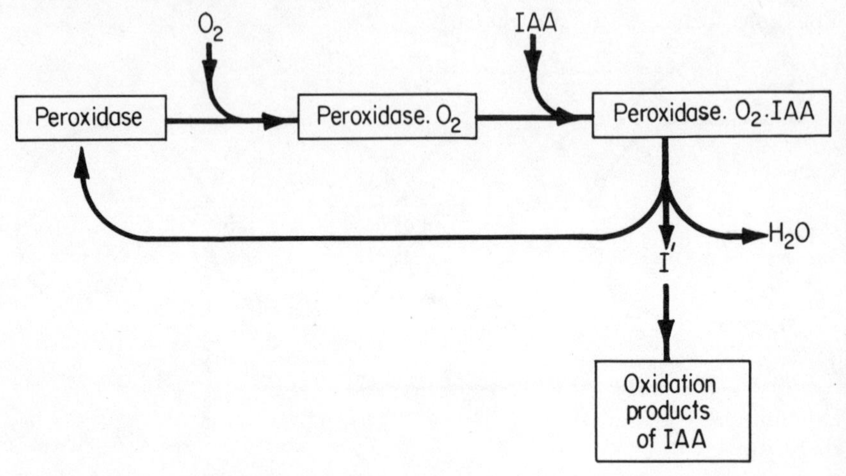

where I′ is an intermediate free radicle.

The free radicle thus produced by the perioxidase, undergoes a spontaneous autocatalytic series of changes leading to the known product 3-methylene-2-oxindole (Hinman & Lang, 1965).

The exact nature of the products of IAA oxidation seems to vary with the source of the enzyme involved. Tang & Bonner (1947) first suggested that the end product was indol-3yl-aldehyde since degradation of IAA did not lead to the loss of the colour reaction of the indole ring (see Chapter II). However the study of the oxidation products by u.v. and i.r. spectrometry and by the use of ^{14}C-labelled IAA has shown that quite other products predominate. For instnace the enzyme from the fungus *Omphalia flavida* yields 3-methyl-3-hydroxy-2-oxindole (Fig. 72 CXCII) as identified by u.v. and i.r. spectra (Stowe *et al.*, 1954; Ray & Thimann, 1955). Pilet (1962) found the same compound to predominate over other products of the enzyme obtained from *Lens culinaris* roots. Other closely related compounds have been produced by enzymes from yeasts and a bacterium (*Escherichia coli*) and from pea tissue (Fukuyama & Moyed, 1964; Tuli & Moyed, 1967); they include 3-hydroxymethyl-2-oxindole (Fig. 72 CXCIII) and 3-methylene-2-oxindole (Fig. 72 CXCIV). In all these products the heterocyclic ring system remains virtually intact. However there is also evidence that the pea enzyme also opens up the pyrrole moiety of the indole ring, yielding ultimately a substance similar to,

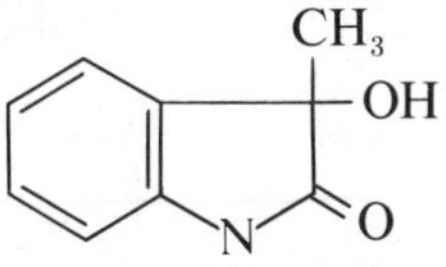

CXCII 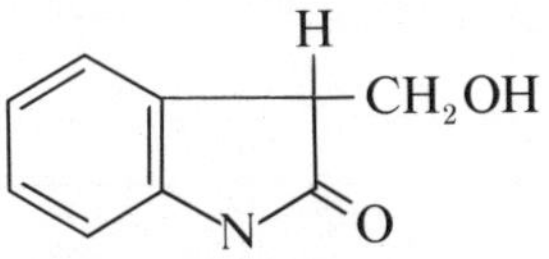 CXCIII

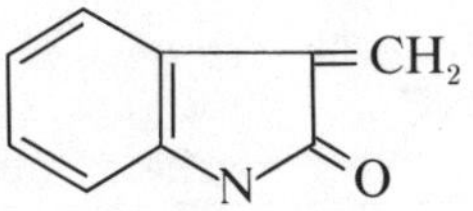

CXCIV 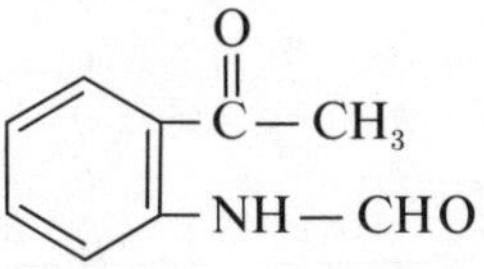 CXCV

Fig. 72. IAA-degradation products.

CXCII 3-Methyl-3-hydroxy-2-oxindole.
CXCIII 3-Hydroxymethyl-2-oxindole
CXCIV 3-Methylene-2-oxindole
CXCV *o*-Formamidoacetophenone.

if not actually identical with *o*-formamidoacetophenone (Fig. 72 CXCV) (Manning & Galston, 1955; Collet, 1968). The probable pathway whereby these various products could be formed from the primary free radicle product of peroxidation have been proposed by Hinman & Lang (1965) (see also survey by Pilet & Gaspar, 1968).

A recent suggestion has involved the metabolite 3-methylene-2-oxindole as an essential factor in the promotion of growth by IAA. It is now recognized that a control of biochemical steps in a biosynthetic sequence can be exercised by a feedback sequence in which the relevant enzymes can be inhibited by its own products. Thus, for example, threonine will inhibit the action of homoserine dehydrogenase, an enzyme catalyzing one of the earlier steps in the biosynthetic sequence which forms threonine. It has been shown, in enzyme preparations and in model systems with the bacterium *Salmonella typhimurium* that 3-methylene-2-oxindole will desensitize such feedback control mechanisms, e.g. by stimulating the growth of a feedback-sensitive mutant of that organism (Tuli & Moyed, 1966). In the same way the production of this potent metabolite of IAA may release feedback control in higher plants and may account for the stimulation of metabolism and growth by IAA in the higher plants (Tuli & Moyed, 1967). The inhibiting effects of IAA could be attributed to the same metabolite since, in higher concentrations it is a potent SH- enzyme inhibitor. Supporting evidence for this hypothesis comes from experiments on pea and mung bean stem segments whose growth is promoted by 3-methylene-2-oxindole at optimal concentrations of the same molarity as those of IAA, although the degree of response was much smaller (Moyed & Tuli, 1968). It is conceivable that such effects might be responsible for a minor part of the IAA response.

(ii) Photo-oxidation of IAA

IAA absorbs ultra-violet light strongly and with an absorption peak at 280 nm. The molecule, thus activated, decomposes *in vitro* (Mills & Schrank, 1954) to produce various substances such as indol-3yl-aldehyde (von Denffer & Fischer, 1952) and indol-3yl-carboxylic acid (Fig. 13 LXVII) (Melchior, 1957). Subsequently the heterocyclic moiety of the indole ring system could be opened with the production of anthranilic acid (Fig. 5 III) (Melchior, 1957). Even the very stable synthetic auxins can be similarly affected and 2,4-D can be degraded to phenols and ultimately, by ring fission, to aliphatic products (Bell, 1956). Although such effects may be responsible for some of the actions of u.v. light on plant growth, by far the greatest interest centres on the action of visible light.

Since IAA does not absorb in the visible spectrum, any photo-decomposition must be associated with radiant energy absorption by a sensitizing pigment. Several pigments have been shown to act in this way (see Gordon, 1954) but the one of most relevence to normal plant growth is the natural enzyme co-factor riboflavin. This pigment will actively sensitize the photo-oxidation of IAA with a maximum activity at its absorption peak in blue light (440 nm) (Galston, 1950). As for u.v. light this action is not specific for IAA since many other compounds including tryptophan (Melchior, 1957), 2,4-D (Bell, 1956) and NAA (Gortner & Kent, 1953) can be destroyed in this way. Photo-oxidation of IAA is also sensitized by plant extracts (Goldacre, 1954) probably because of their riboflavin content, since other plant pigments with absorption bands in the blue end of the spectrum (e.g. β-carotene) are ineffective (Reinert, 1953). The reaction also takes place *in vivo*, where both IAA (Meyer, 1958; Zenk, 1967) and NAA (Leeper *et al.*, 1962) can be photolytically decomposed in plant tissue. The mechanism of this photo-oxidation of auxins is still not completely understood although the products are probably very similar to if not identical with those arising from u.v. action (Meyer, 1958).

It is natural that wide consideration has been given to the role of auxin oxidation, both by IAA-oxidase and by photo-oxidation processes, in the control of plant growth and development. Many theories have been put forward to explain hormone-mediated phenomena in the growth and development of cells, tissues and organs in terms of the direct induction of IAA-oxidase synthesis and the modification of its action by phenolic co-factors and inhibitors; the many actions of light on growth, movement and morphogenesis have also been explained in terms either of direct auxin inactivation or of the induced metabolism of substances modifying IAA-oxidase activity.

The whole subject is complex and still largely controversial and will not be further pursued at this point; it is a subject better discussed in relation to the particular phenomena concerned.

(iii) Conjugation and binding of auxins

Although the existance of an IAA-oxidase in cell-free extracts of plant tissue is well established, there are still discussions as to the extent of its activity in the growing cell, especially since most of the IAA-decarboxylating activity in some excised tissues seems to be associated with the wound surface (Zenk & Müller, 1964). In an attempt to assess its contribution to the metabolism of auxin in cells, Andreae & Good (1955) fed IAA to pea roots, incubated them and then analyzed extracts by paper chromatography. They showed that the major part of the IAA was rapidly *conjugated* with aspartic acid present in the tissue, to yield indol-3yl-acetyl-aspartic acid. This was later shown to occur in several other plant species (Good *et al.*, 1956). Other auxins such as indol-3yl-propionic and butryic acids, NAA and 2,4-D were shown to produce similar conjugates in varying proportions (Andreae & Good, 1957; Andreae, 1967). Nearly all the IAA and NAA taken up by the tissue under these conditions was converted to the aspartyl conjugate (Andreae, 1967).

It was at first thought that these compounds might still exert growth activity, albeit a very small one, but it is now clear that their reported auxin activities came about only after the hydrolytic release of the free auxin in the tissue of the organs used for assay (Kazemie & Klämbt, 1969). Conjugation therefore completely inactivates the auxin, although, unlike oxidative degradation, there is always the possibility of reversal provided the conjugate remains in the tissue. Conjugation can take place with another amino acid, glycine, under the catalytic action of an enzyme octanoate thiokinase from liver in the presence of appropriate co-factors (Zenk, 1960) but its relevence to the plant cell is not clear.

Auxin can also conjugate with sugars; 1-(indol-3yl-acetyl)-β-D-glucose has been produced by incubating leaves of *Colchicum neapolitanum* (Zenk, 1961), wheat coleoptiles (Klämbt, 1961*a*) and pea epicotyl (Zenk, 1964) with IAA. The corresponding naphth-lyl-acetyl- and benzoyl-esters have also been found in pea epicotyl (Zenk, 1962) and wheat coleoptiles (Klämbt, 1961*a*) respectively.

The most interesting characteristic of this conjugation process is that the enzymes responsible for its catalysis are induced by the auxin substrate. This has been clearly shown by Zenk (1962) in pre-incubation studies with pea epicotyl. The simultaneous application of inhibitors of protein synthesis (actinomycin or cycloheximide) prevents the appearance of the conjugating enzyme, showing that the presence of the auxin substrate brings about *de novo* enzyme synthesis (Venis, 1969). The ability to conjugate IAA with glucose is widespread throughout the whole plant kingdom from bacteria to angiosperms, but the corresponding ability with aspartic acid seems to be characteristic only of the more advanced vascular plants (Zenk, 1964).

IAA will also conjugate with cyclitols (see Chapter V) and four esters with myo-inositol have been isolated from mature maize grain. Such conjugates may also contain the sugar arabinose as an additional constituent (Labarga *et al.*, 1965). One has been isolated and positively identified as indol-3yl-acetyl-2-*o*-myo-inositol. This latter compound has a distinct growth-regulating activity in the promotion of callus growth but it is not yet clear whether these conjugates are involved in the actual process of growth control or whether they are storage or transport forms of IAA (Nicholls, 1967).

The physiological significance of conjugation is still a matter for discussion. It can represent a detoxication mechanism of a kind so well known in the animal kingdom where excess doses of foreign and potentially toxic chemicals can be rendered harmless by combination with various amino acids, glucuronic acid, etc. This is most probably true for many plants which are insensitive to the toxic action of high doses of synthetic auxins such as 2,4-D. In combination with oxidative decarboxylation, conjugation in roots can maintain remarkably constant levels of IAA after its transport from the site of application in the shoot apex of pea plants (Morris *et al.*, 1969). Could conjugation participate in similar regulatory functions in normal plants?

However there is an intriguing observation that only compounds that have auxin activity can induce the formation of the conjugation enzymes; homologous antiauxins (e.g. *trans*-cinnamic acid and 2,4-dichlorophenoxyisobutyric acid) do not (Südi, 1966). If the process were purely one of detoxication this sharp differential would not be expected and a more fundamental significance is implied. Furthermore there are those who think that auxin conjugation may be an essential step in the growth process of glucose incorporation into the expanding cell wall (Klämbt,1961*b*). The exact situation has still to be resolved.

There have been reports in the literature that IAA conjugates with proteins, a phenomenon which would be expected if IAA were acting as an essential component of an enzyme system controlling growth. However these cannot be confirmed (Zenk, 1964) and in any case are not likely to be relevent to the control of auxin levels.

(B) OTHER HORMONES

Virtually nothing is known of the details of the breakdown of the gibberellins and the cytokinins in plant cells and we can only guess whether their concentrations can be controlled by such breakdown. There is evidence accumulating that gibberellins can be 'bound' to certain plant metabolites and it is presumed, but not certain, that they are inactive when in this state. The situation is complicated by the possibility of interconversions of gibberellins of widely differing activities in the tissues concerned but it is a plausible hypothesis, supported by ever-increasing evidence, that the control of such interconversions could easily be a major factor in the control of growth and morphogenesis. This is a complex but fascinating field of study only now being uncovered.

Chapter XII

The Mechanisms of Hormone Action

Introduction

The research worker who sets himself the task of elucidating, at the ultimate molecular level, the way in which hormones exert their control on cell growth, has chosen one of the most difficult problems in the whole realm of plant physiology. For fifty years a continuing succession of plant biologists, of very varied interests and eminence and with equally varied approach, have concentrated their time and energies towards its solution which still evades us. The particular difficulty which confronts us is that hormones are extremely versatile in their actions on plant cells, both in the nature of the growth responses they invoke and also in the way they alter biochemical equilibria in those growing cells.

This diversity of activity presents us at once with two major enigmas. The first is whether the different growth responses are the result of separate actions, each in a different biochemical system, or whether only a single system is involved, and the differing responses are conditioned by the particular status of the tissue reacting, and the concentration of the hormone present at the particular site of its controlling action. The latter view, which was first put into a concrete form by Thimann (1936) when he postulated a 'master reaction' as underlying all auxin-induced growth responses, is the one usually favoured. The diversity of hormone effects resides then, either in a series of secondary phenomena arising from the main 'master reaction', or in the modifying effect of cell status or of hormone concentration on it. Speculation as to how changes in hormone concentration at one and the same enzyme surface might switch growth-responses (i.e. from stimulation to inhibition of extension growth) have often been treated mathematically, and auxin response data, acquired under certain growth conditions, have been shown to fit the theoretical formulae tolerably closely (Foster *et al.*, 1952) (see Chapter XI).

The second enigma concerns the extensive biochemical changes that take place in tissues under the action of hormones – changes which have, as we shall see, prompted, each in their turn, suggestions concerning the nature of the 'master reaction'. But which is cause and which is effect of the growth changes produced by the hormones is still, in most cases, an open question and one subject to heated debate. Although there are many facets of hormone action for which mechanisms have been sought, yet there are two major categories of effect at the cell level into which most phenomena usually fall. These are action in the control of cell extension (the main characteristic of auxin action) and action in the control of cell division. However these actions are not always clearly separated in an overall growth response; for example the gibberellins promote main axis extension via effects on both cell extension and cell proliferation and again auxins and cytokinins are both mutually concerned in the control of cell division in callus tissue (see Chapter V).

Most investigations into this problem have been concerned with extension growth in shoots or coleoptiles. In this chapter, therefore, although the broadest possible survey will be given to include any normal hormone activity which may throw light on their ultimate biochemical actions, most of the discussion will deal with extension growth in aerial organs. At concentrations well above those normally met in the plant, auxins and related compounds exert, on plant growth and metabolism, profound disturbances which may result in death of the affected organ or whole plant. The mechanism of these toxic actions will not be dealt with in this chapter.

The Physiology of Cell Extension

Before we can discuss the way in which auxins may accelerate extension growth of cells, we must first consider what processes go on in a cell when it extends.

The living cell is essentially an osmotic system. The central vacuole, containing a variety of dissolved substances, and therefore possessing a definite osmotic potential, is separated from the surrounding protoplasm by a semipermeable membrane. A similar membrane also exists between the protoplasm and the cell-wall, which is usually completely permeable to all non-colloidal molecules. In this osmotic system, the cell-wall plays a very important role because, as water passes into the cell vacuole by osmosis, the resultant increase in volume causes an elastic stretching of the wall and the development of a hydrostatic pressure in the fluid contents of the cell. This hydrostatic pressure, called *turgor pressure*, is exactly balanced by the resistive pressure exerted by the stretched cell-wall on these fluid contents. Naturally this positive hydrostatic pressure tends to force water out of the cell, and in the non-growing cell equilibrium is attained when this pressure rises to a value equal to the osmotic potential difference between the cell-sap and the external solution. In this simplified treatment it has been assumed that osmotic forces are the only purely physical forces acting on the system. Although this is not strictly true, the omission of the other forces does not materially affect the argument.

If a cell, in equilibrium with its surroundings, is to grow, something must happen to disturb the balance of the two forces controlling water movement between cell and environment. The net force, instead of being zero (as at equilibrium), must become positive so that water will be drawn into the cell. This could happen in two ways, either by an increase in the 'absorbing' forces, or by a decrease in the 'expelling' forces. In the first case, the total osmotic potential of the cell might rise by the internal production, and accumulation in the vacuole, of osmotically active solutes. In the second case the expelling forces (i.e. hydrostatic turgor pressure) could be reduced, either by a change in the mechanical properties of the cell-wall, whereby it became more extensible under a given tension, or by an active growth of that wall, which would not need to be accompanied by such changes in properties. Let us now take these possibilities in turn and examine their implications in full.

Firstly it may be assumed that water absorption, as a result of the accumulation of osmotically active solutes, would cause stretching of the cell-wall, with a consequent increase in turgor pressure, and equilibrium would be reached and growth would stop when it had risen to balance the augmented osmotic forces. Continued growth would mean a continued rise in these two pressures until, one would expect, the cell-wall would become so thin that the cell would burst, like a balloon. Of course this does not happen. Thus, the osmotic pressure of the cell-sap does not usually change at all during cell extension (Ursprung & Blum, 1924; Beck, 1941; Burström, 1942*a*) and may actually decrease (Ruge, 1937*b* & *c*; Diehl *et al.*, 1939; Hackett, 1952). The decrease is usually much smaller than would be expected to result from a simple dilution of cell-sap by the entering water, and this means that, during growth, osmotically active material must be accumulating to keep pace with the water absorption. Furthermore the turgor pressure does not build up during cell extension. In root cells Burström (1942*a*) showed long ago that cell turgor stayed more or less constant. However in organs growing in air, e.g. internodes of pea, the turgor may be low in actively growing tissue and actually increase only when growth declines and cells mature (Burström *et al.*, 1967).

The important conclusion we can draw from this is that, as the cell extends and the cell-wall increases in area, elastic and other properties of the cell-wall must also undergo subtle changes if we are to explain the relative constancy of the physical forces causing water uptake. But what is the nature of these changes?

Observations on the walls of extending cells have shown that they are in fact often thicker at the end of the period of extension than they were at the beginning (Green, 1958), and so active formation of new wall material must have kept pace with elongation. During the initial phases of the extension growth of a cell, however, the wall may sometimes get thinner for a time before this active wall synthesis sets in (Ruge, 1937*b*; Diehl *et al.*, 1939; Brown & Sutcliffe, 1950; Green, 1958). Such observations have led to the suggestion that, in the early stages, as a result probably of changes in wall properties, extension is simply by a process of stretching, and that active wall synthesis does not set in until later. If this is so we need to know the exact nature of these stretching processes and one of the most direct sources of relevant information is the study of the mechanical properties of the cell-wall. These changes have been followed in relation to extension growth in several types of cell. During the growth of wheat roots Burström (1942*a* & *b*) immersed samples in hypertonic sugar solutions to reduce turgor to zero. The shrinkage in length observed indicated the degree of elastic stretching of the cell, from which its elastic properties could be calculated. In the early stages of elongation the elastic extension under normal cell turgor conditions increased more rapidly than the cell length, which means that the cell-wall as a whole became more elastic. This corresponds to the phase when cell-wall thickness is decreasing. Subsequently, however, the *absolute* elastic extension reached a constant limit, although the cell length still continued to increase. This actually implies a declining elasticity of the wall as a whole (Frey-Wyssling, 1948), and corresponds to the phase when active cell-wall synthesis is going on.

More direct observations involving sophisticated physical techniques have confirmed this general picture of elastic changes. For example isolated strips of the wall of the giant cells of the alga *Nitella* have been subjected to stretching treatments, observations on extension and elastic recovery being made with a time-lapse micro-ciné camera (Probine & Preston, 1962). These observations showed a significant correlation between the elasticity of the cell-wall and the growth rate of the cell from which it was taken, i.e. the rigidity (elastic modulus) of the wall decreased with growth rate. However the cell-wall is not a perfect elastic solid. When stretched it shows 'give' or 'creep'; in other words it also possesses plasticity, a property which many workers regard as much more important than elasticity in relation to cell-wall extension in growth. This 'creep' under constant tension has also been measured in *Nitella* cells (Probine & Preston, 1962) and its extent is also closely correlated with the growth rate of the cell (see Fig. 73). Furthermore this 'creep' is much more easily induced along the longitudinal axis of the wall (the direction of growth) than along the transverse axis (Probine, 1963).

The usefulness of such overall measurements of cell-wall properties in formulating ideas on the mechanism of wall extension during growth, depends on the assumption that cell-wall structure and its growth behaviour are uniform over the whole area of the extending cell-wall. There have been suggestions that this is not so; for example, Frey-Wyssling (1952) at one time proposed that cells in coleoptiles grow only at their ends where the cell-walls are thinner and thus more easily stretched. But direct observation, using pits or copper oxide dust as markers (Castle, 1955; Wardrop, 1955; Wilson, 1957), has established conclusively that extension growth takes place uniformly over the whole of the length of the cell. This has been confirmed by experiments on the growth of oat coleoptile segments in a medium containing radioactive glucose. Autoradiographs showed no higher concentration of radioactive material at the cell tips than elsewhere, and labelled cellulose appeared to be uniformly distributed in the cell-wall (Wardrop, 1956).

This leads us to a consideration of the molecular architecture of the cell-wall and how this is related to extension growth and these changes in properties.

Chemical and electron-microscope studies have revealed that the structural framework of the cell-wall is constitued of extremely fine fibrils (microfibrils) of pure cellulose of average diameter 5 to 25 nm (see Plate 25). This

Fig. 73. *Top*. Diagram of an apparatus used to study the extensibility of plant cell walls. The sample to be investigated is fixed between two clamps; the upper clamp is immobile and is connected to a mechanism which measures the tension (load) applied to it by the sample and transmits a proportionate electrical signal to a recorder; the lower clamp can be moved slowly down at a constant speed, this stretching the sample. This downward movement is also transmitted as a proportionate signal to the recorder which draws a curve relating load to extension, i.e. a stress/strain curve (right). From such curves the elastic and plastic properties of the sample can be calculated. (From Cleland, 1967*b*; by kind permission of the author.)

Bottom. Graphs showing the relationship between the plasticity of the cell wall, as measured by its 'creep' under a range of stresses, and the growth rate of the cell in *Nitella opaca*. The 'creep' was measured as the percentage extension under a given load between 1 and 100 minutes after the application of the load (vertical axis). The horizontal axis is for stress, expressed as a ratio to a standard stress, calculated as that stress to which the cell would have been subjected under a standard turgor pressure (see Probine & Preston, 1962). This method of expressing stress was necessary to ensure comparability between cells which were of different transverse dimensions. The growth rates of the corresponding cells are marked against the 'creep'/stress curves and against the relevant symbols on the right of the graph. (Graph from Probine & Preston, 1962; by kind permission of the authors.)

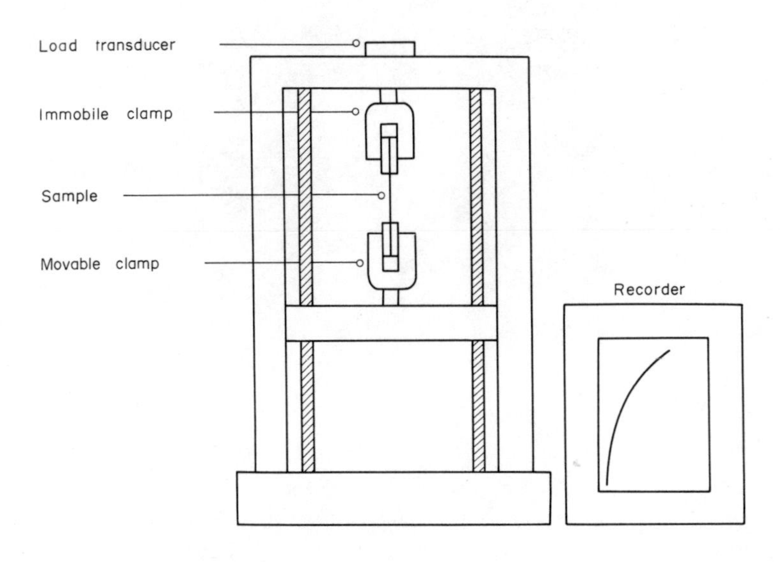
Load transducer
Immobile clamp
Sample
Movable clamp
Recorder

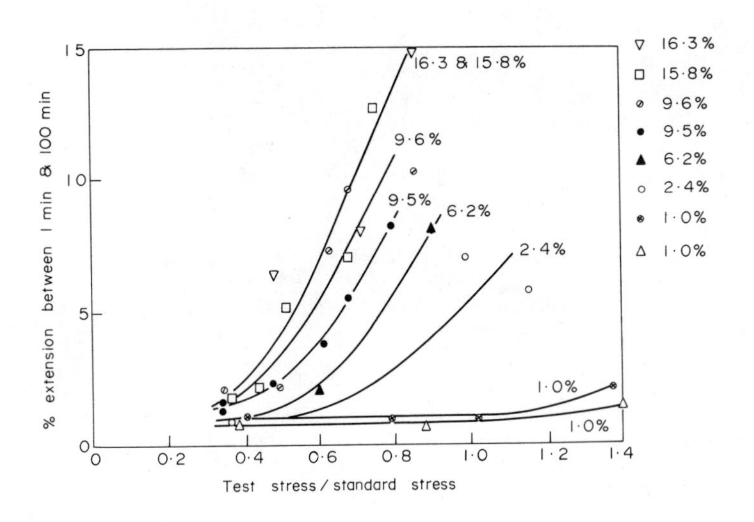
% extension between 1 min & 100 min
Test stress / standard stress
16·3 & 15·8%
9·6%
9·5%
6·2%
2·4%
1·0%
1·0%
▽ 16·3%
□ 15·8%
9·6%
● 9·5%
▲ 6·2%
○ 2·4%
1·0%
△ 1·0%

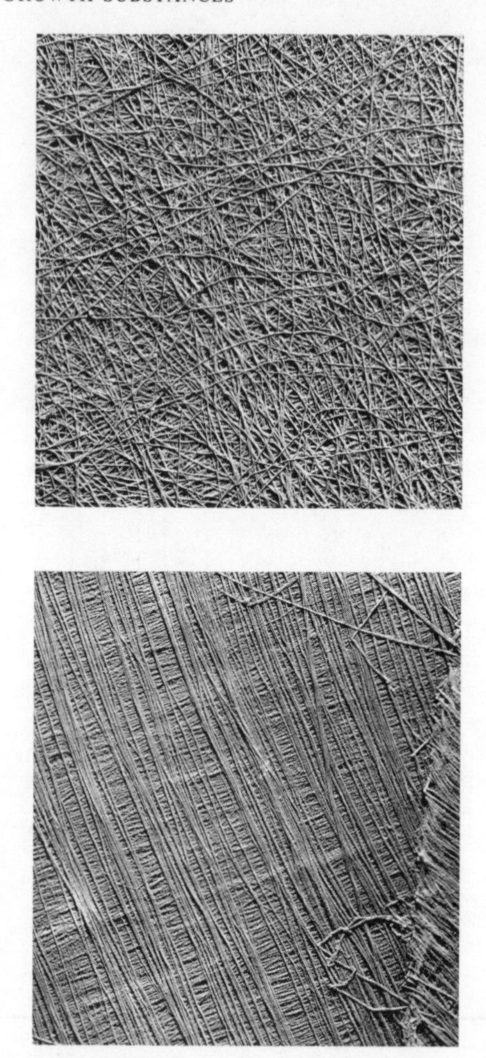

Plate 25. Electron-micrographs of cell-wall preparations.

(A) Wall lamella of a cross wall of *Chaetomorpha melagonium.* Note the random arrangement of the cellulose micro-fibrils; this random arrangement is usually found in the early stages of wall formation. (Magnification × 10,980).

(B) Inner side-wall of *Cladophora rupestris.* Lamellae viewed from the outside. Note the two layers of parallel microfibrils, each with different orientations. This is characteristic of mature cell walls. (Magnification × 10,980). (Photographs by kind permission of Prof. R. D. Preston, University of Leeds. From Frei & Preston, 1961.)

framework, which constitutes about 25 to 30 percent of the dry matter of the growing cell-wall (see Roelofsen, 1959) is embedded in a cementing matrix of largely amorphous material composed of other carbohydrates (hemi-celluloses, and pectic substances) and proteins. Thus cell-wall structure bears a remarkable resemblance to reinforced concrete on a microscopic scale, the micro-fibrils and the interfibrillar matrix serving the same mechanical roles as the steel rods and the encrusting cement respectively serve in building structures. In the newly-formed cell-wall the cellulose microfibrils are more or less randomly orientated, but as the cell-wall begins to thicken they take on a much more highly orientated arrangement, and come to lie largely transverse to the axis of elongation of the cell, forming a shallow helix around the main axis. With increasing wall deposition the direction of this helix may change from time to time giving a kind of crisis-cross basket-work of alternating layers. The two situations are illustrated in Plate 25, which are electron-micrographs of preparations of cell-walls (newly-formed (A) and mature (B) respectively) in which the amorphous matrix has been removed, revealing the cellulose microfibrils. Elastic stretching of these walls can be envisaged as causing a reversible distortion of this basketwork, whereby the 'mesh' is slightly opened up and the cementing material is distorted without changing its bonding to the cellulose microfibrillar framework. As we have mentioned above, such reversible elastic stretching is probably not involved in growth. When plastic flow takes place under tension there must presumably occur a breaking of the cementing bonds between microfibril and matrix so that the mesh is permanently pulled apart.

What exactly happens to the microfibrils and to the matrix when this plastic flow takes place is still a subject for discussion. One might imagine that, as the cell increases in length, the pitch of the microfibrillar helix around the cell would become steeper, just as the direction of the strings of a stretched fish-net would tend to swing around to the direction of the pull as the net is distorted. There is still uncertainty as to whether this is a general characteristic of cell-wall change during growth, although there is good evidence to support such changes. For example as the cell extends more wall material is laid down on the inside surface to maintain wall thickness. The most recently synthesized microfibrils show the same almost transverse orientation to the main axis as was shown in the early stages of wall formation. If reorientation does take place during cell extension then one would expect to find in the relatively mature wall an increasing swing of the orientation towards the direction of the main axis as one proceeds through the cell-wall from the most recently deposited innermost layer to the oldest and outermost layer of the cell-wall nearest the middle lamella. This has been observed by Roelofsen & Houwink (1953) and forms the basis of the 'multi-net' hypothesis of wall extension. In any case, no matter what the nature of these changes, they must ultimately be determined by the changing pattern of binding forces between cellulose microfibrils and cementing matrix, a possible site for auxin action in the control of extension growth. This complexity of change in wall properties, coupled with relative constancy of cell pressure during elongation, affords us little clue to the process or processes initiating the extension and presumably triggered by the hormones controlling it.

Before we turn to consider how auxin studies may throw light on these problems, we must examine the other changes which take place in the extending cell. We have already seen that in the meristematic stage, growth is predominantly characterized by protein synthesis. In extension growth, which follows, much evidence suggests that this protein synthesis still goes on. In the coleoptile of maize, the protein content of the cell continues to increase but at a much slower rate than the cell volume (Frey-Wyssling, 1948). In petals of *Anemone, Tulipa, Gesneria,* and *Dahlia*, extension is accompanied by no detectable protein synthesis (Schumacher & Matthaei, 1955). In dandelion inflorescence stalks, where extension growth takes place in two distinct 'bursts', increase in cell length in the first burst is accompanied by protein synthesis, whereas in the second it is not (Chao, 1947). In broad bean (*Vicia faba*) roots, the protein content of the cell continues to rise as the cell extends (Robinson & Brown, 1952; Morgan & Reith, 1954). There also seems to be a close correlation between protein synthesis and growth induced by auxins in artichoke tuber tissue (Thimann & Loos, 1957). Furthermore this protein synthesis seems undoubtedly to be an essential part of extension growth since its prevention by specific inhibitors such as chloramphenicol also blocks extension growth in *Avena* coleoptiles and the relationship of these two inhibitions to chloramphenicol concentration are closely similar (Noodén & Thimann, 1965).

These overall increases in protein content may however be underestimates of the much greater protein *turnover* of the growing cell, since there is now much evidence of very dramatic changes in the nature of the cell proteins as the cell progresses through its expansion cycle. Studies on the enzyme content of cell protoplasm, calculated per unit of protein, show that the concentrations of many of these cell catalysts may alter a great deal during extension (Robinson & Brown, 1952; Robinson, 1956). The most striking change observed is in invertase, which may increase in concentration by over ten times and then fall away. Analysis of the proteins of extending cells by the technique of gel electrophoresis have shown how complex the changing patterns may be during the extension of the cell.

It is now well established that several enzymes are closely associated with the cell-wall, and this situation has far-reaching implications as regards the possible regulation of wall properties as a necessary process underlying wall extension. Thus ascorbic acid oxidase is mostly found in the cell-wall in tobacco pith callus culture and in the extending cells of barley and maize roots. Other enzymes shown to be present in cell-walls are pectin esterase, peroxidase, acid phosphatase and invertase (for reference details see Newcomb, 1963 and Lamport, 1970). The precise significance of these changes is still obscure, but they indicate that extension growth does not consist merely of wall stretching and water absorption but probably involves, in some way or other, the whole metabolic system of the cell.

Perhaps the most interesting recent discovery is that the predominant protein of the cell-wall is rich in a very characteristic amino acid, hydroxyproline (Dougall & Shimbayashi, 1960; Lamport, 1962). What is exciting is that the same amino acid is characteristic of the structural animal protein collagen, to which it seems to be confined. This has suggested (Lamport & Northcote, 1960) that the hydroxyproline-rich protein found in plants is an integral part of the structure of the primary growing cell-wall; in fact Lamport has suggested that it plays a fundamental role in determining cell-wall properties and may be intimately involved in cell-wall extension processes. For this reason Lamport (1965) has given it the name *extensin* (for further discussion see pp. 372).

Let us now consider the auxins and the experimental evidence for their possible actions in the several systems involved in extension growth.

The Effects of Auxins of Cell Pressures

It was Czaja in 1935 who first suggested that auxins started extension growth by inducing increased osmotic pressure of the cell sap. This theory, based on a few doubtful experiments on the bursting of root hairs, has never received much active support, and the majority of the experimental evidence subsequently accumulated is definitely opposed to it. Auxin application to a variety of growing plant tissues, resulting in increased extension growth, has usually been accompanied by a decrease in osmotic pressure of the cell-sap. In some experiments there was no detectable change in the total of osmotically active solutes in the cell-sap, and so the growth-water entering naturally must have lowered the osmotic pressure proportionately. This is so in potato tissue (van Overbeek, 1944; Hackett, 1952; Brauner, 1954), and in the growing *Avena* coleoptile, where the declining osmotic pressure of the cell-sap is inversely proportional to the increasing length of the cell (Ketellapper, 1953). In other experiments, auxin-induced growth is accompanied by a definite augmentation of osmotically active material in the cell, but this augmentation does not keep pace with cell extension. This is so in young sunflower hypocotyls (Ruge, 1937*b*, 1937*c*; Diehl *et al.*, 1939). In the roots of wheat, where auxin causes a marked inhibition of extension-growth, there is a corresponding increase in osmotically active material per cell (Burström, 1942*b*). Results of this kind suggest that auxin-induced extension-growth is not connected in any direct way with changes in osmotically active solutes. Any such changes are either secondary to the main action of auxins or are the indirect result of growth. Whether osmotic pressures are maintained, or whether they decline or rise, will depend on the relative amounts of water taken in during growth.

But if water is to enter the cell, some net force must be acting on it, and one would expect to find a declining water potential gradient maintained from the environment to the inside of the growing cell, and also some direct relationship between the magnitude of this gradient and the rate of cell extension (i.e. water absorption). The rate at which water can enter a cell, however, and so maintain osmotic equilibrium, would seem to be so great (see p. 362) that only very small water potential differences are necessary to maintain normal growth-rates. No measurable water potential gradient is shown by roots growing in water, and none can be demonstrated in coleoptile sections growing in hypotonic mannitol solutions. A number of investigators, however, have explored the effects of IAA on cell water-potential gradients, and Brauner & Hasman (1949, 1952) studied it in potato tuber tissue. Using a method based on fresh-weight changes in a range of concentrations of sugar solutions, they found that during the first few hours of cell extension, the water potential in IAA was about $\frac{1}{4}$ atmosphere higher than in water. As water absorption proceeded, the water potential rose. The rise was slower in the presence of IAA until, after 48 hours, water was being lost from tissues in the absence of IAA, but was still being taken up in its presence and the water potential of the cell was still $-\frac{1}{4}$ atmosphere. If, as has been suggested above, there is no change in the level of osmotically active materials during cell expansion in this tissue, the maintenance of a water potential gradient by IAA, in spite of increased water absorption, must be *due in large part* to a reduction in wall pressures, i.e. an increase in wall extensibility. In *Avena* coleoptiles during active growth, the water potential gradient is similarly maintained (Ketellapper, 1953). This, coupled with the drop in osmotic concentration of the cell-sap, already mentioned above, also suggests a large drop in wall pressure as growth proceeds.

However the door cannot be completely closed on the possibility of increases in osmotic concentration in the vacuoles of cells of storage tissue directly promoted by auxins. Wain and his colleagues (Wain *et al.*, 1964; Wain, 1967) have shown that 2,4-D will greatly enhance the invertase activity of chicory and Jerusalem artichoke tissue and thereby increase the breakdown of fructosans. Furthermore the response is not evoked by the isomer 3,5-dichlorophenoxyacetic acid, which is inactive as an auxin

(Rutherford *et al.*, 1969). Wain regards the resultant increase in osmotic concentration as being at least partially responsible for the growth response of the tissue to 2,4-D. The fact that this runs counter to all previous osmotic pressure measurements in growing tissue is puzzling and more data are called for.

Auxins and 'Active' Uptake of Material by the Cell

The idea that auxin acts by inducing 'active' water absorption has been popular with some workers. It was set in train by the observations of Reinders (1938) that auxin (IAA at 10 ppm) greatly accelerated the uptake of water by potato tuber tissue and that this uptake was very sensitive to aeration (and thus, presumably, to the supply of metabolic energy from cellular oxidation). The Q_{10} of this induced water uptake was in the region of 2 to 3, and there was also evidence that the uptake was accompanied by a considerably augmented rate of respiration. All this pointed to a water uptake under the control of metabolic rather than of osmotic systems in the cell.

The phenomena have since been confirmed by several workers using both potato and artichoke (*Helianthus tuberosus*) tissue. NAA has been found to be much more effective than IAA (Hackett & Thimann, 1952*a* & *b*). This is probably owing, at least in part, to its greater stability in the tissue and culture solution. Direct observations on gaseous exchange confirmed that much higher rates of oxygen consumption were maintained during the auxin-induced uptake, but the combined effects of auxin and respiratory inhibitors yielded such complex responses that no clear-cut conclusions could be drawn (Hackett & Thimann, 1953). Nevertheless, it was felt that the results still indicated a water uptake depending on the supply of metabolic energy, and that the efficiency with which this energy was used depended in some way on auxin supply. It was, however, pointed out by Levitt (1953), that auxin effects on wall extensibility (see later p. 362) might also be dependent on an energy supply from respiration, and that this too would be blocked by respiratory inhibitors. The 'loosening' of the wall texture might be a metabolic rupture of the linkages between the cellulose micellas, for example by phosphorylations, and the efficiency of this might easily be controlled by auxin supply. Again the promotion of fructose production in artichoke and chicory tissue by 2,4-D treatment (Wain, 1967) could mean increased substrate for metabolic effects on the wall.

The problem has been attacked from another angle by following growth and water uptake in solutions with very high osmotic pressures. Commoner *et al.*, (1943), were the first to claim that IAA could stimulate water uptake by potato tissue from *hypertonic* sucrose solutions, provided certain salts (e.g. potassium chloride and potassium fumarate) were also present. *Avena* coleoptile sections were held to behave similarly (Commoner & Mazia, 1944). In similar solutions but in the absence of auxin, only water *loss* occurred. It was therefore suggested that auxin induced an 'active' *salt* uptake by the tissue, thus reducing its water potential and bringing about water uptake. But such an explanation cannot account for the even greater uptake from distilled water (see above). On the other hand auxins have been shown to have quite definite effects on the uptake of ions by plant tissue. Thus the uptake of potassium and rubidium ions can be stimulated by physiological concentrations of auxins in artichoke tissue (Hanson & Bonner, 1954) pea epicotyl segments (Higinbotham, *et al.*, 1962) and sunflower hypocotyl (Ilan & Reinhold, 1963). In sunflower hypocotyl uptake of lithium ions was unaffected by IAA while that of ammonium ions was inhibited (Ilan & Reinhold, 1963). However the fact that these promotions of ions uptake followed the growth effects and that they, like growth, were inhibited in mannitol solutions of high osmotic pressures (which prevent cell extension) led Ilan & Reinhold to suggest that they were the result not the cause of the growth stimulations. The same may apply to the IAA-augmented sulphate ion uptake by beetroot tissue (Neirinckx, 1968).

There is, however, no real evidence from any subsequent work that auxin induces an active 'pumping' of water into the cell. The claim that Jerusalem artichoke tuber tissue in hypertonic mannitol solution could take up water under the action of IAA (Bonner *et al.*, 1953) was invalidated by the inaccuracy of the methods used to determine the normal osmotic potential of the tissues (Burström, 1953). In isolated coleoptile segments of *Avena*, growth in solutions of the non-penetrating solute mannitol has been independently studied (Pohl, 1953, Ordin *et al.*, 1956) and shown to be proportional to the difference in osmotic potential between tissue and solution. When this is zero, growth ceases. This suggests strongly that osmotic forces are the only ones involved in wall extension, IAA greatly enhances the growth of such sections, but in no way alters the relationship between growth and osmotic-potential differences. This has been interpreted to mean (Pohl, 1953) that auxin does not alter the osmotic forces of the cell but only the speed of entry of water under those forces, i.e. it controls growth by increasing cell permeability to water. This will be discussed in the next section. Further evidence and views against the likelihood of a metabolic water-pump in plant cells are presented in the views of Kramer (1955, 1959) and Galston & Purves (1960).

Growth in *apparently hypertonic* solutions must therefore be explained in terms of internal osmotic adjustment. In normal excised *Avena* coleoptile sections this could be brought about only after absorption of a suitable penetrating solute (e.g. sucrose or NaCl) from the hypertonic medium. This osmotic adjustment by solute uptake goes on independently of auxin (Ordin *et al.*, 1956), and is not therefore an auxin-induced active uptake. Solutions of high concentrations seldom remain hypertonic to intact roots grown in them (see Burström, 1955), as internal adjustment seems to maintain a virtual equality of external and internal osmotic potentials during growth, which then takes place at extremely low cell turgors.

Auxins and the Permeability of the Cell Membranes

In Chapter III it was seen that the deliberations of Veldstra on the relationship between the chemical structure of auxins and their physiological activities led him to postulate physical action at a cell interface. The high lipophilic surface activity of auxins points to the protoplasmic membrane as this interface. Veldstra suggests that adsorption on the membrane might bring about a change in boundary potentials which would affect membrane permeability and thus regulate transport (Veldstra, 1946, 1947). The suggested mechanism is based on the theories of Bungenberg de Jong and his school. (see review by Booij, 1940). The theory was tested by studying the effects of various auxins on the exodiffusion of anthocyanin pigments from the cells of red beet-root tissue, but although auxins in high concentrations undoubtedly increase the permeability of the cell membrane to these pigments, yet no obvious relationship could be found between this effect and the activity of the molecules as auxins (Veldstra & Booij, 1949; Veldstra, 1949). It is doubtful whether, with such high non-physiological concentrations, any useful conclusions could in any case have been drawn concerning the normal function of auxins.

Similarly the leakage of electrolytes from leaf cells of tomato has also been shown to increase in solutions of IAA (Chrominski, 1961) although these effects may not be relevant to growth problems since the concentrations used were probably supra-optimal. Studies of the electrical conductivity of potato tissue also indicate that similar supra-optimal concentrations of IAA and NAA increase permeability of the cells to ions (Meylan & Pilet, 1966). The use of deplasmolysis techniques with onion-scale protoplasts have indicated that low physiological concentrations (10^{-5} to 10^{-8} mol./1.) of IAA increase the permeability of the membrane to both glycerol and urea, whereas high concentrations (10^{-3} to 10^{-4} mol./1.) decrease it (Masuda, 1953). (See Fig. 74). From this it was concluded that auxin affects both the lipoid and the protein phases in the membrane. Later on (Masuda, 1955) a direct correlation was established between the growth rates of coleoptiles under different experimental conditions (i.e. decapitation, IAA application, etc.) and permeability to urea. IAA at about 10ppm has also been shown to increase

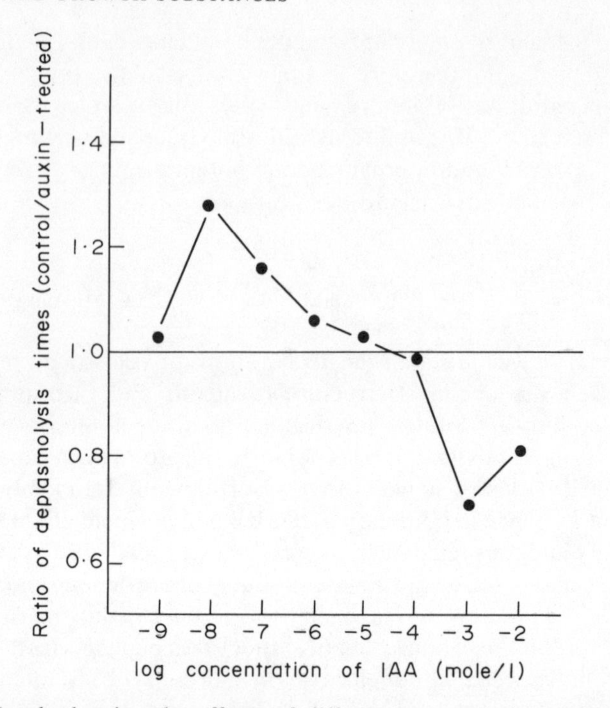

Fig. 74. Graph showing the effects of different concentrations of IAA on the permeability of the cells of onion bulb scale to urea. The effects are expressed as the ratio of the deplasmolysis time in a hypertonic urea solution to the time in the same solution containing IAA. Since the rate of deplasmolysis is a measure of the rate of entry of urea, this ratio will give an indication of the effects of auxin, a value higher than unity representing an increase and one smaller than unity a decrease in permeability. (Data from Masuda, 1953.)

greatly the up-take of amino acids by sunflower hypocotyl (Reinhold & Powell 1958). This was not the result of increased growth, since it also occurred when segment growth was inhibited in hypertonic mannitol solution. The mechanism and significance of this effect remain obscure, but the phenomenon seems unlikely to be due to changes in permeability induced by auxin.

A puzzling effect of IAA has been recorded on the exchange of potassium for sodium ions in leaves of the water plants *Vallisneria* and *Ruppia* (Kawahara & Masuda, 1957; Kawahara & Takada, 1958). These leaves, when placed in KCl solution, show a rapid efflux of sodium ions and this is prevented by physiological concentrations of IAA. However, when the cells are plasmolysed with hypertonic mannitol solutions the auxin stimulates the efflux. A more recent study with the giant cells of *Nitella translucens* showed that IAA at physiological concentrations (2×10^{-5}M) increased the ouabain-sensitive active uptake of ^{42}K by the cell without affecting ^{22}Na uptake (Stolarek, 1968). There were indications that the permeability of the membranes to diffusive influx of ^{42}K was also increased by IAA. These results suggest that auxin action in the movement of ions through membranes is likely to be complex.

A possibly different type of membrane effect has been observed by Sacher & Glasziou (1959). The cell-membranes of excised tissue of bean pods lose their semipermeability after a day or so and liquid containing many solutes exudes into the intercellular spaces; this is prevented by application of auxins at physiological concentrations. This phenomenon seems to be the indirect outcome of auxin effects on the metabolism of these abnormal and possibly senescent cells (see later, p. 390) and which somehow regulates membrane integrity.

Much more attention has been paid to *water* permeability, as the factor through which auxin exerts its controlling effect on water uptake in extension-growth. The idea arose more or less simultaneously from an number of independent workers. Firstly, von Guttenberg & Kröpelin (1947), by measuring deplasmolysis rates of *Rhoeo discolor* and onion-scale protoplasts in IAA, concluded that water permeability was increased by this auxin. Subsequently Pohl (1948) used rates of skrinkage of *Avena* coleoptile sections in hypertonic mannitol solutions as a measure of water permeability and showed that this was raised by IAA. Later on, the effects of both pH and IAA on the water permeability of *Rhoeo discolor* were studied and compared with the actions of similar treatments on the physical properties of the protein gelatin, serving as a physical model of the cell membrane (von Guttenberg & Beythein, 1951; Meinl & von Guttenberg, 1952). Low concentrations of IAA (0.1 to 10 ppm) increased water permeability of both protoplast and gelatin, and also the degree of swelling of gelatin in water. On the other hand, 100 ppm decreased all three. Furthermore, the growth inhibitor coumarin lowered water permeability of protoplasm and reduced the swelling of gelatin. From such results as these, von Guttenberg (1954) concluded that the primary action of auxin was on the protoplasmic membrane, which regulates growth via its permeability to water and perhaps to solutes.

For several reasons this theory is unacceptable. Firstly there are well substantiated results showing that auxin in physiological concentration (1 ppm) may greatly *reduce* the water permeability of onion-scale (Koningsberger, 1947) and of *Avena* coleoptile (Ketellapper, 1953) protoplasts. Then a convincing argument against the theory is that of Levitt (1953), who compared the rate of bulk movement of water into cells during growth ($1{\cdot}7 \times 10^{-3}$ μm/minute) with the rate of movement through the protoplasmic membrane calculated from permeability measurements (0·7 μm/min atm water potential difference). He shows that if water permeability were limiting growth, the water potential gradient necessary to produce the maximum growth-rate would be of the order of only 1/400th atmosphere. Levitt concluded that the relatively slow rate of water movement into the cell during growth could not possibly be under the control of the cell-membrane with its very high permeability to water and suggested that resistance to wall stretching was the real limiting factor. This point of view has been supported experimentally by the demonstration (Thimann & Samuel, 1955; Bonner *et al.*, 1956) that the rate at which heavy water diffuses in and out of a cell is very much greater than the mass inward movement of water induced by auxin.

Remarkable effects which have been regarded as due to the direct action of auxin on protoplast membrane permeability have been demonstrated with isolated protoplasts from fruit tissue (Gregory & Cocking, 1966), roots (Cocking, 1962), cotyledons (Cocking, 1961) and tobacco leaves (Power & Cocking, 1970). Digestion of cell-walls with fungal enzymes allows the preparation of suspensions of large numbers of isolated protoplasts. Auxins such as IAA and 2,4-D in concentrations somewhat above physiological, cause these protoplasts to swell and eventually to burst. This seems to be an action specific to auxins since it is prevented by the antiauxin *trans*-cinnamic acid (Power & Cocking, 1970). Cocking has suggested that the auxin increases the rate of uptake of solutes by the protoplasts; this then entrains an uptake of water leading eventually to protoplast rupture. The exact mechanisms of this effect and its significance for growth control by auxins is still unexplained.

The Effects of Auxin on the Properties of the Cell-Wall

One of the earliest suggested mechanisms of auxin action in extension-growth involved the induction of changes in the physical properties of the cell-wall. This followed on the classical work of Heyn, who studied the relationships between *Avena* coleoptile growth-rates under varying conditions of auxin supply, and correlated with them changes in extensibility of the cell-walls under artificial tension. Observations were made on plasmolysed tissue, to eliminate the complications of cell turgor and its associated tissue tensions. (See review by Heyn, 1940.) Small segments cut from coleoptiles were subject to mechanical bending by the application of known forces. The total amount of bending could be resolved into two components, one which was reversible on removal of the bending force and taken as a measure of cell wall elasticity (elastic extensibility) and the remainder, the permanent curvature, taken as a measure of wall plasticity (plastic extensibility). Growth changes in the coleoptile before and after tip removal (i.e. under changing conditions of internal auxin supply) were more closely correlated with wall plasticity than with wall elasticity; auxin was held to function by making the wall more plastic and thus more capable of acquiring a permanent extension under turgor pressure. Changes in wall elasticity were regarded as *results* of growth.

Since then wall properties have been much studied and much corroborative evidence has now accumulated (see also Fig. 75 A). Ruge (1938*a*, 1937*b*) treated *Helianthus* hypocotyls with auxin pastes and observed immediate increases in wall extensibility which were correlated with the accelerated growth. Crude estimates of wall acidity led to the suggestion that auxin first accumulated in the young cell-wall, there inducing a swelling of the unorientated colloid molecules filling the interstices between the woven fabric of cellulose fibrils. This swelling action of auxin constituted the primary stimulus to extension growth, allowing the altered wall to be stretched under turgor. Claims were made that IAA caused increased swelling of certain natural carbohydrate colloids in water, e.g. apple pectin, salep mucilage, etc., and this implied that auxin action did not involve the participation of the living protoplasm of the cell (Ruge, 1937*c*). We know now that these ideas must be abandoned. Not only is there no substantiating evidence for the initial accumulation of auxin in cell-walls, but its swelling action on 'dead' wall colloids has not been confirmed (Blank & Deuel, 1943).

Nevertheless there is now overwhelming evidence of changes in wall extensibilities of living cells under the action of auxin. In strips of potato tuber tissue exposed to IAA solutions, the total wall extensibility was augmented by 10 percent, while its plastic component rose by 20 percent (Brauner & Hasman, 1949). But these changes failed to occur under conditions of oxygen deficiency, which indicated that they were intimately associated with the respiratory activity of the living protoplasm (Brauner & Hasman, 1952) and were not purely physical actions of auxin. Tagawa & Bonner (1957) using a method modified from that of Heyn, demonstrated that incubation in IAA solution increased the plasticity of the *Avena* coleoptile cell-wall far more than it increased elasticity. Cleland (1958) confirmed these findings in the same material but estimated wall extensibility parameters by measuring length changes in segments after various plasmolysis and deplasmolysis treatments. The increased extensibilities due to auxin were not observed in the presence of the respiratory inhibitor KCN, again confirming the dependence of these changes on the flow of metabolic energy. The reduction of similar auxin-induced changes in wheat coleoptiles by exposure to an atmosphere of nitrogen or to temperatures of 3°C also supports a dependence on metabolic energy (Adamson & Adamson, 1958). The most recent studies have been made on *Avena* coleoptile tissue after boiling in methanol. After rehydration the mechanical properties were analysed by a sensitive Instron stress-strain analyser (Olsen *et al.*, 1965; Cleland, 1967*a*

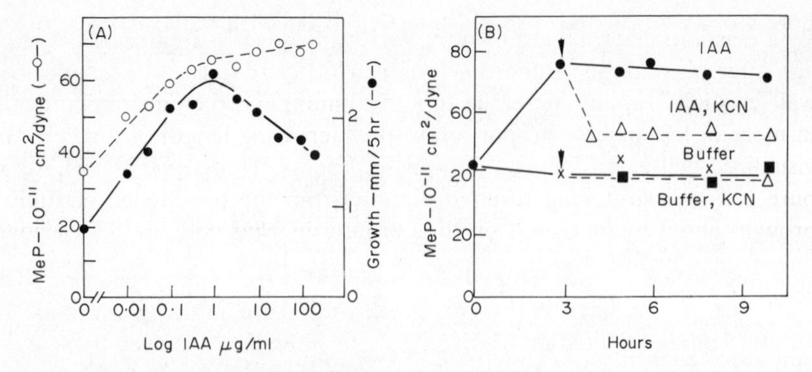

Fig. 75. Graphs illustrating the relationship between the extension growth of the cell and the plastic extensibility of the cell-wall.

(A) Relationship between growth and the modulus of plastic extensibility (MeP) in *Avena* coleoptile segments subjected to various concentrations of IAA. MeP is a measure of the permanent extension of the cell-wall under a given load and was determined on killed de-proteinized segments, i.e. on the matrix of the cell wall alone. It will be seen that up to the optimal concentration of IAA there is a remarkable correlation between growth and plastic extensibility. At supra-optimal concentrations where growth declines, plastic extensibility continues to increase slightly, which is a further indication that the factors controlling this decline in growth rate are very different from those operating in sub-optimal auxin concentrations. (see discussion on p. 307).

(B) Graphs showing the time course of changes in MeP of *Avena* coleoptile cell walls and the effects of IAA and KCN. In buffer alone, where there is little growth, MeP does not change with time either with or without KCN. In IAA, which induced growth, the plasticity of the wall almost doubles. KCN, a respiratory poison, which blocks the growth-stimulating action of auxin, reduces the MeP, i.e. 'stiffens' the cell wall, indicating that wall 'softening' under IAA action depends on a supply of respiratory energy (data of both graphs from Cleland, 1967*b*; by kind permission of the author).

& *b*) (see Fig. 73). Again IAA applied to the living tissue was shown to cause a rapid increase in the plastic extensibility of the cell-wall and this was inhibited by metabolic inhibitors (KCN, iodoacetic acid) to an extent which parallelled their inhibition of extension growth (Cleland, 1967*b*) (see Fig. 75*b*).

Extensive work on cell-wall properties has been done on roots in the laboratory of Burström at Lund, Sweden. These organs offer a different approach to the problem, as their growth is strongly inhibited by auxin concentrations which are optimal for shoot growth. Burström (1942*a* & *b*) made a detailed study of growth and wall extensibility changes in extending epidermal cells of wheat roots throughout the whole grand period of growth.

Wall elasticity determinations were derived from measurements of shrinkage after plasmolysis. Burström found that, at the start of cell extension, wall elasticity rapidly increased to a maximum, and then decreased again in more or less inverse proportion to the increasing length of the cell. He visualized cell extension as a two-phase process. Thus the initial phase of pure cell-wall stretching resulted directly from the heightened elasticity, brought about by increased swelling of intermicellar colloids. The second phase, including the remaining two-thirds of the total extension, was regarded as true wall growth by the insertion of new cellulose microfibrils into the opened 'weave' of the stretching wall. The effect of auxins at growth-inhibitory concentrations was striking. The initial rise in wall elasticity was much steeper and came much earlier, and a corresponding initial cell stretching seemed to take place. But apparently the second phase of wall synthesis was completely prevented. The picture that Burström (1955, 1957) drew of the primary action of auxin in roots and in shoots, was that it set in train the first stage of stretching growth by loosening the intermicellar bonds in the fabric of the cellulose wall, thereby inducing a higher plasticity and allowing extension under cell turgor. Such an action would underlie the growth stimulation of stem and coleoptile tissues by auxins. The inhibiting action of auxins on root growth was held to be on the second phase only, when the persistence of this loosening action prevented the formation of intermicellar linkages, and thus the incorporation of new cellulose microfibrils by intus-susception.

That the cell-wall itself is the site of regulator action in extension growth was further supported by the observation that relatively high concentrations of the non-toxic divalent Ca^{2+} and trivalent Pr^{3+} ions strongly inhibit the growth of coleoptile segments (Bennent-Clark, 1956; Cooil & Bonner, 1957; Tagawa & Bonner, 1957) (see Fig. 76 A) and that this inhibition is reversed by the chelating agent, ethylenediaminetetra-acetic acid (EDTA) which was thought to remove those elements from the cell-wall (Bennet-Clark, 1956). It was proposed that these ions act by the electrovalent binding together by their carboxyl groups of polygalacturonic acids in the interfibrillar cementing material of the cell-wall, thus reducing its extensibility. The wall-plasticizing action of auxin seems to take place even in cells prevented from extending by hypertonic mannitol solutions, since subsequent transfer of such segments to solutions of low osmotic pressure and containing no auxin resulted in almost normal extension-growth (Cleland & Bonner, 1956). The presence of calcium in the pre-treatment mannitol solutions prevented the subsequent elongation in the absence of calcium (Cooil & Bonner, 1957). Furthermore, Ca treatment of segments greatly reduced the elastic and plastic extensibilities of *Avena* coleoptile segments (Tagawa & Bonner, 1957) confirming its modifying action on the inter-

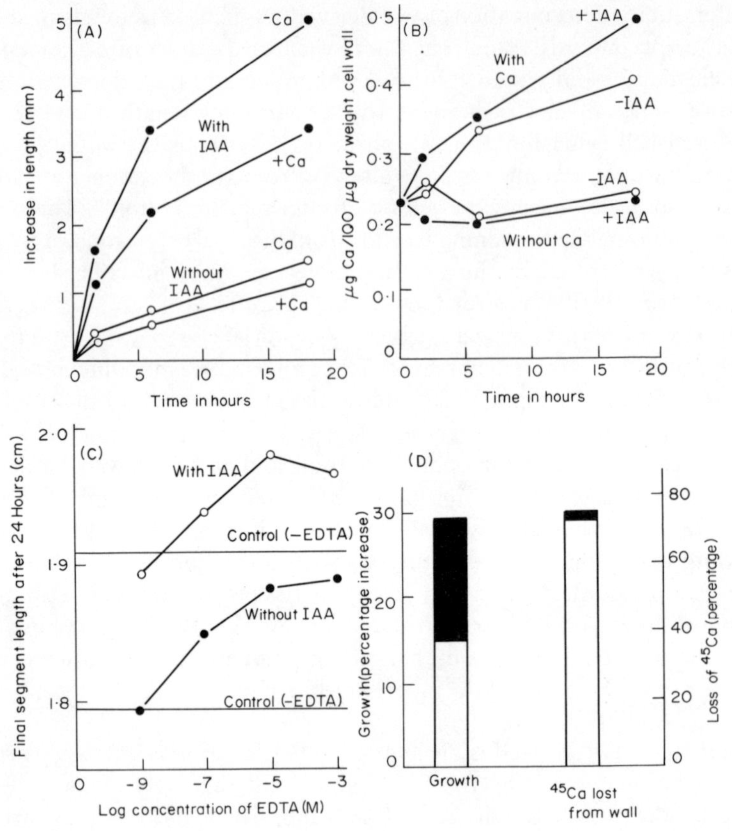

Fig. 76. Graphs illustrating the relationships between wall calcium, cell growth and the actions of auxin.

(A) Progress of growth in 10 mm segments of *Avena* coleoptiles with or without 3·5 mg/1 IAA and with or without 200 mg/1 Ca. Calcium at this concentration inhibits growth and considerably suppresses IAA action.

(B) Progresso of uptake of Ca into the cell walls of the same segments as in (A). IAA would seem to *increase* the uptake of Ca into the walls at the same time as promoting growth, a situation contrary to the Ca-bridge hypothesis. (Data of (A), and (B) from Burling & Jackson, 1965.)

(C) Graphs of the growth responses of whest coleoptile segments to EDTA both with or without IAA at 10^{-5}M. The stimulating effects of the two compounds would seem to be strictly additive. (Data from Heath & Clark, 1960.)

(D) The effect of IAA on the growth and the ease of removal of calcium from the cell wall of the *Avena* coleoptile segment. Segments were first incubated in $2{\cdot}2 \times 10^{-6}$ $^{45}CaCl_2$ for 4 hours to ensure incorporation of Ca into the cell wall. They were then incubated in water or IAA (5 ppm). They were then incubated for 17 hours in 10^{-3}M EDTA with or without IAA in a 1 percent sucrose solution and the loss of ^{45}Ca from the wall measured. Although the IAA incubation just about doubled the growth of the segments, its effect on the ease of removal of Ca from the wall was not significant (data from Cleland, 1960).

micellar forces in the cell-walls. All these actions of Ca were antagonized by K indicating that there we are dealing with an ion exchange phenomenon. Furthermore EDTA was shown to have some auxin activity in that it promoted the extension of coleoptile segments (see Fig. 76 C). All this prompted the formulation of the 'calcium bridge' hypothesis (Bennet-Clark, 1956) in which auxins were regarded as acting via their ability to chelate calcium, thus removing it as an electrovalent linking atom between the chains of pectic acids in the interfibrillar 'cement'.

This theory subsequently came under serious criticism and many different experimental approaches were made to check its validity. Results of studies on calcium exchange failed to support it. *Zea* mesocotyls and *Avena* coleoptiles were incubated with dilute solutions of $^{45}CaCl_2$ and the unbound calcium then removed by washing. Subsequent treatments with IAA had no significant effect on the amount of bound ^{45}Ca which could be extracted either with water or with EDTA solutions (Cleland, 1960) (see Fig. 76 D). Burling & Jackson (1965) have shown that the cell-walls of *Avena* coleoptile segments will accumulate Ca from solutions but that this accumulation is, if anything, *increased* by the presence of IAA whereas the calcium bridge theory would require a reduction of calcium content by IAA (see Fig. 67 B). Both these investigations suggest that auxin action is not exerted via its effect on calcium binding in the cell-walls.

Much more attention has been given to the action of chelating agents in the promotion of cell extension. There has been widespread confirmation that such a promotion does occur. In addition to effects as first reported in coleoptiles, EDTA stimulates growth in lupin hypocotyl segments (Weinstein *et al.*, 1956) and also shows auxin-like patterns in the inhibition of the growth in length of roots (Heath & Clark, 1960; Burström, 1961). Many other chelating agents, of widely varying chemical constitutions have also been claimed to act similarly (Heath & Clark, 1956). They include nitrilotriacetic acid, 8-hydroxyquinoline and diethyldithiocarbamate. Heath & Clark (1960), in factorial experiments involving combinations of these chelating agents with IAA studied their effects on the growth of wheat coleoptile segments and roots and showed that they interact in a remarkable way. Thus in the inhibition of root growth by supra-optimal concentrations of IAA, the action can be largely antagonized by very much smaller concentrations of a chelating agent; the action of the chelating agent is similarly reversed by IAA. This mutual antagonism is entirely removed by non-chelating analogues of the chelating agents. Pilet & Belhanafi (1962) observed similar mutual antagonisms of IAA and 8-hydroxyquinoline in *Lens* roots; the non-chelating analogue 2-hydroxyquinoline showed no such antagonism. Heath & Clark (1960) concluded that the action of auxins and the chelating agents on growth were similar (i.e. both via a chelating or complexing action) but that, because of the strange quantitative characteristics of their interactions, their actions could not be identical. The validity of these conclusions has been disputed by Burström (1963) who could not confirm the mutual antagonisms of IAA and EDTA in root growth and who further claimed that EDTA action on root growth is via the inhibition of cell division whereas that of IAA is on cell extension (Burström & Tullin, 1957)., i.e. two quite separate phases of growth. These discrepancies still remain unresolved. Further misgivings as to the universality of the auxin action of chelating agents arise from their inactivities in the pea-curvature and tomato-leaf epinasty tests for auxins and from the fact that many chelating agents additional to those tested by Health & Clark were inactive, even in the wheat coleoptile cylinder test (Fawcett, Wain *et al.*, 1959).

There are also considerable discrepancies in the characteristics of auxin and chelating agent actions on plant growth. We have already mentioned the differences claimed by Burström for root growth. In addition Buczek (1968) has shown that EDTA, like IAA, induces water uptake of potato tuber tissue but does not induce the correlated increase in respiration characteristic of IAA action. Furthermore the promotion of the growth of sunflower hypocotyl segments by EDTA is reversed by ferric, cobaltous and calcium ions, which have no effects on IAA action (Buczek & Konarzewski, 1968). All this suggests that the mechanism of action of chelating agents is quite different from that of the auxins.

It has been suggested by Ng & Carr (1959) that EDTA action is not due to the sequestration of ions from the cell-wall; it has little or no growth activity at pH values exceeding 6.5 whereas its maximum chelating action occurs at pH 9.

Chemical studies on the chelating properties of auxins have yielded no support for the calcium bridge theory. For example 2, 4-dimethylphenoxyacetic acid, a compound virtually inactive as an auxin, shows the same ability to chelate copper as 2,4-D, a highly active auxin (Armarego *et al.*, 1959). Similarly IAA forms a copper complex which has the same stability as that with acetic acid, which has no auxin properties (Perrin, 1961). All this points to the considerable improbability that auxin action is due to its chelating properties.

One further word needs to be said with regard to the auxin-like action of chelating agents. Thimann & Takahashi (1961) showed the EDTA promoted growth of coleoptiles in the presence of IAA (see also results of Heath & Clark (1960), Fig. 76 C) but not in the presence of NAA or 2,4-D. Since the latter are not destroyed by IAA-oxidase this suggested a possible action of EDTA as a synergist of IAA action by suppressing IAA-oxidase activity via the chelation of the manganese co-factor of the enzyme. Such an inhibition has been demonstrated *in vitro* (Tizio, 1965). However, subsequent work by Tomaszewski & Thimann (1966) showed that EDTA synergism of IAA action takes place even at optimal IAA concentrations, where polyphenols, which also inhibit IAA-oxidase and synergize IAA action, are ineffective or inhibitory. This suggests that EDTA had, in addition to a possible synergism, a further action independant of IAA. What this action is remains a mystery. Taylor & Wain (1966), who have shown that EDTA at concentrations which induce growth in wheat coleoptiles can cause the removal of pectic materials as well as calcium from the cell-walls, regards the action as non-physiological and due to a 'sub-acute toxicity effect' on the cell-wall; but this could hardly be the explanation of the marked effects observed by Heath & Clark (1960) at 10^{-11} molar concentration. The biochemical basis of the auxin-like action of chelating agents remains a mystery.

It is perhaps relevant to note at this point that gibberellic acid, which brings about growth promotion in excised cucumber hypototyl segments comparable with those caused by IAA, has, in contrast to IAA, very little if any effect on the mechanical extensibility of the cell-walls (Cleland *et al.*, 1968). In view of its marked promotion of hydrolase enzyme activity (see later p. 374) the action of gibberellins in extension growth may be via the augmentation of cell osmotica.

Auxins and Cell-Wall Metabolism

Innate in the theories which link the auxin-induced changes in cell-wall properties with the provision of metabolic energy is the suggestion of an action in cell-wall metabolism. An enzymic loosening of the heteropolar bonds between the cellulose microfibrils was first considered as one basic cause of these changes (Burström, 1942*b*). The behaviour of intermicellary substances has since come to the forefront. For instance, it has long been thought that the longitudinal strength and extensibilities of the young cell-wall might reside in these substances, which constitute the continuous phase of this colloidal cell-wall. (The cellulose micellae constitute the discontinuous phase cf. Kerr, 1951). The changes that occur during growth could be caused by shifts in the state of hydration of these carbohydrate gels (Kerr, 1951), and would presumably be enzymatic in nature. Auxins might well be acting in such an enzyme system (van Overbeek, 1952). An association of auxin action with pectin metabolism can be seen in the phenomenon of abscission control. Auxins prevent abscission, which results from changes in the metabolism of the cell-wall and of the middle lamella between the cells of the abscission zone. Such changes are enzymatic, and auxin may inhibit them directly. With all this indirect evidence, pointing to an action of auxin on cell-wall metabolism, what direct evidence is there of such an activity?

Early observations concentrated on the metabolism of the pectic materials. Pectin methylesterase is an enzyme which catalyses the hydrolysis of the methyl ester bonds of pectins to form pectic acids. 2, 4-D, when applied to the base of the lamina of bean leaves, was shown to bring about an increase in the pectin-methylesterase activity of the treated tissue (Neely *et al.*, 1950), correlated with a breakdown of protopectin in the primary cell-walls. Similar increases in pectin methylesterase wore observed in tobacco pith cells growing under the action of auxin in culture (Newcomb, 1954; Bryan & Newcomb, 1954). Studies on the effects of IAA on the rate of incorporation of ^{14}C from labelled sucrose into cell-walls of maize and oat coleoptiles indicated that cellulose synthesis was not affected by auxin, but that the rate of carbon incorporation into pectates and polyuronide hemicellulose was somewhat depressed (Boroughs & Bonner, 1953). On the other hand, in the auxin-induced extension of tobacco pith cells, there was an initial rapid build-up of pectic substance in the extending cell-wall at the expense of soluble uronides. In potato cells expanding under the action of auxins (IAA, NAA) there is a similar shift in the composition of the cell-wall (Carlier & Buffel, 1955).

Impetus was given to the study of pectin metabolism in relation to auxin action by the demonstration by Bennet-Clark (1956) that the methyl donating agent, dimethylthetin chloride, $[(CH_3)_2-\overset{+}{S}-CH_2-COOH]\,Cl^-$ was an active promoter of extension growth of the *Avena* coleoptile. Bennet-Clark suggested that the balance between the free pectic acids and their methyl esters could play a part in the cementing action of the interfibrillar substances. Thus the hydrogen bonding between the carboxyls of the free acids would be much stronger than the van der Waal's forces between the methyl ester groups and methylation would consequently lead to wall loosening, greater plasticity and increased growth. This was checked by studying the effects of IAA on the incorporation of methyl groups into pectins by using ^{14}C-labelled methionine as a donor (Ordin, *et al.*, 1957). IAA was shown to promote this incorporation into the water-soluble pectins but not into the insoluble protopectins (probably closely linked by co-valent bonds to the cellulose of the wall). Further studies (Albersheim & Bonner, 1959), showed that IAA also increased the total synthesis of water-soluble pectins in the wall while having little effect on the water-insoluble fraction, which constituted some 80 percent of the total. Similar increases in the proportion of water-soluble pectins of artichoke tuber tissue under auxin action were recorded by King & Bayley (1963). Since it would be expected that the insoluble pectin bound to the cellulose microfibrils would be the component contributing to wall rigidity, then a decrease in their proportion in the total pectin complement by the augmentation of the soluble component would lead to a loosening of the wall fabric (Bonner, 1960). However Cleland (1963) subsequently showed that ethionine, a competitive analogue of methionine, although completely inhibiting the IAA-induced transfer of methyl groups from methionine to the water-soluble pectin fraction of the cell-wall, did not eliminate the promotion of cell extension by IAA. Furthermore auxin could not cause any measurable increase in the pectin content of the wall in the absence of added sugar, indicating that, if it does act via pectin metabolism, it must do so via the pectins already present and not via a *de novo* synthesis of those materials. Further doubts on the validity of simple methylation of pectins as the basis of auxin action was expressed by Ray (1961) who emphasized that only 20 percent of the wall pectins were soluble and that a small change in the esterification of this fraction should have insignificant effects on wall properties. In addition he demonstrated (Ray, 1958) that most of the water-soluble carbohydrates of growing cell-walls were hemicelluloses and not pectins and these had been neglected in consideration of wall properties.

Consequently in the last few years attention to auxin action in cell-wall metabolism has switched to the hemicelluloses and their hydrolysing enzymes β-1,3- and β-1,6-glucanases. Tanimoto & Masuda (1968) showed that auxin puts up the level of these two enzymes in etiolated barley coleoptiles and pea epicotyls and that β-1,3-glucanase extracted from fungi could induce elongation in barley coleoptile segments. This indicated that hemicellulose linkages, probably with cellulose, in the cell-wall were responsible for cell-wall rigidity. These latter findings were confirmed by Wada *et al.* (1968) who further claimed that cellulase (β-1,4-glucanase) and pectin-methylesterase were without this growth-promoting effect. The effect of β-1,3-glucanase was also studied on diploid strains and mutants of yeast, some of which were capable of growth promotion by IAA. Only those strains which were sensitive to auxin could be induced to expand by the enzyme. The same enzyme also enhanced the growth of pollen tubes (Roggen & Stanley, 1969). Such evidence suggested that the wall-loosening effects of auxin are mediated via the hydrolysis of these glucan linkages and not via pectin metabolism. On the other hand the work of Maclachlan *et al.*, (1968) and Davies & Maclachlan (1968) indicates that cellulase (β-1,4-glucanase) may be the important enzyme concerned. In experiments with pea epicotyls they have shown that IAA causes a marked increase in the specific activity of cellulase, apparently as a result of induced *de novo* synthesis of the enzyme in the cell. Their observations indicate that the clear increase in β-1,3-glucanase activity, caused by IAA is due to the activation of enzymes already present and not to induced synthesis. However there is at present no evidence on which to rule it out as playing a part in wall-loosening.

The role of hydrolyzing enzymes in growth is not universally accepted.

Ruesink (1969) for example has treated *Avena* coleoptile segments with fungal cellulase and, although this increased wall extensibility, it did not promote growth. Cleland (1968) has pointed out that auxin-plasticising effects are prevented by respiratory inhibitors which suggests that *synthetic* processes requiring metabolic energy rather than degradative ones are involved (Baker & Ray, 1965). The next step towards the solution of these problems would be the identification of the enzymes concerned.

There is some evidence that auxin action may be connected with the promotion of cellulose *synthesis*. Thus an enzyme isolated from *Avena* coleoptile cell walls can synthesize cellulose from glucose donors (UDP-glucose). The auxins IAA and 2,4-D had no effect on the enzymes *in vitro* but incubation of segments of living tissue with those auxins increased the synthesis of cellulose from the donor, an effect interpreted as due to an increased production of the cellulose-synthesizing enzyme (cellulose synthetase) (Hall & Ordin, 1968). Similarly in pea stem tissue incubation with IAA stimulates glucose uptake and incorporation and at the same time promotes the activity in that tissue of the enzyme β-glucan-synthetase (UDP-glucose-β-1, 4-glucan-glucosyltransferase) (Abdul-Baki & Ray, 1971).

Another wall component which has received much consideration as a contestant for the role of binding agent in wall-stiffening is the hydroxyproline-rich protein component extensin (see p. 355). Lamport (1965) has advanced a theory of auxin action based on this role. The protein, it was suggested, was linked to galacto-araban residues by glycosidic bonds, the galacto-araban moiety of the complex then acting as the intermediate link to the cellulose microfibrils. The protein moieties were themselves mutually linked via disulphide (-S-S-) bridges, which were responsible for wall 'stiffness'. An action of auxin was proposed on these bridges, breaking them by reduction and the formation of -SH groups. The evidence for this is very indirect and based on some observations that auxins raise NADPH production (Marre & Bianchetti, 1961), the ATP/ADP ratio (Marre & Forti, 1958) and (consequently) -SH levels (Tonzig & Marre, 1961) in growing tissue. Later the existance of a glycosidic linkage with arabinose via the -OH group of hydroxyproline was established in tomato cell-walls (Lamport, 1967). In fact more than 60 percent of the arabinose in the cell-walls of these tomato suspension cultures is linked to hydroxyproline.

This interesting theory has come under criticism. Thus the incorporation of radioactive proline into cell proteins has been followed by autoradiographic techniques and electron microscopy (Israel *et al.*, 1968). From this and other associated experiments Steward has disputed whether the hydroxyproline-rich protein in mainly in the cell-wall and has suggested that it has quite a different role, namely as a structural protein of the *cytoplasm*, permitting the association of individual ribosomes into polysomes, which are the functional units of protein synthesis (see p. 388). However many workers, using somewhat different techniques, have obtained results which support the original observations of Lamport and his colleagues and indicate that a peptido-glycan network ('extensin-complex') is an important structural component of the cell-walls of a great range of plants and that its metabolism may well be involved in wall extension and possibly in the action of hormones upon it. Thus studies on the mechanical properties of the cell-wall have shown that removal of most of the wall protein by the action of a protein-hydrolysing enzyme, pronase, had no effect on the elastic properties of the wall but greatly reduced its plasticity, indicating that wall proteins do contribute to wall stiffening. However the auxin-induced increased in plastic extensibility was not affected by protein removal, indicating that auxin action was on some other component of the cell-wall (Cleland, 1967*b*). Again auxin in the presence of sucrose augments the incorporation of ^{14}C-proline into the hydroxyproline of the cell-walls of *Avena* coleoptile segments, although these effects are not closely correlated with those of auxin on cell extension (Cleland, 1968). A direct connection of auxin mechanism with metabolism in the extensin-complex still remains to be established.

A recent highly speculative theory has been based on the observations that segments of auxin-starved *Helianthus* hypocotyl, which show no growth in buffer solutions at pH 6, can, under *anaerobic* conditions, be stimulated to extend by buffer solutions at pH 4 to the same extent as by auxin treatments in air (Hagar *et al.*, 1971). The response is rapid and can be switched on and off by appropriate pH changes. The effect is prevented by chemicals which make the protoplasmic membranes permeable to protons (hydrogen ions). High-energy phosphates such as ATP can also induce a growth response also under anaerobic conditions and this effect is greatly increased by prior incubation in auxin in air. Such observations led the authors to propose that the wall-softening enzymes are activated by a high proton concentration in the wall 'compartment'. This is maintained by a membrane-bound anisotropic ATP-ase proton pump and the function of auxin is to co-operate with high-energy phosphates in the operation of this pump. Such an effect of auxin on wall enzymes acting purely by way of pH would not be likely to be very specific and this stands in strong contrast to some of the highly specific enzyme activations described in the foregoing paragraphs.

The Action of Hormones in Carbohydrate Metabolism

The stimulation of respiration by auxins (see p. 375) has given rise to the idea that a mobilization of carbohydrate reserves might be the basis of their control of growth. A shift in the polysaccharide-sugar balance, maintaining the constancy of osmotic pressure observed during growth, might be a possible auxin activity. Much data from experiments in which large doses of the synthetic auxins 2, 4-D, MCPA, etc., have been applied to plants, show that, at these levels of incipient toxicity, there occurs gross disturbances in the carbohydrate metabolism of the cell, with the disappearance of polysaccharides and the accumulation of sugars (see Wort, 1964). There is, in addition, a similar dislocation of protein metabolism, but it is most likely that these sub-toxic responses have little to do with the normal auxin control of growth.

Direct observations *in vitro* indicate that any effects on the starch-sugar balance in the cell cannot be explained by a direct action of auxin on the amylases (Smith, 1941; Anker, 1953; Brakke & Nickell, 1952; Pilet & Tubian, 1953). Early suggestions of *in vivo* activation were ill-founded, (see Audus, 1949*b*) but the idea of a reserve-mobilizing action of auxin *in vivo* is still favoured in some circles. Direct observations on tobacco pith cells in tissue culture (Skoog & Robinson, 1950) showed that auxin-induced cell extension was accompanied by a considerable increase in the content of reducing sugars and of sucrose. This was partly, but not entirely, at the expense of starch reserves, and so it was concluded that auxin acted as a catalyst in carbohydrate metabolism. One suggestion has been that mobilization may be regulated by the permeability of internal protoplasmic membranes which may separate enzymes from substrates, and that, accordingly, auxin may act by increasing this permeability (Anker, 1953). The most recent observations of Wain and his colleagues (Wain, 1967) that physiological concentrations of auxin applied to chicory and artichoke tissue augments their content of invertase and other hydrolases has again raised the possibility of a primary action of auxin on the soluble/insoluble carbohydrate balance of the cell.

Whatever may be the situation *vis-à-vis* the auxin, there is little doubt that one of the best-established biochemical actions of the gibberellins is to promote, in a most dramatic manner, the activity of the starch-hydrolysing amylases in the aleurone layers of cereal grains, where they clearly act as reserve-mobilizing hormones. Similar starch hydrolysis has been induced by gibberellic acid in growing tissues such as coleoptiles (for detailed references see Paleg, (1965) and Lang, (1970)). This action is apparently the result of the induction of enzyme synthesis, which may be the basis of the action of gibberellins in promoting cell extension. Thus we have seen that there is no substantial evidence that gibberellins enhance cell-wall plasticity, i.e. act like auxins, and this is paralleled by observations that in pea internodes gibberellin treatments cause no change in the activity levels of cellulase and pectin-esterase, the presumed wall-softening enzymes

(Broughton and McComb, 1971). However such treatments did stimulate amylase and β-fructofuranosidase and these stimulations ran parallel with internode growth. Furthermore the growth-promoting action of gibberellins in pea internodes could be mimicked by glucose injections. In the developing internodes of *Avena* seedlings gibberellic acid applications to excised segments containing intercalary meristems caused a marked promotion of extension and this response was inhibited by kinetin. Running parallel with these growth responses gibberellic acid treatment promoted and then maintained invertase activity in the segments while kinetin counteracted the promotion and seemed actually to hasten the decay of the enzyme (Jones and Kaufman, 1971). All this points to an augmentation of the supply of carbon substrate for general metabolism and wall synthesis in the growing cell as the basis of gibberellin action.

Auxin and Respiratory Processes

Whatever may be the ultimate cell mechanism by which auxin induces water uptake and wall stretching, there is little doubt that the living protoplasm is involved and that an adequate supply of metabolic energy is essential. In fact, the close dependence of auxin activity on an adequate level of oxygen supply to the tissues (Reinders, 1938; Brauner & Hasman, 1952; Hackett & Schneiderman, 1953; Hacket and Thimann, 1953) first suggested the *direct* participation of auxin in enzyme systems concerned with the production of metabolic energy or with the efficiency of its utilization in growth. It was Bonner (1933) who first claimed that auxin from *Rhizopus suinus* stimulated the oxygen uptake of coleoptile sections by 30 percent, and that this was 'a necessary condition of growth promotion'. Subsequently much evidence has accumulated for a small, but apparently direct, effect of auxin on the respiration of growing tissues. In assessing the importance of these results the difficulty has always been to decide how much of the observed increases in respiration have been the *result* of the auxin-induced growth. For example in extending cells active manufacture of protoplasm goes on, and the cell complement of respiratory enzymes should also increase. Some of the increases of respiration claimed had indeed been thus caused. Improved techniques, however, have allowed a direct and more or less immediate effect to be confirmed in several tissues, viz. in coleoptile segments (Anker, 1953; French & Beevers, 1953), in etiolated pea stem segments (Christiansen & Thimann, 1950; Kelly & Avery, 1951), in potato (Hacket & Thimann, 1953) and in artichoke tuber tissue (Hackett & Thimann, 1952*a*).

Attempts to elucidate the nature of this auxin stimulation, and to identify the particular enzyme system responsible, have been based mainly on the use of specific enzyme inhibitors. An early discovery was that extension-growth was very sensitive to respiratory inhibitors and could be suppressed completely by concentration levels which appeared to have little effect on respiration. This led to the hope that the discovery of an inhibitor, which would specifically eliminate the auxin stimulation of respiration, would automatically indicate the enzyme system involved.

The first metabolic system to be pin-pointed in this way was the Krebs organic acid cycle, when Commoner & Thimann (1941) claimed that the auxin stimulation of respiration in the oat coleoptile was dependent on the presence of malic acid, that mono-iodoacetic acid, inhibitor of the Krebs cycle dehydrogenases, was an extremely active inhibitor of extension-growth at concentrations only slightly affecting respiration, and that this inhibition could be neutralized by the Krebs cycle acids. This led to the theory that auxin regulation of growth might be affected by a general stimulation of the supply of metabolic energy by this Krebs acid cycle (Bonner & Thimann, 1950; Thimann, 1951*a*). Unfortunately the various growth effects attributed to the organic acids were later shown to be due to the potassium ion (Cooil, 1952).

The advent of radio-isotopes revivified interest in the possibility of specific effects of auxins on particular stages in respiratory metabolism, in particular in the Kreb's cycle and related biochemical pathways. Early work on the metabolism of acetaldehyde in plant tissues led to the suggestion that auxin might exert a controlling action on the fate of pyruvic acid, the terminal intermediate in the glycolysis sequence (Nance & Shigemura, 1954). This led to a study of the metabolism of ^{14}C-acetate in various tissues, where it was found that IAA and 2,4-D increased the production of $^{14}CO_2$ from acetate and affected the incorporation of acetate-carbon into cell-wall constituents (Perlis & Nance, 1956; Nance, 1958; Fang *et al.*, 1961). Previous to this work Leopold & Guernsey (1953*a*) claimed that IAA could form a thiol ester with coenzyme-A, thus pin-pointing it as the focus of auxin control. Much later it was shown that IAA greatly increased the level of citric acid in artichoke tuber tissue (Rambour, 1968) and also promoted the incorporation of ^{14}C from both acetate and pyruvate into citric and isocitric acids in the *Avena* coleoptile (Sen & Sen Gupta, 1961). This is possibly due to auxin stimulation of the enzyme which catalyses oxalacetic acid condensation with acetyl-coenzyme-A which has been demonstrated in cell-free preparations of *Ricinus* endosperm, *Phaseolus* seedlings and pigs heart (Mitchell & Sarkissian, 1966; Sarkissian, 1966). In studies with mung bean hypocotyls, in which the metabolism of glucose, containing ^{14}C in various positions in the molecule, was followed, it was shown (van Hove & Carlier, 1968) that NAA put up the $^{14}CO_2$ production considerably. The effects of specific inhibitors (malonic acid, mono-iodoacetic acid) on the proportion of the various carbon atoms of the glucose contributing to the total $^{14}CO_2$ suggested that the auxin-stimulating action was operating after the step leading to pyruvate formation but before the decarboxylation of oxalosuccinate. In contrast Kim & Bidwell (1967) found that in pea root tips, IAA and 2,4-D *inhibited* the entry of pyruvate via acetate into the Kreb's cycle, although exogenous acetate oxidation was promoted. They propose that the inhibitions are exerted in the α-lipoic acid cyclic oxidation step which links the pyruvate decarboxylation step to the formation of acetyl-coenzyme-A. Clearly the picture is confusing and calls for much further work.

Another well-studied system was ascorbic acid/ascorbic acid oxidase. The earliest observations by Tonzig & Trezzi (1950) were that ascorbic acid inhibits the growth of *Avena* coleoptiles and pea epicotyls and antagonized the action of IAA. These effects were later attributed to dehydroascorbic acid produced by oxidation in the tissue (Marré *et al.*, 1955). It was also shown that auxin tended to push the balance in the direction of the reduced form of ascorbic (Marré 1954). From this came the theory that auxin acts by inhibiting ascorbic-acid oxidase, keeping ascorbic acid in its reduced form (Marré & Arrigoni, 1955; Tonzig & Marré, 1955). This is in line with the growth-promoting action of chelating agents which inhibit ascorbic acid oxidase by chelating copper. Reduced ascorbic acid was proposed to regulate the state of oxidation of glutathione in the cell, since auxin, at growth-promoting concentrations, could favour glutathione reduction (Marré & Arrigoni, 1957). This was further supported by observations that oxidized glutathione would stimulate growth (Marré & Arrigoni, 1957). Sulphite, at low concentrations also stimulated growth and auxin action in pea tissue and there was a correlated increase in the content of reduced glutathione (Arrigoni, 1960). The key importance of glutathione lay in its activation of dehydrogenases (hence the auxin stimulation of respiration) and other sulphydryl enzymes and its effects on transpeptidase reactions and hence protein synthesis (Marré & Arrigoni, 1957). We have already seen (p. 372) that Lamport has drawn on these ideas to explain the action of auxin in regulating the binding action of extensin. However, not all these claims have been substantiated; for example, in tobacco pith tissue auxin promotes ascorbic acid oxidase activity (Newcomb, 1951) and ascorbic acid can promote coleoptile growth (Chinoy *et al.*, 1957; Recalde & Blesa, 1961). The theory remains an interesting speculation.

The disturbance of the phosphate metabolism of the growing cell results in considerable inhibition of extension growth. This was demonstrated by Bonner (1950*b*) when he showed that arsenate, at concentrations having little effect on respiration, greatly inhibited extension growth in *Avena* coleoptile segments, and that this inhibition was reversed by phosphate. IAA stimulation of respiration was claimed to be prevented by low arsenate concentrations. Even more dramatic effects were produced by the inhibitor 2,4-dinitrophenol (DNP), which may inhibit auxin-induced growth almost completely at concentrations (5 ppm) which considerably stimulate respiration of coleoptile segments (Bonner, 1949*b*; Ketellapper, 1953). It also prevents the IAA stimulation of respiration.

It is known that DNP and arsenate prevent oxidative phosphorylations, i.e. the formation of adenosine triphosphate (ATP) from adenosine diphosphate (ADP) in respiration. The inhibition of growth by DNP and arsenate most probably results from this action, which precludes the provision of the necessary energy as ATP. The ineffectiveness of auxins in the presence of agents that prevent phosphorylations has suggested that auxins may play a part in those systems which couple the energy release in respiratory metabolism to its consumption (from ATP) in growth (Bonner & Bandurski, 1952). Recent evidence has indicated (French & Beevers, 1953; Busse & Kandler, 1956) that IAA stimulates respiration only in growing, not in non-growing tissues. This suggests that in tissues stimulated to grow by IAA, ATP is rapidly used up, thereby raising the level of ADP; since respiration rate is normally limited by the supply of the phosphate-acceptor ADP, the net respiratory turnover will increase. The IAA stimulation of the respiration is thus the *result* of its stimulation of growth and consequent disturbance of the ATP-ADP balance. In tissue where growth cannot be stimulated by auxin, the ATP-ADP balance, and hence the respiration, would not be so affected. If this relationship is so, further study of respiration effects is likely to throw little light on the ultimate auxin mechanism.

However there still remain observations which cannot be so explained. Thus Marré & Forti (1958) claimed that IAA treatment of pea epicotyls induced an immediate rise in the ATP content of the tissue which subsequently declined, presumably as a result of consumption in the subsequent growth processes. Similarly Sen Gupta & Sen (1961) demonstrated that IAA promoted the incorporation of ^{32}P into ADP and ATP in *Avena* coleoptiles and that this was opposed by 2,4-dinotrophenol. Finally there are two contradictory claims for the action of IAA on isolated plant mitochondria; Poljakoff-Mayber (1955) could detect no action on oxidative phosphorylation whereas McDaniel & Sarkissian (1966) observed increases of some 75 percent in the phosphorylative efficiency (P/O ratios) of mitochondria from maize scutellum at IAA concentrations as low as 10^{-10} molar. A direct action of auxin on oxidative metabolism still remains a possibility (see also p. 380).

Theories Derived from Considerations of Structure-Activity Relationships in Hormone Molecules

Much hormone research has been concerned with molecular structure-physiological activity relationships. Many attempts have been made to draw up, from the mass of data, generalized laws from which to predict the nature of the chemical system in which hormones act. The lack of any great specificity in action has led to theories in which hormone molecules, by virtue of a certain 'shape', are attracted by and 'fitted to' protein surfaces by secondary chemical forces. These protein surfaces are probably parts of enzymes, whose activity is consequently modified by the hormones (Veldstra, 1955). In the case of auxins the two-point contact theory of Bonner, the ortho-combination theory of Hansch, and the three-point contact theory of Wain (see Chapter III), favour direct chemical reactions at specific centres on receptor molecules. So far these generalizations from chemical structure have been unable to afford to the plant biochemist and physiologist any indication of the possible identity of these receptor molecules.

We have already seen in Chapter IV that structure/activity relations in the gibberellins have been similarly considered as guide-lines to ultimate mechanisms. In contrast to the active auxins, which have a very wide range of chemical structures and a very low species specificity, the gibberellins have very closely circumscribed structural requirements for activity and very complex patterns of species specificity. We have seen that explanations of structure/activity relations based purely on the efficiency of metabolic conversion to a single active gibberellin are very improbable and that a 'lock and key' relationship of gibberellins at some active enzyme or other molecular site is much more likely. What these molecular sites may be for all hormones will be considered in the last section of this chapter.

Mutual Interactions of Hormones and Interactions with Other Metabolites

It is a little surprising that in view of the favour with which 'coenzyme' theories of auxin action have been regarded, relatively little attention has been paid to the interaction of auxins with other plant metabolites–particularly vitamins and coenzymes. From time to time the metabolic association of IAA with various vitamins, viz. nicotinic acid (Galston, 1949*a*), vitamin B_1 (Kandler, 1953), vitamin K (Hey & Hope, 1951; Duplessy-Graillot, 1954), and ascorbic acid (Tonzig, 1950), have been severally suggested. Only the last vitamin deserves our serious attention.

In 1947, Raadts and Söding reported that dehydroascorbic acid stimulated *Avena* coleoptile growth and concluded that it activated the conversion of endogenous precursor into active auxin. A synergism effect of ascorbic acid with IAA in coleoptile cylinder growth was reported by Scheuermann (1952), who proposed an action involving changes in the redox potential of the growing cell. In contrast to this, Tonzig and his colleagues claimed that, in a large number of physiological responses, ascorbic acid action runs counter to that of IAA. This mutual antagonism was shown in growth and water uptake of a range of tissues such as coleoptiles, hypocotyls and onion epidermis, with associated changes in protoplasmic viscosity (Tonzig & Trezzi, 1954*a*; Tonzig *et al.*, 1952). Ascorbic acid treatment of oat coleoptiles and pea stems increased the amount of free auxin at the expense of that bound to proteins, and it was concluded that the competitive action of ascorbic acid was due to its capacity to remove auxin from auxin-protein complexes, which are the active units in growth (Tonzig & Trezzi, 1954*b*).

All these investigations point to an interaction of ascorbic acid with auxins *outside* the growth system itself; but a much more intimate role for it in extension growth is suggested by the work of Newcomb (1951, 1954) with tobacco pith grown in tissue culture. He found that, under the action of a low IAA concentration, which greatly increased the extension growth of the cells, there was a sevenfold increase in the ascorbic acid oxidase activity of the tissue. Analysis showed that this enzyme was situated in the cell-wall. Phenyl-thiourea inhibited its activity and, at the same time, stopped growth, but did not affect cell respiration. This indicated that the enzyme was specifically associated either with wall extension or with other cell-surface actions such as water uptake, and was not the terminal oxidase of respiration. When it is remembered that ascorbic acid is probably closely related to the pectic substances via the common metabolic intermediate galacturonic acid (Isherwood *et al.*, 1953), a further contact with cell-wall metabolism is indicated. Lamport (1965) has implicated the ascorbic acid oxidase of the cell-wall in his extensin theory of auxin action on wall plasticity (see p. 372). If this wall protein is the predominant ‘cementing’ substance between the cellulose microfibrils and the mutual binding of molecules is via covalent -S-S- linkages, then the reduction of these linkages to form -SH groups should reduce the strength of those linkages to that of hydrogen bonds and thus increase the plasticity of the wall. Although he regards an action of IAA on mitochondrial metabolism as the source of the reductants necessary for this change, he nevertheless suggests that ascorbic acid oxidase in the wall may play a part in ‘disulphide turnover’ in the cell-wall and thus in extension growth.

A metabolite that has received attention as a possible co-factor of auxin in the growth system is coenzyme A (CoA), which is responsible for mediating the transfer of acyl-groups in plant metabolism. Two independent sources of evidence (Leopold & Guernsey, 1953*d*; Siegel & Galston, 1953) suggest that it can form, with IAA and other auxins, a compound similar to those which it forms with the acyl acids. It has been mooted that auxin might regulate growth by controlling a CoA-requiring metabolic reaction (Leopold & Guernsey, 1953*d*). On the other hand, this reaction may simply be the method by which auxins are brought into, and moved about in, the auxin-controlled metabolic system, and may be unrelated to the main action of auxin. This would fit in with the suggestion that some active antiauxins may also form similar compounds with CoA (Leopold & Price, 1956), which would be responsible for introducing them into the growth system.

One of the early approaches to the mechanism of auxin action through its interaction with metabolites, was to consider auxin as the remover of an inhibition rather than as a direct stimulator. For example, Thimann (1951*b*), while stressing the importance of organic acid metabolism in growth, drew attention to the structural similarities between the auxins and malonic acid, the competitive inhibitor of succinic dehydrogenase. He suggested that, in plants, auxins might act in competition with natural inhibitors of the malonic acid type, thus protecting the succinic dehydrogenase, and consequently promoting growth. Again coumarin and the natural inhibitor, chelidonic acid, may be sulphydryl enzyme inhibitors, and auxins may protect important growth-controlling enzymes from their inhibiting action (Leopold *et al*., 1952). Lastly the possibility that auxin may act as a chelating agent indicated that it may promote growth by the removal of harmful metals whose presence in the biochemical systems controlling growth would otherwise depress the activities of these systems to low levels (Heath & Clark, 1956).

All these suggestions have not stood the test of time but there is a much firmer experimental basis for the suggestion that the gibberellins may promote extension growth by suppressing an inhibitor of IAA activity. Such ideas originally stemmed from early observations that the activity of gibberellic acid in promoting extension growth was positively correlated with high endogenous growth rates and thus, by implication, with high IAA content (Hayashi & Murakami, 1953). This was supported by the observation that tryptophan, the precursor of IAA, when applied with gibberellic acid, would significantly augment its activity. This correlation was confirmed by Radley (1958) in basal and apical segments of wheat leaves with high and low endogenous growth rates respectively and by Ng & Audus (1964) in excised segments of *Avena* mesocotyls, whose endogenous growth rates and response to gibberellic acid were greatly promoted by the inclusion of the coleoptilar node, a presumed source of auxin. More direct observations produced additional confirmation; Brian & Hemming (1958) showed that in dwarf pea plants, which are very sensitive to gibberellic acid action, decapitation, and hence removal of the source of endogenous auxin, virtually obliterated the gibberellin response but that the sensitivity was partially recovered by the application of IAA-lanolin paste to the cut surface. Similarly Kuse (1958) showed that gibberellin stimulated the extension of petioles of *Ipomoea batatas* only if the blade were intact but that sensitivity could be restored by applying IAA to the cut surface of a de-bladed petiole.

Subsequently synergistic actions of auxins and gibberellic acid were sought in a variety of tissues. In dwarf pea segments it was demonstrated, not only with IAA, but also with the synthetic auxins NAA and 2,4-D (Brian & Hemming, 1961). Excised lamina joints of rice leaves are senstive to auxins, which bring about an increase in the angle which the blade makes with the sheath, the angle rising from 90° to about 140° at the optimal concentration of 10 ppm with both IAA and NAA (Maeda, 1960). Gibberellic acid treatments greatly increased the sensitivity of this system to auxin, reducing the concentration for maximum response to 1 ppm with NAA and to 3 ppm with IAA (see Fig. 77 A & B). Both gibberellins and auxins can prevent the shedding of young fruit and promote the growth of the flesh tissue. In some species of *Rosa*, particularly *R. spinosissima*, Jackson & Prosser (1959) have demonstrated a dramatic synergism in these responses between gibberellic acid and naphth-lyl-acetamide. The dormant seta of *Pellia epiphylla*, when excised from the thallus, can be induced to extend somewhat by either IAA or gibberellic acid but full normal extension is achieved only when the two hormones are applied together (Asprey *et al.*, 1958).

However, such synergisms have not always been demonstrable. In pea epicotyl they were observed in green but not in etiolated tissue (Galston & McCune, 1961; Hillman & Purves, 1961). Only additive or less than additive effects of the two hormones have been observed in coleoptile and mesocotyl tissue of *Avena* (Recalde *et al.*, 1960; Nitsch & Nitsch, 1961; Ng & Audus, 1964) (see Fig. 77 C), in cucumber hypocotyl (Katsumi *et al.*, 1965) and in the inhibition of root segment growth (Kato, 1958), which suggests that these actions are separate and independent. The action of IAA in the *Avena* curvature test is also unaffected by the presence of gibberellic acid (Kuraishi & Muir, 1964*a*). One possible reason why some workers have been unable to show auxin/gibberellin interactions is that a time factor may be involved. For example Ockerse & Galston (1967) were unable to detect any synergism if IAA were applied to pea stems *before* gibberellic acid but recorded a marked promotion of IAA action by previous pretreatment with gibberellic acid. They suggested that the two hormones act in two separate steps of the growth process, the gibberellic acid step preceding the IAA step.

The interactions of gibberellic acid with antiauxins have also been

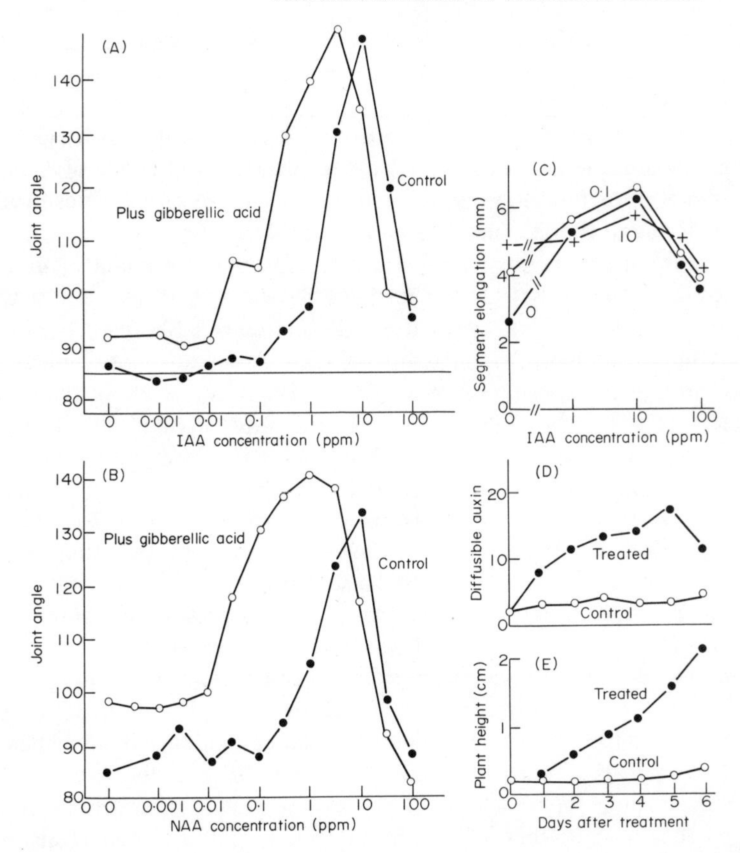

Fig. 77. Selected examples of gibberellin/auxin interactions in the control of growth phenomena.

(A) and (B). Interaction with IAA (A) and NAA (B) in effects on the angle of lamina joints of rice leaves. These joints, excised to include short lengths of blade and sheath, are sensitive to auxins and will 'open', i.e. the angle between the sheath and the blade will increase according to a well-defined concentration/response curve, the maximum opening occurring at the optimal auxin concentration of about 10 ppm. There is a slight response to gibberellic acid on its own, but gibberellic acid dramatically synergises auxin action, the optimum concentration being reduced to 3 ppm for IAA and to 1 ppm for NAA (Data from Maeda, 1960).

(C) Interaction with IAA in the extension growth of 10 mm segments of *Avena* coleoptile. The concentrations of gibberellic acid in ppm are marked on the relevent graphs. Here responses to mixtures of the two hormones are much less than the sums of the responses to the two hormones acting alone. There is no hint of a synergistic action; if anything gibberellic acid at 10 ppm antagonizes the action of low concentrations of IAA (Data from Recalde *et al.*, 1960).

(D) and (E). The effects of gibberellic acid treatment on the levels of diffusible auxin in dwarf pea seedlings. (D) Diffusible auxin levels. (E) Plant growth. (Data from Kuraishi & Muir, 1964*b*.)

studied, the rationale of such experiments being that if gibberellins act via an auxin-controlled system, then suppressing auxin action with antiauxins should correspondingly suppress gibberellin activity. In this area of investigation results are conflicting and interpretations of them diverse. In pea stem segments Kato (1961) showed that *p*-chlorophenoxy-isobutryic acid (PCIB) inhibited IAA action competitively (i.e. is a true antiauxin) but blocked gibberellic acid action completely; he concluded that the two hormones operated in different growth systems. In contrast to this Katsumi *et al.* (1965) claimed that this antiauxin reduced gibberellin action by only a slight amount. Kefford (1962*a*) observed an elimination of gibberellic acid action by PCIB in rice coleoptiles and concluded that auxins and gibberellins acted in the *same* growth system. Ng & Audus (1964) in a study of the mutual interactions of auxins, gibberellic acid and the antiauxins PCIB and α-(naphth-lyl-methylthio)-propionic acid (NMSP) in the growth of *Avena* mesocotyl segments concluded that auxin and gibberellic acid actions were qualitatively similar and therefore probably on the same growth system. Clearly such interaction studies are open to more then one interpretation and have helped little to illuminate the situation.

The first theory to explain auxin/gibberellin interactions was put forward by Brian & Hemming (1958) and was based on comparisons of gibberellin action in intact pea shoots and in excised segments grown with or without auxins. The growth patterns suggested that gibberellin promoted growth by countering an inhibitor of auxin action. The fact that IAA-oxidase in tissues can have the effect of such an inhibitor led to studies of gibberellin action on this enzyme.

Early reports that gibberellic acid can directly inhibit IAA-oxidase action *in vitro* (Pilet, 1957) have not been confirmed (see Kato & Katsumi, 1959) but there are now many reports that it can reduce the concentration of this enzyme in treated tissue. Such reductions of IAA-oxidase levels have been shown in petioles of *Trifolium ochroleucum* (Pilet & Wurgler, 1958) and internodes of *Phaseolus vulgaris* (Pilet & Collet, 1960). Similar reductions in activity have been shown in pea tissue (Galston, 1957; Galston & Warburg, 1959) and were attributed to the induction of an IAA-oxidase inhibitor by gibberellin. From dwarf maize plants Bouillenne-Walrand *et al.*, (1967) isolated such an inhibitor of IAA-oxidase and showed that its concentration in leaves was increased by gibberellin treatment; it was identified as a polyhydroxy-cinnamic acid. A gibberellin repression of peroxidase activity (see Chapter XI) takes place in pea tissue (McCune & Galston, 1959) and in cucumber tissue (Halevy, 1962; Katsumi & Sano, 1968). In the pea tissue there were indications of a shift in substrate specificity of the enzyme, which implies possible selective suppression of one or more peroxidative isoenzymes. In a comparison of tall and dwarf varieties of pea and maize Mc-

Cune & Galston found that peroxidase levels were low in the tall varieties and this was little affected by gibberellic acid treatment. In dwarf varieties the levels of this enzyme were much higher and were considerably reduced by gibberellin applications. It is only fair to record however that in some tissues an influence of gibberellic acid on IAA-oxidase levels could not be detected, e.g. in rice seedlings (Varga & Balint, 1966) and in others, e.g. in the germinating grain of winter wheat (Balduc *et al.*, 1970) and in dark-grown dwarf pea plants (Ockerse and Waber, 1970) gibberellic acid may actually increase the content of the enzyme.

The direct outcome of such a suppression of IAA-oxidase activity should be a rise in endogenous auxin content in gibberellin-treated tissue. This effect had been observed in extracts of *Rhus typhina* tissue (Nitsch, 1957), in the diffusible auxin from pea and sunflower tissue (Kuraishi & Muir, 1962) (see Figs. 77 D & E), from a rosette plant, *Centaurea cyanus* (Kuraishi & Muir, 1963), from savoy cabbage apices (Skytt-Andersen & Muir, 1969) and from tomato flowers (Sastry & Muir, 1963). However gibberellic acid greatly increases the auxin content of both tall and dwarf varieties of maize, irrespective of whether their growth was stimulated or not, which suggests that the two effects are unrelated (Bouillenne & Leyh, 1962).

These augmented auxin levels may not arise from a suppression of IAA-oxidase since no such effects could be detected by Kuraishi & Muir (1964*c*) in dwarf pea stems and gibberellic acid appeared not to inhibit the decarboxylation of IAA-1-^{14}C in *Coleus* (Valdovinos, Ernest & Henry 1967). However it does increase the production of water-soluble auxin from tryptophan in dwarf pea tissue (Kuraishi & Muir, 1964*c*) and in cell-free preparations of *Avena* coleoptiles (Lantican & Muir, 1967; Muir & Lantican, 1968). It enhances the decarboxylation of tryptophan-1-^{14}C in *Coleus* (Valdovinos, Ernest & Perley, 1967), sunflower (Valdovinos & Ernest, 1967) and *Avena* coleoptiles (Valdovinos & Sastry, 1968). The inhibitor dimedon, which blocks auxin synthesis at the indol-3yl-acetaldehyde stage, also blocks gibberellic acid promotion of growth in the coleoptile (Valdovinos & Sastry, 1968). All this points to a gibberellic acid promotion of auxin synthesis via the tryptamine pathway.

It is possible therefore that the enhanced IAA content of gibberellin-treated plants may arise from one or other of two causes, nevertheless there are still some auxin/gibberellin growth interactions which cannot be thus explained. For example one of the major criticisms that have been levelled at this 'auxin sparing' action of gibberellin is that the synergism is evident with the synthetic auxins (see previous data of Jackson & Prosser, Brian & Hemming, and Maeda, pp. 382) which are not metabolized by IAA-oxidase. However the question is not entirely closed since there are reports that gibberellic acid may inhibit the decarboxylation of 2,4-D applied to plants of

Phaseolus vulgaris (Ashton, 1958). Then there are the observations that gibberellic acid may still exert a stimulating action on plant growth even in concentrations of IAA which are supra-optimal. This has been seen in pea epicotyl tissue (Hillman & Purves, 1961) and in *Avena* coleoptile and mesocotyl tissue (Ng & Audus, 1964); an 'auxin-sparing' effect should have increased IAA action and further inhibited growth. Finally of course there are a number of cases in which gibberellic acid and auxin are not synergistic in their action. Thus gibberellic acid may counter the action of auxin in the inhibition of bud growth and the promotion of root formation on cuttings (Kato, 1958); IAA may also virtually eliminate the stimulating effects of gibberellic acid on the growth of the first leaf of *Avena* seedlings (Ng & Audus, 1965). Although high concentrations of the antiauxin PCIB (see p. 384) can inhibit gibberellic acid action in stimulating glucose production by barley endosperm, yet this inhibition is not relieved by the application of auxin (Cleland & McComb, 1965), which indicates that auxin is not involved in the action of gibberellic acid on endosperm tissue.

All these and many other diverse and complex interactions of auxins with the gibberellins, and indeed with other growth-regulating substances, show that there is no simple underlying interdependence of the two hormones. In the first place it is most likely that the gibberellins will produce different metabolic responses in different tissues. On the other hand any particular growth or developmental process may be subject to the regulatory action of a *balance* of several hormones, whose points of action may be quite independent one of the other but which appear to interact because of the obligate mutual association of each component in the total growth system.

Hormones and the Control of Protein Synthesis

The multiplicity of hormone action *in vivo*, and the lack of any clear-cut effects of hormones in physiological concentrations on *in vitro* enzyme systems, has led to the concept of an *in vivo* control of protein, and hence of enzyme synthesis. Thus protein synthesis is a normal concomitant of cell extension. Even in those extending cells where a net accumulation of protein does not seem to occur, a rapid metabolic turn-over of protoplasmic proteins is probably going on with accompanying shifts in the balance of enzymes (Robinson & Brown, 1952, 1954). Such changes undoubtedly play a role in extension-growth, and cell extension hormones may be playing a part in their control. Early work had suggested that auxins have no effect on net protein synthesis. Burström (1951) showed that in roots, under a wide range of growth conditions, which included stimulation with the 'antiauxin' PCIB and inhibition with IAA, the nitrogen content *per cell* remained virtually constant. Again IAA appeared to have no effect on the rate of ^{14}C-labelled-glycine incorporation into the proteins of coleoptile segments of maize or oat grown upon solutions of this amino acid (Boroughs & Bonner, 1953). Nevertheless it is now clear that auxins may regulate the synthesis (or activity) of certain discrete enzymes. For example, auxin treatment of coleoptiles increased their malic and alcohol dehydrogenase activities (Berger & Avery, 1943), a probable outcome of an enhanced enzyme synthesis; in the same tissue it induced the formation of cellulose synthetase (Hall & Ordin, 1968). In tobacco pith in tissue culture auxin may augment the synthesis of ascorbic acid oxidase, of invertase and of pectin methylesterase (Newcomb, 1954) and may act specifically in enhancing a particular isoenzyme of peroxidase (Galston *et al.*, 1968). It may also promote invertase synthesis in sugar-cane tissue (Glasziou *et al.*, 1966). On the other hand IAA can inhibit enzyme synthesis, e.g. invertase in artichoke tuber tissue (Edelman & Hall, 1964) and peroxidase in sugar cane (Glasziou *et al.*, 1968). Clearly these effects of auxin are not expressions of broad modulations of protein synthesis but are manifestations of a much more subtle control of particular enzymes.

We have already seen that one of the most dramatic effects of gibberellin is the induction of the starch-hydrolysing enzyme, α-amylase, in the aleurone layer of cereal grain. Gibberellin is indeed the natural embryo factor which, for a long time, has been known to be responsible for the induction of starch-reserve hydrolysis in the endosperm during normal germination processes. However gibberellic acid also induces other enzymes in the aleurone layer, and these include other carbohydrate hydrolases (MacLeod *et al.*, 1964), proteases (Jacobsen & Varner, 1967) and ribonuclease (Brawerman, 1966).

Ethylene, the proposed hormone which is such a powerful inducer of abscission of plant organs also enhances protein synthesis in stems and petioles of sensitive plants, in particular the synthesis of the cellulose-hydrolyzing enzyme cellulase in the cell layer where separation occurs in bean, cotton and *Coleus* (Abeles & Holm, 1966; Horton & Osborne, 1967).

The undoubted participation of hormones in the control of cell and organ differentiation, as is most strikingly exemplified by the action of gibberellins in the switch from vegetative to reproductive development already described for higher plants in Chapter X, further implies that their control mechanisms involve the regulation of shifts in enzyme patterns which, according to modern concepts, determine the changes in direction of cell differentiation.

It is little wonder then that in the last few years interest in hormone mechanisms has centred on their possible influence on enzyme synthesis and since protein (and hence enzyme) structure is determined by the nucleic acid systems of the cell, most research effort has been directed towards the study of the impact of plant hormones on the nucleic acid metabolism. At this point therefore we should pause to outline what we believe to be the protein synthesis system of the cell, as a base for our further consideration of hormone action.

The chromosomal material which contains the hereditary information and which ultimately therefore determines the exact biochemical and structural organization of the cell is deoxyribonucleic acid (DNA). Basically this molecule consists of a backbone of alternating phosphate and deoxyribose sugar units with side chains each of one of four organic bases (adenine, cytosine, thymine or guanine) attached to each of the sugar units. The whole molecule is organized as a double helix, held together by hydrogen bonding between paired bases, adenine always pairing with thymine and guanine with cytosine. The exact replication of this molecule, on which the continuing and perfect reproduction of the hereditary characteristics of the organism depends, is brought about by a mechanism not yet completely understood; the bonding between the base pairs is broken and the two strands of the double helix come apart, allowing two new double helices to be formed by the building up of two new strands, again with complementary base-pairing, on the old ones that have been thus exposed as templates. This process is known as *replication*.

The genetic information of DNA, which is in the form of the sequence of bases in the helical strands, has to be transmitted to the sites of protein synthesis in the cell. These sites, the ribosomes, are constituted by extremely small two-component granules, 15 to 20 nm in diameter, composed of protein and another nucleic acid, ribonucleic acid (RNA). The chemical structure of this acid is similar to that of DNA but differs in that the sugar is ribose and the base thymine is replaced by another similar base uracil. The characteristics of proteins, and hence of enzymes and their functions, are determined by the sequence in which their component amino acids are linked together in the protein chain and this in turn is ultimately determined by the base sequence in the DNA molecule. This information is transmitted from DNA to the ribosomes by another RNA, called messenger RNA (mRNA), which is a very long single-stranded molecule. This is synthesized, by a process similar to that of the replication of DNA, by base pairing on the separated DNA strands, but this time uracil of RNA pairing with adenine of DNA. The genetic information is thus transferred to the base sequences of the mRNA, which itself is to form the jig on which the proteins are to be formed by the appropriate sequential linking of the twenty amino acids. This transfer of genetic information from DNA to mRNA is known as *transcription.*

This mRNA moves out from the nucleus and becomes linked to the ribosomes with which it will co-operate in the process of protein synthesis.

A third kind of RNA is transfer RNA (tRNA). These molecules are much smaller than those of mRNA and are also single stranded but the strands are folded back on themselves, and assume, by appropriate partial base pairing, variously branched double-helical configurations somewhat reminiscent of clover leaves. These tRNA molecules are specifically concerned with the marshalling of the amino acids on the mRNA template and to this end each of the twenty amino acids has its own exclusive tRNA molecule, with which it links specifically. Furthermore each tRNA has, at one loop end of its folded molecule, three particular unpaired bases, which are specific in their nature and in their sequence to the related amino acid. This triplet of bases is known as the *anticodon.* The genetic information on the mRNA molecule is organized in a sequence of 'words', each word being composed of a triplet of bases, each being called a *codon.* The codon is specific in the nature of its bases and in their sequence for a particular amino acid and is complementary to the base sequence of the corresponding anticodon of the mRNA. In the act of protein synthesis the amino acids are first 'picked up' by their specific tRNA molecules. Then, starting with a specific codon on the mRNA molecule associated with the ribosomes, tRNA molecules with their amino acids attached begin to assemble in sequence, anticodon associating with its specific codon. In similar sequence as they assemble, amino acid is linked to amino acid by the elimination of water between -COOH and $-NH_2$ groups of neighbours and a peptide molecule is put together, the sequence of the amino acids following precisely the sequence of coding triplets of bases on the mRNA. At the same time tRNA is released from the ribosome to 'pick up' further amino acids from the cytoplasm. This stage of protein synthesis is known as *translation.* If hormones are to regulate protein synthesis it is most likely that they will do so at one or more points in the overall sequence, i.e. at the replication, the transcription or the translation stage. We will now consider what evidence there is for hormone action at these stages.

Effects of Hormones on Nucleic Acid Metabolism

(A) LEVELS

The first hint that nucleic-acid metabolism was influenced by hormones came from the work of Skoog on callus cultures of tobacco pith (see Chapter V). Low concentrations of auxin which promoted cell division favoured DNA synthesis while high concentrations caused cell expansion and a correlated increase in RNA. Still higher concentrations blocked both growth and nucleic acid accumulation. Skoog (1954) suggested that the auxin regulated the DNA/RNA ratio in the cells and that this determined the growth pattern, i.e. cell division or cell expansion. Since that time overwhelming evidence has accumulated to show that growth hormones and regulators alter nucleic-acid levels in a very wide variety of organs and tissues. In intact plants the application of gibberellic acid cause an increase in the RNA and DNA levels in dwarf pea (Giles & Myers, 1966) and in *Lens* (Nitsan & Lang, 1966). In *Lemna minor* the cytokinin benzyladenine greatly promoted frond growth and also the synthesis of DNA and RNA. On the other hand abscisic acid inhibits frond growth and nucleic acid synthesis and these suppressive effects are counteracted by benzyladenine (van Overbeek *et al.*, 1967). The synthetic auxin 2,4-D in concentrations above optimal for growth is highly toxic to plants; at these toxic application levels it causes a very great accumulation of nucleic acids and proteins in treated plants, particularly in regions where the 2,4-D stimulates the unco-ordinated proliferation of cells and the consequent disorganization of tissue. The increase in RNA is proportionately much greater than DNA; this dislocation of the DNA/RNA balance could be an important contributing factor in the growth dislocations associated with the toxic action of 2,4-D (see review by Hanson & Slife, 1969). Similar responses to growth substances have been observed by many workers in excised organs and tissues; for example gibberellic acid increases the DNA and RNA content of plugs of potato tuber in the resting state and all three hormone groups, auxins, gibberellins and cytokinins, prevent senescence in a variety of detached leaves and at the same time prevent the decline in DNA and RNA and protein levels which accompany the normal progress of senescence (see reviews by Cherry, 1967*a*; Simon, 1967; Wollgiehn, 1967 and Woolhouse, 1967). The question which now poses itself is whether these effects are general modulations of nucleic acid metabolism or whether they are specific to one or more processes in the over all system.

(B) EFFECTS ON TRANSCRIPTION: THE SYNTHESIS OF RNA

In the last few years a great number of independent investigators have demonstrated that plant growth substances directly affect the synthesis of RNA in experiments which studied the incorporation of radio-tracers such as ^{14}C-labelled bases or ^{32}P into the RNA molecule. Thus auxins at growth-promoting concentrations stimulate such incorporation into the RNA of soybean hypocotyl (Key & Shannon, 1964), wheat coleoptiles (Truelsen & Galston, 1966), pea epicotyls (Davies *et al.*, 1969) oat coleoptile (Tester & Dure, 1967; Masuda & Kamisaka, 1969). These effects seem not to be the indirect result of growth increases since they take place, at least in the oat coleoptile, even when growth is prevented osmotically by a concentrated external mannitol solution (Masuda *et al.*, 1967). These are true auxin effects since they are prevented by antiauxins, e.g. *trans*-cinnamic acid (Masuda & Tanimoto, 1967). Gibberellic acid, which induces enzyme synthesis in barley grains, also stimulates the incorporation of labelled precursors into RNA of the aleurone layer (Chandra & Varner, 1965) and it seems logical that the promotion of RNA synthesis is the essential forerunner of the enzyme production. Auxins, gibberellins and cytokinins all enhance the incorporation of precursors into the RNA of senescing tissue (see p. 390). For auxin this has been shown in pods of *Phaseolus* (Sacher, 1967*a*) and in *Rhoeo* leaves (Sacher 1967*b*); gibberellic acid has a similar action in leaves of *Taraxacum* (Fletcher & Osborne, 1966) and cytokinins do so in many species (see review by Key, 1969). At the same time abscisic acid, which will actively promote the senescence of leaves, inhibits the incorporation of ^{32}P into RNA in leaf discs of *Taraxacum* and *Raphanus* (Wareing *et al.*, 1968). Whether the effects of these hormones on senescence are the direct result of the control of RNA, and hence protein synthesis is still a moot point. For example, senescence is accompanied by a marked increase in the content of enzymes hydrolyzing RNA (i.e. ribonuclease) and proteins (proteases) and it has been shown that kinetin markedly suppressed the activity of these enzymes in senescing barley and tobacco and oat leaves (Srivastava & Ware, 1965; Balz, 1966; Shibaoka & Thimann, 1970) and that of ribonuclease associated with chromatin from those leaves (Srivastava, 1968). A similar suppression of ribonuclease by NAA has been observed in *Rhoeo* leaves and an opposite effect was seen in the case of abscisic acid (de Leo & Sacher, 1969). Since protein and RNA turnover goes on, even in senescing tissue, such suppression of the degradation processes would have the effect of increasing the specific activity of proteins and RNA since less of the newly-formed molecules would be broken down under hormone action. Such results imply that the promotion of RNA synthesis by hormones may only be apparent, the real effect being the suppression of degradation. This possibility has received further experimental support from the results of protein pre-labelling experiments on senescing maize leaves. Such leaves were treated with ^{14}C-labelled leucine which was incorporated into protein, the breakdown of which could be subsequently followed from the loss of ^{14}C. Treatment of such protein-labelled leaves with 6-benzyladenine significantly reduced protein degradation (Tavares & Kende, 1970). A similar situation occurs in relation to abscisic-acid action which inhibits the gibberellin-induced production of α-amylase in barley endosperm, a synthesis probably regulated by RNA. Leshem (1971) has produced evidence to suggest that abscisic acid produces this inhibition not by blocking the synthesis of the appropriate RNA but by inducing the production of ribonuclease.

However a direct effect on RNA synthesis seems likely from studies of isolated plant nuclei where in some species auxins, gibberellins and cytokinins promote the incorporation of radioactive precursors into the RNA molecule. Although some criticisms have been raised that in such experiments, lasting up to 12 hours and carried out under non-sterile conditions, the incorporation of tracer into the RNA of contaminating bacteria may have invalidated the results, a real effect on nuclear RNA synthesis still seems likely from short-term (15 minute) measurements, during which significant bacterial growth could not have occurred (Roychoudhry *et al.*, 1965). Similarly in intact tissue the incorporation of tritium-labelled uridine into nuclear RNA can be followed by the autoradiography of thin sections and microscopic observation to locate the site of incorporation. In such studies bacterial contaminants are ruled out. In this way it has been shown that gibberellic acid promotes and abscisic acid inhibits the synthesis of RNA in the nuclei of dormant 'eyes' of potato tuber (Shih & Rappaport, 1969).

The apparent obligate linkage of the growth effects of hormones to the synthesis of RNA has been shown in many studies with specific inhibitors of RNA transcription. The two most commonly used are actinomycin-D, an antibiotic metabolite of a soil micro-organism which complexes with DNA and prevents its functioning as a template in the synthesis of RNA, and 6-methylpurine, an analogue of the purine bases of RNA, which also specifically blocks its synthesis. A wide range of growing tissues have now been studied and in most, actinomycin-D blocks growth at concentrations which also block the synthesis of RNA, indicating that a continuing production of RNA is essential for growth. In most tissue the effects of actinomycin-D concentrations on the degree of inhibition of growth and RNA synthesis run very closely parallel. With both responses there is a lag after application before inhibitions set in and that for RNA synthesis is shorter than that for growth, suggesting that the inhibition of RNA synthesis is the forerunner of the growth inhibition (Key *et al.*, 1967). This is supported by results of experiments where auxins and actinomycin-D were not applied simultaneously to tissues. In artichoke tuber discs, if actinomycin-D is applied 24 hours after the auxin, there is no inhibition of the growth promotion due to auxin for the next 48 hours (Nooden, 1968). Presumably the RNA essential for growth and induced by auxin treatments in the first 24 hours persists over the 48 hour period and its action is not affected by actinomycin-D.

The increase in the plasticity of the cell-wall, which we have seen as the probable immediate cause of the promoted extension (see p. 362) is also prevented by actinomycin-D and again the degree of blockage runs parallel with that of the inhibition of RNA synthesis. This holds for *Avena* coleoptiles, where both inhibitions are only partial and for soybean hypocotyl and maize mesocotyl where inhibitions are complete (Coartney 1967). However sometimes concentrations of actinomycin-D, which partially inhibit *gross* RNA synthesis, have no effect on endogenous growth but inhibit the growth stimulating action of auxin (see Key, 1969, p. 463). Such results are very puzzling but tempt the suggested explanation that there may be types of RNA specific for growth and auxin action and that the synthesis of these RNA species may be differentially affected by actinomycin-D. Results of Masuda (1967), again with artichoke tuber tissue, support the suggestions. Investigations of the effects of the auxin (2,4-D), gibberellic acid and kinetin on growth showed that pretreatment of the discs for 15 hours with gibberellic acid caused a much greater response to 2,4-D applied subsequently than if the gibberellic acid had been applied with the 2,4-D after 15 hours pretreatment with water. This prompted the suggestion of the involvement of two processes in the expansion growth of tuber cells; the first takes place during the pretreatment 'ageing' in the absence of auxin, is promoted by gibberellins and establishes a 'cell state' which allows subsequent auxin-induced expansion; the second is the auxin-induced expansion itself. Masuda found that the base analogue 8-azaguanine would inhibit auxin-induced growth if applied during the ageing period but had little direct effect if applied with auxin in the expansion period. With actinomycin-D the situation was reversed, little inhibition occurring in the ageing phase but a much more marked action being seen in the expansion phase. Finally a study was made of the effects of 2,4-D and actinomycin-D on the various kinds of RNA synthesised during the ageing and expansion periods. Extracted RNA was separated into its major components (in terms of molecular size) by chromatographic fractionation on columns of methylated albumin on kieselguhr (MAK). 2,4-D was shown to promote the synthesis of both soluble and ribosomal RNA whereas actinomycin-D inhibited the synthesis of ribosomal RNA only. From these and other similar results, e.g. on yeast cell expansion (Shimoda *et al.*, 1967) Masuda concluded that the preparatory phase for auxin-induced cell expansion in some way depended on the lighter type of RNA (soluble RNA) while the cell expansion itself was linked to the ribosomal RNA synthesis. Now this is a highly speculative theory but it does point to the great complexity of the situation and to the possibility of specificity in the RNA relationships with growth and hormone action.

Direct attempts to identify the species of RNA most affected by hormone action, by a combination of the labelling of the synthesized RNA with radioactive atoms and their fractionation by MAK column chromatography and differential centrifugation, have not as yet clarified the situation. The synthesis of all types of RNA seems to be promoted by hormones, e.g. by auxins in the *Avena* coleoptile (Hamilton *et al.*, 1965), soybean hypocotyl (Ingle & Key, 1965; Key *et al.*, 1967) (see Fig. 78), pea epicotyl (Trewavas, 1968) etc. However certain preferential actions have been observed, predominantly on ribosomal RNA. This was shown for the action of IAA and 2,4-D in excised soybean hypocotyl tissue where the stimulatory effects of low concentra-

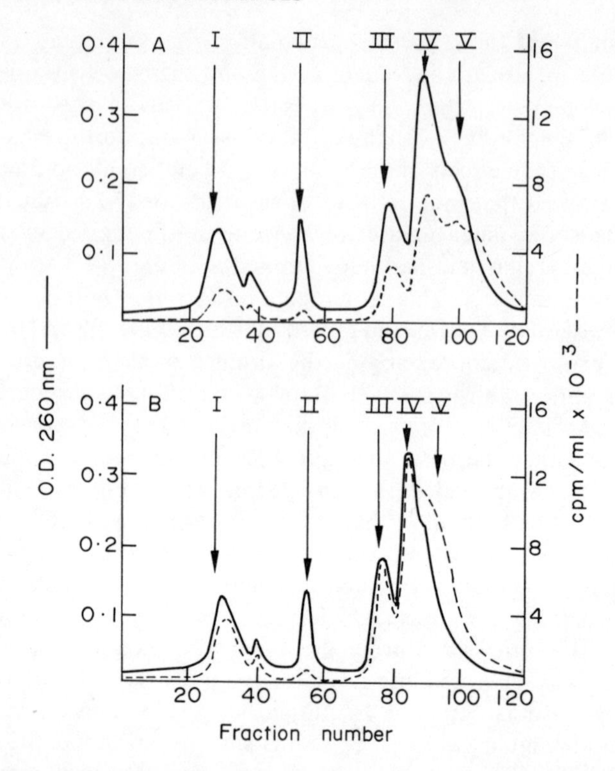

Fig. 78. The effects of 2,4-D treatment (10 μg/ml) on the synthesis of RNA in the elongating hypocotyl of soybean. The hypocotyl was treated for 8 hours with ^{32}P-orthophosphate. The RNA was then extracted, the various fractions separated by gel-filtration (MAK columns) and the total RNA (solid lines; estimated by optical density at 260 nm) and the newly-synthesized RNA (dotted lines; estimated from radioactivity) determined.

(A) Control plants untreated with 2,4-D. (B) 2,4-D-treated plants. The nature of the various fractions are as follows:

I = Soluble RNA
II = DNA
III = 18S-RNA
IV = 25S-RNA
V = D-RNA

(From Key & Ingle, 1968; by kind permission of the authors.)

tions and the inhibitory effects of high ones ran parallel (Key & Shannon, 1964; Ingle & Key, 1965). Such preferential synthesis has been shown to take place in the nucleus (presumably in the nucleolus) of this tissue (Cherry, 1967*b*). A most interesting observation is due to Trewavas (1968). The synthesis of the protein on the messenger-RNA template involves the simultaneous operation of a number of ribosomes on one and the same messenger-

RNA molecule. Aggregates of such ribosomes are called polysomes and Trewavas has found that the incubation of pea epicotyl tissue with IAA increases the number of polysomes that can be isolated. This resulted from an increased synthesis of ribosomes rather than an increased production of messenger RNA; an absolute increase in the number of ribosomes per unit volume of cytoplasm of tobacco-pith cells in culture has also been observed to arise from auxin treatment in electron microscope studies (Nitsch, 1968).

However the possibility of a specific effect on messenger RNA cannot be ignored. Undoubtedly auxin does promote the synthesis of messenger RNA and in pea internode tissue a very rapid labelling, predominantly of messenger RNA, took place when the tissue was incubated in orthophosphate-^{32}P in the presence of 2,4-D (Masuda, 1968). 2,4-D does greatly increase the synthesis of an RNA in soybean hypocotyl tissue, which closely resembled messenger RNA but which is characterised by being rich in the nucleotide adenosine monophosphate (AMP) (Tester & Dure, 1967; Key *et al.*, 1967; Key & Ingle, 1968). Moreover Key & Ingle (1968) regard this AMP-rich RNA as being specifically related to the promoting action of auxin in cell elongation. They base this hypothesis on the fact that the base analogue 5-fluorouracil markedly inhibits the synthesis of most RNA fractions (including ribosomal RNA) at concentration which have no effect on extension growth or on the synthesis of the AMP-rich RNA fraction. Furthermore the action of actinomycin-D on auxin-induced extension runs strictly parallel with the suppression of AMP-rich RNA-synthesis. They regard the effects of auxin on gross RNA synthesis as being related to the regulation of cell division, not cell elongation.

However the role even of this apparently specific RNA in cell elongation could be called into question on the grounds of discrepancies in the speed of action of auxin on growth and nucleic acid synthesis. Thus changes in RNA under auxin action are usually regarded as taking hours to become established (Key, 1964). However auxin acceleration of extension growth can be detected within a few minutes (Ray & Ruesink, 1962; Nissl & Zenk, 1969; Burström *et al.*, 1970) and the acceleration of streaming in *Avena* coleoptile cells in an even shorter time (Sweeney, 1941). Using a new and sophisticated technique involving the measurement of the resonance frequency of segments of pea internodes and roots subjected to forced mechanical oscillations, Burström *et al.* (1970) have shown that changes in the elastic modulus of cell-walls can be demonstrated within 2–3 minutes of auxin treatment. This suggests that auxin action is much more likely to be an immediate physical one (e.g. on the properties of cell membranes, see p. 359) rather than chemical ones in a complex biosynthetic sequence involving nucleic acids. However in bacteria the activation of the gene and the consequent synthesis of messenger RNA and protein takes about 2 minutes

(Kaempfer & Magasanik, 1967) and there is one report (Masuda & Kamisaka, 1969) that a stimulated incorporation of ^{14}C-labelled uracil into RNA of *Avena* coleoptile cells can be detected in 10 minutes. The point cannot therefore be settled as yet.

The possible mechanisms of the above actions of hormones on RNA production will be considered later but one possible complicating phenomenon should be mentioned at this point. This is that in the promotion of the synthesis of the enzyme invertase in sugar-cane tissue by NAA and gibberellic acid, the synthesis of new RNA seems to play no part (Gayler & Glasziou, 1969). Thus tissue treated with actinomycin-D or 6-methylpurine to block RNA synthesis still responded to hormones by producing more enzyme. The suggestion is made that these hormones act in this system by stabilizing the RNA already synthesized and thus indirectly maintaining enzyme synthesis which declines in control untreated tissue.

(C) HORMONES AND DNA REPLICATION

We have already seen in the pioneering work of Skoog that low concentrations of auxins favour DNA synthesis in cultured pith tissue. Similar responses to auxins have been shown in the promotion of the incorporation of ^{32}P and thymine-2-^{14}C into the DNA of coconut milk nuclei and of pea internodes (Roychoudhury & Sen, 1964). However most attention in this area has been given to gibberellic acid, whose action is undoubtedly associated with DNA synthesis. Such a conviction comes from the results of experiments with specific inhibitors of DNA synthesis, particularly 5-fluorodeoxyuridine (FUDR), an analogue of a DNA nucleotide. Thus Nitsen & Lang (1965) showed that the gibberellic acid-stimulated elongation of lettuce hypocotyl could be inhibited to the extent of 50 per cent by FUDR concentrations as low as 3×10^{-7}M. This inhibition could be completely reversed by the base thymidine. Both cell elongation and cell division seemed to be inhibited. Similar results were obtained with lentil hypocotyl where extension growth was almost exclusively due to cell elongation (Lang & Nitsen, 1967). Gibberellic acid also stimulated DNA synthesis and there was a correlated increase in ribosomal RNA content which was presumably dependant on the DNA synthesis; both effects were prevented by FUDR. Lang & Nitsen saw in this response a possible distinction between the actions of gibberellins and auxins on cell extension, the first being operated through DNA replication (on which subsequent RNA synthesis could depend) and the second on RNA transcription on the DNA template. These ideas were supported by observations on the responses of the *Avena* coleoptile to gibberellic acid and to auxin. In the initial phase of coleoptile growth, cell extension is most sensitive to gibberellins whereas at a later stage cell extension is most sensitive to auxin (Wright, 1961). Lang & Nitsen showed that the first gibberellin-sensitive phase was inhibited by FUDR, which was ineffective on the auxin-sensitive phase. A similar inhibition of dark-induced cell extension (etiolation) by FUDR has been demonstrated in *Kalanchoë daigremontianum* and *Sinapis alba* (Bopp, 1967). FUDR also blocks the promotive action of 2,4-D and its synergism by gibberellic acid on the growth of artichoke tuber tissue and also the correlated synthesis of DNA and RNA (Kamisaka & Masuda, 1970). It seems therefore that cell extension may also be dependant on DNA replication, at least in certain phases of growth or under certain conditions.

The situation is clouded however since it is not certain how much cell division, and hence associated nuclear division (mitosis) and DNA replication, were taking place in the excised tissue used for experimentation. It is just possible (see Key, 1969, p. 465) that gibberellic acid, which does induce mitotic activity in stems, might thus stimulate indirectly the growth of other cells in their extension phase, by some factor arising from the stimulated mitotic activity. Even so the situation remains complex since it has been shown in soybean seedlings deprived of their roots, thus removing a possible major source of endogenous gibberellins and cytokinins (see p. 316), that meristematic activity, extension growth and DNA synthesis were all inhibited. The application of gibberellic acid stimulated DNA synthesis in the apex and also restored extension growth but FUDR inhibited this DNA synthesis without inhibiting growth or gibberellin action (Holm & Key, 1969). The exact relationship between DNA replication, extension growth and gibberellin action still remains obscure.

It is perhaps important to mention at this point that the natural growth inhibitor, abscisic acid, seems also to operate via DNA synthesis. Studies with the small water-weed *Lemna minor* (van Overbeek *et al.*, 1967) have shown that its growth inhibition can be overcome by the cytokinin benzyladenine, but not by auxin or gibberellin. The correlated inhibition of DNA synthesis can also be overcome by benzyladenine. These workers suggest possible opposing actions of abscisic acid and of cytokinins on the configuration, and hence activity, of the enzyme DNA-polymerase, which controls DNA replication. In the aleurone layer of the barley grain abscisic acid similarly prevents the gibberellin induction of the enzyme amylase and this is partially overcome by further gibberellin application, (Chrispeels & Varner, 1966). However these authors were of the opinion that the action of both gibberellin and abscisic acid in this system were likely to be at the levels either of transcription or of translation. From kinetic considerations and comparisons with the action of FUDR and cycloheximide, an inhibitor of peptide-bond formation in protein synthesis, the action of abscisic acid in the inhibition of bean stems has been regarded as taking place at the level of translation (Walton *et al.*, 1970).

(D) THE MOLECULAR MECHANISMS OF HORMONE ACTION IN NUCLEIC ACID AND PROTEIN METABOLISM

There is now no doubt in the minds of biologists that *all* the information necessary for any aspect of the control of growth and development in the organism is present in the DNA of all cells. At any particular phase of cell development only a portion of this information is being called upon to regulate the synthesis of the specific enzymes needed for that phase. Thus during the course of development it is believed that the genes – the DNA templates – are 'switched on' and 'switched off' in an appropriate sequence by the operation of repressor molecules which mask the template and effectively prevent their employment as 'jigs' for the synthesis of RNA. The strongest candidates for this role of gene repressors are the histones, the basic protein molecules which are essential components of the chromatin, the DNA-containing material of the chromasomes. The experimental evidence for such a rôle in plant tissues is mainly due to Bonner and his colleagues (see Bonner, 1965). Thus chromatin extracted from plant cells, in the presence of appropriate enzymes, will serve as templates for the *in vitro* synthesis of messenger RNA if the nucleotide 'building bricks' are supplied. This messenger RNA, in the presence of ribosomes, amino acids, the requisite transfer RNA molecules and the appropriate enzymes will synthesize proteins *in vitro*. By using these systems Bonner has shown that the chromatin extracted from pea cotyledons will 'programme' the synthesis of proteins with a high content of globulins, which are characteristic of the storage proteins of those organs. On the other hand chromatin from pea embryos 'programmed' the synthesis of proteins which contained no significant amounts of globulin, again resembling the situation in the living cell from which the chromatin had been taken. Obviously different genes, and therefore different parts of the total chromatin, were being used for the RNA transcription in these two preparations. Later Bonner showed that removal of the histone from the chromatin of the apical bud nuclei greatly increased RNA synthesis and that the protein synthesized on this messenger RNA had the high globulin content of the cotyledon protein. It seemed therefore that the DNA coded for globulin was repressed in the embryo and apical bud chromatin by a masking histone.

Criticisms have been raised against such an hypothesis of histone action because, it has been pointed out, the variety of histones known to occur is too small to block, with the necessary degree of specificity, the almost astronomical number of genes involved. However Huang & Bonner (1966) have demonstrated a histone-RNA complex in chromatin and it is possible therefore that the RNA component of this complex may confer the necessary specificity to allow it to act as the postulated suppressor molecule.

One of the possible actions of hormones then could be to promote the de-repression of the gene, particularly where hormone-directed development is concerned (e.g. the flowering switch, p. 246). For this it would seem most probable that the hormones would have to interact in some chemical way with the DNA-repressor complex, causing it to dissociate and thus expose the appropriate segment of the DNA for template functioning. But we then find ourselves faced with the question of specificity, of why only some genes are 'turned on' and not all. With IAA this is a real problem but we have already seen indications of some specificity in the score or so gibberellin molecules, whose actions differ from plant to plant and process to process (see Chapter IV). Nevertheless there is some experimental evidence for such a de-repression.

Thus Fellenberg (1966) has shown that the induction of tumours in *Kalanchoë daigremontianum* by *Agrobacterium tumefaciens* is inhibited by histones prepared from calf thymus gland, an effect which is not operating via the growth or virulence of the infecting organism. This suggests a suppression of the genes responsible for the release of unorganized tumour growth. He has also shown (Fellenberg, 1969) that during the induction of root formation by IAA in pea epicotyls there is a change in the properties of the nucleoproteins (chromatin) and the DNA extracted from it. Both show a considerable reduction in their melting points under the action of IAA during the first 48 hours of treatment. This he suggests is due to the loosening of the bonds that bind DNA to histones and of the individual strands of the DNA helix to each other; this is presented as evidence for the gene de-repressing theory of hormone action. In studies of the binding action of histones prepared from pea plants on a variety of auxins Venis (1968) has shown that, although overall binding is low, there is a definite correlation between the degree of binding of particular auxins and their physiological activities. There were indications that such binding might produce conformational changes in histone molecules, and this could be the basis of the gene de-repressing action of auxins.

Again gibberellic acid can act as a dormancy-breaking agent in seeds by inducing the synthesis of enzymes necessary for germination. In hazel (*Corylus avellana*) seeds this action of gibberellin has been well established. Chromatin isolated from hazel seeds can be used as a template for *in vitro* RNA synthesis. Incubation of such chromatin with excess RNA polymerase (from the bacterium *Escherichia coli*) and the necessary nucleotides will produce RNA, the rate of production of which has been taken to indicate the 'template availability' of the chromatin. Treatment of hazel seeds with gibberellic acid has been shown to increase this 'template availability' of the chromatin extracted from it (Jarvis *et al.*, 1968) a response which is compatible with the gene-de-repression hypothesis. On the other hand gibberellic acid appears to augment the RNA-polymerase content of chromatin extracted from the internodes of light-grown dwarf pea plants without affecting the DNA-template availability (McComb *et al.*, 1970).

Direct evidence of the chemical interaction of hormones and nucleic acids is conflicting. There seems to be no indication whatever that the gibberellins combine chemically or even are associated physically with any nucleic acid, despite the evidence of these gene-de-repressing actions. There is some evidence that IAA can become attached to RNA. For example ^{14}C-labelled IAA, fed to pea stem tissue, seems to form a 'complex' with RNA, from which unchanged IAA can be subsequently recovered (Bendaña *et al.*, 1965). However IAA-degradation products are undoubtedly utilized in the synthesis of new RNA, a situation which somewhat clouds the issue. More data are needed before the existance of a 'functioning' RNA-IAA complex can be regarded as established.

Another possible action of growth substances could be in the actual transcription itself (see Fig. 79). We have already mentioned the theory of van Overbeek *et al.*, (1967) in which abscisic acid and cytokinins were proposed as controlling growth via the activity of DNA polymerase. A similar hormone action might control overall transcription via the activity of RNA polymerase, although again no great specificity of action would be expected. The activity of this enzyme in hazel seed chromatin preparations has been shown to be higher from gibberellin-treated than from control material (Jarvis *et al.*, 1968). The same situation was shown in soybean hypocotyl treated with 2,4-D (O'Brian *et al.*, 1968).

When we turn to cytokinins the evidence points to their action at the translation level by chemical association with RNA, in this case transfer-RNA. As we have seen cytokinins are adenine derivatives and are thus close relatives of the nucleic acid bases. Recently such base derivatives with high cytokinin activity have been isolated from transfer-RNA of several organisms. Thus the extremely active cytokinin 6-(γ,γ-dimethylallylamino)-purine (isopentyladenine or IPA) has been shown to be the naturally-occurring base in the nucleotide neighbouring the anticodon triplet in the transfer RNAs specific for serine and tyrosine in yeast (Biemann *et al.*, 1966; Zachau *et al.*, 1966; Madison *et al.*, 1967). It also occurs in the transfer RNA of the bacterium *Corynebacterium fascians* (Matsubara *et al.*, 1968). Another bacterium, *Escherichia coli* contains the active cytokinin 6-(3-methyl-2-butenylamino)-2-methylthio-9-β-D-ribofuranosylpurine in transfer RNA (Burrows *et al.*, 1968). Again ribosylzeatin, a nucleotide derived from the natural cytokinin zeatin (see p. 146) has been obtained by hydrolysis of transfer RNA of immature maize grain, garden peas and spinach (Hall, 1967, 1968). Since it appears that the nature of the nucleotides neighbouring the anticodon triplet in transfer RNA may play a part in the proper functioning of this molecule in protein synthesis (Fittler & Hall, 1966) perhaps in its linking

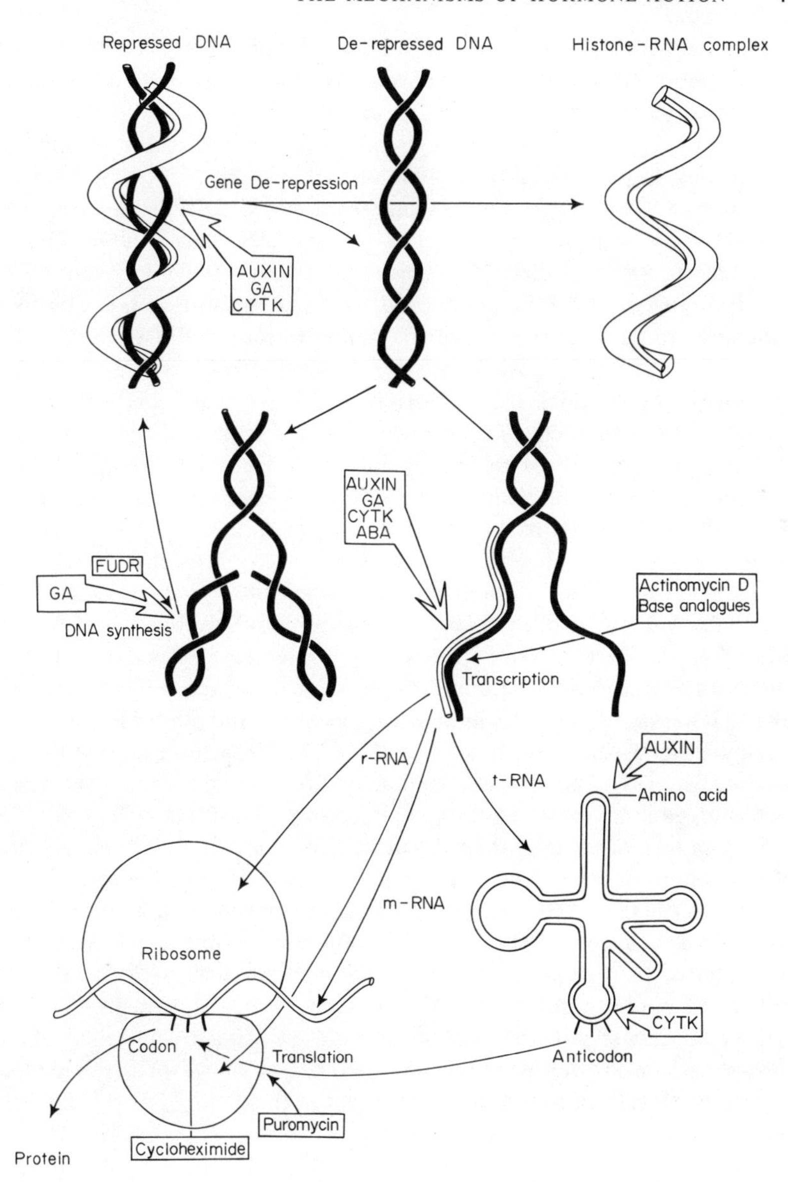

Fig. 79. Diagram to illustrate the possible sites of action of plant growth substances in the control of nucleic acid and protein synthesis in plants (for further explanation see text).

up with the messenger RNA codon, it seems also possible that cytokinins, as units in some but not all transfer RNA molecules, may regulate the incorporation of particular amino acids into proteins. The question now arises – is the physiological activity of cytokinins, as demonstrable by external applications, to be derived from such a mechanism?

In support of the suggestion there are several records of the incorporation of labelled cytokinins, e.g. benzylaminopurine-^{14}C into soluble RNA of treated tissue. In some there is evidence that the cytokinin is incorporated as such (Fox & Chen, 1967) and this could mean that the basis of its action could be the determination of the nature of some of the transfer RNA produced and hence of the proteins subsequently synthesized. However it is recognized that usually nucleotides of the normal bases are first incorporated into transfer RNA and then subsequently modified (e.g. into molecules with cytokinin properties) (see review by Borek & Srinivasan, 1966), and such a biosynthesis would invalidate the above explanation of applied cytokinin action. Furthermore there is no apparent correlation between the degree to which different cytokinins can be incorporated into transfer RNA and their physiological activities. Thus 6-benzylaminopurine was incorporated into soluble RNA of soybean callus but 6-benzylamino-9-methylpurine was not, although both were physiologically active (Kende & Tavares, 1968). In detached tobacco leaves kinetin-8-^{14}C was incorporated into various fractions of RNA and 6-benzylaminopurine was not, although both were active in delaying senescence (Richmond *et al.*, 1970). This would suggest that the activities of cytokinins are not dependant on their incorporation into transfer RNA; indeed it has been suggested that, on the contrary, transfer RNA may serve as an intermediate in the biosynthesis of cytokinins and act as a source for maintaining proper levels of this growth regulator in the cell (Chong-Maw Chen & Hall, 1969). The above is only an outline of the work done in recent years on the cytokinins of RNA and their possible significance in the regulation of protein synthesis. For a complete review of the position up to 1970 the reader is referred to the article by Skoog & Armstrong (1970).

From all this the reader will be in no doubt that the mechanisms of hormone action at the molecular level are still subjects for active debate. There is no dearth of problems for workers in this field.

Bibliography

ABDUL-BAKI, A. A. & RAY, P. M. (1971). Regulation by auxin of carbohydrate metabolism involved in cell-wall synthesis by pea stem tissue. *Pl. Physiol. Lancaster,* **47**, 537–44.

ABDUL-WAHAB, A. S. & RICE, E. L. (1967). Plant inhibition by Johnson grass and its possible significance in old-field succession. *Bull. Torrey bot. Club*, **94**, 486–97.

ABELES, F. B. (1966). Auxin stimulation of ethylene evolution. *Pl. Physiol. Lancaster,* **41**, 585–8.

ABELES, F. B. (1967). Inhibition of flowering in *Xanthium pennsylvanicum* Walln. by ethylene. *Pl. Physiol. Lancaster,* **42**, 608–609.

ABELES, F. B. & HOLM, R. E. (1966). Enhancement of RNA synthesis, protein synthesis and abscission by ethylene. *Pl. Physiol. Lancaster*, **41**, 1337–42.

ÅBERG, B. (1950). On auxin antagonists and synergists in root growth. *Physiologia Pl.*, **3**, 447–61.

ÅBERG, B. (1952). On the growth regulating effects of some 1-naphthyl- and 2,4-dichlorophenoxy- derivatives without carboxyl groups. *Physiologia Pl.*, **5**, 567–74.

ÅBERG, B. (1953). Studies on plant growth regulators. VIII On optically active plant growth regulators. *K. LantbrHögsk. Annlr.*, **20**, 241–95.

ÅBERG, B. (1958). Studies on plant growth regulators. XIV Some indole derivatives. *K. LantbrHögsk. Annlr.*, **24**, 375–95.

ÅBERG, B. (1959). Studies on plant growth regulators. XV The naphthylacetic and the α-naphthylpropionic acids. *K. LantbrHögsk. Annlr.*, **25**, 221–39.

ÅBERG, B. (1961). Studies on plant growth regulators. XVIII Some β-substituted acrylic acids. *K. LantbrHögsk. Annlr.*, **27**, 99–123.

ÅBERG, B. (1963). Studies on plant growth regulators. XIX Phenylacetic acid and related substances. *K. LantbrHögsk. Annlr.*, **29**, 3–43.

ÅBERG, B. (1965). Studies on plant growth regulators. XXI Some naphthoxy compounds. *K. LantbrHögsk. Annlr.*, **31**, 333–53.

ÅBERG, B. (1967). Studies on plant growth regulators. XXIII Further mono-

substituted phenoxyacetic acids. *K. LantbrHögsk. Annlr.*, **33**, 625–41.
Åberg, B. (1969). Plant growth regulators. XXV Some ring-substituted phenylacetic acids. *K. LantbrHögsk. Annlr.*, **35**, 935–51.
Åberg, B. & Johansson, I. (1969). Studies on plant growth regulators. XXIV Some phenolic compounds. *K. LantbrHögsk. Annlr.*, **35**, 3–27.
Abrams, G. J. von, 1953. Auxin relations of a dwarf pea. *Pl. Physiol. Lancaster.*, **28**, 443–56.
Adamson, D. & Adamson, H. (1958). Auxin action on coleoptiles in the presence of nitrogen and low temperature. *Science, N. Y.*, **128**, 532–3.
Addicott, F. T., Carns, H. R., Cornforth, J. W., Lyon, L. J., Millborrow, B. V., Ohkuma, K., Ryback, G., Smith, O. E., Thiessen, W. E. & Wareing, P. F. (1968). Abscisic acid: a proposal for the redesignation of abscisin II (dormin). In Wightman, F. & Setterfield, G. (eds.) *Biochemistry and Physiology of Plant Growth Substances*. Runge Press, Ottawa, pp. 1527–9.
Addicott, F. T. & Lyon, J. L. (1969). Physiology of abscisic acid and related substances. *A. Rev. Pl. Physiol.*, **20**, 139–64.
Akkerman, A. M. & Veldstra, H. (1947). The chemical nature of Köckermann's blastocholine from *Lycopersicon esculentum*. *Recl. Trav. chim. Pays-Bas Belg.*, **66**, 411–412.
Albersheim, P. & Bonner, J. (1959). Metabolism and hormonal control of pectic substances. *J. biol. Chem.*, **234**, 3105–3108.
Al-Omary, S. A. (1968). A preliminary study of the relationship of mineral nutrition to auxin content of *Zebrina pendula* Schnitzl. *Annls. Physiol. vég. Univ. Bruxelles.*, **13**, 109–35.
Anderson, J. D. & Moore, T. C. (1967). Biosynthesis of (–)-kaurene in cell-free extracts of immature pea seeds. *Pl. Physiol. Lancaster.*, **42**, 1527–34.
Andreae, W. A. (1967). Uptake and metabolism of indoleacetic acid, naphthaleneacetic acid and 2,4-dichlorophenoxyacetic acid by pea root segments in relation to growth inhibition during and after auxin application. *Can. J. Bot.*, **45**, 737–53.
Andreae, W. A. & Andreae, S. R. (1953). Studies on indoleacetic acid metabolism. I The effect of methyl umbelliferone, maleic hydrazide and 2,4-D on indoleacetic acid oxidation. *Can. J. Bot.*, **31**, 426–37.
Andreae, W. A. & Good, N. E. (1955). The formation of indoleacetylaspartic acid in pea seedlings. *Pl. Physiol. Lancaster*, **30**, 380–2.
Andreae, W. A. & Good, N. E. (1957). Studies in 3-indoleacetic acid metabolism. IV Conjugation with aspartic acid and ammonia as processes in the metabolism of carboxylic acids. *Pl. Physiol. Lancaster*, **32**, 566–72.
Anker, L. (1953). *The effect of indoleacetic acid and other growth-promoting substances on the endogenous respiration of the Avena coleoptile*. Doctoral thesis, University of Utrecht.
Armarego, W. L. F., Canny, M. J. & Cox, S. F. (1959). Metal-chelating properties of plant-growth substances. *Nature, Lond.*, **183**, 1176–7.
Arrigoni, O. (1960). The interaction between sulphite and auxin in stimulating growth. *Ital. J. Biochem.*, **9**, 84–97.

ASEN, S. & EMSWELLER, S. L. (1962). Hydroxycinnamic acids and their glucose esters in hybrids of *Lilium* species and their relation to germination. *Phytochemistry*, **1**, 169–74.
ASHTON, F. M. (1958). The interaction of gibberellic acid and 2,4-D. *Proc. 16th. Western Weed Control Conf.*
ASPREY, G. F., BENSON-EVANS, K. & LYON, A. G. (1958). Effect of gibberellin and indoleacetic acid on seta elongation in *Pellia epiphylla*. *Nature, Lond.*, **181**, 1351.
ATSMON, D., LANG, A. & LIGHT, E. N. (1968). Contents and recovery of gibberellins in monoecious and gynoecious cucumber plants. *Pl. Physiol. Lancaster.*, **43**, 806–810.
AUDUS, L. J. (1949*a*). The biological detoxication of 2,4-dichlorophenoxy-acetic acid in soil. *Pl. Soil*, **2**, 31–36.
AUDUS, L. J. (1949*b*). The mechanism of auxin action. *Biol. Rev.*, **24**, 51–93.
AUDUS, L. J. (1951). The biological detoxication of hormone herbicides in soil. *Pl. Soil*, **3**, 170–92.
AUDUS, L. J. (1954). Auxin antagonists and synergists. A critical approach. *New Phytol.*, **53**, 461–9.
AUDUS, L. J. (1961). Special problems of synthetic auxins: Metabolism and mode of action. *Encyclopaedia of Plant Physiology*. Springer, Berlin, Göttingen, Heidelberg, Vol. 14, pp. 1055–83.
AUDUS, L. J. (1967). The transport of growth regulators in plants. (*VI Symp. int. Agrochim. Varenna.*) *Agrochimica*, **11**, 309–29.
AUDUS, L. J. & QUASTEL, J. H. (1947). Toxic effects of amino acids and amines on seedling growth. *Nature, Lond.*, **160**, 222.
AUDUS, L. J. & SHIPTON, M. E. (1952). 2,4-dichloranisole-auxin interactions in root growth. *Physiologia Pl.*, **5**, 430–55.
AUDUS, L. J. & THRESH, R. (1953). A method of plant growth-substance assay for use in paper partition chromatography. *Physiologia Pl.*, **6**, 451–65.
AUDUS, L. J. & THRESH, R. (1956*a*). The effects of synthetic growth-regulator treatments on the level of free endogenous growth-substances in plants. *Ann. Bot.*, **20**, 439–59.
AUDUS, L. J. & THRESH, R. (1956*b*). The effects of synthetic growth substances on the level of endogenous auxins in plants. In Wain, R. L. and Wightman, F. (eds.), *The Chemistry and Mode of Action of Plant Growth Substances*. Butterworths, London, pp. 248–52.
AUNG, L. H. & DE HERTOGH, A. A. (1967). Gibberellin-like substances in non-cold and cold-treated tulip bulbs (*Tulipa* spp.). *Abstr. 6th int. Congr. Pl. Growth Substances.*, p. 6.
AVELLA, T., DINANT, M. & GASPAR, TH. (1966). Action des acides *o*- et *p*-hydroxybenzöique dur la destruction de l'acide β-indolylacétique par la catalase de foie de boeuf.*Bull. Soc. r. Sci. Liège*, **35**, 307–314.
AVERY, G. S. (1935). Differential distribution of a phytohormone in the developing leaf of *Nicotiana* and its relation to polarised growth. *Bull. Torrey bot. Club*, **62**, 313–30.

AVERY, G. S., BURKHOLDER, P. R. & CREIGHTON, H. B. (1937*a*). Production and distribution of growth hormone in shoots of *Aesculus* and *Malus* and its probable role in stimulating cambial activity. *Am. J. Bot.*, **24**, 51–8.

AVERY, G. S., BURKHOLDER, P. R. & CREIGHTON, H. B. (1937*b*). Growth hormone in terminal shoots of *Nicotiana* in relation to light. *Am. J. Bot.*, **24**, 666–73.

BAKER, B. & RAY, P. M. (1965). Relation between effects of auxin on cell wall synthesis and cell elongation. *Pl. Physiol. Lancaster*, **40**, 360–8.

BAKER, H. G. (1966). Volatile growth inhibitors produced by *Eucalyptus globulus. Madroño*, **18**, 207–10.

BALDEV, B. & LANG, A. (1965). Control of flower formation by growth retardants and gibberellin in *Samolus parviflorus*, a long-day plant. *Am. J. Bot.*, **52**, 408–17.

BALDEV, B., LANG, A. & AGATEP, A. (1965). Gibberellin production in pea seeds developing in excised pods: Effect of growth-retardant AMO-1618. *Science, N. Y.*, **147**, 155–7.

BALZ, H. P. (1966). Intrazelluläre Lokalisation und Funktion von hydrolytischen Enzymen bei Tabak. *Planta*, **70**, 207–36.

BANBURY, G. H. (1955). Physiological studies in the Mucorales. II The zygotropism of zygophores of *Mucor mucedo. J. exp. Bot.*, **6**, 235–44.

BARKSDALE, A. W. (1963). The role of hormone A during sexual conjugation in *Achlya ambisexualis. Mycologia*, **55**, 627–32.

BARKSDALE, A. W. (1967). The sexual hormones of the fungus *Achlya. Ann. N. Y. Acad. Sci.*, **144**(1), 313–9.

BARLOW, H. W. B., HANCOCK, C. R. & LACEY, H. J. (1957). Studies on extension growth in coleoptile sections. I The influence of age of coleoptile upon the response of sections to IAA. *Ann. Bot.*, **21**, 257–71.

BARNES, M. F., LIGHT, E. N. & LANG, A. (1969). The action of plant growth retardants on terpenoid biosynthesis. *Planta.*, **88**, 172–82.

BARRETT, D. W. A. & GEORGE, E. F. (1969). The auxin activity of thioglycollic acid derivatives of β-diketones. *Physiologia Pl.*, **22**, 18–24.

BASSALIK, K. (1934). Zur 'Auximon' Frage. *Acta Soc. Bot. Pol.*, **11**, 583–660.

BASTIN, M. (1964). The active enzymatic centre of indoleacetic acid peroxidation. *Bull. Soc. r. Sci. Liège.*, **34**, 678–83.

BASZYNSKI, T. (1967). The effect of vitamin E on flower initiation in *Calendula officinalis* grown in short day. *Naturwissenschaften*, **54**, 339–40.

BAUTZ, E. (1953). Einwirkung verschiedener Bodentypen und Bodenextrakte auf die Keimung von *Picea excelsa. Z. Bot.*, **41**, 41–84.

BAYLISS, W. M. & STARLING, E. (1904). The chemical regulation of the secretory process. *Proc. R. Soc. B.*, **73**, 310–22.

BECK, W. A. (1941). Production of solutes in growing epidermal cells. *Pl. Physiol. Lancaster*, **16**, 637–42.

BECQUEREL, P. & ROUSSEAU, J. (1941). Sécrétions par les racines du Lin d'un substance specifique toxique pour une nouvelle culture de cette plante. *C. r. hebd. Séanc. Acad. Sci. Paris.*, **213**, 1028–30.

BELL, G. R. (1956). On the photochemical degradation of 2,4-dichlorophenoxyacetic acid and structurally related compounds in the presence and absence of riboflavin. *Bot. Gaz.*, **118**, 133–6.

BENDAÑA, F. E., GALSTON, A. W., KAUR-SAWHNEY, R. & PENNY, P. J. (1965). Recovery of labelled ribonucleic acid following administration of labelled auxin to green pea stem sections. *Pl. Physiol. Lancaster*, **40**, 977–83.

BENEDICT, H. M. (1941). The inhibitory effect of dead roots on the growth of brome grass. *J. Am. Soc. Agron.*, **33**, 1108–1109.

BENNET-CLARK, T. A. (1956). Salt accumulation and mode of action of auxin. A preliminary hypothesis. In Wain, R. L. and Wightman, F. (eds.), *The Chemistry and Mode of Action of Plant Growth Substances*. Butterworths, London, pp. 284–91.

BENNET-CLARK, T. A. & KEFFORD, N. P. (1954). The extension growth–time relationships for *Avena* coleoptile sections. *J. exp. Bot.*, **5**, 292–304.

BENNET-CLARK, T. A., TAMBIAH, M. S. & KEFFORD, N. P. (1952). Estimation of plant growth substances by partition chromatography. *Nature, Lond.*, **169**, 452–3.

BENNET, J. P. & SKOOG, F. (1938). Premilinary experiments on the relation of growth promoting substances to the rest period in fruit trees. *Pl. Physiol. Lancaster*, **13**, 219–35.

BENTLEY, J. A. (1950*a*). An examination of the method of auxin assay using the growth of isolated sections of *Avena* coleoptile in test solutions. *J. exp. Bot.*, **1**, 201–213.

BENTLEY, J. A. (1950*b*). Growth-regulating effect of certain organic compounds. *Nature, Lond.*, **165**, 449.

BENTLEY, J. A. (1958). The naturally-occurring auxins and inhibitors. *A. Rev. Pl. Physiol.*, **9**, 47–80.

BENTLEY, J. A. (1961*a*). Extraction and purification of auxins. *Encyclopaedia of Plant Physiology*. Springer, Berlin, Göttingen, Heidelberg, Vol. 14, pp. 501–520.

BENTLEY, J. A. (1961*b*). Chemistry of native auxins. *Encyclopaedia of Plant Physiology*. Springer, Berlin, Göttingen, Heidelberg, Vol. 14, pp. 485–500.

BENTLEY, J. A. & BICKLE, A. S. (1952). Studies on plant growth hormones. II Further biological properties of 3-indolylacetonitrile. *J. exp. Bot.*, **3**, 406–423.

BENTLEY, J. A. FARRER, K. R., HOUSLEY, S., SMITH, G. F. & TAYLOR, W. C. (1956). Some chemical and physiological properties of 3-indolylpyruvic acid. *Biochem. J.*, **64**, 44–9.

BENTLEY, J. A. & HOUSLEY, S. (1952). Studies on plant growth hormones. I Biological activities of 3-indolylacetaldehyde and 3-indolylacetonitrile. *J. exp. Bot.*, **3**, 393–405.

BENTLEY-MOWAT, J. A. (1966). Activity of gibberellins A_1 to A_9 in the *Avena* first-leaf bioassay and location after chromatography. *Ann. Bot.*, **30**, 165–71.

BERGER, J. & AVERY, G. S. (1943). The mechanism of auxin action. *Science*, **98**, 454–5.

BERRIE, A. M. M., PARKER, W., KNIGHT, B. A. & HENDRIE, M. R. (1968).

Studies on lettuce seed germination. I Coumarin-induced dormancy. *Phytochemistry*, **7**, 567–73.
BEYER, E. M. & MORGAN, P. W. (1969*a*). Time sequence of the effect of ethylene in transport, uptake and decarboxylation of auxin. *Pl. Cell Physiol.*, **10**, 787–99.
BEYER, E. M. & MORGAN, P. W. (1969*b*). Ethylene modification of an auxin pulse in cotton stem sections. *Pl. Physiol. Lancaster*, **44**, 1690–4.
BIALE, J. B. (1960). Respiration of fruits. *Encyclopaedia of Plant Physiology*. Springer, Berlin, Göttingen, Heidelberg, Vol. **12**(2), pp. 536–92.
BIALE, J. B., YOUNG, R. E. & OLMSTEAD, A. J. (1954). Fruit respiration and ethylene production. *Pl. Physiol. Lancaster*, **29**, 168–74.
BIEMANN, K., TSUNAKAWA, S., SONNEBICHLER, J., FELDMAN, H., DÜTTING, D. & ZACHAU, H. G. (1966). Struktur eines ungewöhnlichen Nucleosids aus Serinspezifischer Transfer-ribonucleinsäuren. *Angew. Chem.*, **78**, 600.
BIGOT, M. C. (1968). Action d'adénines substituées sur la synthèse des bétacyanines dans la plantule d'*Amaranthus caudatus*. Possibilité d'un test biologique de dosage des cytokinines. *C.r.hebd. Séanc. Acad. Sci. Paris*, **266**, 349–52.
BINKS, R., MACMILLAN, J. & PRYCE, R. J. (1969). Plant hormones. VIII Combined gas chromatography-mass spectrometry on the methyl esters of gibberellins A_1 to A_{24} and their trimethylsilyl ethers. *Phytochemistry*, **8**, 271–84.
BIRCH, A. J., RICKARDS, R. W. & SMITH, H., (1958). The biosynthesis of gibberellic acid. *Proc. chem. Soc.*, pp. 192–3.
BISWAS, P. K., PAUL, K. B. & HENDERSON, J. H. M., (1966). Effect of *Chrysanthemum* plant extract on flower initiation in short day plants. *Physiologia Pl.*, **19**, 875–82.
BLAAUW-JANSEN, G. (1959). The influence of red and far-red light on growth and phototropism of the *Avena* seedling. *Acta bot.neerl.*, **8**, 1–39.
BLACK, L. M. (1945). A virus tumour disease of plants. *Am. J. Bot.*, **32**, 408–415.
BLANK, F. & DEUEL, H. E. (1943). Der Einfluss von Hetero-auxin auf die Quellung von Membransubstanzen. *Vierteljahrschr. naturf. Ges. Zürich*, **88**, 161.
BLOMMAERT, K. L. J. (1955). The significance of auxins and growth inhibiting substances in relation to winter dormancy of the peach tree. *Sci. Bull. S. Afr. Dept. Agric.* No. 368.
BLUM, U. & RICE, E. L. (1969). Inhibition of symbiotic nitrogen fixation by gallic and tannic acids and possible roles in old-field succession. *Bull. Torrey bot. Club*, **96**, 531–41.
BODE, H. R. (1940). Über die Blattausscheidungen des Wermuts und ihre Wirkung auf andere Pflanzen. *Planta*, **30**, 567–89.
BOLDUC, R. J., CHERRY, J. H. & BLAIR, B.O. (1970). Increase in indoleacetic oxidase activity of winter wheat by cold treatment and gibberellic acid. *Pl. Physiol. Lancaster*, **45**, 461–4.
BONNER, D. M. (1938). Relation of environment and of the physical properties of synthetic growth substances to the growth reaction. *Bot. Gaz.*, **100**, 200–14.

BONNER, J. (1932). The production of growth substance by *Rhizopus suinus*. *Biol. Zbl.* **52**, 565–82.
BONNER, J. (1933). The action of the plant growth hormone. *J. gen. Physiol.* **17**, 63–76.
BONNER, J. (1946). The role of organic matter, especially manure, in the nutrition of rice. *Bot. Gaz.*, **108**, 267–9.
BONNER, J. (1949*a*). Further experiments on flowering in *Xanthium*. *Bot. Gaz.*, **110**, 625–7.
BONNER, J. (1949*b*). Limiting factors and growth inhibitors in the growth of the *Avena* coleoptile. *Am. J. Bot.*, **36**, 323–32.
BONNER, J. (1950*a*). The role of toxic substances in the interactions of higher plants. *Bot. Rev.*, **16**, 51–65.
BONNER, J. (1950*b*). Arsenate as a selective inhibitor of growth substance action. *Pl. Physiol. Lancaster*, **25**, 181–4.
BONNER, J. (1960). The mechanical analysis of auxin-induced growth. *Beih. Z. schweiz. Forstv.*, **30**, 141–59.
BONNER, J. (1965). *The Molecular Biology of Development*. Clarendon Press, Oxford.
BONNER, J. & BANDURSKI, R. S. (1952). Studies of the physiology, pharmacology and biochemistry of the auxins. *A. Rev. Pl. Physiol.*, **3**, 59–86.
BONNER, J., BANDURSKI, R. S. & MILLERD, A. (1953). Linkage of respiration to auxin-induced water absorption. *Physiologia Pl.*, **6**, 511–22.
BONNER, J. & FOSTER, R. J. (1955*a*). The growth-time relationships of the auxin-induced growth in *Avena* coleoptile sections. *J. exp. Bot.*, **6**, 293–302.
BONNER, J. & FOSTER, R. J. (1955*b*). The kinetics of auxin-induced growth. In Wain R. L. & Wightman, F. (eds.), *The Chemistry and Mode of Action of Plant Growth Substances*. Butterworths, London, pp. 295–309.
BONNER, J. & GALSTON, A. W. (1947). The physiology and biochemistry of rubber formation in plants. *Bot. Rev.*, **13**, 543–96.
BONNER, J., HEFTMANN, E. & ZEEVAART, J. A. D. (1963). Suppression of floral induction by inhibitors of steroid biosynthesis. *Pl. Physiol. Lancaster*, **38**, 81–8.
BONNER, J. & LIVERMAN, J. (1952–3). Hormonal control of flower initiation. In *Growth and Differentiation in Plants*. Iowa State College Press, pp. 283–303.
BONNER, J., ORDIN, L. & CLELAND, R. (1956). Auxin-induced water uptake. In Wain, R. L. and Wightman, F. (eds.), *The Chemistry and Mode of Action of Plant Growth Substances*. Butterworths, London, pp. 260–70.
BONNER, J. & THURLOW, J. (1949). Inhibition of photoperiodic induction in *Xanthium* by applied auxin. *Bot. Gaz.*, **110**, 613–24.
BONNER, W. D. Jr. & THIMANN, K. V. (1950). Studies on the growth and inhibition of isolated plant parts. III The action of some inhibitors concerned with pyruvate metabolism. *Am. J. Bot.*, **37**, 66–75.
BOOIJ, H. L. (1940). *The protoplasmic membrane regarded as a complex system*. Doctoral Thesis, University of Leiden.

BOPP, M. (1967). Hemmung der Streckungswachstums etiolierter Sprosseachsen durch FUDR. *Z. PflPhysiol.*, **57**, 173–87.

BOPP, M. (1968). Control of differentiation in fern-allies and bryophytes. *A. Rev. Pl. Physiol.*, **19**, 361–80.

BOREK, E. & SRINIVASAN, P. R. (1966). The methylation of nucleic acids. *A. Rev. Biochem.*, **35**, 275–98.

BORER, K., HARDY, R. J., LINDSAY, W. S. & SPRATT, D. A. (1966). Ethylene-bis-nitrourethane: Investigations of the plant growth stimulant action of analogous compounds. *J. exp. Bot.*, **17**, 378–89.

BORGSTRÖM, G. (1939). Theoretical suggestions regarding the ethylene responses of plants and observations on the influence of apple emanations. *K. fysiogr. Sällsk. Lund Förhandl.*, **9**, 135–74.

BÖRNER, H. (1959). The apple-replant problem. I The excretion of phlorizin from apple root residues. *Contrib. Boyce Thompson Inst.*, **20**, 39–56.

BÖRNER, H. (1960). Liberation of organic substances from higher plants and their role in the soil-sickness problem. *Bot. Rev.*, **26**, 393–424.

BOROUGHS, H. & BONNER, J. (1953). Effects of indoleacetic acid on metabolic pathways. *Archs Biochem. Biophys.*, **46**, 279–90.

BORTHWICK, H. A. (1959). Photoperiodic control of flowering. In *Photoperiodism. Am. Ass. Adv. Sci. Wash.*, pp. 275–87.

BORTHWICK, H. A. & HENDRICKS, S. B. (1961). Effect of radiation on growth and development. *Encyclopaedia of Plant Physiology*. Springer, Berlin, Göttingen, Heidelberg, Vol. 16, pp. 299–330.

BOSE, T. K. & NITSCH, J. P. (1970). Chemical alteration of sex expression in *Luffa acutangula*. *Physiologia Pl.*, **23**, 1206–11.

BOTTELIER, H. P. (1934). Über den Einfluss äusserer Faktoren auf die Protoplasmaströmung in der *Avena*-Koleoptile. *Rec. Trav. bot. néerl.*, **31**, 474–582.

BOTTOMLEY, W. B. (1914). The bacterial treatment of peat. *J. R. Soc. Arts.*, **62**, 373–80.

BOTTOMLEY, W. B. (1917). Some effects of organic growth-promoting substances (auximones) on the growth of *Lemna minor* in mineral culture. *Proc. R. Soc. B.*, **89**, 481–507.

BOTTOMLEY, W. B. (1920). The effect of nitrogen-fixing organisms and nucleic acid derivatives on plant growth. *Proc. R. Soc. B.*, **91**, 83–95.

BOTTOMLEY, W., KEFFORD, N. P., ZWAR, J. A. & GOLDACRE, P. L. (1963). Kinin activity from plant extracts. I Biological assay and sources of activity. *Aust. J. biol. Sci.*, **16**, 395–406.

BOTTOMLEY, W., SMITH, H. & GALSTON, A. W. (1966). Flavonoid complexes in *Pisum sativum*. III The effect of light on the synthesis of kaempferol and quercitin complexes. *Phytochemistry*, **5**, 117–23.

BOUGHEY, A. S., MUNRO, P. E., MEIKLEJOHN, J., STRANG, R. M. & SWIFT, M. J. (1964). Antibiotic reactions between African savanna species. *Nature, Lond.*, **203**, 1302–1303.

BOUILLENNE, M. & LEYH, C. (1962). The auxin (IAA) metabolism of *Zea mays* L (Phinney dwarf varieties) treated with gibberellic acid. *Meded. LandbHoogesch. OpzoekStat. Staat Gent.*, **27**, 1353–70.

Bouillenne-Walrand, M. Leyh, C., Bastin, M. & Gaspar, T. (1967). Extraction, dosage, analyse chromatographique et charactérisation des effecteurs auxines-oxidasiques des feuilles de Zea mays (variété naine) traité par l'acide gibbérellique. *Bull. Soc. r. bot. Belge.*, **100**, 153–62.
Boysen-Jensen, P. (1911). Le transmission de l'irritation phototropique dans *l'Avena. K. Dansk. Vidensk. Selskab. Förhandl.*, **3**, 1–24.
Boysen-Jensen, P. (1936). Über die Verteilung des Wuchsstoffes in Keimstengeln und Wurzeln während die phototropischen und geotropischen Krümmung. *K. Dansk. Vidensk. Selskab. Biol. Med.*, **13**, 1–31.
Brakke, M. K. & Nickell, L. G. (1952). Lack of effect of plant growth regulators on the action of alpha-amylase secreted by virus tumour tissue. *Bot. Gaz.*, **113**, 482–4.
Brauner, L. (1954). The initial effect of heteroauxin on the water uptake by potato tissue. *Rep. Commun. 8th. int. bot. Congr. Sect.*, **11**, 155–7.
Brauner, L. & Hasman, M. (1949). Über den Mechanismus der Heteroauxinwirkung auf die Wasseraufnahne von pflanzlichen Speichergewebe. *Istanb. Univ. Fen. Fac. Mecm.*, **12**, 57–71.
Brauner, L. & Hasman, M. (1952). Weitere Untersuchungen über den Wirkungs-mechanismus des Heteroauxins bei der Wasseraufnahme von Pflanzenparanchymen. *Protoplasma*, **41**, 302.
Brawerman, G. (1966). Nucleic acids associated with chloroplasts of *Euglena gracilis*. In Goodwin, T. W. (ed.), *Biochemistry of Chloroplasts.* Academic Press, London, Vol. I, pp. 301–17.
Brian, P. W. (1949). The production of antibiotics by microorganisms in relation to biological equilibria in the soil. *Symp. Soc. exp. Biol.*, **3**, 357–72.
Brian, P. W. (1951). Antibiotics produced by fungi. *Bot. Rev.*, **17**, 357–430.
Brian, P. W. (1954). The use of antibiotics for control of plant diseases caused by bacteria and fungi. *J. appl. Bact.*, **17**, 142.
Brian, P. W. (1957). The ecological significance of antibiotic production. *Symp. Soc. gen. Microbiol.*, **7**, 168–88.
Brian, P. W. (1966). The gibberellins as hormones. *Int. Rev. Cytol.*, **19**, 229–66.
Brian, P. W., Grove, J. F. & MacMillan, J. (1960). The gibberellins. *Fortschr. chem. organ. Naturstoffe.*, **18**, 350–433.
Brian, P. W., Grove, J. F. & Mulholland, T. P. C. (1967). Relationships between structure and growth-promoting activity of the gibberellins and some allied compounds in four test systems. *Phytochemistry*, **6**, 1475–99.
Brian, P. W. & Hemming, H. G. (1955). The effect of gibberellic acid on shoot growth of pea seedlings. *Physiologia Pl.*, **8**, 669–81.
Brian, P. W. & Hemming, H. G. (1957). The effect of maleic hydrazide on the growth response of plants to gibberellic acid. *Ann. appl. Biol.*, **45**, 489–97.
Brian, P. W. & Hemming, H. G. (1958). Complementary action of gibberellic acid and auxins in pea internode extension. *Ann. Bot.*, **22**, 1–17.
Brian, P. W. & Hemming, H. G. (1961). Interaction of gibberellic acid and auxin in extension growth of pea stems. In *Plant Growth Regulation.* Iowa State University Press, pp. 645–56.

BRIAN, P. W., HEMMING, H. G. & LOWE, D. (1964). Comparative potency of nine gibberellins. *Ann. Bot.*, **28**, 369–89.
BRIAN, P. W., HEMMING, H. G. & RADLEY, M. (1955). A physiological comparison of gibberellic acid with some auxins. *Physiologia Pl.*, **8**, 899–912.
BRIAN, P. W., PETTY, J. H. P. & RICHMOND, P. T. (1959). Extended dormancy in deciduous plants treated in autumn with gibberellic acid. *Nature, Lond.*, **184**, 69.
BRIAN, R. C. (1967). Action of plant growth regulators. IV Adsorption of unsubstituted and 2,6-dichloro-aromatic acids to oat monolayers. *Pl. Physiol. Lancaster*, **42**, 1209–14.
BRIGGS, D. E. (1966). Gibberellin-like activity of helminthosporol and helminthosporic acid. *Nature. Lond.*, **210**, 418–9.
BRIGGS, W. R. (1963). Red-light, auxin relationships and the phototropic responses of corn and oat coleoptiles. *Am. J. Bot.*, **50**, 196–207.
BRIGGS, W. R. & RAY, P. M. (1956). An auxin-activating system involving tyrosinase. *Pl. Physiol. Lancaster*, **31**, 165.
BROUGHTON, W. J. & MCCOMB, A. J. (1971). Changes in the pattern of enzyme development in gibberellin-treated pea internodes. *Ann. Bot.*, **35**, 213–28.
BROWN, D. S. GRIGGS, W. H. & IWAKIRI, B. T. (1960). The influence of gibberellin on resting pear buds. *Proc. Am. Soc. hort. Sci.*, **76**, 52–8.
BROWN, R. & EDWARDS, M. (1944). The germination of the seed of *Striga lutea*. I Host influence and the progress of germination. *Ann. Bot.*, **8**, 131–48.
BROWN, R., GREENWOOD, A. D., JOHNSON, A. W., LANSDOWN, A. R., LONG, A. G. & SUNDERLAND, N. (1952). The *Orobanche* germination factor. III Concentration of the factor by counter-current distribution. *Biochem. J.*, **52**, 571–4.
BROWN, R., JOHNSON, A. W., ROBINSON, E. & TODD, A. R. (1949). The stimulant involved in the germination of *Striga hermonthica*. *Proc. R. Soc. B.*, **136**, 1–12.
BROWN, R., ROBINSON, E. & JOHNSON, A. W. (1949). The effect of D-xyloketose and certain root exudates on extension growth. *Proc. R. Soc. B.*, **136**, 577–91.
BROWN, R. & SUTCLIFFE, J. F. (1950). The effects of sugar and potassium on extension growth in the root. *J. exp. Bot.*, **1**, 88–113.
BROYER, T. C. & STOUT, P. R. (1959). The macronutrient elements. *A. Rev. Pl. Physiol.*, **10**, 277–300.
BRUCE, M. I. & ZWAR, J. A. (1966). Cytokinin activity of some substituted ureas and thioureas. *Proc. R. Soc. B.*, **165**, 245–65.
BRUINSMA, J. (1962). The effect of 4,6-dinitro-*o*-cresol (DNOC) on growth development and yield of winter rye (*Secale cereale* L). *Weed Res.*, **2**, 73–89.
BRUINSMA, J. & PATIL, S. S. (1963). The effect of 3-indoleacetic acid, gibberellic acid and vitamin E on flower initiation in unvernalised Petkus winter rye plants. *Naturwissenschaften*, **50**, 505.
BRYAN, W. H. & NEWCOMBE, E. H. (1954). Stimulation of pectin methyl-

esterase activity of cultured tobacco pith by indoleacetic acid. *Physiologia Pl.*, 7, 290–8.
BUCZEK, J. (1968). Comparative investigations on the effect of IAA and EDTA on water uptake and respiration of potato discs during and after the resting period. *Acta Soc. Bot. Pol.*, **37**, 111–118.
BUCZEK, J. & KONARZEWSKI, Z. (1968). Effect of metal ions on IAA- and EDTA-induced elongation of sunflower hypocotyl sections and on the absorption of water. *Acta Soc. Bot. Pol.*, **37**, 245–54.
BUKOVAC, M. J. & NAKAGAWA, S. (1967). Comparative potency of gibberellins in inducing parthenocarpic fruit growth in *Malus sylvestris* Mill. *Experientia*, **23**, 865.
BUKOVAC, M. J. & WITTWER, S. H. (1957). Gibberellin and higher plants. II Induction of flowering in biennials. *Q. Bull. Mich. St. Univ. agric. Exp. Stn.*, **39**, 650–60.
BÜNNING, E. (1960). Circadian rhythms and the time measurement in photoperiodism. *Cold Spring Harb. Symp. quant. Biol.*, **25**, 249–56.
BÜNNING, E. (1967). *The Physiological Clock*. Longmans, Springer, New York.
BURG, S. P. (1962). The physiology of ethylene formation. *A. Rev. Pl. Physiol.*, **13**, 265–302.
BURG, S. E. & BURG, E. A. (1965). Ethylene action and the ripening of fruits. *Science, N. Y.*, **148**, 1190–5.
BURG, S. P. & BURG, E. A. (1966). The interaction between auxin and ethylene and its role in plant growth. *Proc. natl. Acad. Sci. U.S.A.*, **55**, 262–9.
BURG, S. P. & BURG, E. A. (1967). Molecular requirements for the biological activity of ethylene. *Pl. Physiol. Lancaster*, **42**, 144–52.
BURG, S. P. & BURG, E. A. (1968). Auxin-stimulated ethylene formation; its relationship to auxin-inhibited growth, root geotropism and other plant processes. In Wightman, F. and Setterfield, G. (eds.) *Biochemistry and Physiology of Plant Growth Substances*. Runge Press, Ottawa, pp. 1275–94.
BURGEFF, H. (1924). Untersuchungen über Sexualität und Parasitismus bei Mucorinëen. *Bot. Abh.*, **4**, 5–135.
BURLING, E. & JACKSON, W. T. (1965). Changes in calcium levels in cell walls during elongation of oat coleoptile sections. *Pl. Physiol. Lancaster*, **40**, 138–41.
BURNETT, D. & AUDUS, L. J. (1964). The use of fluorimetry in the estimation of naturally-occurring indoles in plants. *Phytochemistry*, **3**, 395–415.
BURNETT, D., AUDUS, L. J. & ZINSMEISTER, H. D. (1965). Growth substances in the roots of *Vicia faba*. II. *Phytochemistry*, **4**, 891–904.
BURROWS, W. J., ARMSTRONG, D. J., SKOOG, F., HECHT, S. M., BOYLE, J. T. A., LEONARD, N. J. & OCCOLOWITZ, J. (1968). Cytokinins from soluble RNA of *Escherichia coli*: 6-(3-methyl-2-butenylamino)-2-methylthio-9-β-D-ribofuranosylpurine. *Science, N. Y.*, **161**, 691–3.
BURROWS, W. J. & CARR, D. J. (1969). Effects of flooding the root system of sunflower plants on the cytokinin content of the xylem sap. *Physiologia Pl.*, **22**, 1105–1112.

BURSTRÖM, H. (1942*a*). Die osmotischen Verhältnisse während das Streckungswachstum der Wurzel. *K. LantbrHögsk. Annlr.*, **10**, 1.
BURSTRÖM, H. (1942*b*). The influence of heteroauxin on cell growth and root development. *K. LantbrHögsk. Annlr.*, **10**, 209.
BURSTRÖM, H. (1950). Studies on growth and metabolism of roots. IV Position and negative auxins effects on cell elongation. *Physiologia Pl.*, **3**, 277–92.
BURSTRÖM, H. (1951). Studies in the growth and metabolism of roots. V Cell elongation and dry matter content. *Physiologia Pl.*, **4**, 199–207.
BURSTRÖM, H. (1953). Growth and water absorption of *Helianthus* tuber tissue. *Physiologia Pl.*, **6**, 685–91.
BURSTRÖM, H. (1955). Zur Wirkungsweise chemischer Regulatoren des Wurzelwachstums. *Bot. Notiser.*, **108**, 400.
BURSTRÖM, H. (1957). Auxin and the mechanism of root growth. *Symp. Soc. exp. Biol.*, **11**, 44–62.
BURSTRÖM, H. (1961). Growth action of EDTA in light and darkness. *Physiologia Pl.*, **14**, 354–77.
BURSTRÖM, H. (1963). Growth regulation by metals and chelates. *Adv. bot. Res.*, **1**, 73–100.
BURSTRÖM, H. & HANSEN, B. A. M. (1956). Root growth effects of indan, indene and thionaphthene derivatives. *Physiologia Pl.*, **9**, 502–514.
BURSTRÖM, H., SJÖBERG, B. & HANSEN, B. A. M. (1956). The plant growth activity of phenoxythioacetic acids. *Acta agric. Scand.*, **6**, 155–77.
BURSTRÖM, H. & TULLIN, V. (1957). Observations on chelates and root growth. *Physiologia Pl.*, **10**, 406–17.
BURSTRÖM, H. G., UHRSTRÖM, I. & OLAUSSON, B. (1970). Influence of auxin on Young's modulus in stems and roots of *Pisum* and the theory of changing the modulus in tissues. *Physiologia Pl.*, **23**, 1223–33.
BURSTRÖM, H., UHRSTRÖM, I. & WURSCHER, R. (1967). Growth, turgor, water potential and Young's modulus in pea internodes. *Physiologia Pl.*, **20**, 213–31.
BURTON, M. O., SOWDEN, F. J. & LOCHHEAD, A. G. (1954). Studies on the isolation and nature of the 'terregens factor'. *Can. J. Biochem. Physiol.*, **32**, 400–406.
BUSSE, M. & KANDLER, O. (1956). Über die Wirkung der β-indolylessigsäure auf den Stoffwechsel von *Avena* Koleoptilen. *Planta*, **46**, 619–42.
BUSTINZA, F. (1954). Antibiotics from lichens. *Rep. Commun. 8th. int. bot. Congr. Sect.*, **24**, 57–67.
BUTCHER, D. N. (1963). The presence of gibberellins in excised tomato roots. *J. exp. Bot.*, **14**, 272–80.
BUTENKO, R. G., & BUSKAKOV, Y. A. (1961). [On the mechanism of the effect of maleic hydrazide on plants.] In Russian. *Fiziol. Rast.*, **7**, 323–9. (From *Biol. Abstr.*, **36**, 42317, 1961.)
BUTLER, W. L., HENDRICKS, S. B. & SIEGELMAN, H. W. (1964). Purification and properties of phytochrome. In Goodwin, T. W. (ed.), *Biochemistry of Plant Pigments*. Pergamon Press, London, pp. 197–210.

BUTLER, W. L., MORRIS, K. H., SIEGELMAN, H. W. & HENDRICKS, S. B. (1959). Detection, assay and preliminary purification of the pigment controlling photoresponsive development of plants. *Proc. natl. Acad. Sci. U.S.A.*, **45**, 1703–8.

CALLAGHAN, J. J., HAAB, W., EVANS, J. T., ANDROSKY, C. F. & STAHL, H. D. (1965). The effect of maleic hydrazide on the growth of three uracil-requiring strains of *Saccharomyces cerevisae* UR_1, UR_2 and UR_3. *Proc. Penn. Acad. Sci.*, **39**, 297–304.

CARLIER, A. & BUFFEL, K. (1955). Polysaccharide changes in the cell-walls of water-absorbing potato tuber tissue in relation to auxin action. *Acta bot. neerl.*, **4**, 551–64.

CARLISLE, D. B., OSBORNE, D. J., ELLIS, P. E. & MOOREHOUSE, J. E. (1963). Reciprocal effects of insect- and plant-growth substances. *Nature, Lond.*, **200**, 1230.

CARR, D. J. (1967). The relationship between florigen and the flower hormones. *Ann. N. Y. Acad. Sci.*, **144**(1), 305–12.

CARR, D. J., McCOMB, A. J. & OSBORNE, L. D. (1957). Replacement of the requirement for vernalisation in *Centaurium minus* Moench by gibberellic acid. *Naturwissenschaften*, **44**, 428–9.

CARR, D. J., & REID, D. M. (1968). The physiological significance of the synthesis of hormones in roots and their export to the shoot systems. In Wightman, F. & Setterfield, G. (eds.), *Biochemistry and Physiology of Plant Growth Substances*. Runge Press, Ottawa, pp. 1169–85.

CARTWRIGHT, P. M., SYKES, J. T. & WAIN, R. L. (1956). The distribution of natural hormones in germinating seeds and seedling plants. In Wain, R. L. & Wightman, F. (eds.), *The Chemistry and Mode of Action of Plant Growth Substances*. Butterworths, London, pp. 32–9.

CASTLE, E. S. (1955). The mode of growth of epidermal cells of the *Avena* coleoptile. *Proc. natl. Acad. Sci. U.S.A.*, **41**, 197–8.

CATARINO, F. M. (1964). Some effects of kinetin on sex expression in *Bryophyllum crenatum* Bak (LSDP). *Port. Acta Biol*, **8**, 267–84.

CATHEY, H. M. (1964*a*). The physiology of growth-retarding chemicals. *A. Rev. Pl. Physiol.*, **15**, 271–302.

CATHEY, H. M. (1964*b*). Relation of phosphon structure to its growth-retarding activity. *Phyton, B. Aires*, **21**, 203–8.

CATHEY, H. M. (1965). Plant selectivity in response to variation in the structure of AMO–1618. *Phyton, B. Aires*, **22**, 19–24.

CATHEY, H. M., & STUART, N. W. (1961). Comparative plant growth-retarding activity of AMO-1618, Phosphon and CCC. *Bot. Gaz.*, **123**, 51–7.

CAVÉ, A., DEYRUP, J. A., GOUTAREL, R., LEONARD, N. J. & MONSEUR, X. G. (1962). Identité de la triacanthine, de la togholamine et de la chidlovine. *Ann. Pharm. fr.*, **20**, 285–92.

CAVELL, B. D. & MACMILLAN, J. (1967). Plant hormones. VI Isolation of (—)-kaur-16-en-19-oic acid from the mycelium of *Gibberella fujikuroi*. *Phytochemistry*, **6**, 1151–4.

CAVELL, B. D., MACMILLAN, J., PRYCE, R. J. & SHEPPARD, A. C. (1967). Plant hormones. V Thin-layer and gas-liquid chromatography of the gibberellins; direct identification of the gibberellins in a crude extract by gas-liquid chromatography. *Phytochemistry*, **6**, 867–74.

CHADWICK, A. V. & BURG, S. P. (1970). Regulation of root growth by auxin-ethylene interaction. *Pl. Physiol. Lancaster*, **45**, 192–200.

CHAILAKHYAN, M. KH. (1936). *C. r. (Doklady.) Acad. Sci. U.S.S.R.*, **3**, 442–7. (From Murneek and Whyte, 1948.)

CHAILAKHYAN, M. KH. (1957). [Effects of gibberellin on growth and flowering in plants.] In Russian. *Dokl. Akad. Nauk. SSSR.* **117**, 1077–80.

CHAILAKHYAN, M. KH. (1958). Hormonale Faktoren des Pflanzenblühens. *Biol. Zbl.*, **77**, 641–62.

CHAILAKHYAN, M. KH. (1961). Principles of ontogenesis and physiology of flowering in higher plants. *Can. J. Bot.*, **39**, 1817–41.

CHAILAKHYAN, M. KH. (1968). Internal factors of plant flowering. *A. Rev. Pl. Physiol.*, **19**, 1–36.

CHAILAKHYAN, M. KH., ASEEVA, I. V. & KHLOPENKOVA, L. P. (1958). Gibberellin-like substances formed by soil yeasts. *Dokl. Akad. Nauk. SSSR (Biol. Sci. Sect.)*, **123**, 274–6.

CHAILAKHYAN, M. KH. & BUTENKO, R. G. (1957). [Movement of assimilates of leaves to shoots under different photoperiodic conditions of leaves.] In Russian. *Fiziol. Rast.*, **4**, 450–62.

CHAILAKHYAN, M. KH. & BUTENKO, R. G. (1959). [Influence of adenine and kinetin on the differentiation of flower buds in isolated apices of *Perilla*.] In Russian, *Dokl. Akad. Nauk. SSSR.*, **129**, 224–7.

CHAILAKHYAN, M. KH. & KHLOPENKOVA, L. P. (1961). [Influence of growth substances and nucleic acid metabolites on growth and flowering of photoperiodically-induced plants.] In Russian. *Dokl. Akad. Nauk. SSSR.*, **141**, 1497–1500.

CHAKRAVARTI, S. C. & PILLAI, V. N. K. (1955). Studies in auxin-vernalisation relations. I The effects of certain synthetic auxins and their antagonists on the vernalisation of *Brassica campestris* L. *Phyton, B. Aires*, **5**, 1–17.

CHANDRA, G. R. & VARNER, J. E. (1965). Gibberellic acid-controlled metabolism of RNA in aleurone cells of barley. *Biochim. biophys. Acta.*, **108**, 583–92.

CHAO, M. D. (1947). Growth of the dandelion scape. *Pl. Physiol. Lancaster*, **22**, 393–406.

CHERRY, J. H. (1967*a*). Nucleic acid metabolism in ageing cotyledons. *Symp. Soc. exp. Biol.*, **21**, 247–68.

CHERRY, J. H. (1967*b*). Nucleic acid biosynthesis in seed germination: influence of auxin and growth-regulating substances. *Ann. N. Y. Acad. Sci.*, **144**(1), 154–68.

CHESTERS, C. G. C. & STREET, H. E. (1948). Studies in plant nutrition. I The effect of some organic supplements on the growth of lettuce in sand culture. *Ann. appl. Biol.*, **35**, 443–59.

CHIN, T. Y. & LOCKHART, J. A. (1965). Translocation of applied gibberellin in bean seedlings. *Am. J. Bot.*, **52**, 828–33.
CHINOY, J. J., GROVER, R. & SIROHI, G. S. (1957). A study of the interaction of ascorbic acid and indole-3-acetic acid in the growth of the *Avena* coleoptile. *Physiologia Pl.*, **10**, 92–9.
CHOLODNY, N. (1936). Hormonisation of grains. *Dokl. Akad. Nauk. SSSR.*, **3**, 349.
CHONG-MAW CHEN & HALL, R. H. (1969). Biosynthesis of N^6-(Δ^2-isopentenyl) adenosine in the transfer ribonucleic acid of cultured tobacco pith tissue. *Phytochemistry*, **8**, 1687–95.
CHOUARD, P. (1937). Action combinée de la folliculine et de la durée d'éclairement sur la floraison des reines-marguerites. *C. r. Séanc. Soc. Biol.*, **126**, 509–512.
CHOUARD, P. (1957). Diversité des méchanisms des dormances de la vernalisation et du photoperiodisme revelée notamment par l'action de l'acide gibbérellique. *Bull. Soc. bot. fr.*, **104**, 51–64.
CHOUARD, P. (1960). Vernalisation and its relation to dormancy. *A. Rev. Pl. Physiol.*, **11**, 191–238.
CHOUARD, P. & POIGNANT, P. (1951). Recherches préliminaire sur la vernalisation en présence d'inhibiteurs de germination et de respiration. *C. r. hebd. Séanc. Acad. Sci. Paris*, **232**, 103–5.
CHRISPEELS, M. J. & VARNER, J. E. (1966). Inhibition of gibberellic acid-induced formation of α-amylase by abscisin II. *Nature, Lond.*, **212**, 1066–7.
CHRISTIANSEN, G. S. & THIMANN, K. V. (1950). The metabolism of stem tissue during growth and its inhibition. II Respiration and ether-soluble materials. *Archs Biochem.*, **26**, 248–59.
CHROMINSKI, A. (1961). Effect of plant growth regulators on membrane permeability of tomato leaf cells. *Acta Soc. Bot. Pol.*, **30**, 267–70.
CHVOJKA, L., VEREŠ, K. & KOZAL, J. (1961). [Effect of kinins on the growth of apple tree buds and on the incorporation of P^{32}.] In Polish. *Biologia Pl.*, **3**, 140–7.
CLARK, H. E. & KERNS, K. R. (1942). Control of flowering with phytohormones. *Science, N. Y.*, **95**, 536–7.
CLARK, W. G. (1938). Electrical polarity and auxin transport. *Pl. Physiol. Lancaster*, **13**, 529–52.
CLARKE, G. C. & WAIN, R. L. (1963). Studies in plant growth-regulating substances XVIII. Chloromethylphenoxyacetic acids and chloromethylphenylglycines. *Ann. appl. Biol.*, **51**, 453–8.
CLELAND, R. (1958). A separation of auxin-induced cell-wall loosening into its plastic and elastic components. *Physiologia Pl.*, **11**, 599–609.
CLELAND, R. (1960). Effect of auxin upon loss of calcium from cell walls. *Pl. Physiol. Lancaster*, **35**, 581–4.
CLELAND, R. (1963). Independance of effects of auxin on cell wall methylation and elongation. *Pl. Physiol. Lancaster*, **38**, 12–18.
CLELAND, R. (1965). Evidence on the site of action of growth retardants. *Pl.*

Cell Physiol., **6**, 7–15.
CLELAND, R. (1967*a*). Extensibility of isolated cell walls. Measurement and changes during cell elongation. *Planta*, **74**, 197–209.
CLELAND, R. (1967*b*). Auxin and the mechanical properties of the cell wall. *Ann. N. Y. Acad. Sci.* **144**(1), 3–18.
CLELAND, R. (1968*a*). Auxin and wall extensibility: Wall-loosening process, reversibility of auxin-induced. *Science, N. Y.*, **169**, 192–4.
CLELAND, R. (1968*b*). Hydroxyproline formation and its relation to auxin-induced cell elongation in the *Avena* coleoptile. *Pl. Physiol. Lancaster.*, **43**, 1625–30.
CLELAND, R. & BONNER, J. (1956). The residual effect of auxin on the cell wall. *Pl. Physiol. Lancaster*, **31**, 350–4.
CLELAND, R. & MCCOMBS, N. (1965). Gibberellic acid: Action in barley endosperm does not require endogenous auxin. *Science, N. Y.*, **150**, 497–8.
CLELAND, R., THOMPSON, M. L., RAYLE, D. L. & PURVES, W. K. (1968). Differences in effects of gibberellins and auxins on wall extensibility in cucumber hypocotyls. *Nature, Lond.*, **219**, 510–11.
COARTNEY, J. S., MORRE, D. J. & KEY, J. L. (1967). Inhibition of RNA synthesis and auxin-induced cell-wall extensibility and growth by Actinomycin D. *Pl. Physiol. Lancaster*, **42**, 434–9.
COCKING, E. C. (1961). The action of indolyl-3-acetic acid on isolated protoplasts of tomato cotyledons. *Biochem. J.*, **82**, 12–13,P.
COCKING, E. C. (1962). Action of growth substances, chelating agents and antibiotics on isolated root protoplasts. *Nature, Lond.*, **193**, 998–9.
COLLET, G. F. (1968). Intra- and extra-cellular metabolism of IAA in aseptic or non-aseptic culture of *Pisum sativum* root tips. *Can, J. Bot.*, **46**, 969–78.
COMMONER, B., FOGEL, S. & MULLER, W. H. (1943). The mechanism of auxin action. The effect of auxin on water absorption by potato tuber tissue. *Am. J. Bot.*, **30**, 23–8.
COMMONER, B. & MAZIA, D. (1944). The mechanism of auxin action. The effect of auxin and the C_4-acids on salt and water absorption in *Avena* coleoptile and potato tuber tissue. *Am. J. Bot.*, **31**, i (Supplement).
COMMONER, B. & THIMANN, K. V. (1941). On the relation between growth and respiration in the *Avena* coleoptile. *J. gen. Physiol.*, **24**, 279–96.
CONRAD, K. (1962). Über geschlechtsgebundene Unterscheide im Wuchsstoffgehalt männlicher and weiblicher Hanfpflanzen. *Flora*, **152**, 68–73.
COOIL, B. J. (1952). Relationships of certain nutrients, metabolites and inhibitors to growth in the *Avena* coleoptile. *Pl. Physiol. Lancaster*, **27**, 49–69.
COOIL, B. J. & BONNER, J. (1957). The nature of growth inhibition by calcium in the *Avena* coleoptile. *Planta*, **48**, 696–723.
COOK, A. H. & POLLOCK, J. R. A. (1952). Chemical aspects of malting.I Removal of free amino acids and sugars from barley during steeping. *J. Inst. Brew.*, **58**, 325.
COOK, A. H. & POLLOCK, J. R. A. (1954). Chemical aspects of malting.VI

Presence of phenolic acids, including vanillic acid, in barley steeping liquors and barley. *J. Inst. Brew.*, **60**, 300.
COOKE, A. R. (1954). Changes in free auxin content during the photoinduction of short-day plants. *Pl. Physiol. Lancaster*, **29**. 440–4.
COOKE, A. R. & RANDALL, D. I. (1968). 2-Haloethanephosphonic acids as ethylene-releasing agents for the induction of flowering in pineapple. *Nature, Lond.*, **218**, 974–5.
COOPER, W. S. & STOESZ, A. D. (1931). The subterranean organs of *Helianthus scaberrimus. Bull. Torrey bot. Club*, **58**, 67–72.
CORNFORTH, J. W., MILBORROW, B. V., RYBACK, G. & WAREING, P. F. (1965). Identity of sycamore 'dormin' with abscisin II. *Nature, Lond.*, **205**, 1269–70.
CROCKER, W., HITCHCOCK, H. E. & ZIMMERMAN, P. W. (1935). Similarities in the effects of ethylene and the plant auxins. *Contrib. Boyce Thompson Inst.*, **7**, 231–48.
CROSBY, D. G. & VLITOS, A. (1958). n-Butyl 5-chloro-2-pyrimidoxyacetate—a plant growth regulator analogue. *Science, N. Y.*, **128**, 480.
CROSBY, D. G. & VLITOS, A. J. (1959). Growth substances from Maryland Mammoth tobacco; long chain alcohols. *Contrib. Boyce Thompson Inst.*, **20**, 283–92.
CROSS, B. E. (1954). Gibberellic acid. Part I. *J. chem. Soc.*, 4670–6.
CROSS, B. E., GALT, R. H. B. & HANSON, J. R. (1964). Recent work on the gibberellins. I The biosynthesis of the gibberellins. In *Régulateurs naturels de la croissance végétale. Colloq. int. Centre natl. Rech. sci.*, No. 123, pp. 264–72.
CROSS, B. E., GROVE, J. F., MACMILLAN, J., MOFFATT, J. S., MULHOLLAND, T. P. C., SEATON, J. C. & SHEPPARD, N. (1959). A revised structure for gibberellic acid. *Proc. chem. Soc. Lond.*, 302–3.
CROZIER, A., AOKI, H. & PHARIS, R. P. (1969). Efficiency of countercurrent distribution, Sephadex-GlO and silicic acid partition chromatography in the purification and separation of gibberellin-like substances from plant tissue. *J. exp. Bot.*, **20**, 786–95.
CROZIER, A., BOWEN, D. H., MACMILLAN J., REID, D. M. & MOST, B. H. (1971). Characterisation of gibberellins from dark-grown *Phaseolus coccineus* seedlings by gas-liquid chromatography and combined gas-chromatography-mass spectrometry. *Planta*, **97**, 142–54.
CROZIER, A., KUO, C. C., DURLEY, R. C. & PHARIS, R. P. (1970). The biological activities of 26 gibberellins in nine plant bioassays. *Can. J. Bot.*, **48**, 867–77.
CURRY, G. M. & WASSINK, E. C. (1956). Photoperiodic and formative effects of various wavelength regions in *Hyoscyamus niger* as influenced by gibberellic acid. *Meded. LandbHoogesch. Wageningen*, **56**, 1–8.
CURTIS, J. T. & COTTAM, G. (1950). Antibiotic and autotoxic effects in prairie sunflower. *Bull. Torrey bot. Club*, **77**, 187–91.
CURTIS, P. J., & CROSS, B. E. (1954). Gibberellic acid. A new metabolite from

the culture filtrates of *Gibberella fujikuroi*. *Chemy Ind.*, pp. 1066.
CZAJA, A. TH. (1935). Wurzelwachstum, Wuchsstoff und die Theorie der Wuchsstoffwirkung. *Ber. dt. bot. Ges.*, **53**, 221.
CZAPEK, F. (1903). Antifermente im Pflanzenorganismus. *Ber. dt. bot. Ges.*, **21**, 229–42.
CZYGAN, F. C. (1962). Blütenbildung bei *Lemna minor* nach Zusatz von Oestrogenen. *Naturwissenschaften*, **49**, 285–6.
DAINTY, J. (1969). Fundamentals of water movement. *Adv. Sci.*, **25**, 404–8.
DAVIES, D. D., GIOVANELLI, J. & AP REES, T. (1964). *Plant biochemistry*. Botanical monographs No. 3, Blackwell, Oxford.
DAVIES, D. D., PATTERSON, B. D. & TREWAVAS, A. J. (1969). Studies on the mechanism of action of indoleacetic acid. In *Plant growth regulators*. *Monograph* No. 31, *Soc. chem. Ind.* pp. 208–23.
DAVIES, E. & MACLACHLAN, G. A. (1968). Effects of indoleacetic acid on intracellular distribution of β-glucanase activities in the pea epicotyl. *Archs Biochem. Biophys.*, **128**, 595–600.
DAVIS, E. F. (1928). The toxic principle of *Juglans nigra* as identified with juglone and its toxic effects on tomato and alfalfa plants. *Am. J. Bot.*, **15**, 620.
DAVIS, L. A., HEINZ, D. E. & ADDICOTT, F. T. (1968). Gas-liquid chromatography of trimethylsilyl derivatives of abscisic acid and other plant hormones. *Pl. Physiol. Lancaster*, **43**, 1389–94.
DECANDOLLE, A. P. (1882). *Physiologie végétale*, vol. **3**, p. 1474.
DEDOLPH, R. R. (1962). Effect of benzthiazole-2-oxyacetic acid on flowering and fruiting of papaya. *Bot. Gaz.*, **124**, 75–8.
DE LEO, P.—*see* LEO, P. DE
DELEUIL, G. (1950–1). (Quoted by Guyot, 1954).
DEL MORAL, R.—*see* MORAL, R. DEL
DENFFER, D. VON, BEHRENS, M. & FISCHER, A. (1952). Papierelektrophoretische Trennung von Indolderivaten aus Pflanzenextrakten. *Naturwissenschaften*, **39**, 258.
DENFFER, D. VON & FISCHER, A. (1952). PapierchromatographischerNachweis des β-Indolaldehyds in photolytische zersetzter IES-Lösung. *Naturwissenschaften*, **23**, 549–50.
DENFFER, D. VON & GRÜNDLER, H. (1950). Über wuchsstoffinduzierte Blühhemmung bei Langtagspflanzen. *Biol. Zbl.*, **69**, 272–82.
DENNIS, D. T., UPPER, C. D., LEW, F. & WEST, C. A. (1964). The biosynthesis of gibberellin-related diterpenoids in the endosperm of *Echinocystis macrocarpa* Greene. *Pl. Physiol. Lancaster*, **39**, Suppl. xxvi.
DENNIS, D. T. UPPER, C. D. & WEST, C. A. (1965). An enzyme site of inhibition of gibberellin biosynthesis by AMO-1618 and other plant growth retardants. *Pl. Physiol. Lancaster*, **40**, 948–52.
DENNIS, D. T. & WEST, C. A. (1967). Biosynthesis of gibberellins. III The conversion of (−)-kaurene to (−)-kauren-19-oic acid in endosperm of *Echinocystis macrocarpa* Greene. *J. biol. Chem.*, **242**, 3293–300.
DENNY, F. E. (1924). Effect of ethylene on the respiration of lemons. *Bot.*

Gaz., **77**, 322–9.
DE WAARDE, J.—*see* WAARDE, J. DE
DE ZEEUW, D.—*see* ZEEUW, D. DE
DIEHL, J. M., GORTER, CH. J., VAN ITERSON, G. Jr. & KLEINHOONTE, A. (1939). The influence of growth hormone on hypocotyls of *Helianthus* and the structure of their cell-walls. *Recl. Trav. bot. néerl.*, **36**, 709.
DOLK, H. E. & THIMANN, K. V. (1932). Studies on the growth hormone of plants I. *Proc. natl. Acad. Sci. U.S.A.*, **18**, 30–46.
DÖPP, W. (1950). Eine die Antheridienbildung bei Farnen förderende Substanz in den Prothallien von *Pteridium aquilinum*. *Ber. dt. bot. Ges.*, **63**, 139–47.
DÖRFFLING, K. & BÖTTGER, M. (1968). Transport von Abscisinsäure in Explanten, Blatsteil- und Internodialsegmenten von *Coleus rhenaltianus*. *Planta*, **80**, 299–308.
DOUGAL, D. K. & SHIMBAYASHI, K. (1960). Factors affecting growth of tobacco callus tissue and its incorporation of tyrosine. *Pl. Physiol. Lancaster*, **35**, 396–404.
DUBUY, H. D. & OLSEN, R. A. (1940). The relation between respiration, protoplasmic streaming and auxin transport of *Avena* coleoptiles, using a polarographic respirometer. *Am. J. Bot.*, **27**, 401–13.
DUPLESSY-GRAILLOT, G. (1954). Interaction du chlorohydrate de l'amino-3-méthyl-4-naphtol et de l'acide 2,4-dichloro-phénoxy-acétique sur des fragments de tubercules de topinambour et de tissus de carottes cultivés *in vitro*. *C. r. Séanc. Soc. Biol.*, **148**, 975.
EAGLES, C. F. & WAREING, P. F. (1963). Dormancy regulators in woody plants. *Nature, Lond.*, **199**, 874–5.
EAGLES, C. F. & WAREING, P. F. (1964). The role of growth substances in the regulation of bud dormancy. *Physiologia Pl.*, **17**, 697–709.
EBERHARDT, F. & MARTIN, P. (1957). Das Problem der Wurzelausschiedungen und seiner Bedeutungen für die gegenseitiger Beeinflussung höherer Pflanzen. *Z. PflKrankh. PflPath. PflSchutz.*, **64**, 193–205.
EDELMAN, J. & HALL, M. A. (1964). Effect of growth hormone on the development of invertase associated with cell-walls. *Nature, Lond.*, **201**, 296–7.
EL-ANTABLY, H. M. M. & WAREING, P. F. (1966). Stimulation of flowering in certain short-day plants by abscisin. *Nature, Lond.*, **210**, 328–9.
EL-ANTABLY, H. M. M., WAREING, P. F. & HILLMAN, J. (1967). Some physiological responses to D,L-abscisin (Dormin). *Planta*, **73**, 74–90.
ELIASSON, L. (1959). Inhibition of the growth of wheat roots in nutrient solutions by substances exuded from the roots. *K. LantbrHögsk. Annlr.*, **25**, 285–93.
ELLINGER, A. (1905). Ueber die Constitution der Indolgruppe im Eiweiss. II Mittheilung: Synthese der Indol-3-propionsäure. *Ber. dt. chem. Ges.*, **38**, 2884.
ELSON, G. W., JONES, D. F., MACMILLAN, J. & SUTER, P. J. (1964). Plant hormones. IV Identification of the gibberellins of *Echinocystis macrocarpa* Greene by thin layer chromatography. *Phytochemistry*, **3**, 93–101.

ENDE, H. VAN DEN (1967). Sexual factor of Mucorales. *Nature, Lond.*, **215**, 211–2.
ENDO, T. (1968). Indoleacetic oxidase activity of horseradish and other plant peroxidase isoenzymes. *Pl. Cell Physiol.*, **9**, 333–41.
ENGELSMA, G. & MEIJER, G. (1965). The influence of light of different spectral regions on the synthesis of phenolic compounds in gherkin seedlings in relation to photomorphogenesis. I Biosynthesis of phenolic compounds. *Acta bot. neerl.*, **14**, 54. II Indoleacetic acid oxidase activity and growth. *Acta bot. neerl.*, **14**, 73.
EPSTEIN, S. S. & MANTEL, N. (1968). Hepatocarcinogenicity of the herbicide maleic hydrazide following a parenteral administration to infant swiss mice. *Int. J. Cancer*, **3**, 325–35.
ERDMANN, N. & SCHWIEWER, U. (1971). Tryptophan-dependent indoleacetic acid biosynthesis from indole, demonstrated by double-labelling experiments. *Planta*, **97**, 135–41.
ERDTMAN, H. & NILSSON, G. (1949). Synthetic plant hormones. I Sulphur analogues of some phenoxy acetic acids. *Acta Chim. scand.*, **3**, 901–3.
EULER, H. VON (1946). Ergone und Anti-ergone des Wachstums. *Svensk. kem. Tidskr.*, **53**, 180.
EVANS, L. T. (1964). Inflorescence mutation in *Lolium temulentum* L. VI Effects of some inhibitors of nucleic acid, protein and steroid biosynthesis. *Aust. J. biol. Sci.*, **17**, 24–35.
EVANS, L. T. (1966). Abscisin II. Inhibitory effect on flower induction in a long-day plant. *Science, N. Y.*, **151**, 107–8.
EVANS, L. T. (1969). *The Induction of Flowering; Some Case Histories.* MacMillan, Melbourne.
EVENARI, M. (1949). Germination inhibitors. *Bot. Rev.*, **15**, 153–94.
EVENARI, M. (1961). Chemical influences of other plants. *Encyclopaedia of Plant Physiology*, Springer, Berlin, Göttingen, Heidelberg, vol. **16**, pp. 691–736.
FANG, S. V., TEENY, F. & BUTTS, J. S. (1961). Effect of 2,4-D on the utilisation of labelled acetate by bean leaf and stem tissue. *Pl. Physiol. Lancaster*, **36**, 192–6.
FAWCETT, C. H. (1961). Indole auxins. *A. Rev. Pl. Physiol.*, **12**, 345–68.
FAWCETT, C. H., OSBORNE, D. J., WAIN, R. L. & WALKER, R. D. (1953). Studies on plant growth regulating substances. VI Side-chain structure in relation to growth-regulating activity in the aryloxyalkylcarboxylic acids. *Ann, appl. Biol.*, **40**, 232–43.
FAWCETT, C. H., PASCAL, R. M., PYBUS, M. B., TAYLOR, H. F., WAIN, R. L. & WIGHTMAN, F. (1959). Plant growth-regulating activity in homologous series of ω-phenoxyalkanecarboxylic acids and the influence of ring-substitution on their breakdown by β-oxidation within plant tissue. *Proc. R. Soc. B.*, **150**, 95–119.
FAWCETT, C. H., TAYLOR, H. F., WAIN, R. L. & WIGHTMAN, F. (1956). The degradation of certain phenoxy acids, amides and nitriles within plant

tissue. In *The Chemistry and Mode of Action of Plant Growth Substances*. Butterworths, London, pp. 187–94.

FAWCETT, C. H., WAIN, R. L. & WIGHTMAN, F. (1955). Studies on plant growth-regulating substances. VIII The growth-promoting activity of certain aryloxy- and arylthio-alkanecarboxylic acids. *Ann. appl. Biol.*, **43**, 342–54.

FAWCETT, C. H., WAIN, R. L. & WIGHTMAN, F. (1956). Plant growth-regulating activity in certain carboxylic acids not possessing a ring structure. *Nature, Lond.*, **178**, 972–4.

FAWCETT, C. H., WAIN, R. L. & WIGHTMAN, F. (1959). An examination of a suggested relationship between chelating properties and plant growth-regulating activity. *Proc. int. bot. Congr.*, **2**, 113.

FELLENBERG, G. (1966). Versuche zur Hemmung pflanzlicher Tumoren mit Histon. *Z. PflPhysiol.*, **56**, 446–52.

FELLENBERG, G. (1968). Veränderung des Nucleoproteids unter dem Einfluss von Auxin und Ascorbinsäure bei der Wurzelneubildung an Erbsenepikotyl. *Planta*, **84**, 324–38.

FERENCZY, L. (1956). Occurrence of antibacterial compounds in seeds and fruits. *Acta biol. hung.*, **6**, 317–23.

FILIPPENKO, I. A. (1940). Inhibition of developmental processes in vernalised plants that have suffered partial anaerobiosis. *Dokl. Akad. Nauk. SSSR*, **28**, 167–9.

FITTING, H. (1909). Die Beeinflussing der Orchideenblüten durch die Bestäubung und durch andere Umstände. *Z. Bot.*, **1**, 1–86.

FITTLER, F. & HALL, R. H. (1966). Selective modification of yeast seryl-t-RNA and its effect on the acceptance and binding functions. *Biochem. biophys. Res. Commun.* **25**, 441–6.

FLAIG, W. (1965). Effect of lignin degradation products on plant growth. In *Isotopes and Radiation in Soil–Plant Nutrition Studies. Int. Atomic. Energy Agency, Vienna*, pp. 3–19.

FLAIG, W., SAALBACH, E. & SCHOBINGER, U. (1960). Zur Kenntnis der Huminsäuren. XIX Mitteilung über den Einfluss von Kaltwasserextrakten aus unterscheidlich lange gerottenem Weizenstroh auf das Anfangswachstum und die Nahrstoffaufnahme von Roggenkeimpflanzen. *Z. Pfl-Ernahr. Düng. Bodenk.*, **88**, 232–6.

FLETCHER, R. A. & OSBORNE, D. J. (1966*a*). A simple bioassay for gibberellic acid. *Nature, Lond.*, **211**, 743–4.

FLETCHER, R. A. & OSBORNE, D. J. (1966*b*). Gibberellin as a regulator of protein and ribonucleic acid synthesis during senescence in leaf cells of *Taraxacum officinale*. *Can. J. Bot.*, **44**, 739–45.

FLETCHER, R. A. & ZALIK, S. (1964). Effect of light on growth and free indoleacetic acid content of *Phaseolus vulgaris*. *Pl. Physiol. Lancaster*, **39**, 328–31.

FOGG, G. E. (1968). *Photosynthesis*. English Universities Press, London.

FÖRSTER, H., WIESE, L. & BRAUNITZER, G. (1956). Über das agglutinierendwirkende Gynogamon von *Chlamydomonas eugametos*. *Z. Naturf.*, **11b**, 315–7.

FOSTER, R. J., McRAE, D. H. & BONNER, J. (1952). Auxin-induced growth inhibitions, a natural consequence of two-point attachment. *Proc. natl. Acad. Sci. U.S.A.*, **38**, 1014–1022.

FOX, E. J. & MILLER, C. O. (1959). Factors in corn steep water promoting growth of plant tissues. *Pl. Physiol. Lancaster*, **34**, 577–9.

FOX, J. E. & CHEN, C. M. (1967). Characterisation of labelled ribonucleic acid from tissue grown on ^{14}C-containing cytokinins. *J. biol. Chem.*, **242**, 4490–4.

FOX, L. R., PURVES, W. K. & NAKADA, H. I. (1965). The role of horseradish peroxidase in indole-3-acetic acid oxidation. *Biochemistry*, **4**, 2754–63.

FRANKLAND, B. & WAREING, P. F. (1960). Effect of gibberellic acid on hypocotyl growth of lettuce seedlings. *Nature, Lond.*, **185**, 255–6.

FREDGA, A. & ÅBERG, B. (1965). Stereoisomerism in plant growth regulators of the auxin type. *A. Rev. Pl. Physiol.*, **16**, 53–72.

FREI, E. & PRESTON, R. D. (1961). Cell wall organisation and wall growth in the filamentous green algae *Cladophora* and *Chaetomorpha*. I The basic structure and its formation. *Proc. R. Soc. B.*, **154**, 70–94.

FRENCH, R. C. & BEEVERS, H. (1953). Respiratory and growth responses induced by growth regulators and allied compounds. *Am. J. Bot.*, **40**, 660–6.

FREY–WYSSLING, A. (1948). The growth in surface of the plant cell wall. *Growth Symposium*, **12**, 151–69.

FREY-WYSSLING, A. (1952). Growth of plant cell walls. *Symp. Soc. exp. Biol.*, **6**, 322–8.

FROESCHEL, P. (1954). Inhibiting substances in lower plants. *Natuurwet. Tijdschr.*, **35**, 70–5.

FROESCHEL, P. (1956). Hemmstoffe und Wachstum. *Acta bot. neerl.*, **5**, 264–7.

FUJITA, T., KAWAZU, K., MITSUI, T & KATO, J. (1966). Studies on plant growth regulators XVIII. Plant growth activity of α-alkylphenylacetic acids. *Agric. biol. Chem.*, **30**, 1277–9.

FUJITA, T., KAWAZU, K., MITSUI, T., KATSUMI, M. & KATO, J. (1966). Studies on plant growth regulators XIX. Plant growth activity of cyclic homologues of the hydro-l-naphthoic acids. *Agric. biol. Chem.*, **30**, 1280–4.

FUKUYAMA, T. T. & MOYED, H. S. (1964). Inhibition of cell growth by photo-oxidation products of indole-3-acetic acid. *J. biol. Chem.*, **239**, 2392.

FULTS, J. H. & JOHNSON, G. A. (1950). A cumulative fluorescent chemical found in certain plants treated with 2,4-D identified as scopoletin (6-methoxy-7-hydroxycoumarin). *Proc. 8th Western Weed Control Conf. Denver*, pp. 108.

FURUYA, M., GALSTON, A. W. & STOWE, B. B. (1962). Isolation from peas of co-factors and inhibitors of indolyl-3-acetic acid oxidase. *Nature, Lond.*, **193**, 456–7.

GALSTON, A. W. (1947). The effects of 2,3,5-tri-iodobenzoic acid on the growth and flowering of the soya bean plant. *Am. J. Bot.*, **34**, 356–60.

GALSTON, A. W. (1949*a*). Indoleacetic-nicotinic acid interactions in the etiolated pea plant. *Pl. Physiol. Lancaster*, **24**, 577–86.
GALSTON, A. W. (1949*b*). Transmission of the floral stimulus in soya bean. *Bot. Gaz.*, **110**, 495–501.
GALSTON, A. W. (1950). Riboflavin, light and the growth of plants. *Science, N. Y.*, **111**, 619–24.
GALSTON, A. W. (1957). Studies on indoleacetic acid oxidase and its inhibition in light-grown peas. *Pl. Physiol. Lancaster*, **32**, Suppl. xxi.
GALSTON, A. W. & BAKER, R. S. (1951). Studies in the physiology of light action III. Light activation of a flavoprotein enzyme by reversal of a naturally-occurring inhibition. *Am. J. Bot.*, **38**, 190–5.
GALSTON, A. W. & BAKER, R. S. (1953). Studies in the physiology of light action. V Photoinductive alteration of auxin metabolism in etiolated pea. *Am. J. Bot.*, **40**, 512–16.
GALSTON, A. W., BONNER, J. & BAKER, R. S. (1953). Flavoprotein and peroxidase as components of the indoleacetic oxidase system of peas. *Archs Biochem.* **49**, 456–70.
GALSTON, A. W. & CHEN, H. R. (1965). Auxin activity of isatin and oxindole-3-acetic acid. *Pl. Physiol. Lancaster*, **40**, 699–705.
GALSTON, A. W. & HILLMAN, W. S. (1961). The degradation of auxin. *Encyclopaedia of Plant Physiology*. Springer, Berlin, Göttingen, Heidelberg, vol. **14**, 647–70.
GALSTON, A. W., LAVEE, S. & SIEGEL, B. Z. (1968). The induction and repression of peroxidase isoenzymes by 3-indoleacetic acid. In Wightman F. and Setterfield, G. (eds.), *The Physiology and Biochemistry of Plant Growth Substances*. Runge Press, Ottawa, pp. 455–72.
GALSTON, A. W. & MCCUNE, D. C. (1961). An analysis of gibberellin-auxin interaction and its possible metabolic basis. In *Plant Growth Regulation*. Iowa State University Press, pp. 611–26.
GALSTON, A. W. & PURVES, W. K. (1960). The mechanism of auxin action. *A. Rev. Pl. Physiol.*, **11**, 230–76.
GALSTON, A. W. & WARBURG, H. (1959). An analysis of auxin-gibberellin interaction in pea stem tissue. *Pl. Physiol. Lancaster,* **34**, 16–22.
GALUN, E., IZHAR, S. & ATSMON, D. (1965). Determination of relative auxin content in hermaphrodite and andomonoecious *Cucumis sativus* L. *Pl. Physiol. Lancaster*, **40**, 321–6.
GALUN, E., JUNG, Y. & LANG, A. (1963). Morphogenesis of floral buds of cucumber cultured *in vitro*. *Devl. Biol.*, **6**, 370–87.
GANDAR, J.-C. & NITSCH, J. P. (1959). Activités comparées de quelques auxines dérivés de l'acids phenylacétique. *Bull. Soc. bot. Fr.*, **106**, 441–5.
GANE, R. (1934). Production of ethylene by some ripening fruit. *Nature, Lond.*, **134**, 1008.
GARB, S. (1961). Differential growth-inhibitors produced by some plants. *Bot. Rev.*, **27**, 422–43.
GARNER, W. W. & ALLARD, H. A. (1920). Effect of relative length of day and

night and other factors of the environment on growth and reproduction in plants. *J. agric. Res.*, **18**, 580.

GARRETT, S. D. (1950). Ecology of the root-inhabiting fungi. *Biol. Rev.*, **25**, 220–54.

GASSNER, G. (1910). Beobachtungen und Versuche über den Anbau und die Entwicklung von Getriedpflanzen im subtropischen Klima. *Jb. Vereinig. angew. Bot.*, **8**, 95–163.

GAUTHERET, R. J. (1939). Sur la possibilité de réaliser la culture indéfinie des tissus de tubercules de carotte. *C. r. hebd. Séanc. Acad. Sci. Paris*, **208**, 118–20.

GAUTHERET, R. J. (1942). Hétéro-auxines et cultures de tissus végétaux. *Bull. Soc. Chim. biol.*, **24**, 13–47.

GAUTHERET, R. J. 1950. Le cancer végétal. *Endeavour*, **9**, 21–5.

GAUTHERET, R. J. (1955*a*). Sur la variabilité des propriétés physiologiques des cultures de tissus végétaux. *Rev. gén. Bot.*, **62**, 1–107.

GAUTHERET, R. J. (1955*b*). The nutrition of plant tissue culture. *A. Rev. Pl. Physiol.*, **6**, 433–84.

GAYLER, K. R. & GLASZIOU, K. T. (1969). Plant enzyme synthesis: hormonal regulators of invertase and peroxidase synthesis in sugar cane. *Planta*, **84**, 185–94.

GEISSMAN, T. A., VERBISCAR, A. J., PHINNEY, B. O. & CRAGG, G. (1966). Studies on the biosynthesis of gibberellins from (−)-kaurenoic acid in cultures of *Gibberella fujikuroi*. *Phytochemistry*, **5**, 933–47.

GILES, K. W. & MYERS, A. (1966). The effects of gibberellic acid and light on RNA, DNA and growth of the three basal internodes of dwarf and tall peas. *Phytochemistry*, **5**, 193–6.

GIRARDIN, J. P. L. (1864). Einfluss des Leuchtgases auf die Promenaden- und Strassenbäume. *J. Fortschr. agrik. Chem.*, **7**, 199.

GLASZIOU, K. T., GAYLER, K. R. & WALDRON, J. C. (1968). Effect of auxin and gibberellic acid on the regulation of enzyme synthesis in sugar cane stem tissue. In Wightman, F. and Setterfield G. (eds.), *Physiology and Biochemistry of Plant Growth Substances*. Runge Press, Ottawa, pp. 433–2.

GLASZIOU, K. T., WALDRON, J. C. & BULL, T. A. (1966). Control of invertase synthesis in sugar cane. Loci of auxin and glucose effects. *Pl. Physiol. Lancaster*, **41**, 282–8.

GMELIN, R. (1964). Occurrence, isolation and properties of glucobrassicin and neo-glucobrassicin. In *Régulateurs naturels de la criossance végétale. Colloq. int. Centre natl. Rech. sci.*, No. 123, pp. 159–67.

GMELIN, R. & VIRTANEN, A. I. (1961). Glucobrassicin, der Precursor von SCN^-, 3-Indolylacetonitril und Ascorbigen in *Brassica oleracea* Species. *Ann. Acad. Sci. fenn. Ser. A. II. Chem.*, **107**, 1–25.

GOESCHL, J. D., PRATT, H. K. & BONNER, B. A. (1967). An effect of light on the production of ethylene and the growth of the plumular portion of etiolated pea seedlings. *Pl. Physiol. Lancaster*, **42**, 1077–80.

GOESCHL, J. D., RAPPAPORT, L. & PRATT, H. K. (1966). Ethylene as a factor regulating the growth of pea epicotyls subjected to physical stress. *Pl. Physiol. Lancaster*, **41**, 877–84.

Goldacre, P. L. (1949). On the mechanism of action of 2,4-dichlorophenoxyacetic acid. *Aust. J. Sci. Res*, **B.2**, 154–6.

Goldacre, P. L. (1954). The photochemical inactivation of indoleacetic acid sensitised by non-protein components of plant tissues. *Aust, J. biol. Sci.*, **7**, 225–50.

Goldacre, P. L. & Bottomley, W. (1959). A kinin in apple fruitlets. *Nature, Lond.*, **184**, 555–6.

Goldacre, P. L., Galston, A. W. & Weintraub, R. L. (1953). The effect of substituted phenols on the activity of the indoleacetic acid oxidase of peas. *Archs Biochem. Biophys.*, **43**, 358–73.

Goldsmith, M. H. M. (1966). Movement of indoleacetic acid in coleoptiles of *Avena sativa* L. II Suspension of polarity by total inhibition of the basipetal transport. *Pl. Physiol. Lancaster*, **41**, 15–27.

Goldsmith, M. H. M. (1967*a*). Separation of transit of auxin from uptake: Average velocity and reversible inhibition by anaerobic conditions. *Science, N. Y.*, **156**, 661–3.

Goldsmith, M. H. M. (1967*b*). Movement of pulses of labelled auxin in corn coleoptiles. *Pl. Physiol. Lancaster*, **42**, 258–63.

Goldsmith, M. H. M. (1968). Comparison of aerobic and anaerobic movement of 3-indoleacetic acid in coleoptiles of oats and corn. In Wightman, F. and Setterfield, G. (eds.), *Biochemistry and Physiology of Plant Growth Substances*. Runge Press, Ottawa, pp. 1037–50.

Good, N. E., Andreae, W. A. & Ysselstein, M. W. H. van (1956). Studies on 3-indoleacetic acid metabolism. II Some products of the metabolism of exogenous indoleacetic acid in plant tissues. *Pl. Physiol. Lancaster*, **31**, 231–5.

Gordon, S. A. (1953). Physiology of hormone action. In Loomis, W. E. (ed.), *Growth and Differentiation in Plants*. Iowa State College Press, Chap. 13.

Gordon, S. A. (1954). Occurrence, formation and inactivation of auxins. *A. Rev. Pl. Physiol.*, **5**, 341–78.

Gordon, S. A. (1956). The biogenesis of natural auxins. In Wain R. L. and Wightman, F. (eds.), *The Chemistry and Mode of Action of Plant Growth Substances*. Butterworths, London, pp. 65–75.

Gordon, S. A. (1961). The biogenesis of auxin. *Encyclopaedia of Plant Physiology*. Springer, Berlin, Göttingen, Heidelberg, vol. 14, pp. 620–46.

Gordon, S. A. & Moss, R. A. (1958). The activity of S-(carboxymethyl)-dimethyldithiocarbamate as an auxin. *Physiologia Pl.*, **11**, 208–14.

Gordon, S. A. & Sanchez-Nieva, F. (1949). The biosynthesis of auxin in the vegetative pineapple. I Nature of active auxin. II The precursors of indoleacetic acid. *Archs Biochem.*, **20**, 256–66 and 267–85.

Gordon, S. A. & Weber, R. P. (1951). Colorimetric estimation of indoleacetic acid. *Pl. Physiol. Lancaster*, **23**, 192–5.

GOREN, R. & TOMER, E. (1971). Effects of seselin and coumarin on growth, indoleacetic acid oxidase and peroxidase, with special reference to cucumber (*Cucumis sativa* L.) radicles. *Pl. Physiol. Lancaster*, **47**, 312–6.
GORIS, A. & BOURIQUET, R. (1953). Action de l'hydrazide maléique sur les tissus de carotte cultivés *in vitro*. *Bull. Soc. Chim. biol.*, **35**, 1401–8.
GORTER, C. J. (1961). Dwarfism of peas and the action of gibberellic acid. *Physiologia Pl.*, **14**, 332–43.
GORTNER, W. A. & KENT, M. (1953). Indoleacetic acid oxidase and an inhibitor in pineapple tissue. *J. biol. Chem.*, **204**, 593–603.
GORTON, B. S., SKINNER, C. G. & EAKIN, R. E. (1957). Activity of some 6-(substituted) purines on the development of the moss *Tortella caespitosa*. *Archs Biochem. Biophys.*, **66**, 493–6.
GOTTLIEB, D. & SHAW, P. D. (1967). *Antibiotics*. Vol. V. *Mechanism of action*. Springer, Berlin.
GRACE, N. H. (1939). Physiological activity of a series of naphthyl acids. *Can. J. Res. C.*, **17**, 247–55.
GRAEBE, J. E. (1968). Biosynthesis of kaurene, squalene and phytoene from mevalonate-2-^{14}C in a cell-free system from pea fruits. *Phytochemistry*, **7**, 2002–20.
GRAY, R. & BONNER, J. (1948). Structure determination and synthesis of a plant growth inhibitor, 3-acetyl-6-methoxy-benzaldehyde, found in the leaves of *Encelia farinosa*. *J. Am. chem. Soc.*, **70**, 1249–53.
GREEN, P. B. (1958). Structural characteristics of developing *Nitella* internodal cell-walls. *J. biophys. biochem. Cytol.*, **4**, 505–16.
GREENHAM, C. G. (1958). α-Fluoro derivatives of 2,4-dichlorophenoxyacetic acid. *Aust. J. Sci.*, **20**. 212–3.
GREGORY, D. W. & COCKING, E. C. (1966). Studies on isolated protoplasts and vacuoles. II The action of growth substances. *J. exp. Bot.*, **17**, 68–77.
GREGORY, F. G. (1948). The control of flowering in plants. *Symp. Soc. exp. Biol.*, **2**, 75–103.
GREGORY, F. G. & HANCOCK, C. R. (1955). The rate of transport of natural auxin in woody shoots. *Ann. Bot.*, **19**, 451–65.
GREGORY, F. G. & PURVIS, O. N. (1938). Studies in the vernalisation of cereals. III. The use of anaerobic conditions in the analysis of the vernalising effect of low temperatures during germination. *Ann. Bot.* **2**, 753–64.
GREULACH, V. A. (1953). Notes on the starch metabolism of plants treated with maleic hydrazide. *Bot. Gaz.*, **114**, 480–1.
GROSSBARD, E. (1952). Antibiotic production by fungi on organic manures and in soil. *J. gen. Microbiol.*, **6**, 295.
GRÜMMER, G. (1955). *Die gegenseitige Beeinflussung höherer Pflanzen-Allelopathie*. Fischer, Jena.
GRÜMMER, G. (1964). Die Schädigung von Raps durch Queckenrhizome. *Naturwissenschaften*, **51**, 366.
GRUNWALD, C., MENDEZ, J. & STOWE, B. B. (1968). Substrates for optimum gas chromatographic separation of indolic methyl esters and the resolution

of components of methyl-3-indolepyruvate solutions. In Wightman, F. and Setterfield G. (eds.), *Biochemistry and Physiology of Plant Growth Substances*. Runge Press, Ottawa, pp. 163–71.

GRUNWALD, C., VENDRELL, M. & STOWE, B. B. (1967). Evaluation of gas and other chromatographic separations of indole methyl esters. *Analyt. Biochem.*, **20**, 484–94.

GUENZI, W. D. & McCALLA, T. M. (1961). Phenolic acids in oats, wheat, sorghum and corn residues and their phytotoxicity. *Agron. J.*, **58**, 303–4.

GUMIŃSKI, S. & GUMIŃSKA, Z. (1953). Le fondement chimique de l'activitée physiologique analogue de l'humus et des extraits aqueux preparés de feuilles de quelquesune espèces végétales. *Acta Soc. Bot. Pol.*, **22**, 771–85.

GUMIŃSKI, S. & SULEJ, J. (1967). La relation entre les certaines qualités physicochimique des composés humique et leur activité biologique. *Acta Soc. Bot. Pol.*, **36**, 109–22.

GUSTAFSON, F. G. (1946). Influence of external and internal factors on growth hormone in green plants. *Pl. Physiol. Lancaster*, **21**, 49–62.

GUTTENBERG, H. VON (1954). Der Einfluss von Wirkstoffen auf die Plasmapermeabilität. *Rep. Commun. 8th. int. bot. Congr., Sect.* 11, 158–60.

GUTTENBERG, H. VON & BEYTHEIN, A. (1951). Über den Einfluss von Wirkstoffen auf die Wasserpermeabilität des Protoplasmas. *Planta*, **40**, 36–69.

GUTTENBERG, H. VON, EIFLER, I. & NEHRING, G. (1953). Die Wuchsstoffe von *Coleus, Cucumis* und *Helianthus. Planta*, **42**, 209.

GUTTENBERG, H. VON & KROPELIN, L. (1947). Über den Einfluss des Heteroauxins auf das Laminargelenk von *Phaseolus. Planta*, **35**, 257–80.

GUTTENBERG, H. VON & ZETSCHE, K. (1956). Der Einfluss des Lichtes auf die Auxinbildung und den Auxintransport. *Planta*, **48**, 99–134.

GUTTRIDGE, C. G. & THOMPSON, P. A. (1964). The effects of gibberellins on growth and flowering of *Fragaria* and *Duchesnea*. *J. exp. Bot.*, **15**, 631–46.

GUYOT, L. (1954). Répercussions sur l'équilibre fongique de profondeur et l'équilibre phanerogamique de surface. *Rep. Commun. 8th. int. bot. Congr. Sect.* 24, pp. 47–52.

GUYOT, L., BECKER, Y., MASSENOT, M. & MONTIGUT, J. (1951). Les excrétions racinaire toxique chez les végétaux. *Bull. tech. Inform. Serv. Ingénieurs agric.*, **59**, 346.

HAAGEN SMIT, A. J., DANDLIKER, W. B., WITTWER, S. H. & MURNEEK, A. E. (1946). Isolation of 3-indoleacetic acid from immature corn kernels. *Am. J. Bot.*, **33**, 118–20.

HAAGEN SMIT, A. J., LEACH, W. D. & BERGEN, W. R. (1942). The estimation, isolation and identification of auxins in plant materials. *Am. J. Bot.*, **29**, 500–6.

HAAGEN SMIT, A. J. & WENT, F. W. (1935). A physiological analysis of the growth substance. *Proc. K. ned. Acad. Wet.*, **38**, 852–7.

HABERLANDT, G. (1913). Zur Physiologie der Zellteilung. *Sber. preuss. Akad. Wiss.*, pp. 318–45.

HABERLANDT, G. (1914). Zur Physiologie der Zellteilung. *Sber. preuss. Akad. Wiss.*, pp. 1096–1111.
HACKETT, D. P. (1952). The osmotic change during auxin-induced water uptake by potato tissue. *Pl. Physiol. Lancaster*, **27**, 279–92.
HACKETT, D. P. & SCHNEIDERMAN, H. A. (1953). Terminal oxidases and growth in plant tissues. I The terminal oxidase mediating growth of *Avena* coleoptile and *Pisum* stem sections. *Archs Biochem. Biophys.*, **47**, 190–204.
HACKETT, D. P. & THIMANN, K. V. (1952*a*). The effect of auxin on growth and respiration of artichoke tissue. *Proc. natl. Acad. Sci. U.S.A.*, **38**, 770–5.
HACKETT, D. P. & THIMANN, K. V. (1952*b*). The nature of auxin-induced water uptake by potato tissue. *Am. J. Bot.*, **39**, 553–60.
HACKETT, D. P. & THIMANN, K. V. (1953). The nature of auxin-induced water uptake by potato tissue. II The relation between respiration and water absorption. *Am. J. Bot.*, **40**, 183–8.
HAGAR, A., MENZEL, H. & KRAUSS, A. (1971). Versuche und Hypothese zur Primärwirkung des Auxins beim Strekungswachstums. *Planta*, **100**, 47–75.
HALEVY, A. H. (1962). Inverse effects of gibberellin and AMO-1618 on growth, catalase and peroxidase activity in cucumber seedlings. *Experientia*, **18**, 74.
HALEVY, A. H. (1963). Interaction of growth-retarding compounds and gibberellin on indoleacetic acid oxidase and peroxidase in cucumber seedlings. *Pl. Physiol. Lancaster*, **38**, 731–7.
HALEVY, A. H. & SHILO, R. (1970). Promotion of growth and flowering and increase in content of endogenous gibberellins in *Gladiolus* plants treated with the growth retardant CCC. *Physiologia Pl.*, **23**, 820–7.
HALEVY, A. H. & WITTWER, S. H. (1965). Growth promotion in the snapdragon by CCC, a growth retardant. *Naturwissenschaften*, **52**, 310.
HALL, M. A. & ORDIN, L. (1968). Auxin-induced control of cellulose synthetase activity in *Avena* coleoptile sections. In Wightman F. and Setterfield G. (eds.), *Biochemistry and Physiology of Plant Growth Substances*. Runge Press, Ottawa, pp. 659–71.
HALL, R. H. (1967). An N^6-(alkyl) adenosine in the S-RNA of *Zea mays*. *Ann. N. Y. Acad. Sci.*, **144**(1), 258–9.
HALL, R. H. (1968). Cytokinins in the transfer-RNA: their significance to the structure of t-RNA. In Wightman F. and Setterfield G. (eds.), *Biochemistry and Physiology of Plant Growth Substances.* Runge Press, Ottawa, pp. 47–56.
HAMILTON, R. H., KILVILAAN, A. & MCMANUS, J. M. (1960). Biological activity of tetrazole analogues of indole-3-acetic acid and 2,4-dichlorophenoxyacetic acid. *Pl. Physiol. Lancaster*, **35**, 136–40.
HAMILTON, T. H., MOORE, R. J., RUMSEY, A. F., MEANS, A. R. & SCHRANK, A. R. (1965). Stimulation of synthesis of ribonucleic acid in sub-apical sections of *Avena* coleoptile by indole-3-acetic acid. *Nature, Lond.*, **208**, 1180–3.
HAMNER, K. C. & BONNER, J. (1938). Photoperiodism in relation to hormones as factors in floral initiation and development. *Bot. Gaz.*, **100**, 388–431.

HAMNER, K. C. & NANDA, K. K. (1956). A relationship between applications of indoleacetic acid and the high-intensity-light reaction of photoperiodism. *Bot. Gaz.*, **118**, 13–18.
HANCOCK, C. R. & BARLOW, H. W. B. (1953). The assay of growth substances by a modified straight growth method. *Rep. E. Malling Res. Stn.*, pp. 88.
HANSCH, C. & MUIR, R. M. (1950). The ortho effect in plant growth-regulators. *Pl. Physiol. Lancaster*, **25**, 389–93.
HANSCH, C., MUIR, R. M. & METZENBERG, R., Jr. (1951). Further evidence for a chemical reaction between plant growth-regulators and a plant substrate. *Pl. Physiol. Lancaster*, **26**, 812–21.
HANSEN, B. A. M. (1954). The physiological classification of 'shoot auxins' and 'root auxins'. I and II. *Bot. Notiser.*, **3**, 230–68 and 318–25.
HANSON, J. B. & BONNER, J. (1954). The relationship between salt and water uptake in Jerusalem artichoke tuber tissue. *Am. J. Bot.*, **41**, 702–9.
HANSON, J. B. & SLIFE, F. W. (1969). Role of RNA metabolism in the action of auxin-herbicides. *Residue Rev.*, **25**, 59–67.
HANSON, J. R. & WHITE, A. F. (1968). Studies in terpenoid biosynthesis. II The biosynthesis of steviol. *Phytochemistry*, **7**, 595–7.
HANSON, J. R. & WHITE, A. F. (1969). The oxidative modification of the kaurenoid ring B during gibberellin biosynthesis. *Chem. Commun.*, pp. 410–2.
HARADA, H. (1962). *Étude des substances naturelles de criossance en relation avec la floraison. Isolement d'une substance de montaison.* Doctoral Thesis, University of Paris.
HARADA, H. & NITSCH, J. P. (1959). Changes in endogenous growth substances during flower development. *Pl. Physiol. Lancaster*, **34**, 409–15.
HARADA, H. & NITSCH. J. P. (1964). Isolement et propriétés physiologique d'une substance de montaison. In *Regulateurs naturels de la croissance végétale. Colloq, int. Centre natl. Rech. sci.* No. 123, 599–609.
HARDER, R. & BÜNSOW, R. (1956). Einfluss des Gibberellins auf die Blütenbildung bei *Kalanchoë blossfeldiana. Naturwissenschaften*, **45**, 544.
HARLEY, J. L. (1948). Mycorrhiza and soil ecology. *Biol. Rev.*, **23**, 127–58.
HARLEY, J. L. (1969). *Biology of Mycorrhiza.* Leonard Hill, London, 2nd edition.
HARLEY-MASON, J. H. & ARCHER, A. A. P. G. (1958). Use of *p*-dimethylaminocinnamaldehyde as a spray reagent for indole derivatives on paper chromatograms. *Biochem. J.*, **69**, 60.P.
HARNISH, W. N., BERG, L. A. & LILLY, V. G. (1964). Factors in lima bean and hemp seed required for oospore formation by species of *Phytophthora*. *Phytopathology*, **54**, 895.
HARPER, D. B. & WAIN, R. L. (1969). Studies on plant growth-regulating substances. XXX The plant growth-regulating activity of substituted phenols. *Ann. appl. Biol.*, **64**, 395–407.
HARTLEY, R. D., HILL, T. A., PEGG, G. F. & THOMAS, G. G. (1969). Solvent

and chemical impurities as sources of gibberellin-like growth-promoting activity. *J. exp. Bot.*, **20**, 276–87.
HARTMANN, M. (1955). Sex problems in algae, fungi and protozoa. A critical account following the review of R. A. Lewin. *Am. Nat.*, **89**, 321.
HARTSEMA, A. M. (1961). Influence of temperature on flower formation and flowering of bulbous and tuberous plants. In *Encyclopaedia of Plant Physiology*. Springer, Berlin, Göttingen, Heidelberg, vol. 16, pp. 123–67.
HASHIMOTO, T., IKAI, T. & TAMURA, S. (1968). Isolation of (+)-abscisin II from dormant aerial tubers of *Discorea batatas*. *Planta*, **78**, 89–92.
HASHIMOTO, T., SAKURAI, A. & TAMURA, S. (1967). Physiological activities of helminthosporol and helminthosporic acid. I Effects on growth of intact plants. *Pl. Cell Physiol.*, **8**, 23–34.
HASHIMOTO, T. & TAMURA, S. (1967*a*). Physiological activities of helminthosporol and helminthopsoric acid. II Effects on excised plant parts. *Pl. Cell Physiol.*, **8**, 35–45.
HASHIMOTO, T. & TAMURA, S. (1967*b*). Physiological activities of helminthosporol and helminthosporic acid. III Effects on seed germination. *Pl. Cell Physiol.*, **8**, 197–200.
HAVINGA, E. & NIVARD, R. J. F. (1948). Ultraviolet absorption spectra and stereochemical structure of plant growth substances of the *cis*-cinnamic acid type and of stilboestrol. *Recl. Trav. chim. Pays-Bas Belg.*, **67**, 846–54.
HAYASHI, T. & MURAKAMI, Y. (1953). Biochemistry of the 'Bakanae' fungus. XXIX The physiological action of gibberellins (5). The effect of gibberellin on straight growth of etiolated pea epicotyl sections. *J. agric. Chem. Soc. Japan*, **27**, 675.
HEACOCK, R. A., WAIN, R. L. & WIGHTMAN, F. (1958). Studies on plant growth regulating substances XII. Polycyclic acids. *Ann. appl. Biol.*, **46**, 352–65.
HEATH, O. V. S. & CLARK, J. E. (1956). Chelating agents as plant growth substances. A possible clue to the mode of action of auxin. *Nature, Lond.*, **177**, 1118–21.
HEATH, O. V. S. & CLARK, J. E. (1960). Chelation and auxin action. I A study of the interaction of 3-indoleacetic acid and synthetic chelating agents as affecting the growth of wheat roots and coleoptile sections. *J. exp. Bot.*, **11**, 167–87.
HEATHERBELL, D. A., HOWARD, B. H. & WICKEN, A. J. (1966). The effect of growth retardants on the respiration and coupled phosphorylation of preparations from etiolated pea seedlings. *Phytochemistry*, **5**, 635–42.
HECHT, S. M., LEONARD, N. J., SCHMITZ, R. Y. & SKOOG, F. (1970*a*). Cytokinins: Synthesis and growth-promoting activity of 2-substituted compounds in the N^6-isopentenyladenine and zeatin series. *Phytochemistry*, **9**, 1173–80.
HECHT, S. M., LEONARD, N. J., SCHMITZ, R. Y. & SKOOG, F. (1970*b*). Cytokinins: Influence of side-chain planarity of N^6-substituted adenines and

adenosines on their activity in promoting cell growth. *Phytochemistry*, **9**, 1907–1913.

HEINRICH, G. & HOFFMEISTER, H., (1967). Ecdyson als Begleitsubstanz des Ecdysterons in *Polypodium vulgare* L. *Experientia*, **23**, 995.

HELGESON, J. P. & LEONARD, N. J. (1966). Cytokinins: Identification of compounds isolated from *Corynebacterium fascians*. *Proc. natl. Acad. Sci. U.S.A.*, **56**, 60–3.

HELLSTRÖM, N. (1953). An attempt to explain the interaction of auxin and antiauxin in root growth by an adsorption mechanism. *Acta Chem. scand.*, **7**, 461–8.

HEMBERG, T. (1949*a*). Significance of growth-inhibiting substances and auxins for the rest period of the potato tuber. *Physiologia Pl.*, **2**, 24–36.

HEMBERG, T. (1949*b*). Growth-inhibiting substances in terminal buds of *Fraxinus*. *Physiologia Pl.* **2**, 37–44.

HEMBERG, T. (1961). Biogenous inhibitors. In *Encyclopaedia of Plant Physiology*. Springer, Berlin, Göttingen, Heidelberg, vol. 14, 1162–84.

HENBEST, H. B., JONES, E. R. H. & SMITH, G. F. (1953). Isolation of a new plant hormone, 3-indolylacetonitrile. *J. chem. Soc.*, **776**, 3796–3801.

HENDERSCHOTT, C. H. & BAILEY, L. F. (1955). Growth-inhibiting substances of extracts of dormant flower buds of peach. *Proc. Am. Soc. hort. Sci.*, **65**, 85–92.

HENDERSCHOTT, C. H. & WALKER, D. R. (1959*a*). Identification of a growth inhibitor from extracts of dormant peach flower buds. *Science N. Y.*, **130**, 798.

HENDERSCHOTT, C. H. & WALKER, D. R. (1959*b*). Seasonal fluctuations in quantity of growth substances in resting peach flower buds. *Proc. Am. Soc. hort. Sci.*, **74**, 121–9.

HENDRICKS, S. B. & BORTHWICK, H. A. (1964). The physiological functions of phytochrome. In Goodwin, T. A. (ed.), *Biochemistry of Plant Pigments*. Pergamon Press, Oxford, pp. 405–36.

HENDRIX, J. W. (1964). Sterol induction of reproduction and stimulation of growth of *Pythium* and *Phytophthora*. *Science N. Y.*, **144**, 1028–9.

HERTEL, R., EVANS, M. L., LEOPOLD, A. C. & SELL, H. M. (1969). The specificity of the auxin transport system. *Planta*, **85**, 238–49.

HERTEL, H. & LEOPOLD, A. C. (1963). Versuche zur Analyse des Auxintransportes in der Koleoptile von *Zea mays* L. *Planta*, **59**, 535–62.

HERTER, C. A. (1909). Note on the occurrence of skatol and indol in the wood of *Celtis reticulosa* (Miquel). *J. biol. Chem.*, **5**, 489–92.

HESLOP-HARRISON, J. (1948). *Studies on flowering and the phenomenon of sex intergradation and reversal in some angiospermic families*. Doctoral Thesis, Queen's University, Belfast.

HESLOP-HARRISON, J. (1956). Auxins and sexuality in *Cannabis sativa*. *Physiologia Pl.*, **9**, 588–97.

HESLOP-HARRISON, J. (1957). The experimental modification of sex expression in flowering plants. *Biol. Rev.*, **32**, 38–90.

HESLOP-HARRISON, J. (1961). The experimental control of sexuality and inflorescence structure in *Zea mays* L. *Proc. Linn. Soc. Lond.*, **172**, 108–23.
HESLOP-HARRISON, J. (1963). Sex expression in flowering plants. In *Meristems and Differentiation. Brookhaven Symp. Biol.*, **16**, 109–25.
HESLOP-HARRISON, J. (1964). The control of flower differentiation and sex expression. In *Régulateurs naturels de la croissance végétale. Colloq. int. Centre natl. Rech. sci.*, No. 123, 649–64.
HESLOP-HARRISON, J. & HESLOP-HARRISON, Y. (1957*a*). Studies on flowering-plant growth and organogenesis. II The modification of sex expression in *Cannabis sativa* by carbon monoxide. *Proc. R. Soc. Edinb. B.*, **66**, 424–34.
HESLOP-HARRISON, J. and HESLOP-HARRISON, Y. (1957*b*). The effect of carbon monoxide on sexuality in *Mercurialis ambigua-* L.fils. *New Phytol.*, **56**, 352–5.
HESS, D. (1958). Die Regulatoren des Streckungswachstums bei *Streptocarpus wendlandii* Utrecht und ihre Veränderung während der Blühinduktion. *Planta*, **50**, 504–25.
HEY, G. L. & HOPE, P. P. (1951). A new theory of the action of plant hormones. *The Grower*, **35**, Nos. 2–4.
HEYN, A. N. J. (1940). The physiology of cell elongation. *Bot. Rev.*, **6**, 515–74.
HIGINBOTHAM, N., PRATT, M. J. & FOSTER, R. J. (1962). Effects of calcium, indoleacetic acid and distance from stem apex on potassium and rubidium absorption by excised segments of etiolated pea epicotyl. *Pl. Physiol. Lancaster*, **37**, 203–14.
HILLMAN, W. S. (1967). The physiology of phytochrome. *A. Rev. Pl. Physiol.*, **18**, 301–24.
HILLMAN, W. S. & GALSTON, A. W. (1957). Inductive control of indoleacetic acid oxidase by red and near infra-red light. *Pl. Physiol. Lancaster*, **32**, 129–35.
HILLMAN, W. S. & GALSTON, A. W. (1961). The effect of external factors on auxin content. *Encyclopaedia of Plant Physiology*. Springer, Berlin, Göttingen, Heidelberg, vol. **14**, pp. 683–702.
HILLMAN, W. S. & PURVES, W. K. (1961). Does gibberellin act through an auxin-mediated mechanism? In *Plant Growth Regulation*. Iowa State University Press, pp. 589–600.
HINMAN, R. L, & LANG, A. (1965). Peroxidase-catalysed oxidation of indole-3-acetic acid. *Biochemistry*, **4**, 144.
HIRONO, Y., OGAWA, Y. & IMAMURA, S-I. (1960). Eine neue Methode für Gibberellin-Test bei einem Zwergmutanten von *Pharbitis nil* Chois. *Pl. Cell Physiol.*, **1**, 81–9.
HOFFMAN, O. L., FOX, S. W. & BULLOCK, M. W. (1952). Auxin-like activity of systematically substituted indoleacetic acid. *J. biol. Chem.*, **196**, 437–41.
HOFINGER, M., GASPAR, Th. & DARIMONT, E. (1970). Occurrence, titration and enzymatic degradation of 3-(3-indolyl)-acrylic acid in *Lens culinaris* Med. extracts. *Phytochemistry*, **9**, 1757–61.
HOLDSWORTH, M. & NUTMAN, P. S. (1947). Flowering response in a strain of *Orobanche minor*. *Nature, Lond.*, **160**, 223.

HOLM, R. E. & KEY, J. L. (1969). Hormone regulation of cell elongation in the hypocotyl of rootless soybeans: An evaluation of the role of DNA synthesis. *Pl. Physiol. Lancaster*, **44**, 1295–1302.

HOLST, U. -B. (1971). Some properties of inhibitor-β from *Solanum tuberosum* compared to abscisic acid. *Physiologia Pl.*, **24**, 392–6.

HOPKINS, F. G. & COLE, S. W. (1903). A contribution to the chemistry of proteins. II The constitution of tryptophane and the action of bacteria upon it. *J. Physiol.*, **29**, 451.

HORTON, R. F. & OSBORNE, D. J. (1967). Senescence, abscission and cellulase activity in *Phaseolus vulgaris*. *Nature, Lond.*, **214**, 1086–8.

HOUSLEY, S. (1961). Kinetics of auxin-induced growth. *Encyclopaedia of Plant Physiology*. Springer, Berlin, Göttingen, Heidelberg, vol. 14, 1007–43.

HOUSLEY, S., BENTLEY, J. A. & BICKLE, A. S. (1954). Studies on plant growth hormones. III Application of enzyme reaction kinetics to cell elongation in the *Avena* coleoptile. *J. exp. Bot.*, **5**, 373–88.

HOUSLEY, S. & TAYLOR, W. C. (1958). Studies on plant growth hormones. *J. exp. Bot.*, **9**, 458–71.

HOVE, C. VAN & CARLIER, A. (1968). The influence of 1-naphthylacetic acid on growth and glycolysis-Krebs cycle pathway as affected by malonic acid, monoiodoacetic acid and sodium fluoride. *Z. PflPhysiol.*, **58**, 395–401.

HUANG, R.-C. C. & BONNER, J. (1966). Histone-bound RNA, a component of native nucleohistone. *Proc. natl. Acad. Sci. U.S.A.*, **54**, 960–7.

HUFFMAN, C. W., GODAR, E. M., OHKI, K. & TORGESON, D. C. (1968). Synthesis of hydrazine derivatives as plant growth inhibitors. *Agric. Fd. Chem.*, **16**, 1041–6.

HUGHES, C. & SPRAGG, S. P. (1958). The inhibition of mitosis by the reaction of maleic hydrazide with sulphydryl groups. *Biochem. J.*, **70**, 205–12.

HULL, H. M., WENT, F. W. & YAMADA, N. (1954). Fluctuations in sensitivity of the *Avena* test due to air pollutants. *Pl. Physiol. Lancaster*, **29**, 182–7.

HUMPHREYS, E. C. (1963). Effects of (2-chloroethyl)trimethylammonium chloride on plant growth, leaf area and net assimilation rate. *Ann. Bot.*, **27**, 517–32.

HUMPHREYS, E. C., WELBANK, P. J. & WITTS, K. J. (1965). Effect of CCC (chlorcholine chloride) on growth and yield of spring wheat in the field. *Ann. appl. Biol.*, **56**, 351–61.

HURST, H. M. & BURGES, N. A. (1967). Lignin and humic acids. In McLaren, A. D. & Peterson, G. H. (eds.), *Soil Biochemistry*. Dekker, New York, pp. 260–86.

HUSSEY, G. & GREGORY, F. G. (1954). The effect of auxin on the flowering behaviour of winter barley and Petkus rye. *Pl. Physiol. Lancaster*, **29**, 292–6.

ILAN, I. & REINHOLD, L. (1963). Analysis of the effects of indole-3- acetic acid on the uptake of monovalent cations. *Physiologia Pl.*, **16**, 596–603.

INGERSOLL, R. B. & SMITH, O. E. (1971). Transport of abscisic acid. *Pl. Cell. Physiol.*, **12**, 301–9.

INGHAM, G. (1950). Effect of materials absorbed from the atmosphere in maintaining soil fertility. *Soil Sci.*, **70**, 205–12.

INGLE, J. & KEY, J. L. (1965). A comparative evaluation of the synthesis of DNA-like RNA in excised and intact plant tissue. *Pl. Physiol. Lancaster*, **40**, 1212–9.

IRVINE, V. C. (1938). Studies in growth-promoting substances as related to X-radiation and photoperiodism. *Univ. Colorado Studies*, **26**, 69–70.

ISHERWOOD, F. A., CHEN, Y. T. & MAPSON, L. W. (1953). Synthesis of L-ascorbic acid in plants and animals. *Nature, Lond.*, **171**, 348.

ISOGAI, Y., OKAMOTO, T. & KOIZUMI, T. (1964). Isolation of growth-regulating substances from *Phaseolus* seedlings. In *Régulateurs naturels de la croissance végétale. Colloq. int. Centre natl. Rech. sci.*, No. **123**, 143–58.

ISOGAI, Y., OKAMOTO, T. & KOIZUMI, T. (1967). Studies on plant growth regulators. I Isolation of indole-3-acetamide, phenylacetamide and indole-3-carboxaldehyde from etiolated seedlings of *Phaseolus. Chem. pharm. Bull.*, **15**, 151–8.

ISRAEL, H. W., SALPETER, M. M. & STEWARD. F. C. (1968). The incorporation of radioactive proline into cells. Interpretations based on radio-autography and electron microscopy. *J. Cell. Biol.*, **39**, 698–715.

ITO, S. (1932). Primary outbreak of the important diseases of the rice-plant and common treatment for their control. *Rep. Hokkaido natl. agric. Exp. Stn.*, No. 28.

JABLONSKI, J. R. & SKOOG, F. (1954). Cell enlargement and cell division in excised tobacco pith tissue. *Physiologia Pl.*, **7**, 16–24.

JACKSON, G. A. D. & BLUNDELL, J. B. (1964). Relative effectiveness of different gibberellins in the induction of parthenocarpic development on *Rosa. Nature, Lond.*, **202**, 1027–8.

JACKSON, G. A. D. & PROSSER, M. V. (1959). The induction of parthenocarpic development in *Rosa* by auxin and gibberellic acid. *Naturwissenschaften*, **46**, 407.

JACOBS, S. E. & DADD, A. H. (1959). Antibacterial substances and their role in the infection of sweet peas by *Corynebacterium fascians* (Tilford) Dowson. *Ann. appl. Biol.*, **47**, 666–72.

JACOBS, W. P. (1951). Auxin relationships in an intercalary meristem: further studies on the gynophore of *Arachis hypogaea. Am. J. Bot.*, **38**, 307–10.

JACOBS, W. P. (1961). The polar movement of auxin in the shoots of higher plants: Its occurrence and physiological significance. In *Plant Growth Regulation*, Iowa State University Press, pp. 397–409.

JACOBS, W. P. (1968). Hormonal regulation of leaf abscission. *Pl. Physiol. Lancaster*. **43**, 1480–95.

JACOBS, W. P. & CASE, D. B. (1965). Auxin transport, gibberellin and apical dominance. *Science, N. Y.*, **148**, 1729–31.

JACOBSEN, J. V. & VARNER, J. E. (1967). Gibberellic acid-induced synthesis of protease by isolated aleurone layers of barley. *Pl. Physiol. Lancaster*, **42**, 1596–1600.

JAFFE, M. J. (1970). Evidence for the regulation of phytochrome-mediated processes in bean roots by the neurohumor, acetylcholine. *Pl. Physiol. Lancaster*, **46**, 768–77.

JAMES, C. S. & WAIN, R. L. (1969). Studies on plant growth regulating substances. XXVIII Halogen-substituted benzoic acids. *Ann. appl. Biol.*, **63**, 205–10.

JAMIESON, W. D. & HUTZINGER, O. (1970). Identification of simple naturally occurring indoles by mass spectrometry. *Phytochemistry*, **9**, 2029–36.

JARVIS, B. C., FRANKLAND, B. & CHERRY, J. H. (1968). Increased nucleic acid synthesis in relation to the breaking of dormancy in hazel seed by gibberellic acid. *Planta*, **83**, 257–66.

JEFFERYS, E. G. & BRIAN, P. W. (1954). Are antibiotics produced in soil? *Rep. Commun. 8th. int. bot. Congr.*, Sect. 24, pp. 71–2.

JEFFREYS, E. G. & HEMMING, H. G. (1953). Fungistasis in soils. *Nature, Lond.*, **172**, 1953.

JEPSON, J. B. (1958). Indoles and related Ehrlich reactors. In *Chromatographic Techniques*, Heinemann Medical Books, London.

JOHNSON, S. P. & GREULACH, V. A. (1953). Some effects of maleic hydrazide on the growth of *Avena* seedlings. *J. Elisha Mitchell Sci. Soc.*, **69**, 177–81.

JOHNSTON, A. M. (1962). The relationship of structure to activity in plant-growth-regulating compounds. 2. Growth-regulating activity of substituted amides of 2,4-dichlorophenoxyacetic acid. *Biochem. J.*, **82**, 425–8.

JONES, D. F. (1964). Examination of the gibberellins of *Zea mays* and *Phaseolus multiflorus* using thin-layer chromatography. *Nature, Lond.*, **202**, 1309–10.

JONES, D. F., MACMILLAN, J. & RADLEY, M. (1963). Plant hormones. III Identification of gibberellic acid in immature barley and immature grass. *Phytochemistry*, **2**, 307–14.

JONES, E. R. H., HENBEST, H. B., SMITH, G. F. & BENTLEY, J. A. (1952). 3-Indolylacetonitrile; a naturally occurring plant growth hormone. *Nature, Lond.*, **169**, 485.

JONES, K. C. (1969). Similarities between gibberellins and related compounds in inducing acid phosphatase and reducing sugar release from barley endosperm. *Pl. Physiol. Lancaster*, **44**, 1695–700.

JONES, R. A. & KAUFMAN, P. B. (1971). Regulation of growth and invertase activity by kinetin and gibberellic acid in developing *Avena* internodes. *Physiologia Pl.*, **25**, 198–203.

JONES, R. L. (1967). Agar diffusion technique for estimating gibberellin production by plant organs: the discrepancy between extractable and diffusible gibberellins of peas. *Abstr. 6th. int. Conf. Pl. Growth Substances*, pp. 38.

JONES, R. L., METCALFE, T. P. & SEXTON, W. A. (1949). The relationship between the constitution and effect of chemical compounds on plant growth. I 2-phenoxyethylamine derivatives. *Biochem. J.*, **45**, 143–9.

JONES, R. L., METCALFE, T. P. & SEXTON, W. A. (1951). The relationship between the constitution and the effect of chemical compounds in plant growth. 3. Chlorinated benzaldehydes and benzoic acids. *Biochem. J.*, **48**, 422–5.

JONES, R. L. & PHILLIPS, I. D. J. (1966). Organs of gibberellin synthesis in light-grown sunflower plants. *Pl. Physiol. Lancaster*, **41**, 1381–6.

JONES, R. L. & VARNER, J. E. (1967). The bioassay of gibberellins. *Planta*, **72**, 155–61.

JÖNSSON, A. (1955). Synthetic plant hormones. VIII Relationship between chemical structure and plant growth activity in the arylalkyl-, aryloxyalkyl- and indolealkylcarboxylic acid series. *Svensk. kem. Tidskr.*, **67**, 166–87.

JÖNSSON, A. (1961). Chemical structure and growth activity of auxins and antiauxins. *Encyclopaedia of Plant Physiology*, Springer, Berlin, Göttingen, Heidelberg, vol. 14, pp. 959–1006.

JOST, L. & REISS, L. (1936). Zur Physiologie der Wuchsstoffe. III. *Z, Bot.*, **31**, 65.

JULIA, M. & BAILLARGÉ, M. (1953). Facteurs de croissance des plantes. III Sur l'acide carboxyméthoxy-2-naptalène-1-acétique et l'acide carboxyméthoxy-2-chloro-5-phénylacétique. *Bull. Soc. chim. Fr.*, **20**, 640–3.

JULIA, H. & BAILLARGÉ, M. (1954). Facteurs de croissance végétale VI. Sur des acides aryloxysucciniques et des corps apparentés. *Bull. Soc. chim. Fr.*, **21**, 470–3.

KAEMPFER, R. O. R. & MAGASANIK, B. (1967). Mechanism of β-galactosidase induction in *E. coli*. *J. molec. Biol.*, **27**, 475–94.

KAINDL, K. (1951). Zur Wirkungsweise von Wuchs- und Hemmstoffe II. *Biochim. biophys. Acta.*, **6**, 395–405.

KAINDL, K. (1954). Biophysikalische Analyse der Konzentrations-Wirkungskurven von Wirkstoffen (insbesondere Zellstrekungswuchsstoffen). *Monatshefte für Chem.*, **85**, 985–1002.

KAINDL, K. (1955). The action-concentration curves of mixtures of growth-promoting and growth-inhibiting substances. In Wain, R. L. & Wightman, F. (eds.), *The Chemistry and Mode of Action of Plant Growth Substances*. Butterworths, London, pp. 159–64.

KAISER, W. (1967). Der Einfluss epiphytischer Bakterien auf den Gehalt extrahierbarer IES bei *Zea mays*. *Wiss. Z. Univ. Rostock.*, **16**, 467–8.

KALDEWEY, H. (1957). Wachstumsverlauf, Wuchsstoffbildung und Nutationsbewegungen von *Fritillaria meleagris* L. im Laufe der Vegetationsperiode. *Planta*, **49**, 300–44.

KALDEWEY, H. (1965). Wuchsstofftransport, Temperatur und Pflanzenalter. *Ber. dt. bot. Ges.*, **78**, 128.

KALDEWEY, H. & STAHL, E. (1964). Die quantative Auswertung dünnschichtchromatographisch getrennter Auxine im Avena-tageslichttest nach Söding (*Avena*-silicagel-Krümmungstest). *Planta*, **62**, 22–38.

KALDEWEY, H., WAKHLOO, J. L., WEIS, A. & JUNG, H. (1969). The *Avena* geocurvature test. *Planta*, **84**, 1–10.

KALDEWEY, H., WEIS, A. & WÁKHLOO, J. L. (1969). ($\pm$)-Abscisic acid effect on IAA-2-^{14}C transport in *Avena* coleoptiles. *Abstr. 11th. int. bot. Congr. Seattle.* 106.

Kamisaka, S. & Masuda, Y. (1970). Auxin-induced growth of *Helianthus* tuber tissue. V Role of DNA synthesis in hormonal control of growth. *Physiologia Pl.*, **23**, 343–50.
Kandler, O. (1953). Untersuchungen über die Wirkung von 2,4-Dichlorophenoxyessigsäure, Natriumfluorid und Vitamin B_1 auf den Stoffwechsel *in vitro* kultivierter Maiswurzeln. *Planta*, **42**, 304–48.
Kang, B. C. & Ray, P. M. (1969). Ethylene and carbon dioxide as mediators in the response of the bean hypocotyl hook in light and auxin. *Planta*, **87**, 206–16.
Kang, B. G., Yocum, C. S., Burg, S. P. & Ray, P. M. (1967). Ethylene and carbon dioxide: mediation of hypocotyl hook-opening response. *Science, N. Y.*, **156**, 958–9.
Kaper, J. M. (1957). [*On the breakdown of tryptophan by Agrobacterium tumefaciens.*] In Dutch. Doctoral Thesis, University of Leiden.
Kasembe, J. N. R. (1967). Phenotypic restoration of fertility in a male-sterile mutant by treatment with gibberellic acid. *Nature, Lond.*, **215**, 668.
Kato, J. (1958). Studies on the physiological effect of gibberellin. II On the interaction of gibberellin with auxins and growth inhibitors. *Physiologia Pl.*, **11**, 10–15.
Kato, J. (1961). Studies on the physiological effect of gibberellin. VI Interactions of gibberellin with antiauxins. *Mem. Coll. Sci. Univ. Kyoto. B*, **28**, 119–29.
Kato, J. & Katsumi, M. (1959). Studies on the physiological effect of gibberellin. V Effect of gibberellic acid and gibberellin A on the activity of indoleacetic acid oxidase. *Mem. Coll. Sci. Univ. Kyoto, B.*, **26**, 53–60.
Kato, J. & Katsumi, M. (1967). Pseudogibberellin A_1 as an inhibitor of the GA_3-induced growth of rice seedlings. *Planta*, **74**, 194–6.
Kato, J., Purves, W. K. & Phinney, B. O. (1962). Gibberellin-like substances in plants. *Nature, Lond.*, **196**, 687–8.
Kato, J., Shiotany, Y., Tamura, S. & Sakura, A. (1964). A new plant growth-promoting substance: Helminthosporol. *Naturwissenschaften*, **51**, 341.
Katsumi, M., Phinney, B. O. & Purves, W. K. (1965). The role of gibberellin and auxin in cucumber hypocotyl growth. *Physiologia Pl.*, **18**, 462–73.
Katsumi, M. & Sano, H. (1968). Relationship of IAA-oxidase activity to gibberellin- and IAA-induced elongation of light-grown cucumber seedlings. *Physiologia Pl.*, **21**, 1348–55.
Katznelson, H., Lochhead, A. G. & Timonin, M. I. (1948). Soil microorganisms and the rhizosphere. *Bot. Rev.*, **14**, 542–87.
Katznelson, H., Rouatt, J. W. & Payne, T. M. B. (1954). Liberation of amino acids by plant roots in relation to disiccation. *Nature, Lond.*, **174**, 1110.
Katznelson, H., Rouatt, J. W. & Payne, T. M. B. (1955). The liberation of amino acids and reducing compounds by plant roots. *Pl. Soil.*, **7**, 35–48.
Katznelson, H., Sirois, J. C. & Cole, S. E. (1962). Production of a gibberellin-like substance by *Arthrobacter globiformis. Nature, Lond.*, **196**, 1012–3.
Kawahara, A. & Masuda, Y. (1957). Auxin-induced sodium efflux by *Valisneria* leaves. *J. Inst. Polytech. Osaka City Univ. Ser. D.* **8**, 89–98.

KAWAHARA, A. & TAKADA, H. (1958). Further experiments on auxin-induced Na efflux by *Valisneria* and *Ruppia* leaves. *J. Inst. Polytech. Osaka City Univ. Ser. D*, **9**, 19–26.

KAZEMIE, M. & KLÄMBT, D. (1969). Untersuchungen zur Aufnahme von Naphthalin-l-essigsäure und ihre Asparaginsäure-Kongugates in Weisen-koleoptilegewebe. *Planta*, **89**, 76–81.

KEFFORD, N. P. (1962). Auxin-gibberellin interaction in rice coleoptile elongation. *Pl. Physiol. Lancaster*, **37**, 380–6.

KEFFORD, N. P. (1962*b*). The inactivity of 1-docosanol in some plant growth tests in relation to the auxin of Maryland Mammoth tobacco. *Aust. J. biol. Sci.*, **15**, 304–311.

KEFFORD, N. P. & KASO, O. H. (1966). A potent auxin with unique chemical structure, 4-amino-3,5,6-trichloropicolinic acid. *Bot. Gaz.*, **127**, 159–63.

KEITT, G. W. Jr. & BAKER, R. A. (1966). Auxin activity of substituted benzoic acids and their effects on polar auxin transport. *Pl. Physiol. Lancaster*, **41**, 1561–9.

KELLY, S. & AVERY, G. S. (1951). The age of pea tissue and other factors influencing the respiratory response to 2,4-dichlorophenoxyacetic acid and dinitro compounds. *Am. J. Bot.*, **38**, 1–5.

KENDE, H. (1964). Preservation of chlorophyll in leaf sections by substances obtained from root exudate. *Science N. Y.*, **145**, 1066–7.

KENDE, H. (1965). Kinetin-like factors in the root exudate of sunflowers. *Proc. natl. Acad. Sci. U.S.A.*, **53**, 1302–9.

KENDE, H. & LANG, A. (1964). Gibberellins and light inhibition of stem growth in peas. *Pl. Physiol. Lancaster.*, **37**, 435–40.

KENDE, H. & TAVARES, J. E. (1968). On the significance of cytokinin incorporation into RNA. *Pl. Physiol. Lancaster*, **43**, 1244–8.

KENNY, G., SUDI, J. & BLACKMAN, G. E. (1969). The uptake of growth substances. XIII Differential uptake of indol-3yl-acetic acid through epidermal and cut surfaces of etiolated stem segments. *J. exp. Bot.*, **20**, 820–40.

KENT, M. & GORTNER, W. A. (1951). Effect of pre-illumination on the response of split pea stems to growth substances. *Bot. Gaz.*, **112**, 307–11.

KENTEN, R. H. (1953). The oxidation of phenylacetaldehyde by plant saps. *Biochem. J.*, **55**, 359.

KENTEN, R. H. (1955). The oxidation of indolyl-3-acetic acid by waxpod bean root sap and peroxidase systems. *Biochem. J.*, **59**, 110–21.

KENTZER, T. (1967). [Studies on the role of endogenous gibberellins during the process of vernalisation of winter wheat.] In Polish. *Acta Soc. Bot. Pol.*, **36**, 7–22.

KENTZER, T. & LIBBERT, E. (1961). Blokade der Gibberellinsäure-transport in Hypocotylsegmenten durch Trijodbenzoesäure, zugleich ein neuer Agarblocktest auf Gibberelline. *Planta*, **56**, 23–7.

KERK, G. J. M. VAN DER, RAALTE, M. H. VAN, SLIJPSTEIJN, A. K. & VEEN, R. VAN DER (1955). A new type of plant growth-regulating substances. *Nature, Lond.*, **176**, 308–10.

KERR, T. (1951). Growth and the structure of the primary wall. In Skoog. F. (ed.), *Plant Growth Substances*, Univ. Wisconsin Press, pp. 37–42.
KETELLAPPER, H. J. (1953). *The mechanism of the action of indole-3-acetic acid on the water absorption by* Avena *coleoptile sections*, Doctoral Thesis, University of Utrecht. (*Acta bot. neerl.*, **2**, 387.)
KETRING, D. L. & MORGAN, P. W. (1971). Physiology of oil seed. II Dormancy release in Virginia-type peanut seeds by plant growth regulators. *Pl. Physiol. Lancaster*, **47**, 488–92.
KEY, J. L. (1964). Ribonucleic acid and protein synthesis as essential processes for cell elongation. *Pl. Physiol. Lancaster*, **39**, 365–70.
KEY, J. L. (1969). Hormones and nucleic acid metabolism. *A. Rev. Pl. Physiol.*, **20**, 449–74.
KEY, J. L., BARNETT, N. M. & LIN, C. Y. (1967). RNA and protein biosynthesis and the regulation of cell elongation by auxin. *Ann. N. Y. Acad. Sci.*, **144**(1), 49–62.
KEY, J. L. & INGLE, J. (1968). RNA metabolism in response to auxin. In Wightman, F. & Setterfield, G. (eds.), *Biochemistry and Physiology of Plant Growth Substances*. Runge Press, Ottawa, pp. 711–22.
KEY, J. L. & SHANNON, J. C. (1964). Enhancement by auxin of ribonucleic acid synthesis in excised soybean hypocotyl tissue. *Pl. Physiol. Lancaster*, **39**, 360–4.
KHALIFAH, R. A., LEWIS, L. N. & COGGINS, C. W. Jr. (1963). New natural growth-promoting substances in young *Citrus* fruits. *Science, N.Y.*, **142**, 399–400.
KHALIFAH, R. A., LEWIS, L. N. & COGGINS, C. W., Jr. (1965). Gradient elution column chromatography systems for the separation and identification of gibberellins. *Analyt. Biochem.*, **12**, 113–118.
KHALIFAH, R. A., LEWIS, L. N., COGGINS, C. W., Jr. & RADLICK, P. C. (1964). Fluorimetric, chromatographic and spectronic evidence for the nonindolic nature of *Citrus* auxin. *J. exp. Bot.*, **16**, 511–18.
KHUDAIRI, A. & HAMNER, K. C. (1951). (Ex. Bonner and Liverman, 1953).
KIDD, F. & WEST, C. (1933). Effects of ethylene and of apple vapours on the ripening of fruit. *G. Brit. Dept. Sci. Ind. Res. Fd. Invest. Bd. Rep., 1932*, pp. 55–8.
KIERMAYER, O. (1956). Eine einfach Arbeitsweise für den Koleoptilzylindertest. *Planta*, **47**, 527–31.
KIERMAYER, O. (1961). Über die formativen Wirksamkeit der 2,3,5-Trijodbenzoësäure (TIBA) im vegitativen und generativen Bereich von *Solanum. Öst. bot. Z.*, **108**, 102–56.
KIM, W. K. & BIDWELL, R. G. S. (1967). The effects of indoleacetic acid and 2,4-dichlorophenoxyacetic acid on intermediary metabolism of ^{14}C-labelled organic acids by pea root tips. *Can. J. Bot.*, **45**, 1789–96.
KIM, W. K. & GREULACH, V. A. (1963). A comparative study of some influences of maleic hydrazide and 5-fluorouracil on the metabolism of *Chlorella pyrenoidosa. Phyton, B. Aires*, **20**, 127–36.

King, N. J. & Bayley, S. T. (1963). A. chemical study of the cell walls of Jerusalem artichoke tuber tissue under different growth conditions. *Can. J. Bot.*, **41**, 1141–53.

Kirk, S. C. & Jacobs, W. P. (1968). Polar movement of indole-3-acetic acid-^{14}C in roots of *Lens* and *Phaseolus*. *Pl. Physiol. Lancaster*, **43**, 675–82.

Klämbt, D. (1967). Nachweis eines Cytokinins aus *Agrobacterium tumefaciens* und sein Vergleich mit dem Cytokinin aus *Corynebacterium fascians*. *Wiss. Z. Univ. Rostock Math.-Naturwiss. Reihe*, **16** (4/5), 623–5.

Klämbt, D. (1968). Cytokinine aus *Helianthus annuus*. *Planta*, **82**, 170–8.

Klämbt, D., Thies, G. & Skoog, F. (1966). Isolation of cytokinins from *Corynebacterium fascians*. *Proc. natl. Acad. Sci. U.S.A.*, **56**, 52–9.

Klämbt, H.-D. (1961*a*). Wachstumsinduktion der Indol-3-essigsäure und der Benzoësaüre. *Planta*, **56**, 618–31.

Klämbt, H.-D. (1961*b*). Wachstumsinduktion und Wuchsstoffmetabolismus im Weizenkoleoptilzylinder. II. *Planta*, **56**, 618–31.

Klebs, G. (1906). Über künstliche Metamorphosen. *Abh. naturf. Ges. Halle*, **25**, 135–294.

Klein, W. H., Withrow, R. B. & Elstad, V. B. (1956). Response of the hypocotyl hook of bean seedlings to radient energy and other factors. *Pl. Physiol. Lancaster*, **31**, 289–94.

Klippart, J. H. (1858). An essay on the origin, growth, diseases, varieties, etc. of the wheat plant. *A. R. Ohio State Bd. Agric*, **12**, 562–816.

Knight, B. E. A., Taylor, H. F. & Wain, R. L. (1969). Studies on plant growth-regulating substances. XXIX The plant growth-retarding properties of certain ammonium, phosphonium and sulphonium halides. *Ann. appl. Biol.*, **63**, 211–23.

Knight, L. L., Rose, R. C. & Crocker, W. (1910). Effect of various gases and vapours upon etiolated seedlings of sweet pea. *Science (Lancaster, Pa.) N.S*, **31**, 635–6.

Knipe, D. & Herbel, C. H. (1966). Germination and growth of some semi-desert grassland species treated with aqueous extracts from creosotebush. *Ecology*, **47**, 775–81.

Knoop, F. (1904). Der Abbau aromatischerer Fettsäuren im Tierkörper. *Beitr. chem. physiol. Path*, **6**, 150.

Knott, J. E. (1934). Effect of localised photoperiod on spinach. *Proc. Am. Soc. hort. Sci.*, **31**, Suppl., 152–4.

Knudson, L. (1920). The secretion of invertase by plant roots. *Am. J. Bot.* **7**, 371–9.

Knudson, L. & Smith, R. S. (1919). Secretion of amylase by plant roots. *Bot. Gaz.*, **68**, 460–6.

Knypl, J. S. (1970). Inhibition of chlorophyll synthesis by growth-retarding chemicals and coumarin in detached cotyledons of pumpkin. *Biochim. Physiol. Pflanzen.*, **161**, 1–13.

Köckemann, A. (1934). Über ein keimungshemmende Substanz in fleischigen Früchten. *Ber. dt. bot. Ges.*, **52**, 523.

KOEPELI, J. B., THIMANN, K. V. & WENT, F. W. (1938). Phytohormones: Structure and physiological activity. I. *J. biol. Chem.*, **122**, 763–80.
KÖGL, F. (1937). Wirkstoffprinzip und Pflanzenwachstum. *Naturwissenschaften*, **25**, 465.
KÖGL, F. & HAAGEN SMIT, A. J. (1931). Über die Chemie des Wuchsstoffs. *Proc. K. ned. Akad. Wet.*, **34**, 1411–16.
KÖGL, F., HAAGEN SMIT, A. J. & ERXLEBEN, H. (1934). Über ein neues Auxin ('Heteroauxin') aus Harn. *Hoppe-Seyl. Z.*, **228**, 90–103.
KÖGL, F. & KOSTERMANS, D. G. F. R. (1935). Über die Konstitutions-Spezifität des Heteroauxin. *Hoppe-Seyl. Z.*, **235**, 201–6.
KÖGL, F. & VERKAAIK, B. (1944). Über die Antiden der α-(β-Indolyl) propionsäure und ihre verschieden starke physiologische Wirksamkeit. XXXVIII Über pflanzliche Wachstumsstoffe. *Hoppe-Seyl. Z.*, **280**, 167–76.
KOJIMA, H., YAHIRO, M. & ETO, T. (1957). On the influence of auxins upon vernalisation. *J. Fac. Agric. Kyushu Univ.*, **11**, 25–35.
KOLLER, D. (1959). Germination. *Scient. Am.*, (*April*), 75–84.
KÖNIG, K. H. (1968). N, N-dimethylhydraziniumsalze als neue Wachstumsregulatoren. *Naturwissenschaften*, **55**, 217–9.
KONINGSBERGER, V. J. (1947). [Concerning the primary action of growth substances of the auxin type.] In Dutch. *Meded. K. Vlaamse Akad. Wet. Lett. schone Kunst. Belg.*, **9**, 5.
KOSHIMIZU, J. (1967). Isolation of a new cytokinin from immature yellow lupin seeds. *Agric. biol. Chem.*, **31**, 795–801.
KOSHIMIZU, K., FUJITA, T., MITSUI, T. & KATO, J. (1960). Studies on plant growth substances. XIII Plant growth activity of substituted l-naphthoic acid derivatives. *Bull. agric. Chem. Soc. Japan*, **24**, 221–5.
KOSHIMIZU, K., FUKUI, H., INUI, M., OGAWA, Y. & MITSUI. T. (1968). Gibberellin A_{23} in immature seed of *Lupinus luteus*. *Tetrahedron Lett.*, **9**, 1143–7.
KOSHIMIZU, K., FUKUI, H., KUSAKI, T., MITSUI. T. & OGAWA, Y. (1966). A new C_{20} gibberellin in immature seeds of *Lupinus luteus*. *Tetrahedron Lett.*, **22**, 2459–63.
KOSHIMIZU, K., INUI, M., FUKUI, H. & MITSUI, T. (1968). Isolation of (+)-abscisyl-β-D-glucopyranoside from immature fruit of *Lupinus luteus*. *Agric. biol. Chem. (Tokyo).*, **32**, 789–91.
KOTTE, W. (1922). Wurzelmeristem in Gewebekultur. *Ber. dt. bot. Ges.*, **40**, 269–72.
KÖVES, E. (1957). Papierchromatographische Untersuchungen der ätherlöslichen keimungs- and wachstumshemmenden Stoffe der Haferspelze. *Acta biol. hung.*, **3**, 179–87.
KÖVES, E. & VARGA, M. (1958). Growth inhibiting substances in rice straw. *Acta biol. hung.*, **4**, 13–16.
KRAMER, P. J. (1955). Water uptake of plant cells and tissues. *A. Rev. Pl. Physiol.*, **6**, 253–69.
KRAMER, P. J. (1959). Transpiration and the water economy of plants. In

Steward, F. C. (ed.), *Plant Physiology*. Academic Press, New York and London, vol. 2, pp. 607–726.
KRAUS, E. J. & KRAYBILL. H. R. (1918). Vegetation and reproduction with special reference to the tomato. *Bull. Ore. agric. Exp. Stn.*, No. 149.
KREKULE, J. (1961). Application of some inhibitors in studying the physiology of vernalisation. *Biologia Pl. (Prague)*, **3**, 107–14.
KREWSON, C. F., WOOD, J. W., WOLFE, W. C., MITCHELL, J. W. & MARTH, P. C. (1959). Synthesis and biological activity of some quaternary ammonium and related compounds that suppress plant growth. *Agric. Fd. Chem.*, **7**, 264.
KRIBBEN, F. J. (1952). Die Blütenbildung von *Orobanche* in Abhängigkeit von der Entwicklungsphase des Wirtes. *Ber. dt. bot. Ges.*, **64**, 353–5.
KRUYT, W. (1954). A study in connection with the problem of hormonisation of seeds. *Acta bot. neerl.*, **3**, 1–82.
KULAEVA, O. N. (1962). The effect of roots on leaf metabolism in relation to the action of kinetin on leaves. *Fiziol. Rast.*, **9**, 229–39.
KULESCHA, Z. (1955). Action de l'hydrazide maléique sur la teneur en auxine des tissus de Topinambour cultivés en présence de diverses substances de division. *Acta bot, neerl.*, **4**, 404–409.
KULESCHA, Z. & GAUTHERET. R. J. (1948). Sur l'élaboration de substances de croissance par trois types de cultures de tissus de Scorsonère: cultures normales, cultures de Crown-gall et cultures accoutumées a l'hétéro-auxine. *C. r. hebd. Séanc. Acad. Sci. Paris.*, **227**, 292–4.
KURAISHI, S. (1959). Effect of kinetin analogues on leaf growth. *Sci. Pap. Coll. gen. Educn. Univ. Tokyo*, **9**, 67–104.
KURAISHI, S. & MUIR, R. M. (1962). Increase in diffusible auxin after treatment with gibberellin. *Science N. Y.*, **137**, 760–1.
KURAISHI, S. & MUIR, R. M. (1963). Diffusible auxin increase in a rosette plant treated with gibberellin. *Naturwissenchaften*, **50**, 337–8.
KURAISHI, S. & MUIR. R. M. (1963). Mode of action of growth retarding chemicals. *Pl. Physiol. Lancaster*, **38**, 19–24.
KURAISHI, S. & MUIR. R. M. (1964*a*). The relationships of gibberellin and auxin in plant growth. *Pl. Cell Physiol.*, **5**, 61–70.
KURAISHI, S. & MUIR. R. M. (1964*b*). Paper chromatographic study of diffusible auxin. *Pl. Physiol. Lancaster*, **39**, 23–8.
KURAISHI, S. & MUIR, R. M. (1964*c*). The mechanism of gibberellin action in pea. *Pl. Cell Physiol.*, **5**, 259–72.
KURAISHI, S. & YAMAKI, T. (1964). A simplified *Avena* curvature test using cut coleoptiles. *Bot. Mag., Tokyo*, **77**, 199–205.
KUROSAWA, E., (1926). [Experimental studies on the secretion of the 'bakanae' fungus on rice plants.] In Japanese. *Trans. nat. Hist. Soc. Formosa*, **16**, 213–27.
KUROSAWA, E. (1926). Experimental studies on the secretion of *Fusarium heterosporum* on rice plants. *Trans. nat. Hist. Soc. Formosa*, **20**, 218–39.
KUSE. G. (1953). Effect of 2,3,5-tri-iodobenzoic acid on the growth of the lateral bud and on tropism of petiole. *Mem. Coll. Sci. Univ. Kyoto, Ser. B*, **20**, 207.

KUSE, G. (1958). Necessity of auxin for the growth effect of gibberellin. *Bot. Mag., Tokyo*, **71**, 152–9.
KUTÁČEK, M. (1967). Indolderivate in Pflanzen der Familie Brassicaceae. *Wiss. Z. Univ. Rostock Math,-Naturwiss.*, (1/5), **16** 417–26.
KUTÁČEK, M. & GALSTON, A. W. (1968). The metabolism of ^{14}C-labelled isatin and anthranilate in *Pisum* stem sections. *Pl. Physiol. Lancaster*, **43**, 1793–8.
KUTÁČEK, M. & KEFELI, V. I. (1968). The present knowledge of indole compounds in plants of the Brassicaceae family. In Wightman, F. & Setterfield, G. (eds.), *Biochemistry and Physiology of Plant Growth Substances.* Runge Press, Ottawa, pp. 127–52.
KUTÁČEK, M., MAŠEV, N., OPLIŠTILOVÁ, K. & BULGAKOV, R. (1966). The influence of gamma radiation on the biosynthesis of indoles and gibberellins in barley. The action of zinc on the restitution of growth substances levels in irradiated plants. *Biologia Pl. (Prague)*, **8**, 152–63.
KUTÁČEK, M. & PROCHÁZKA, Z. (1964). Méthodes de détermination et d'isolement des composés indoliques chez les crucifères. In *Régulateurs naturels de croissance végétale, Colloq. int. Centre natl. Rech. Sci.*, No. **123**, 445–56.
LAAN, P. A. VAN DER (1934). Der Einfluss von Aethylen auf die Wuchsstoffbildung bei *Avena* und *Vicia. Recl. Trav. bot. néerl.*, **31**, 691–742.
LABARGA, C., NICHOLLS, P. B. & BANDURSKI, R. S. (1966). A partial characterisation of indolacetylinositol from *Zea mays. Biochem. biophys. Res. Commun.*, **20**, 641–6.
LABORIE, M.-E. (1963). Contribution à l'étude des actions de la gibbérelline et du chlorure de chlorocholine sur le métabolisme des pigments foliaires. *Annls. Physiol. vég. Paris*, **5**, 89–113.
LAGERSTEDT, H. B. & LANGSTON, R. G. (1966). Transport of kinetin-8-^{14}C in petioles. *Physiologia Pl.*, **19**, 734–40.
LAIBACH, F. (1952). Wuchsstoff und Blütenbildung. *Beitr. Biol. Pfl.*, **29**, 129–41.
LAIBACH, F. & KRIBBEN, F. J. (1950). Der Einfluss von Wuchsstoff auf die Bildung männlicher und weiblicher Blüten bei einer monözischen Pflanze (*Cucumis sativus* L.). *Ber. dt. bot. Ges.*, **62**, 53.
LAIBACH, F. & KRIBBEN, F. J. (1951). Die Bedeutung des Wuchsstoff für die Bildung und Geschlechtsbestimmung der Blüten. *Beitr. Biol. Pfl.*, **28**, 131.
LAMPORT, D. T. A. (1962). Hydroxyproline of primary cell walls. *Fedn. Proc. Fedn. Am. Soc. exp. Biol.*, **21**, 398.
LAMPORT, D. T. A. (1965). The protein component of primary cell walls. *Adv. bot. Res.* **2**, 151–218.
LAMPORT, D. T. A. (1967). Hydroxyproline-*o*-glucoside linkage of the plant cell wall glycoprotein extensin. *Nature, Lond.*, **216**, 1322–4.
LAMPORT, D. T. A. (1970). Cell wall metabolism. *A. Rev. Pl. Physiol.*, **21**, 235–70.
LAMPORT, D. T. A. & NORTHCOTE, D. H. (1960). Hydroxyproline in primary cell walls of higher plants. *Nature, Lond.*, **188**, 665–6.
LANE, F. E. & BAILEY, L. F. (1964). Isolation and characterisation studies

on the β-inhibitor in dormant buds of the silver maple (*Acer saccharinum* L.) *Physiologia Pl.*, **17**, 91–9.

LANG, A. (1956*a*). Stem elongation in a rosette plant, induced by gibberellic acid. *Naturwissenschaften*, **43**, 257–8.

LANG, A. (1956*b*). Bolting and flowering in biennial *Hyoscyamus niger*, induced by gibberellin. *Pl. Physiol. Lancaster*, **31**, Suppl. xxxv.

LANG, A. (1957). The effect of gibberellin upon flower formation. *Proc. natl. Acad. Sci. U.S.A.*, **43**, 709–17.

LANG, A. (1960). Gibberellin-like substances in photoinduced and vegetative plants. *Planta*, **54**, 498–504.

LANG, A. (1961). Auxins in flowering. *Encyclopaedia of Plant Physiology*. Springer, Berlin, Göttingen, Heidelberg, vol. 14, pp. 909–50.

LANG, A. (1965). Physiology of flower initiation. *Encyclopaedia of Plant Physiology*, Springer, Berlin, Göttingen, Heidelberg, vol. 15(1), pp. 1380–1536.

LANG, A. (1966). Intercellular regulation in plants. *Symp. Soc. devl. Biol.* No. 25, Academic Press, New York & London.

LANG, A. (1970). Gibberellins: Structure and metabolism. *A. Rev. Pl. Physiol.*, **21**, 537–70.

LANG, A. & LIVERMAN, J. L. (1954). The role of auxin in the photoperiodic response of long-day plants. *Rep. Commun. 8th. int. bot. Congr., Sect.* **11**, pp. 330–1.

LANG, A. & NITSEN, J. (1967). Relations among cell growth, DNA synthesis and gibberellin action. *Ann. N. Y. Acad. Sci.*, **144**(1), 180–90.

LANG, A., SANDOVAL, J. A. & BERDI, A. (1957). Induction of bolting and flowering in *Hyoscyamus* and *Samolus* by a gibberellin-like material from a seed plant. *Proc. natl. Acad. Sci. U.S.A.*, **43**, 960–4.

LANTICAN, B. P. & MUIR. R. M. (1967). Isolation and properties of the enzyme system forming indoleacetic acid. *Pl. Physiol. Lancaster*, **42**, 1158–60.

LARSEN, P. (1936). Über einen wuchsstoffinaktivierenden Stoff aus *Phaseolus*-Keimpflanzen. *Planta*, **25**, 311–4.

LARSEN, P. (1940). Untersuchungen über den thermolabilen wuchsstoffoxydierenden Stoff in *Phaseolus*-Keimpflanzen. *Planta*, **30**, 673–82.

LARSEN, P. (1944). 3-Indoleacetaldehyde as a growth hormone in higher plants. *Dansk bot. Ark.*, **11**, 11–132.

LARSEN, P. (1949). Conversion of indolacetaldehyde to indoleacetic acid in excised coleoptiles and in coleoptile juice. *Am. J. Bot.*, **36**, 32–41.

LARSEN, P. (1950). Quantitative relationships in the enzymatic conversion of indole- and naphthalene-acetaldehyde to auxins. *Am. J. Bot.*, **37**, 680.

LARSEN, P. (1951). Enzymatic conversion of indole acetaldehyde and naphthalene acetaldehyde to auxins. *Pl. Physiol. Lancaster*, **26**, 697–707.

LARSEN, P. (1955*a*). Nomenclature of plant growth substances. *Pl. Physiol. Lancaster*, **30**, 190–1.

LARSEN, P. (1955*b*). Growth substances in higher plants. In Paech K. and Tracey, M. V. (eds.), *Modern Methods of Plant Analysis*, Springer, Berlin, vol. 3, pp. 565–625.

LARSEN, P. (1961). Biological determination of natural auxins. *Encyclopaedia of Plant Physiology*. Springer, Berlin, Göttingen, Heidelberg, vol. 14, 521–82.
LARSEN, P. (1967). The biogenesis of indole auxins. *Wiss. Z. Univ. Rostock, Math.-Naturwiss*. **16**, (4/5), 431–8.
LARSEN, P. & RAJAGOPAL, R. (1964). The activity of indole-3-acetaldehyde in relation to that of indole-3-acetic acid in various auxin tests. In *Régulateurs naturels de la croissance végétale, Colloq. int. Centre natl. Rech. Sci.*, No. 123, 221–34.
LAZER, L., DALZIEL, A. M., BAUMGARTNER, W. E., DAHLSTROM, R. V. & MORTON, B. J. (1961). Determination of residual tritium-labelled gibberellic acid in potatoes, grapes and products derived from barley by isotopic dilution techniques. *Adv, Chem.*, **28**, 116–21.
LEAL, J. A., FRIEND, J. & HOLLIDAY, P. (1964). A factor controlling sexual reproduction in *Phytophthora., Nature, Lond.*, **203**, 545–6.
LEAPER, J. M. F. & BISHOP, J. R. (1951). Relation of halogen position to physiological properties in the mono-, di- and trichlorophenoxyacetic acids. *Bot. Gaz.*, **112**, 250–8.
LEEPER, R. W., GOWING, D. P. & STEWART, W. S. (1962). Decarboxylation of alpha-naphthaleneacetic acid by pineapple leaves in sunlight. *J. appl. Radiation Isotopes*, **13**, 399–402.
LEIKE, H. (1967). Beeinflussung des Auxintransports durch Kinetin. *Wiss. Z. Univ. Rostock, Math.-Naturwiss.*, **16**(4/5), 501–2.
LEO, P. DE, DALESSANDRO, G. & ARRIGONI, O. (1968). [Effects of growth retardants on oxidative phosphorylation and on cyclic photophosphorylation.] In Italian. *Nuovo G. bot. ital.*, **102**, 73–80.
LEO, P. DE & SACHER, J. A. (1969). Control of RNAase activity and RNA and protein synthesis during senescence of *Rhoeo* leaf sections. *Abstr. 11th. int. bot. Congr. Seattle.*
LEONARD, N. J., HECHT, S. M., SKOOG, F. & SCHMITZ, R. Y. (1969). Cytokinins: Synthesis, mass spectra and biological activities of compounds related to zeatin. *Proc. natl. Acad. Sci. U.S.A.*, **63**, 175–82.
LEOPOLD, A. C. (1963). The polarity of auxin transport. In *Meristems and Differentiation. Brookhaven Symp. Biol.*, No. 16, 218–34.
LEOPOLD, A. C. & FUENTE, R. K. DELA (1968). A view of polar auxin transport. In Vardar, Y. (ed.), *Transport of Plant Hormones*, North-Holland, pp. 24–47.
LEOPOLD, A. C. & GUERNSEY, F. S. (1953*a*). Modification of floral initiation with auxins and temperatures, *Am. J. Bot.* **40**, 603–7.
LEOPOLD, A. C. & GUERNSEY, F. S. (1953*b*). Flower initiation in Alaska pea. I Evidence as to the role of auxin. *Am. J. Bot.*, **40**, 46–50.
LEOPOLD, A. C. & GUERNSEY, F. S. (1953*c*). Auxin polarity in the *Coleus* plant. *Bot. Gaz.*, **115**, 147–54.
LEOPOLD, A. C. & GUERNSEY, F. S. (1953*d*). A theory of auxin action involving coenzyme A. *Proc. natl. Acad. Sci. U.S.A.*, **39**, 1105.
LEOPOLD, A. C. & GUERNSEY, F. S. (1954). Flower initiation in the Alaska pea. II Chemical vernalisation. *Am. J. Bot.*, **41**, 181–5.

LEOPOLD, A. C. & HALL, O. F. (1966). Mathematical model of polar auxin transport. *Pl. Physiol. Lancaster*, **41**, 1476–80.
LEOPOLD, A. C. & KLEIN, W. H. (1952). Maleic hydrazide as an anti-auxin. *Physiologia Pl.*, **5**, 91–9.
LEOPOLD, A. C. & LAM, S. L. (1961). Polar transport of three auxins. In Klein, R. M. (ed.), *Plant Growth Regulation*. Iowa State Univ. Press, pp. 411–18.
LEOPOLD, A. C. & LAM, S. L. (1962). The auxin transport gradient. *Physiologia Pl.*, **15**, 631–8.
LEOPOLD, A. C. & PRICE, C. A. (1956). The influence of growth substances upon sulphydryl compounds. In Wain, R. L. & Wightman, F. (eds.), *The Chemistry and Mode of Action of Plant Growth Substances*. Butterworths, London, pp. 271–83.
LEOPOLD, A. C., SCOTT, F. T., KLEIN, W. H. & RAMSTAD, E. (1952). Chelidonic acid and its effects on plant growth. *Physiologia Pl.*, **5**, 85–90.
LEOPOLD, A. C. & THIMANN, K. V. (1949). The effect of auxin on flower initiation. *Am. J. Bot.*, **36**, 342–7.
LERNER, H. R., MAYER, A. M. & EVENARI, M. (1959). The nature of germination inhibitors present in dispersal units of *Zygophyllum dumosum* and *Trigonella arabica*. *Physiologia Pl.*, **12**, 245–50.
LESHEM, Y. (1971). Abscisic acid as a ribonuclease promoter. *Physiologia Pl.*, **24**, 85–9.
LETHAM, D. S. (1963*a*). Zeatin, a factor inducing cell division isolated from *Zea mays*. *Life Sci.*, No. 8, pp. 569–73.
LETHAM, D. S. (1963*b*). Regulators of cell division in plant tissues. I Inhibitors and stimulants of cell division in developing fruits; their properties and activity in relation to the cell division period. *N. Z. J. Bot.*, **1**, 336–50.
LETHAM, D. S. (1964). Isolation of a kinin from plum fruitlets and other tissues. In *Régulateurs naturels de la croissance végétale, Colloq, int. Centre natl. Rech. Sci.*, No. 123, pp. 109–117.
LETHAM, D. S. (1966*a*). Purification and probable identity of a new cytokinin in sweet corn extracts. *Life Sci.*, **5**, 551–4.
LETHAM, D. S. (1966*b*). Isolation and identity of a third cytokinin in sweet corn extracts. *Life. Sci.*, **5**, 1999–2004.
LETHAM, D. S. (1967*a*). Regulators of cell division in plant tissues. V A comparison of the activities of zeatin and other cytokinins in five bioassays. *Planta*, **74**, 228–42.
LETHAM, D. S. (1967*b*). Chemistry and physiology of kinetin-like compounds. *A. Rev. Pl. Physiology*, **18**, 349–64.
LETHAM, D. S. (1968). A new cytokinin bioassay and the naturally occurring cytokinin complex. In Wightman, F. & Setterfield G. (eds.), *Biochemistry and Physiology of Plant Growth Substances*. Runge Press, Ottawa., pp. 19–31.
LETHAM, D. S. & MILLER, C. O. (1965). Identity of kinetin-like factors from *Zea mays*. *Pl. Cell Physiol.*, **6**, 355–9.
LETHAM, D. S., SHANNON, J. S. & MCDONALD, I. R. (1964). The structure of zeatin, a factor inducing cell division. *Proc. chem. Soc.*, 230–1.

LETHAM, D. S. & WILLIAMS, M. W. (1969). Regulators of cell division in plant tissue. VIII The cytokinins of apple fruit. *Physiologia Pl.*, **22**. 925–36.
LEVITT, J. (1953). Further remarks on the thermodynamics of active (non-osmotic) water absorption. *Physiologia Pl.*, **6**, 240–52.
LEWIN, R. A. (1954). Sex in unicellular algae. In *Sex in Microorganisms. Am. Ass, Adv. Sci. Wash.*, pp. 100.
LEWIS, L., KHALIFAH, R. A. & COGGINS, C. W. Jr. (1965). The existence of the non-indolic citrus auxin in several plant families. *Phytochemistry*, **4**, 203–5.
LIBBERT, E. & GERDES, I. (1964). Können Gibberelline an tropischen Krümmung beteiligt sein? *Planta*, **61**, 245–58.
LIBBERT, E., KAISER, W. & KUNERT, R. (1969). Interaction between plants and epiphytic bacteria regarding their auxin metabolism. VI The influence of the epiphytic bacteria on the content of extractible auxins in the plant. *Physiologia Pl.*, **22**, 432–9.
LIBBERT, E. & LÜBKE, H. (1960). Physiologische Wirkungen des Scopoletins. V Scopoletin und Strekungswachstum von Koleoptilzylindern. *Flora*, **149**, 95–105.
LIBBERT, E., WICHNER, S., DUERST, E., KAISER, W., KUNERT, R., MANIKI, A., MANTEUFFEL, R., RIECKE, E. & SCHRÖDER, R. (1968). Auxin content and auxin synthesis in sterile and non-sterile plants with regard to the influence of epiphytic bacteria. In Wightman, F. and Setterfield, G. (eds.), *Biochemistry and Physiology of Plant Growth Substances*, Runge Press, Ottawa, pp. 213–230.
LIBBERT, E., WICHNER, S., SCHIEWER, U., RISCH, H. & KAISER, W. (1966). The influence of epiphytic bacteria on auxin metabolism. *Planta*, **68**, 327–34.
LIEBERMAN, M., KUNISHI, A., MAPSON, L. W. & WARDALE, D. A. (1966). Stimulation of ethylene production in apple tissue slices by methionine. *Pl. Physiol. Lancaster*, **41**, 376–82.
LIEBERMAN, M. & MAPSON, L. W. (1962). Fatty acid control of ethane production by sub-cellular particles from apples and its possible relationship to ethylene biosynthesis. *Nature, Lond.*, **195**, 1016–7.
LIEBERMAN, M. & MAPSON, L. W. (1964). Genesis and biogenesis of ethylene. *Nature, Lond.*, **204**, 343–5.
LINCOLN, R. G., CUNNINGHAM, A., CARPENTER, B. H., ALEXANDER, J. & MAYFIELD, D. L. (1966). Florigenic acid from fungal culture. *Pl. Physiol. Lancaster*, **41**, 1079–80.
LINCOLN, R. G., MAYFIELD, D. L. & CUNNINGHAM, A. (1961). Preparation of a floral initiating extract from *Xanthium*. *Science N. Y.*, **133**, 756.
LINSER, H. (1938). Zur Methodik der Wuchsstoffbestimmung. *Planta*, **28**, 227–56.
LINSER, H. (1951). Versuche zur chromatographischen Trennung pflanzlicher Wuchsstoffe. *Planta*, **39**, 377–401.
LINSER, H. (1954). Chemische Konstitution und Zellstrekungswirkung verschiedener Stoffe. *Mheft. Chem.*, **85**, 196.
LINSER, H. & KAINDL, K. (1951). The mode of action of growth substances and growth inhibitors. *Science N. Y.*, **114**, 69–70.

LINSER, H. & KIERMAYER, O. (1956). Fluoressensanalytischer Nachweis von Indolkörpern insbesondere Indol-3-acetonitril auf Papierchromatogrammen. *Biochim. biophys. Acta.*, **21**, 382.

LINSKENS, H. F. & KNAPP, R. (1955). Über die Ausscheidung von Aminosäuren in seinem und gemischten Beständen verschiedener Pflanzenarten, *Planta*, **45**, 106–117.

LIPE, N. W. & CRANE, J. C. (1966). Dormancy regulation in peach seeds. *Science N. Y.*, **153**, 541–2.

LIVERMAN, J. L. & LANG, A. (1956). Induction of flowering in long-day plants by applied indoleacetic acid. *Pl. Physiol. Lancaster*, **31**, 147–50.

LIVINGSTONE, B. E. (1907). Further studies on the properties of unproductive soils. *Bull. U. S. Divn. Soils.*, No. 36.

LLOYD, R. F., SKINNER, C. G. & SHIVE, W. (1967). 4-Substituted-pteridines, analogues of kinetin. *Can. J. Chem.*, **45**, 2213–6.

LOCHHEAD. A. G. (1958). Soil bacteria and growth-promoting substances. *Bact. Rev.*, **22**, 145–53.

LOCHHEAD, A. G. & BURTON, M. O. (1953). An essential bacterial growth factor produced by microbial synthesis. *Can. J. Bot.*, **31**, 7–22.

LOCHHEAD, A. G. & THEXTON, R. H. (1951). Vitamin B_{12} as a growth factor for soil bacteria. *Nature, Lond.*, **167**, 1034.

LOCHHART, J. A. (1957). Studies on the organ of production of the natural gibberellin factor in higher plants. *Pl. Physiol. Lancaster*, **32**, 204–7.

LOCKHART, J. A. (1959). Control of stem growth by light and gibberellic acid. In Withrow, R. B. (ed.), *Photoperiodism and Related Phenomena in Plants and Animals. Am. Ass. Adv. Sci. Wash.*, pp. 217–221.

LOCKHART. J. A. (1962). Kinetic studies of certain antigibberellins. *Pl. Physiol. Lancaster*, **37**, 759–64.

LOCKHART, J. A. & DEAL, P. H. (1960). Prevention of red light inhibition of stem growth in the Cucurbitaceae by Gibberellin A_4. *Naturwissenschaften*, **47**, 141–2.

LOEFFLER, J. E. & OVERBEEK, J. VAN (1964). Kinin activity in coconut milk. In *Régulateurs naturels de la croissance végétale, Colloq. int. Centre natl. Rech. Sci.*, No. 123, 77–82.

LOEHWING, W. F. (1937). Root interactions of plants. *Bot. Rev.*, **3**, 195–239.

LONA, F. (1949). L'induzione fotoperiodica di foglie staccate. *Boll. Soc. ital. Biol. sper.*, **25**, 761–3.

LONA, F. (1955). Gibberellic acid induces germination of seeds of *Lactuca scariola* in the dark-inhibition phase. *L'Ateneo parmense*, **27**, 641–4.

LONA, F. (1956). [The action of gibberellic acid on the stem growth of some herbaceous plants under controlled external conditions.] In Italian. *Nuovo G. bot. ital.* **63**, 61–76.

LONA, F. & BOCCI, A. (1956). [The vegetative and reproductive development of some long-day plants in relation to the action of gibberellic acid.] In Italian. *Nuovo G. bot. ital.*, **68**, 469–86.

LONA, F. & BOCCI, A. (1957). [The morphogenesis and organogenic effects

produced by kinetin on herbaceous plants under controlled environmental conditions.] In Italian. *Nuovo G. bot. ital.*, **64**, 236–46.
LONA, F. & FIORETTI, L. (1962). [Growth and flowering of short- and long-day plants in relation to the diverse activities of the gibberellins ($A_1 - A_9$). In Italian. *Annali. Bot.* **27**, 313–22.
LÖVE, A. & LÖVE, D. (1945). Experiments on the effects of animal sex hormones on dioecious plants. *Ark. Bot.*, **32** A, 1–60.
LOVEYS, B. R. & WAREING, P. F. (1971). The red-light-controlled production of gibberellin in etiolated wheat leaves. *Planta*, **98**, 109–116.
LUCKWILL, L. C. (1952). Growth-inhibiting and growth-promoting substances in relation to the dormancy and after-ripening of apple seed. *J. hort. Sci.*, **27**, 53–67.
LUCKWILL, L. C. & WOODCOCK, D. (1956). Relationship of molecular structure of some naphthyloxy compounds and their biological activity as plant growth regulating substances. In Wain, R. L. and Wightman, F. (eds.), *The Chemistry and Mode of Action of Plant Growth Substances.* Butterworths, London, pp. 195–204.
LUNDEGÅRDH, H. & STENLID, G. (1944). On the exudation of nucleotides and flavones from living roots. *Ark. Bot.*, **31** A, 1–27.
LYON, J. L. & SMITH, O. E. (1966). Effects of gibberellins on abscission in cotton seedling explants. *Planta*, **69**, 347–56.
LYSENKO, T. D. (1928), *Trud. Azerbaidzan Op. Stn.*, No. 3, p. 168. (From Murneek and Whyte, 1948.)
MACHLIS, L. (1958). A procedure for the purification of sirenin. *Nature, Lond.*, **181**, 1790–1.
MACHLIS, L. (1966). Sex hormones in fungi. In Ainsworth, G. C. & Sussman, A. S. (eds.), *The Fungi.* Academic Press, New York & London, vol. 2, chap. 13.
MACHLIS, L., NUTTING, W. H. & RAPOPORT, H. (1968). The structure of sirenin. *J. Am. chem. Soc.*, **90**, 1674–6.
MACLACHLAN, G. A., DAVIES, E. & FAN, D. F. (1968). Induction of cellulase by 3-indoleacetic acid. In Wightman, F. & Setterfield, G., (eds.), *Biochemistry and Physiology of Plant Growth Substances.* Runge Press, Ottawa, pp. 443–54.
MACLACHLAN, G. A. & WAYGOOD, E. R. (1956). Kinetics of the enzymatically catalysed oxidation products of indoleacetic acid. *Pl. Physiol. Lancaster*, **30**, 225–31.
MACLEOD, A. M., DUFFUS, J. H. & JOHNSTON, C. S. (1964). Development of hydrolytic enzymes in germinating grain. *J. Inst. Brew.*, **70**, 521–8.
MACMILLAN, J. (1968). Direct identification of gibberellins in plant extracts by gas chromatography-mass spectrometry. In Wightman, F. and Setterfield, G. (eds.), *Biochemistry and Physiology of Plant Growth Substances.* Runge Press, Ottawa, pp. 101–107.
MACMILLAN, J. & PRYCE, R. J. (1968). Phaseic acid, a putative relation of abscisic acid from seed of *Phaseolus multiflorus. Chem. Commun.*, 124–6.

MacMillan, J., Seaton, J. C. & Suter, P. J. (1959). A new plant growth-promoting acid, gibberellin A_5 from the seed of *Phaseolus multiflorus*. *Proc. chem. Soc., Lond.*, pp. 325.
MacMillan, J., & Suter, P. J. (1958). The occurrence of gibberellin A_1 in higher plants. Isolation from the seed of runner bean (*Phaseolus multiflorus*). *Naturwissenschaften*, **45**, 46.
MacNicol, P. K. & Reinert, J. (1963). Untersuchungen über Auxinoxydasen II. Auftrennung und IES-Oxydaseactivität der multiplen Peroxydasen des etiolierten Erbsensprosses. *Z. Naturf.*, **18** b, 572–9.
Madison, J. T., Everett, G. A. & Kung, H. (1967). Oligonucleotides from yeast tyrosine transfer ribonucleic acid. *J. biol. Chem.*, **242**, 1318.
Maeda, E. (1960). Interaction of gibberellin and auxins in lamina joints of excised rice leaves. *Physiologia Pl.*, **13**, 214–26.
Mahadevan, S. & Thimann, K. V. (1964). Nitrilase II. Substrate specificity and possible mode of action. *Archs Biochem. Biophys.*, **107**, 62–8.
Mann, H. H. & Barnes, T. W. (1953). The mutual effect of rye grass and clover when grown together. *Ann. appl. Biol.*, **40**, 556.
Mann, J. D. & Jaworski, E. G. (1970). Minimising loss of indoleacetic acid during purification of plant extracts. *Planta*, **92**, 285–91.
Manning, D. T. & Galston, A. W. (1955). On the nature of the enzymatically catalysed oxidation products of indoleacetic acid. *Pl. Physiol. Lancaster*, **30**, 225–31.
Marcelle, R. & Sironval, C. (1964). Action du Phosphon D et de la gibbérelline sur la croissance et la floraison de *Bryophyllum tubiflorum*. Harv. *Physiol. vég.*, **2**, 409–18.
Marinos, N. G. (1957). Responses of *Avena* coleoptile sections to high concentrations of auxins. *Aust. J. biol. Sci.*, **10**, 147–63.
Marré, E. (1954). [Researches on the physiology of ascorbic acid X. Quantitative changes in ascorbic acid in *Avena* coleoptiles and in segments of the internode of *Pisum* treated with indoleacetic acid.] In Italian. *Atti. Accad. naz. Lincei Rc. Cl. sci. fis. mat. nat.*, **16**, 758–63.
Marré, E. & Arrigoni, O. (1955). Ulteriori ricerche sull'azione inibente dell'auxina nei confronti dell'ossidasi dell'acido ascorbico. *Atti. Accad. naz. Lincei Rc. Cl. sci. fis. mat. nat.*, **18**, 539–47.
Marré. E. & Arrigoni, O. (1957). Metabolic reactions of auxins. I The effect of auxins on glutathione and the effects of glutathione on the growth of isolated plant parts. *Physiologia Pl.*, **10**, 289–304.
Marré, E. & Bianchetti, R. (1961). Metabolic responses to auxins. VI The effect of auxin on the oxidation-reduction state of triphosphopyridine nucleotide. *Biochim. biophys. Acta.*, **48**, 583–5.
Marré, E. & Forti, G. (1958). Metabolic responses to auxins. II The effect of auxin on ATP level as related to the auxin-induced respiration increase. *Physiologia Pl.*, **11**, 36–47.
Marré, E., Laudi, G. & Arrigoni, O. (1955). [Researches on the physiology of ascorbic acid. XV Inhibitory action of dehydroascorbic acid on the

dehydrogenase activity of plant enzyme preparations.] In Italian. *Atti. Accad. naz. Lincei Rc. Cl. sci. fis. mat. nat.*, **19**, 460–5.
MARTIN, J. P. (1950). Effect of various leaching treatments on growth of orange seedlings in old citrus soils. *Soil Sci.*, **69**, 433–42.
MARUMO, S., ABE, H., HATTORI, H. & MUNAKATA, K. (1968). Isolation of a novel auxin, methyl 4-chloroindoleacetate from immature seeds of *Pisum sativum*. *Agric. biol. Chemy.*, **32**, 117–8.
MARUMO, S. & HATTORI, H. (1970). Isolation of D-4-chlorotryptophan derivatives from immature seeds of *Pisum sativum*. *Planta*, **90**, 208–11.
MARUSHIGE, K. & MARUSHIGE, Y. (1962). Effects of 8-azaguanine and ethionine on floral initiation and development in seedlings of *Pharbitis nil* Chois. *Bot. Mag., Tokyo*, **75**, 270–2.
MASUDA, Y. (1953). Über den Einfluss von Auxin auf die Stoffpermeabilität des Protoplasmas. *Bot. Mag., Tokyo*, **66**, 785–6.
MASUDA, Y. (1955). Über den Einfluss von Auxin auf die Stoffpermeabilität das Protoplasmas II Mitteilung. Harnstoffpermeabilität des Protoplasmas von *Avena* Koleoptile. *Bot. Mag., Tokyo*, **68**, 180–3.
MASUDA, Y. (1967). Auxin-induced expansion growth of Jerusalem artichoke tuber tissue in relation to nucleic acid and protein metabolism. *Ann. N. Y. Acad. Sci.*, **144** (1), 68–80.
MASUDA, Y. (1968). Requirement of RNA biosynthesis for the auxin-induced elongation of oat coleoptile and pea stem cells. In Wightman, F. and Setterfield, G. (eds.), *Biochemistry and Physiology of Plant Growth Substances*, Runge Press, Ottawa, pp. 699–710.
MASUDA, Y. & KAMISAKA, S. (1969). Rapid stimulation of RNA biosynthesis by auxin. *Pl. Cell. Physiol.*, **10**, 79–86.
MASUDA, Y. & TANIMOTO, E. (1967). Effect of auxin and antiauxin on the growth and RNA synthesis of etiolated pea internodes. *Pl. Cell Physiol.*, **8**, 459–65.
MASUDA, Y., TANIMOTO, E. & WADA, S. (1967). Auxin-stimulated RNA synthesis in oat coleoptile cells. *Physiologia Pl.*, **20**, 713–9.
MATELL, M. (1953). *Stereochemical Studies on Plant Growth Substances.* Alquist & Wikseus, Uppsala.
MATSUBARA, S., ARMSTRONG, D. J. & SKOOG, F. (1968). Cytokinins in tRNA of *Corynebacterium fascians*. *Pl. Physiol. Lancaster*, **43**, 451–3.
MAVRODINEANU, R., SANFORD, W. W. & HITCHCOCK, A. E. (1955). Use of fluorescence for the estimation of substances separated on paper by partition chromatography. *Contrib. Boyce Thompson Inst.*, **18**, 167–72.
MAYER, A. M. & POLJAKOFF-MAYBER, A. (1963). *The Germination of Seeds*, Pergamon Press, Oxford.
MAYR, H. R. & PRESOLY, E. (1963). Untersuchungen an mit Chlorcholinchlorid (CCC) behandelten Weisenpflanzen. Anatomisch-morphologische Ergebnisse. *Z. Acker- u. PflBau.*, **118**, 109–24.
McCALLA, T. M. & HASKINS, F. A. (1964). Phytotoxic substances from soil microorganisms and crop residues. *Bact. Rev.*, **28**, 181

McComb, A. J. (1961). 'Bound' gibberellin in mature runner bean seeds. *Nature, Lond.*, **192**, 575–6.
McComb, A. J. (1964). The stability and movement of gibberellic acid in pea seedlings. *Ann. Bot.*, **28**, 669–87.
McComb, A. J. (1967). The control by gibberellic acid of stem elongation and flowering in biennial plants of *Centaurium minus* Moench. *Planta*, **76**, 242–51.
McComb, A. J. & Carr, D. J. (1958). Evidence from a dwarf pea bioassay of naturally occurring gibberellins in the growing plant. *Nature, Lond.*, **181**, 1548–9.
McComb, A. J., McComb, J. A. & Duda, C. T. (1970). Increase of ribonucleic acid polymerase activity associated with chromatin from internodes of drawf pea plants. *Pl. Physiol. Lancaster*, **46**, 221–3.
McCready, C. C. (1963). Movement of growth regulators in plants. I Polar transport of 2,4-dichlorophenoxyacetic acid in segments from the petiole of *Phaseolus vulgaris*. *New Phytol.*, **62**, 3–18.
McCready, C. C. (1968*a*). The acropetal movement of auxin through segments excised from petioles of *Phaseolus vulgaris*. In Vardar, Y. (ed.), *Transport of Plant Hormones*, North Holland, Amsterdam, pp. 108–29.
McCready, C. C. (1968*b*). The polarity of auxin movement in segments excised from petioles of *Phaseolus vulgaris* L. In Wightman F. and Setterfield, G. (eds.), *Biochemistry and Physiology of Plant Growth Substances*. Runge Press, Ottawa, pp. 1005–23.
McCready, C. C. & Jacobs, W. P. (1963). Movement of growth regulators in plants. II Polar transport of radioactivity from indoleacetic acid-(^{14}C) and 2,4-dichlorophenoxyacetic acid-(^{14}C) in petioles of *Phaseolus vulgaris*. *New Phytol.*, **62**, 19–34.
McCready, C. C., Osborne, D. J. & Black, M. K. (1965). Promotion by kinetin of the polar transport of two auxins. *Nature*, *Lond.*, **208**, 1065–7.
MacCune, D. C. & Galston, A. W. (1959). Inverse effects of gibberellin on peroxidase activity and growth in dwarf strains of pea and corn. *Pl. Physiol. Lancaster*, **34**, 416–8.
McDaniel, R. G. & Sarkissian, I. V. (1966). Enhancement of oxidation and phosphorylation of maize scutellum mitochondria by physiological concentrations of indoleacetic acid. *Physiologia Pl.*, **19**, 187–93.
McDonald, J. J., Leonard, N. J. Schmitz, R. Y. & Skoog, F. (1971). Cytokinins: Synthesis and biological activity of ureidopurines. *Phytochemistry*, **10**, 1429–39.
McElroy, W. D. & Mason, A. (1954). Mechanism of action of micronutrient elements in enzyme systems. *A. Rev. Pl. Physiol.*, **5**, 1–30.
McIlrath, W. J. (1950). Response of the plant to maleic hydrazide. *Am. J. Bot.*, **37**, 816–8.
McMorris, T. C. & Barksdale, A. W. (1967). Isolation of sex hormones from the water mould *Achlya bisexualis*. *Nature, Lond.*, **215**, 320–1.
McPherson, J. K. & Muller, C. H. (1969). Allelopathic effects of *Adenostoma fasciculata*, 'Chamise', in the California chaparral. *Ecol. Monogr.*, **39**, 177–98.

McRae, D. H. & Bonner, J. (1952). Diortho substituted phenoxyacetic acids as antiauxins. *Pl. Physiol. Lancaster*, **27**, 834–8.
McRae, D. H. & Bonner, J. (1953). Chemical structure and antiauxin activity, *Physiologia Pl.*, **6**, 485–510.
McRae, D. H., Foster, R. J. & Bonner, J. (1953). Kinetics of auxin activity. *Pl. Physiol. Lancaster*, **28**, 343–55.
Mees, G. C. (1965). The plant-growth stimulating properties of ethylene-bis-nitrourethane and related compounds. *J. exp. Bot.*, **16**, 48–58.
Meinl, G. & Guttenberg, H. von (1952). Über den Einfluss von Wirkstoffen auf die Permeabilität des Protoplasmas, III und IV Mitteilungen. *Planta*, **41**, 167–89.
Melchers, G. (1937). Die Wirkung von Genen, tiefen Temperaturen und blühenden Propfpartneren auf die Blühreife von *Hyoscyamus niger*. *Biol. Zbl.*, **57**, 568–614.
Melchers, G. (1939). Die Blühhormone. *Ber. dt. bot. Ges.*, **57**, 29–48.
Melchers, G. & Lang, A. (1941). Weitere Untersuchungen zur die Frage der Blühhormone. *Biol. Zbl.*, **61**, 16–39.
Melchior, G. H. (1957). Über den Abbau von Indolderivaten. I Photolyse durch ultraviolettes Licht. *Planta*, **50**, 262–90.
Meyer, J. (1958). Die photolytischen Abbauprodukte der 3-Indolessigsäure und ihre physiologische Wirkung auf das Wachstum der *Avena*-Koleoptile. *Z. Bot.*, **46**, 125–60.
Meyer, J. & Pohl, R. (1956). Neue Erkenntniss zum Problem der phototropen Krümmung bei höheren Pflanzen. *Naturwissenschaften*, **43**, 5.
Meylan, S. & Pilet, P.-E. (1966). Traitement auxinique et variation de la conductivité d'un tissu végétale. *Physiol. vég.*, **4**, 221–36.
Michener, H. D. (1942). Dormancy and apical dominance in potato tubers. *Am. J. Bot.*, **29**, 558–68.
Michniewicz, M. & Kamienska, A. (1964). Flower formation induced by kinetin and vitamin E treatment in the cold-requiring plant (*Cichorium intybus* L.) grown under non-inductive conditions. *Naturwissenschaften*, **51**, 295–6.
Michniewicz, M. & Kamienska, A. (1965). Flower formation induced by kinetin and vitamin E treatment in long-day plant (*Arabidopsis thaliana*) grown in short day. *Naturwissenachaften*, **52**, 623.
Michniewicz, M. & Lang, A. (1962). Effect of nine different gibberellins on stem elongation and flower formation in cold-requiring and photoperiodic plants grown under non-inductive conditions. *Planta*, **58**, 549–63.
Milborrow, B. V. (1967). The identification of (+)-abscisin II [(+)-dormin] in plants and measurements of its concentration. *Planta*, **76**, 93–113.
Milborrow, B. V. (1968). Identification and measurement of (+)-abscisic acid in plants. In Wightman F. and Setterfield G. (eds.), *Biochemistry and Physiology of Plant Growth Substances*. Runge Press, Ottawa, pp. 1531–45.
Miller, C. O. (1955). Kinetin and kinetin-like compounds. In Linskens, H. F. and Tracey, M. V. (eds.), *Modern Methods of Plant Analysis*. Springer, Berlin, Göttingen, Heidelberg, vol. 6, 194–202.

MILLER, C. O. (1961*a*). A kinetin-like compound in maize. *Proc. natl. Acad. Sci., U.S.A.*, **47**, 170–4.
MILLER, C. O. (1961*b*). Kinetin and related compounds in plant growth. *A. Rev. Pl. Physiol.*, **12**, 395–408.
MILLER, C. O. (1967). Zeatin and zeatin riboside from a mycorrhizal fungus. *Science, N. Y.*, **157**, 1055–7.
MILLER, C. O., SKOOG, F., OKUMURA, F. S., SALTZA, M. H. VON & STRONG, F. M. (1956). Isolation, structure and synthesis of kinetin, a substance promoting cell division. *J. Am. chem. Soc.*, **78**, 1375–80.
MILLER, C. O., SKOOG, F., von SALTZA, M. H. & STRONG, F. M. (1955). Kinetin, a cell division factor from deoxyribonucleic acid. *J. Am. chem. Soc.*, **77**, 1392.
MILLS, K. S., & SCHRANK, A. R. (1954). Electrical and curvature responses of the *Avena* coleoptile to unilateral ultraviolet irradiation. *J. cell. comp. Physiol.*, **43**, 39–65.
MITCHELL, E. D. & TOLBERT, N. E. (1968). Isolation from sugar beet fruit and characterisation of *cis*-4-cyclohexane-1,2-dicarboximide as a germination inhibitor. *Biochemistry, N. Y.*, **7**, 1019–25.
MITCHELL, J. W. & BRUNSTETTER, B. C. (1939). Colorimetric methods for the quantitative estimation of indole-3-acetic acid. *Bot. Gaz.*, **100**, 802–16.
MITCHELL, J. W., SMALE, B. C. & PRESTON, W. H. (1959). New plant regulators that exude from roots. *J. agric. Fd. Chem.*, **7**, 841–3.
MITCHELL, J. W., SKAGGS, D. P. & ANDERSON, W. P. (1951). Plant growth stimulating hormones in immature bean seeds. *Science, N. Y.*, **114**, 159.
MITCHELL, J. W., WIRWILLE, J. W. & WEIL, L. (1949). Plant growth-regulating properties of some nicotinium compounds. *Science N. Y.*, **110**, 252–4.
MITCHELL, K. K. & SARKISSIAN, I. V. (1966). Effect of indol-3yl-acetic acid on the citrate-condensing reaction by preparations of root and shoot tissue. *J. exp. Bot.*, **17**, 839–43.
MOEWUS, F. (1948). Ein neuer quantativer Test für pflanzliche Wuchsstoff. *Naturwissenschaften*, **4**, 124.
MOEWUS, F. (1950*a*). Die Bedeutung von Farbstoffen bei den Sexualprozessen der Algen und Blütenpflanzen. *Angew. Chem.*, **62**, 496–502.
MOEWUS, F. (1950*b*). Über ein Blastokolin der *Chlamydomonas*-Zygoten. *Naturforschung*, **5b**, 196–202.
MOEWUS, F. & BANERJEE, B. (1951). Über die Wirkung von *cis*-Zimtsäure und einigen isomeren Verbindungen auf *Chlamydomonas*-Zygoten. *Naturforschung*, **6b**, 210.
MOEWUS, F., MOEWUS, L. & SCHADER, E. (1951). Vorkommen und Bedeutung von Blastokolinen in fleischigen Früchten. *Naturforschung*, **6b**, 261–70.
MOEWUS, F. & SCHADER, E. (1951). Die Wirkung von Coumarin und Parasorbinsäure auf das Austreiben von Kartoffelknollen. *Naturforschung*, **6b**, 112–115.
MOORE, T. C. (1969). Comparative net biosynthesis of indoleacetic acid

from tryptophan in cell-free extracts of different parts of *Pisum sativum* plants. *Phytochemistry*, **8**, 1109–20.
MOORE, T. C. & SHANER, C. A. (1968). Synthesis of indoleacetic acid from tryptophan via indolepyruvic acid in cell-free extracts of pea seedlings. *Archs Biochem. Biophys.*, **127**, 613–21.
MORAL., R. DEL & MULLER, C. H. (1969). Fog drip: a mechanism of toxin transport from *Eucalyptus globulus*. *Bull. Torrey bot. Club*, **96**, 467–75.
MOREL, G. (1946). Action de l'acide pantothénique sur la croissance des tissus d'Aubépine. *C. r. hebd. Séanc. Acad. Sci., Paris*, **223**, 116–8.
MORGAN, C. & REITH, W. S. (1954). The compositions and quantitative relations of protein and related fractions in developing root cells. *J. exp. Bot.*, **5**, 119–35.
MORGAN, D. G. (1964). Influence of α-naphthylphthalamic acid on the movement of indole-3-acetic acid in plants. *Nature, Lond.*, **201**, 476–7.
MORGAN, D. G. (1965). Über die wachstumsfördernden Eigenschaften von Äthylen-bis-Nitrourethan und verwandten Verbindungen. *Ber. dt. bot. Ges.*, **78**, 143.
MORGAN, D. G. & SODING, H. (1958). Über der Wirkungsweise von Phthalsäure-mono-α-Naphthylamid (PNA) auf das Wachstum der Haferkoleoptile. *Planta*, **52**, 235–49.
MORGAN, P. W. & GAUSMAN, H. W. (1966). Effects of ethylene on auxin transport, *Pl. Physiol. Lancaster*, **41**, 45–52.
MORGAN. P. W. & HALL, W. C. (1962). Effect of 2,4-dichlorophenoxyacetic acid on the production of ethylene by cotton and grain sorghum. *Physiologia Pl.*, **15**, 420–7.
MORGAN, P. W. & HALL, W. C. (1964). Accelerated release of ethylene by cotton following application of indolyl-3-acetic acid. *Nature, Lond*, **201**, 99.
MORI, S., KUMAZAWA, K. & MITSUI, S. (1965). Stimulation of release of reducing sugar from the endosperms of rice seed by helminthosporol. *Pl. Cell Physiol.*, **6** 571–4.
MORRIS, D. A., BRIANT, R. E. & THOMPSON, P. G. (1969). The transport and metabolism of ^{14}C-labelled indoleacetic acid in intact pea seedlings. *Planta*, **89**, 178–97.
MORTENSEN, J. L. & HIMES, F. L. (1964). In Bear, F. (ed.), *Chemistry of the Soil*, 2nd edition, Reinhold, New York.
MOSHKOV, B. S. (1936). Die Rolle der Blätter in der photoperiodischen Reaktion der Pflanze. *Bull. appl. Bot. gen. Plant Breed., Ser. A*, **17**, 25–30.
MOST, B. H. (1968). A sugarcane spindle bioassay for gibberellins and its use in detecting diffusible gibberellin from sugarcane. In Wightman, F. & Setterfield, G. (eds.), *Biochemistry and Physiology of Plant Growth Substances*. Runge Press, Ottawa, pp. 1619–33.
MOST, B. H., WILLIAMS, J. C. & PARKER, K. J. (1968). Gas chromatography of cytokinins. *J. Chromatog.*, **38**, 136–8.
MOWAT, J. A. (1963). Gibberellin-like substances in algae. *Nature, Lond.*, **200**, 453–5.

MOYED, H. S., & TULI, V. (1968). The oxindol pathway of 3-indolacetic acid metabolism and the action of auxins. In Wightman, F. and Setterfield, G. (eds.), *Biochemistry and Physiology of Plant Growth Substances*. Runge Press, Ottawa, pp. 289–300.

MUDD, J. B. & ZALIK, S. (1958). The metabolism of indole by tomato plant tissues and extracts. *Can. J. Bot.*, **36**, 467–72.

MUIR, R. M. & HANSCH, C. (1953). On the mechanism of action of growth regulators. *Pl. Physiol. Lancaster*, **28**, 218–32.

MUIR, R. M., HANSCH, C. & GALLUP, A. H. (1949). Growth regulation by organic compounds. *Pl. Physiol. Lancaster*, **24**, 359–66.

MUIR, R. M. & LANTICAN, B. P. (1968). Purification and properties of the enzyme system forming indoleacetic acid. In Wightman F. and Setterfield, G. (eds.), *Biochemistry and Physiology of Plant Growth Substances*. Runge Press, Ottawa, pp. 259–72.

MULLER, C. H. (1953). The association of desert annuals with shrubs. *Am. J. Bot.*, **40**, 53–60.

MULLER, C. H. (1965). Inhibitory terpenes volatilized from *Salvia* shrubs. *Bull. Torrey bot. Club*, **92**, 38–45.

MULLER, C. H. (1966). The role of chemical inhibitors (allelopathy) in vegetational composition. *Bull. Torrey bot. Club*, **93**, 332–51.

MULLER, C. H. (1968). The role of allelopathy in the evolution of vegetation. In *Biochemical evolution. Proc. 29th. annu. biol. Colloq.*, Oregon State University Press, pp. 13–31.

MULLER, C. H. (1970). Phytotoxins as plant habitat variables. In Steelink, C. & Runeckles, V. C. (eds.), *Recent Advances in Phytochemistry*. Meredith Corp. New York, vol. 3, pp. 105–21.

MULLER, C. H. & MORAL, R. DEL (1966). Soil toxicity induced by terpenes from *Salvia leucophylla*. *Bull. Torrey bot. Club*, **93**, 130–7.

MÜLLER, D. G. (1968). Versuche zur Charakterisierung einer Sexual-Lockstoffes bei der Braunalgen *Ectocarpus saliculosus*. *Planta,* **81**, 160–8.

MULLER, W. H., LORBER, P., HALEY, B. & JOHNSON, K. (1969). Volatile growth inhibitors produced by *Salvia leucophylla*: effect on oxygen uptake by mitochondrial suspensions. *Bull. Torrey bot. Club*, **96**, 89–96.

MUMFORD, F. E., SMITH, D. H. & CASTLE, J. E. (1961). An inhibitor of indoleacetic acid oxidase from pea tips. *Pl. Physiol. Lancaster*, **36**, 752–6.

MURAKAMI, Y. (1966). [Bioassay of gibberellins using rice endosperm and some problems on its application.] In Japanese. *Bot. Mag., Tokyo*, **79**, 315–27.

MURAKAMI, Y. (1968*a*). [A new rice-seedling test for gibberellins, 'microdrop method', and its use for testing extracts of rice and morning glory.] In Japanese. *Bot. Mag. (Tokyo)*, **81**, 33–43.

MURAKAMI, Y. (1968*b*). Gibberellin-like substances in roots of *Oryza sativa, Pharbitis nil* and *Ipomoea batatas* and the site of their synthesis in the plant. *Bot. Mag. (Tokyo)*, **81**, 334–43.

MURASHIGE, T. & SKOOG, F. (1962). A revised medium for rapid growth and bioassays with tobacco tissue cultures. *Physiologia Pl.* **15**, 473–97.

MURNEEK, A. E. & WHYTE, R. O. (1948). *Vernalisation and photoperiodism*, Chronica Botanica Publ., Waltham, Massachussetts.

NÄF, U. (1959). Control of antheridium formation in the fern species *Anemia phyllitidis*. *Nature, Lond.*, **184**, 798–800.

NÄF, U. (1960). On the control of antheridium formation in the fern species *Lygodium japonicum*. *Proc. Soc. exp. Biol. Med.*, **105**, 82–6.

NÄF, U. (1962). Developmental physiology of lower archegoniates. *A. Rev. Pl. Physiol.* **13**, 507–32.

NÄF, U. (1968). On separation and identity of fern antheridogens. *Pl. Cell Physiol.*, **8**, 27–33.

NAKAMURA, T., TAKAHASHI, N., MATSUI, M. & HWANG, Y.-S. (1966). Activity of synthesized auxin B lactone as the plant growth regulator of auxin type. *Pl. Cell Physiol.* **7**, 693–4.

NAKAYAMA, S., TOBITA, H. & OKUMURA, F. S. (1962). Antagonism of kinetin and far-red light on β-indoleacetic acid in the flowering of *Pharbitis* seedlings. *Phyton, B. Aires*, **19**, 43–8.

NANCE, J. F. (1958). Effect of indoleacetic acid on the utilisation of acetate-1-C^{14} by pea stem slices. *Pl. Physiol. Lancaster*, **33**, 93–8.

NANCE, J. F. & SHIGEMURA, Y. (1954). Studies on acetaldehyde evolution and utilisation. *Am. J. Bot.*, **41**, 829–32.

NANDA, K. K., TOKY, K. L. & SAWHNEY, S. (1970). Seasonal variation in GA_3 effects on flowering of *Impatiens balsamina*, a qualitative short day plant. *Physiologia Pl.*, **23**, 1085–8.

NAPP-ZINN, K. 1963. Über den Einfluss von Genen und Gibberellinen auf die Blütenbildung von *Arabidopsis thaliana*. *Ber. dt. bot. Ges.*, **76**, 77–89.

NAQVI, S. M. & GORDON, S. A. (1967). Auxin transport in *Zea mays* coleoptiles. II Influence of light on the transport of indoleacetic acid-2-^{14}C. *Pl. Physiol. Lancaster*, **42**, 138–43.

NASON, A. 1950. Effect of zinc deficiency on the synthesis of tryptophan by *Neurospora* extracts. *Science, N. Y.*, **112**, 111–112.

NAVEZ, A. E. (1933). Growth-promoting substance and illumination. *Proc. natl. Acad. Sci., U.S.A.*, **19**, 636–8.

NAYLOR, A. W. (1953). Reactions of plants to photoperiod. In *Growth and Differentiation in Plants*, Iowa State College Press, Chap. 9.

NAYLOR, A. W. & DAVIS, E. A. (1951). Respiration response of root tips to maleic hydrazide. *Bull. Torrey bot. Club*, **78**, 73.

NEELY, P. M. & PHINNEY, B. O. (1957). The use of mutant dwarf-1 of maize as a quantitative bioassay for gibberellin activity. *Pl. Physiol. Lancaster*, **32**, Suppl. xxxi.

NEELY, W. B., BALL, C. D., HAMNER, C. L. & SELL, H. M. (1950). Effect of 2,4-dichlorophenoxyacetic acid on the invertase, phosphorylase and pectin methoxylase activity in the stems and leaves of the red kidney bean plants. *Pl. Physiol. Lancaster*, **25**, 525–7.

NEGI, S. S. & OLMO, H. P. (1966). Sex conversion in a male *Vitis vinifera* L. by a kinin. *Science N. Y.*, **152**, 1624–5.
NEIRINCKX, L. J. A. (1968). Influence de quelques substances de croissance sur l'absorption du sulfat par le tissu radiculaire de beterave rouge (*Beta vulgaris* L. ssp. *vulgaris* var. *rubra* L.). *Annls Physiol. vég. Univ. Bruxelles*, **13**, 83–108.
NELJUBOW, D. (1901). Über die horizontale Nutation der Stengel von *Pisum sativum* und einiger anderer Pflanzen. *Beih. bot. Zbl.*, **10**, 128–38.
NEŠKOVIĆ, M. & CULAFIĆ, L. (1968). The influence of light on the content of growth substances in pea shoots. I. Effect of red light on the extractable indole auxins. *Bull. Inst. Jard. bot. Univ. Beograd.*, **3**, 255–62.
NEWCOMB, E. H. (1951). Effect of auxin on ascorbic acid oxidase activity in tobacco pith cells. *Proc. Soc. exp. Biol. Med.*, **76**, 504–509.
NEWCOMB, E. H. (1954). The use of cultured tissue in a study of the metabolism controlling cell enlargement. *Colloq. int. Union biol. Sci.*, **20**, 195–214.
NEWCOMB, E. H. (1963). Cytoplasm–cell wall relationships. *A. Rev. Pl. Physiol.*, **14**, 43–64.
NEWMAN, I. A. (1965). Distribution of indole-3-acetic acid labelled with carbon-14 in *Avena*. *Nature, Lond.*, **205**, 1336.
NG, E. K. & AUDUS, L. J. (1964). Growth-regulator interactions in the growth of the shoot system of *Avena sativa* seedlings. I The growth of the first internode. *J. exp. Bot.*, **15**, 67–95.
NG, E. K. & AUDUS, L. J. (1965). Growth-regulator interactions in the growth of the shoot system of *Avena sativa* seedlings. II The growth of the first leaf and the coleoptile. *J. exp. Bot.*, **16**, 107–27.
NG, E. K. & CARR, D. J. (1959). Effects of pH on the activity of chelating agents and auxins in cell extension. *Physiologia Pl.*, **12**, 275–87.
NICHOLLS, P. B. N. (1962). *On the Growth of the Barley Apex*, Doctoral thesis, University of Adelaide, Australia. (Quoted by Paleg, L. G. (1965). *A. Rev. Pl. Physiol.*, **16**, 290–322.)
NICHOLLS, P. B. (1967). Isolation of indole-3-acetyl-2-*o*- myoinositol. *Planta*, **72**, 258–64.
NICHOLLS, P. B. N. PALEG, L. G. (1963). A barley endosperm bioassay for gibberellins. *Nature, Lond.*, **199**, 823–4.
NICKELL, L. G. (1955). Effects of antigrowth substances in normal and atypical plant growth. In *Antimetabolites and Cancer, Am. Ass. Adv. Sci.*
NICKELL, L. G. (1959). Antimicrobial activity of vascular plants. *Econ. Bot.*, **13**, 281–318.
NICKELL, L. G. & FINLAY, A. C. (1954). Antibiotics and their effects on plant growth. *Agric. Fd. Chem.*, **2**, 178–82.
NICKELL, L. G. & GORDON, P. N. (1960). Effects of tetracycline antibiotics, their degradation products, derivatives and some synthetic analogues on the growth of certain plant systems. *Antimicrobial Agents Annual*, Plenum Press, New York, pp. 588–94.
NICKELL, L. G. & KORTSCHAK, H. P. (1964). Arginine; its role in sugar cane

growth. *Hawaiian Planters Record*, **57**, 230.

NICKERSON, W. J. & THIMANN, K. V. (1943). The chemical control of conjugation in Zygosaccharomyces, II. *Am. J. Bot.*, **30**, 94–101.

NICOL, H. (1934). The derivation of the nitrogen of crop plants, with special reference to associated growth. *Biol. Rev.* **9**, 383–410.

NICOLAS, D. J. D. (1961). Minor mineral nutrients. *A. Rev. Pl. Physiol.*, **12**, 63–90.

NIEDERGANG-KAMIEN, E. & LEOPOLD, A. C. (1957). Inhibition of polar auxin transport. *Physiologia Pl.*, **9**, 60–73.

NIEDERGANG-KAMIEN, E. & LEOPOLD, A. C. (1959). The inhibition of transport of indoleacetic acid by phenoxyacetic acids. *Physiologia Pl.*, **12**, 776–85.

NIEDERGANG-KAMIEN, E. & SKOOG, F. (1956). Studies on polarity and auxin transport in plants. I Modification of polarity and auxin transport by triiodobenzoic acid. *Physiologia Pl.*, **9**, 60–73.

NIELSEN, N. (1930). Untersuchungen über einen neuen wachstumsregulierenden Stoff: Rhizopin. *Jb. wiss. Bot.*, **73**, 125–91.

NINNEMANN, H., ZEEVAART, J. A. D., KENDE, H. & LANG, A. (1964). The plant growth retardant CCC as inhibitor of gibberellin biosynthesis in *Fusarium moniliforme*. *Planta*, **61**, 229–35.

NISSL, D. & ZENK, M. H. (1969). Evidence against induction of protein synthesis during auxin-induced elongation of *Avena* coleoptiles. *Planta*, **89**, 323–41.

NITSAN, J. & LANG, A. (1965). Inhibition of cell division and cell elongation in higher plants by inhibitors of DNA synthesis. *Devl. Biol.*, **12**, 358–76.

NITSAN, J. & LANG, A. (1966). DNA synthesis in the elongating nondividing cells of the lentil epicotyl and its promotion by gibberellin. *Pl. Physiol. Lancaster*, **41**, 965–70.

NITSCH, C. (1967). L'induction *in vitro* de la floraison chez *Plumbago indica* L. *Bull. Soc. fr. Physiol. vég.*, **13**, 119–36.

NITSCH, C. (1968). Effects of growth substances on the induction of flowering of a short-day plant *in vitro*. In Wightman, F. & Setterfield, G. (eds.), *Biochemistry and Physiology of Plant Growth Substances.* Runge Press, Ottawa, pp. 1385–9.

NITSCH, C & NITSCH, J. P. (1967). The induction of flowering *in vitro* in stem segments of *Plumbago indica* L. II The production of reproductive buds. *Planta*, **72**, 371–84.

NITSCH, C. & NITSCH, J. P. (1969). Floral induction in a short-day plant, *Plumbago indica* L. by 2-chloroethanephosphonic acid. *Pl. Physiol. Lancaster*, **44**, 1747–8.

NITSCH, J. P. (1953). The physiology of fruit growth. *A. Rev. Pl. Physiol.*, **4**, 199–236.

NITSCH, J. P. (1957). Growth responses of woody plants to photoperiodic stimuli. *Proc. Am. Soc. hort. Sci.* **70**, 512–25.

NITSCH, J. P. (1965). Physiology of flower and fruit development. *Encyclo-*

paedia of plant physiology, Springer, Berlin, Göttingen, Heidelberg, **15**(1), pp. 1537–1647.
NITSCH, J. P. (1967*a*) Progress in the knowledge of natural plant growth regulators. *Ann. N.Y. Acad. Sci.*, **144**(1), 279–94.
NITSCH, J. P. (1967*b*). Towards a biochemistry of flowering and fruiting: contributions of the *'in vitro'* technique. *Rep. 17th. int. hort. Congr.*, **3**, 291–308.
NITSCH, J. P. (1968). Studies on the mode of action of auxins, cytokinins and gibberellins at the subcellular level. In Wightman, F. & Setterfield, G. (eds.), *Biochemistry and Physiology of Plant Growth Substances.* Runge Press, Ottawa, pp. 563–80.
NITSCH, J. P. (1969). Natural cytokinins. In *Plant Growth Regulators. Soc. chem. Indust. Monogr.*, **31**, 111–23.
NITSCH, J. P. & NITSCH, C. (1956). Studies on the growth of coleoptile and first internode sections. A new sensitive straight-growth test for auxins. *Pl. Physiol. Lancaster*, **31**, 94–111.
NITSCH, J. P. & NITSCH, C. (1959). Photoperiodic effects in woody plants: evidence for the interplay of growth-regulating substances. In *Photoperiodism and Related Phenomena in Plants and Animals. Am. Ass. Adv. Sci. Wash.*, pp. 225–42.
NITSCH, J. P. and NITSCH, C. (1961). Synergistes naturels des auxines et des gibbérellines. *Bull. Soc. bot. Fr.*, **108**, 349–62.
NITSCH, J. P. & NITSCH, C. (1962*a*). Activités comparées de noef gibbérellines sur trois tests biologiques. *Ann. Physiol. vég.*, **4**, 85–97.
NITSCH, J. P. & NITSCH, C. (1962*b*). Composés phénoliques et croissance végétale. *Ann. Physiol. vég.*, **4**, 211–225.
NITSCH, J. P. & NITSCH, C. (1965). Néoformation des fleurs *in vitro* chez une espèce de jours courts: *Plumbago indica* 1. *Ann. Physiol. vég*, **7**, 251–6.
NOBÉCOURT, P. (1939). Sur la pérennité et l'augmentation de volume des cultures de tissus végétaux, *C. r. Séanc. Soc. Biol.*, **130**, 1270–1.
NOODÉN, L. D. (1968). Studies on the role of RNA synthesis in auxin-induction of cell enlargement. *Pl. Physiol. Lancaster*, **43**, 140–50.
NOODÉN, L. D. & THIMANN, K. V. (1965). Inhibition of protein synthesis and auxin-induced growth by chloramphenicol. *Pl. Physiol. Lancaster*, **40**, 193–201.
NORMAN, A. G. (1955). The effect of polymyxin on plant roots. *Archs Biochem. Biophys.*, **58**, 461–77.
NORMAN, A. G. (1959*a*). Inhibition of root growth and cation uptake by antibiotics. *Proc. Soil. Sci. Soc. Am.*, **23**, 368–70.
NORMAN, A. G. (1959*b*). Terramycin and plant growth. *Agron. J.*, **47**, 585–7.
NORMAN, A. G. (1960). The action of duramycin on plant roots. *Proc. Soil Sci. Soc. Am.*, **24**, 109–11.
NORRIS, R. F. (1966). Effect of (2-chloroethyl)trimethylammonium chloride

on the level of endogenous indole compounds in wheat seedlings. *Can. J. Bot.*, **44**, 675–84.

NUTILE, G. E. (1945). Inducing dormancy in lettuce seeds with coumarin. *Pl. Physiol. Lancaster*, **20**, 433–42.

O'BRIEN, T. J., JARVIS, B. C., CHERRY, J. H. & HANSON, J. B. (1968). Enhancement by 2,4-dichlorophenoxyacetic acid of chromatin RNA polymerase in soybean hypocotyl tissue. *Biochem. biophys. Acta.*, **169**, 35–43.

OCKERSE, R. & GALSTON, A. W. (1967). Gibberellin-auxin interaction in pea stem elongation. *Pl. Physiol. Lancaster*, **42**, 47–54.

OCKERSE, R. & WABER, J. (1970). The promotion of indole-3-acetic acid oxidation in pea buds by gibberellic acid treatment. *Pl. Physiol. Lancaster*, **46**, 821–4.

ODHNOFF, C. (1961). The influence of boric acid and phenylboric acid on the root growth of bean (*Phaseolus vulgaris*). *Physiologia Pl.*, **14**, 187–220.

OGAWA, Y. (1961). Über die Wirkung von Kinetin auf die Blüten-Bildung von *Pharbitis nil* Chois. *Pl. Cell Physiol.*, **2**, 343–59.

OGAWA, Y. (1965). Extracts from animals promote enzyme production in embryoless rice endosperm. *Proc. Jap. Acad.*, **41**, 850–3.

OHKUMA, K., LYON, J. L., ADDICOTT, F. T. & SMITH, O. E. (1963). Abscisin II, an abscission-accelerating substance from young cotton fruit. *Science N. Y.*, **142**, 1592–3.

OLSEN, A. C., BONNER, J. & MORRÉ, D. J. (1965). Force extension analysis of *Avena* coleoptile cell walls. *Planta*, **66**, 126–34.

ORDIN, L., APPLEWHITE, T. H. & BONNER, J. (1956). Auxin-induced water-uptake by *Avena* coleoptile sections. *Pl. Physiol. Lancaster*, **31**, 44–53.

ORDIN, L., CLELAND, R. & BONNER, J. (1957). Methyl esterification of cell wall constituents under the influence of auxin. *Pl. Physiol. Lancaster*, **32**, 216–20.

OSBORNE, D. J. (1952). A synergistic interaction between 3-indolylacetonitrile and 3-indolylacetic acid. *Nature, Lond.*, **170**, 210.

OSBORNE, D. J. (1968). A theoretical model for polar auxin transport. In Vardar, Y. (ed), *Transport of Plant Hormones*. North-Holland, Amsterdam, pp. 97–107.

OSBORNE, D. J. & BLACK, M. K. (1964). Polar transport of a kinin, benzyladenine. *Nature, Lond.*, **201**, 97.

OSBORNE, D. J., BLACKMAN, G. E., NOVOA, S., SUDZUKI, F. & POWELL, R. G. (1955). The physiological activity of 2,6-substituted phenoxyacetic acids. *J. exp. Bot.*, **6**, 392–408.

OSBORNE, D. J., HORTON, R. F., & BLACK, M. K., (1968). Senescence in excised petiole segments: the relevence to auxin and kinetin transport. In Vardar, Y. (ed.), *Transport of Plant Hormones*. North-Holland, Amsterdam, pp. 79–96.

OSBORNE, D. J. & MCCALLA, D. R. (1961). Rapid bioassay for kinetin and kinins using senescing leaf tissue. *Pl. Physiol. Lancaster*, **36**, 219–21.

OSBORNE, D. J. & MCCREADY, C. C. (1965). Transport of the kinin N_6-benzyladenine: non-polar or polar? *Nature, Lond.*, **206**, 678–80.
OSBORNE, D. J. & WAIN, R. L. (1951). Plant growth-regulating activity in certain aryloxyalkylcarboxylic acids. *Science, N. Y.*, **114**, 92.
OSVALD, H. (1947). [Equipment of plants in the struggle for space.] In Swedish. *Pl. Husbandry.*, **2**, 288–303.
OSVALD, H. (1949). Root exudates and seed germination. *K. LantbrHögsk. Annlr.*, **16**, 789–96.
OVERBEEK, J. VAN (1933). Wuchsstoff, Lichtwachstumsreaktion und Phototropismus bei *Raphanus. Recl. Trav. bot. néerl.*, **30**, 538–626.
OVERBEEK, J. VAN (1944). Auxin, water uptake and osmotic pressure in potato tissue. *Am. J. Bot.*, **31**, 265–9.
OVERBEEK, J. VAN (1950). Plant hormones and other regulatory factors. In *Agricultural Chemistry*, van Nostrand, London, vol. 1, Chap. 22.
OVERBEEK, J. VAN (1951) Use of growth substances in tropical agriculture. In Skoog, F. (ed.), *Plant Growth Substances*. University of Wisconsin Press, pp. 225–44.
OVERBEEK, J. VAN (1952). Agricultural application of growth regulators and their physiological basis. *A. Rev. Pl. Physiol.*, **3**, 87–108.
OVERBEEK, J. VAN (1956). Absorption and translocation of plant regulators. *A. Rev. Pl. Physiol.*, **7**, 355–72.
OVERBEEK, J. VAN (1961). New theory on the primary mode of auxin action. In *Plant Growth Regulation*. Iowa State University Press, pp. 449–61.
OVERBEEK, J. VAN & DOWDING, L. (1961). Inhibition of gibberellin action by auxin. In *Plant Growth Regulation*. Iowa State University Press, pp. 657–63.
OVERBEEK, J. VAN, LOEFFLER, J. E. & MASON, M. I. R. (1967). Dormin (Abscisin II), inhibitor of plant DNA synthesis. *Science, N. Y.*, **156**, 1497–9.
OVERLAND, L. (1966). The role of allelopathic substances in the 'smother crop' barley. *Am. J. Bot.*, **53**, 423–32.
PAÁL, A. (1919). Ueber phototropische Reizleitung. *Jb. wiss. Bot.*, **58**, 406–58.
PALEG, L. G. (1960). Physiological effects of gibberellic acid. I On carbohydrate metabolism and amylase activity of barley endosperm. *Pl. Physiol. Lancaster*, **35**, 293–9.
PALEG, L. G. (1965). Physiological effects of gibberellins. *A. Rev. Pl. Physiol.*, **16**, 291–322.
PALEG, L. G., ASPINAL, D., COOMBE, B. & NICHOLLS, P. B. N. (1964). Physiological effects of gibberellic acid. VI Other gibberellins in three test systems. *Pl. Physiol. Lancaster*, **39**, 286–90.
PALEG, L., KENDE, H., NINNEMANN, H. & LANG, A. (1965). Physiological effects of gibberellic acid. VIII Growth retardants on barley endosperm. *Pl. Physiol. Lancaster*, **40**, 165–9.
PARIS, D. & DUHAMET, L. (1953). Action d'un mélange d'acides aminés et des vitamines sur la proliferation des cultures de tissus de Crown Gall de

Scorsonère; comparison avec l'action du lait de Coco. *C. r. hebd. Séanc. Acad. Sci. Paris*, **236**, 1690–2.

Parish, R. W. (1969*a*). The effect of light on peroxidase synthesis and indoleacetic acid oxidase inhibitors in coleoptiles and first leaves of wheat. *Z. PflPhysiol.*, **60**, 90–7.

Parish, R. W. (1969*b*). The effects of various aromatic compounds on indoleacetic acid oxidase before and after treatment with peroxidase and hydrogen peroxide. *Z. PflPhysiol.*, **60**, 296–306.

Parker, M. W., Hendricks, S. B., Borthwick, H. A. & Scully, N. J. (1946). Action spectrum for photoperiodic control of floral initiation of short day plants. *Bot. Gaz.*, **108**, 1–25.

Parks, J. M. & Rice, E. L. (1969). Effects of certain plants of old-field successions on the growth of blue-green algae. *Bull. Torrey bot. Club*, **96**, 345–60.

Patrick, Z. A. (1955). The peach replant problem in Ontario. II Toxic substances from microbial decomposition of peach root residues. *Can. J. Bot.*, **33**. 461–86.

Patrick, Z. A. (1971). Phytotoxic substances associated with the decomposition in soil of plant residues. *Soil. Sci.*, **111**, 13–18.

Patrick, Z. A., Toussoun, T. A. & Snyder, W. C. (1963). Phytotoxic substances in soils associated with decomposition of plant residues. *Phytopathology*, **53**, 152–61.

Paulet, P. & Nitsch, J. P. (1964). La néoformation de fleurs sur culture *in vitro* de racines de *Cichorium intybus*: étude physiologique. *Ann. Physiol. vég.*, **6**, 333–45.

Pavlinova, O. A., Prasolova, M. F. & Ivanova, E. A. (1967). [Questions of the mechanisms of the effect of maleic acid hydrazide on the growth and sugar accumulation in the roots of sugar beet.] In Russian. *Fiziol. Rast.*, **14**, 992–6.

Perley, J. E. & Stowe, B. B. (1966). On the ability of *Taphrina deformans* to produce indoleacetic acid from tryptophan by way of tryptamine. *Pl. Physiol. Lancaster*, **41**, 234–7.

Perlis, J. B. & Galston, A. W. (1955). Studies on the peroxidase of etiolated pea seedlings and their induction by indoleacetic acid. *Pl. Physiol. Lancaster*, **30**, (Suppl.) xii.

Perlis, J. B. & Nance, J. F. (1956). Indoleacetic acid and the utilisation of radioactive pyruvate and acetate by wheat roots. *Pl. Physiol. Lancaster*, **31**, 451–5.

Perrin, D. D. (1961). Improbability of the chelation hypothesis of auxin action. *Nature, Lond.*, **191**, 253.

Petersen, E. L. & Naylor, A. W. (1953). Some metabolic changes in tobacco stem tips accompanying maleic hydrazide treatment and the appearance of frenching symptoms. *Physiologia Pl.*, **6**, 816.

Phatak, S. C., Wittwer, S. H., Honma, S. & Bukovac, M. J. (1966). Gibberellin-induced anther and pollen development in a stamenless tomato mutant. *Nature, Lond.*, **209**, 635–6.

PHELPS, R. H. & SEQUEIRA, L. (1967). Synthesis of indoleacetic acid via tryptamine by a cell-free system from tobacco terminal buds. *Pl. Physiol. Lancaster*, **42**, 1161–3.
PHILLIPS, D. A. & TORREY, J. G. (1970). Cytokinin production by *Rhizobium japonicum*. *Physiologia Pl.*, **23**, 1057–63.
PHILLIPS, I. D. J. (1962). Some interactions of gibberellic acid with naringenin (5,7,4′-trihydroxyflavanone) in the control of dormancy and growth in plants. *J. exp. Bot.*, **13**, 213–26.
PHILLIPS, I. D. J. & JONES, R. L. (1964). Gibberellin-like activity in the bleeding sap of root systems of *Helianthus annuus* detected by a new pea epicotyl assay and other methods. *Planta*, **63**, 269–78.
PHILLIPS, I. D. J. & WAREING, P. F. (1959). Studies in dormancy of sycamore. II The effect of daylength on the natural growth-inhibitor content of the shoot. *J. exp. Bot.*, **10**, 504–14.
PHINNEY, B. O. (1956). Growth responses of single-gene dwarf mutants in maize to gibberellic acid. *Proc. natl. Acad. Sci. U.S.A.*, **42**, 185–9.
PHINNEY, B. O. & SPECTOR, C. (1967). Genetics and gibberellin production in the fungus *Gibberella fujikuroi*. *Ann. N. Y. Acad. Sci.*, **144**(1), 204–10.
PHINNEY, B. O. & WEST, C. A. (1960). Gibberellins as native plant growth regulators. *A. Rev. Pl. Physiol.*, **11**, 411–36.
PHINNEY, B. O. & WEST, C. A. (1961). Gibberellins and plant growth. *Encyclopaedia of Plant Physiology*. Springer, Berlin, Göttingen, Heidelberg, vol. 14, pp. 1185–227.
PHINNEY, B. O., WEST, C. A., RITZEL, M. & NEELY, P. M. (1957). Evidence for 'gibberellin-like' substances from flowering plants. *Proc. natl. Acad. Sci. U.S.A.*, **43**, 398–404.
PHOUPHAS, C. & GORIS, A. (1952). Sur la modification des taux de saccharose et d'inuline sur l'influence de l'hydrazide maléique dans les tissus de Topinambour cultivés *in vitro*. *C. r. hebd. Séanc. Acad. Sci. Paris*, **234**, 2002–4.
PICKERING, S. U. (1917). The effect of one plant on another. *Ann. Bot.*, **31**, 181–97.
PIENIĄŻEK, J. (1964). Kinetin-induced breaking of dormancy in 8-month old apple seedlings of 'Antonovka' variety. *Acta agrobot.*, **16**, 157–69.
PIENIĄŻEK, J. & GROCHOWSKA, M. J. (1967). The role of the natural growth inhibitor (Abscisin II) in apple seed germination and the changes in the content of phenolic substances during stratification. *Acta. Soc. Bot. Pol.*, **36**, 579–87.
PILET, P.-E. (1954). Croissance et rhizogénèse des racines de plantules vernalisées. *Rev. gén. Bot.*, **61**, 637–64.
PILET, P.-E. (1957*a*). Action of maleic hydrazide on *in vivo* auxin destruction. *Physiologia Pl.*, **10**, 791–3.
PILET, P.-E. (1957*b*). Action des gibberellines sur l'activité auxin-oxydasique de tissus cultivés *in vitro*. *C. r. hebd. Séanc. Acad. Sci. Paris*, **245**, 1327.
PILET, P.-E. (1961). *Les Phytohormones de Croissance*. Masson, Paris.

PILET, P.-E. (1962). L'hydroxy-3-méthyl-3-oxo-2-indole, un produit possible du catabolisme de l'acide β-indolylacétique. *Bull. Soc. chim. Biol.*, **44**, 875–6.
PILET, P.-E. (1965*a*). Action of gibberellic acid on auxin transport. *Nature, Lond.*, **208**, 1344–5.
PILET, P.-E. (1965*b*). Action de la kinetin sur le transport de l'acide β-indolylacétique marqué par du radiocarbone. *C. r. hebd. Séanc. Acad. Sci. Paris*, **260**, 4053.
PILET, P.-E. & BELHANAFI, A. (1962). Chélation et croissance: effets des 8- et 2-hydroxyquinoléines. *C. r. hebd. Séanc. Acad. Sci. Paris*, **254**, 3416–8.
PILET, P.-E. & COLLET, G. (1960). Étude du nanisme. I Action de l'acide gibbérellique sur la croissance et la destruction *in vitro* des auxines. *Bull. Soc. bot. Suisse*, **70**, 180–94.
PILET, P.-E. & GALSTON, A. W. (1955). Auxin destruction, peroxidase activity and peroxide genesis in the roots of *Lens culinaris*. *Physiologia Pl.*, **8**, 888–98.
PILET, P.-E. & GASPAR, TH. (1968). *Le catabolisme auxinique. Monographie de Physiologie Végétale. No. 1,* Masson, Paris.
PILET, P.-E., GUERN, J. & HUGON, E. (1967). Sur le transport de la 6-benzylaminopurine-α^{14}C. *Physiol. vég.*, **5**, 261–70.
PILET, P.-E. & MARGOT, L. (1955). Applications d'hétéroauxine et d'hydrazide maléique continués dans le lanoline, sur les racines du *Lens culinaris* Med., et répercussions sur leur croissance, leur rhizogénèse et leur morphologie. *Bull. Soc. bot. Suisse*, **65**, 47–59.
PILET, P.-E. & SIEGENTHALER, P. A. (1959). Gradients biochimique radiculaire. I Auxines et réserves azotées. *Bull. Soc. bot. Suisse*, **69**, 58.
PILET, P.-E. & TUBIAN, G. (1953). Auxines et amidon. III Étude *in vitro* de l'action des auxines sur l'amylose. *Bull. Soc. vaud. Sci. nat.*, **65**, 391.
PILET, P.-E. & WURGLER, W. (1958). Action des gibbérellines sur la croissance et l'activité auxines-oxydasique du *Trifolium ochroleucum*. Hudson. *Bull. Soc. bot. Suisse*, **68**, 54–63.
PINCUS, G. & THIMANN, K. V. (1948). *The Hormones*. Academic Press, New York, Chapters 2 and 3.
PLEMPEL, M. (1963). Die chemischen Grundlage der Sexualreaktion bei Zygomycetes. *Planta*, **59**, 492–508.
PLUIJGERS, C. W. & KERK, G. J. M. VAN DER (1961). Plant growth-regulating activity of S-carboxymethyl-N,N-dimethyldithiocarbamate and related compounds. *Recl. Trav. chim. Pays-Bas. Belg.*, **80**, 1089–1100.
POHL, R. (1948). Ein Beitrag zur Analyse der Streckungswachstums der Pflanze. *Planta*, **36**, 230–61.
POHL, R. (1953). Zur Reaktionsweise des Wuchsstoffes bei der Zellstreckung. *Z. Bot.*, **41**, 343.
POLJAKOFF-MAYBER, A. (1955). The effect of IAA on the oxidative activity of mung-bean mitochondria. *J. exp. Bot.*, **6**, 321–7.
POLLARD, J. K., SHANTZ, E. M. & STEWARD, F. C. (1961). Hexitols in coconut

milk. Their role in the nurture of dividing cells. *Pl. Physiol. Lancaster*, **36**, 492–501.
PONT LEZICA, R. F. (1965). Gibberellins in *Rudbeckia bicolor* grown under different photoperiods. *Bull. Soc. r. Sci. Liège.*, **34**, 49–55.
POPOLOCKAJA, K. L. (1961). [On the mechanism of the action of maleic hydrazide on plants.] In Russian. *Izv. Akad. Nauk. SSSR Ser. biol.*, **2**, 250–5. (From *Biol. Abstr.*, **39**, 20294, 1962.)
PORTER, W. L. & THIMANN, K. V. (1959). Molecular and functional complementarity of auxins and phosphatides. *Abstr. 9th. int. bot. Congr.*, University of Toronto Press, pp. 305.
PORTER, W. L. & THIMANN. K. V. (1965). Molecular requirements for auxin action. I Halogenated indoles and indoleacetic acid. *Phytochemistry*, **4**, 229–43.
POST, L. C. (1959). [*The isolation of a new plant growth hormone of the indole type.*] In Dutch. Doctoral thesis, University of Utrecht.
POWELL, L. E. (1964). Preparation of indole extracts from plants for gas chromatography and spectrofluorimetry. *Pl. Physiol. Lancaster*, **39**, 836–42.
POWER, J. B. & COCKING, E. C. (1970). Isolation of leaf protoplasts. Macromolecule uptake and growth substance response. *J. exp. Bot.*, **21**, 64–70.
PRAMER, D. (1958). The persistance and biological effects of antibiotics in soil. *Appl. Microbiol.*, **6**, 221–4.
PRESTON, W. H. Jr. & LINK, C. B. (1958). Dwarfed progeny produced by plants treated with several quaternary ammonium compounds. *Pl. Physiol. Lancaster*, **33**, (Suppl.) xlix.
PRESTON, W. H., MITCHELL, J. W. & REEVE, W. (1954). Movement of alpha-methoxy-phenylacetic acid from one plant to another through their root systems. *Science N. Y.*, **119**, 437–8.
PRINGLE, R. B., NÄF, U. & BRAUN, A. C. (1960). Purification of a specific inducer of the male sex organ in certain fern species. *Nature, Lond.*, **186**, 1066–7.
PROBINE, M. C. (1963). The plant cell wall. *Tuatara*, **11**, 115–41.
PROBINE, M. C. & PRESTON, R. D. (1962). Cell growth and the structure of the wall in internodal cells of *Nitella opaca*. II Mechanical properties of the walls. *J. exp. Bot.*, **13**, 111–27.
PROCHÁZKA, Z. & KOŘÍSTEK, S. (1951). A proof of the existance of a bound form of ascorbic acid in cabbage by paper chromatography. *Coll. Czech. chem. Comm.*, **16**, 65–8.
PROCHÁZKA, Z., SANDA, V. & ŠORIN, F. (1957). On the structure of ascorbigen. *Coll. Czech. chem. Comm.*, **22**, 654–6.
PROEBSTING, E. L. & GILMORE, A. E. (1940). The relation of peach root toxicity to the re-establishing of peach orchards. *Proc. Am. Soc. hort. Sci.*, **38**, 21–6.
PROVASOLI, L., HUTNER, S. H. & PINTER, L. J. (1951). Destruction of chloroplasts by streptomycin. *Cold Spring Harb. Symp. quant. Biol.*, **16**, 113.
PRYCE, R. J. & MACMILLAN, J. (1967). A new gibberellin in the seed of

Phaseolus multiflorus. Tetrahedron Lett., No. 42, 4173–5.
PRYCE, R. J., MACMILLAN, J. & MCCORMACK, A. (1967). The identification of bamboo gibberellin in *Phaseolus multiflorus* by combined gas-chromatography-mass spectrometry. *Tetrahedron Lett.*, No. 49. 5009–11.
PURVIS, O. N. (1961). The physiological analysis of vernalisation. *Encyclopaedia of Plant Physiology*, Springer, Berlin, Göttingen, Heidelberg, vol. 16, pp. 77–122.
PURVIS, O. N. & GREGORY, F. G. (1937). Studies in the vernalisation of cereals. I A comparative study of vernalisation of winter rye by low temperature and by short days. *Ann. Bot.*, **1**, 569–92.
PURVIS, O. N. & GREGORY, F. G. (1952). Studies in vernalisation of cereals. XII The reversibility by high temperature of the vernalised condition in Petkus winter rye. *Ann. Bot.*, **16**, 1–21.
PYATT, F. B. (1967). The inhibitory influence of *Peltigera canina* on the germination of graminaceous seeds and the subsequent growth of the seedlings. *The Bryologist*, **70**, 326–9.
PYBUS, M. B., SMITH, M. S., WAIN, R. L. & WIGHTMAN, F. (1959). Studies on plant growth-regulating substances. XII Chloro- and methyl-substituted phenoxyacetic and benzoic acids. *Ann. appl. Biol.*, **47**, 173–81.
PYBUS, M. B., WAIN, R. L. & WIGHTMAN, F. (1958). Studies on plant growth-regulation substances. XIV Chloro-substituted phenylacetic acids. *Ann. appl. Biol.*, **47**, 593–600.
RAADTS, E. & SÖDING, H. (1947). Über den Einfluss der Askorbinsäure auf die Auxinaktivierung. *Naturwissenschaften*, **11**, 344–6.
RAALTE, M. H. VAN (1951). Interaction of indole and hemi-auxins with indoleacetic acid in root formation. I and II. *Proc. K. ned. Acad. Wet.*, **54**, 21–9 & 117–25.
RADLEY, M. (1956). Occurrence of substances similar to gibberellic acid in higher plants. *Nature, Lond.*, **178**, 1070–1071.
RADLEY, M. (1958). The distribution of substances similar to gibberellic acid in higher plants. *Ann. Bot.*, **22**, 297–307.
RADLEY, M. (1969). The effect of endosperm on the formation of gibberellins in barley embryo. *Planta*, **86**, 218–23.
RADLEY, M. & DEAR, E. (1958). Occurrence of gibberellin-like substances in the coconut. *Nature, Lond.*, **182**, 1098.
RAI, V. K. & LALORAYA, M. M. (1967). Effect of different gibberellins on the growth of the hypocotyl. *Physiologia Pl.*, **20**, 879–85.
RAJAGOPAL, R. (1968). Occurrence and metabolism of indoleacetaldehyde in certain plant tissues under aseptic conditions. *Physiologia Pl.*, **21**, 378–85.
RAMBOUR, S. (1968). Action de l'acide β-indolelacétique sur la teneur en acides organiques de fragments de tissus de Topinambour cultivés *in vitro. C. r. hebd. Séanc. Acad. Sci. Paris*, **266**, 1120–2.
RAMSTAD, E. (1953). Über das Vorkommen und die Verbreitung von Chelidonsäure in einigen Pflanzenfamilien. *Helv. pharmacol. Acta*, **28**, 45–57.
RAPER, J. R. (1951). Chemical regulation of sexual processes in fungi. In

Skoog, F. (ed.), *Plant Growth Substances*, University of Wisconsin Press, pp. 301–13.
RAPER, J. R. (1952). Chemical regulation of sexual processes in the thallophyta. *Bot. Rev.*, **18**, 447–545.
RAPER, J. R. (1957). Hormones and sexuality in lower plants. In *The biological action of growth substances. Symp. Soc. exp. Biol.*, **11**, 143–65.
RAWITSCHER-KUNKEL, E. & MACHLIS, L. (1962). The hormonal integration of sexual reproduction in *Oedogonium. Am. J. Bot.*, **49**, 177–83.
RAY, P. M. (1958). Composition of cell walls of *Avena* coleoptile. *Pl. Physiol.* **33**, (Suppl.) xlvii.
RAY. P. M. (1960). The distruction of indoleacetic acid III. Relationships between peroxidase action and indoleacetic acid oxidation. *Archs Biochem. Biophys.*, **87**, 19–30.
RAY, P. M. (1961). Hormonal regulation of plant cell growth. In Bonner, D. M. (ed.), *Control Mechanisms in Cellular Processes*, Ronald Press, New York, pp. 185–212.
RAY, P. M. & RUESINK, A. W. (1962). Kinetic experiments on the nature of the growth mechanism on oat coleoptile cells. *Devl. Biol.*, **4**, 377–97.
RAY. P. M. & THIMANN, K. V. (1955). Steps in the oxidation of indoleacetic acid. *Science N. Y.*, **122**, 187–8.
RAYLE, D. L. & PURVES, W. K. (1967*a*). Isolation and identification of indole-3-ethanol (tryptophol) from cucumber seedlings. *Pl. Physiol. Lancaster*, **42**, 520–24.
RAYLE, D. L. & PURVES, W. K. (1967*b*). Conversion of indole-3-ethanol to indole-3-acetic acid in cucumber seedling shoots. *Pl. Physiol. Lancaster*, **42**, 1091–3.
RAYLE, D. L. & PURVES, W. K. (1968). Studies on 3-indolethanol in higher plants. In Wightman, F. & Setterfield, G. (eds.), *Biochemistry and Physiology of Plant Growth Substances*. Runge Press, Ottawa, pp. 153–61.
RAYNER, M. C. & NEILSON JONES, W. (1946). *Problems in Tree Nutrition*. Faber and Faber, London.
RECALDE, L. & BLESA, A. C. (1961). [Contributions to the study of growth in excised sections of coleoptiles. IV Action of ascorbic acid on growth.] In Spanish. *Anal. Edafol. Agrobiol.*, **20**, 119–28.
RECALDE, L., VERDEJO, G. & BLESA, A. C. (1960). The effect of gibberellic acid on the growth of oat coleoptile sections. *Phyton, B. Aires*, **14**, 55–60.
REDEMANN, C. T., RAPPAPORT, L. & THOMPSON, R. A. (1968). Phaseolic acid: a new plant growth regulator from bean seed. In Wightman, F. & Setterfield, G, (eds.), *Biochemistry and Physiology of Plant Growth Substances*, Runge Press, Ottawa, pp. 109–24.
REED, D. J. (1965). Tryptamine oxidation by extracts of pea seedlings: Effect of growth retardant β-hydroxyethylhydrazine. *Science, N. Y.*, **148**, 1097–9.
REED, D. L. (1968). Tryptophan decarboxylation in cell-free extracts of Alaska pea epicotyls. In Wightman, F. & Setterfield, G. (eds.), *Biochemistry*

and Physiology of Plant Growth Substances. Runge Press, Ottawa, pp. 243–58.
REED. D. J., MOORE, T. C. & ANDERSON, J. D. (1965). Plant growth retardant B-995, a possible mode of action. *Science, N. Y.*, **148**, 1469–71.
REID, D. M. & BURROWS, W. J. (1968). Cytokinin and gibberellin-like activity in the spring sap of trees. *Experientia*, **24**, 189–90
REID, D. M. & CARR, D. J. (1967). Effects of a dwarfing compound on the production and export of gibberellin-like substances by root systems. *Planta*, **73**, 1–11.
REID, D. M. & CLEMENTS, J. B. (1968). RNA and protein synthesis; prerequisites of red light-induced gibberellin synthesis. *Nature, Lond.*, **219**, 607–9.
REID, D. M., CROZIER, A. & HARVEY, B. M. R. (1969). The effect of flooding on the export of gibberellins from the root to the shoot. *Planta*, **89**, 376–9.
REINDERS, D. E. (1938). The process of water uptake by discs of potato tissue. *Proc. K. ned. Akad. Wet. Sec. Sci.*, **41**, 820.
REINERT, J. (1953). Über die Wirkung von Riboflavin und Carotin beim Phototropismus von Avena-Koleoptilen und bei anderen pflanzlichen Lichtreizreaktionen. *Z. Bot.*, **41**, 103–22.
REINHOLD, L. & POWELL, R. G. (1958). The stimulatory effect of indole-3-acetic acid on the uptake of amino acids by tissues of *Helianthus annuus*. *J. exp. Bot.*, **9**, 82–96.
RICE, E. L. (1964). Inhibition of nitrogen-fixing and nitrifying bacteria by seed plants. *Ecology*, **45**, 824–37.
RICE, E. L. (1965*a*). Inhibition of nitrogen-fixing and nitrifying bacteria by seed plants. II Characterisation and identification of inhibitors. *Physiologia Pl.*, **18**, 255–68.
RICE, E. L. (1965*b*). Inhibition of nitrogen-fixing and nitrifying bacteria by seed plants. IV The inhibitors produced by *Ambrosia elatior* and *A. psilostachya*. *Southwestern Naturalist*, **10**, 248–55.
RICE, E. L. (1967). Inhibition of nitrogen-fixing and nitrifying bacteria by seed plants. V Inhibitors produced by *Bromus japonicus* Thunb. *Southwestern Naturalist*, **12**, 97–103.
RICE, E. L. (1968). Inhibition of nodulation of inoculated legumes by pioneer plant species from abandoned fields. *Bull. Torrey bot. Club*, **95**, 346–58.
RICE, E. L., PENFOUND, W. T. & ROHRBAUGH, L. M. (1960). Seed dispersal and mineral nutrition in succession in abandoned fields in central Oklahoma. *Ecology*, **41**, 224–8.
RICHMOND, A., BACK, A. & SACHS, B. (1970). A study of the hypothetical role of cytokinins in completion of tRNA. *Planta*, **90**, 57–65.
RICHMOND, A. & LANG, A. (1957). Effect of kinetin on protein content and survival of detached *Xanthium* leaves. *Science, N. Y.*, **125**, 650–1.
RIDDELL, J. A., HAGEMAN, H. A., J'ANTHONY, C. M. & HUBBARD, W. L. (1962). Retardation of plant growth by a new group of chemicals. *Science, N.Y.*, **136**, 39.

RIDDLE, V. M. & MAZELIS, M. (1964). A role for peroxidase in biosynthesis of auxin. *Nature. Lond.*, **202**, 391–2.
RIDLEY, V. W. (1923). Some principles involved in the handling of fruits. *Fruit Dispatch. N. Y.*, **8**, 523–5.
RIGAUD, J. (1970). La biosynthése de l'acide indolyl-3-acétique en liaison avec le métabolisme du tryptophol et de l'indolyl-3-acétaldéhyde chez *Rhizobium. Physiologia Pl.*, **23**, 171–8.
RISHBETH, J. (1950). Observations on the biology of *Fomes annosus* with particular reference to East Anglian pine plantations. I. The outbreak of disease and ecological status of the fungus. *Ann. Bot.*, **14**, 365–83.
ROBBINS, W. J. (1922). Cultivation of excised root tips and stem tips under sterile conditions. *Bot. Gaz.*, **73**, 376–83.
ROBBINS, W. J. & KAVANAGH, F. (1942). Hypoxanthine, a growth factor of *Phycomyces. Proc. natl. Acad. Sci U.S.A.*, **28**, 65.
ROBERTS, R. H. (1951). The induction of flowering with a plant extract. In Skoog, F. (ed.), *Plant Growth Substances*, University of Wisconsin Press, pp. 347–50.
ROBINSON, E. (1956). Proteolytic enzymes in growing root cells. *J. exp. Bot.*, **7**, 296–305.
ROBINSON, E. & BROWN, R. (1952). The development of the enzyme complement in growing root cells. *J. exp. Bot.*, **3**, 356–74.
ROBINSON, E. & BROWN, R. (1954). Enzyme changes in relation to cell growth in excised root tissue. *J. exp. Bot.*, **5**, 71–8.
ROBINSON, P. M. & WAREING, P. F. (1964). Chemical nature and biological properties of an inhibitor varying with photoperiod in sycamore (*Acer pseudoplatanus*). *Physiologia Pl.*, **17**, 314–23.
RODIONOVA, N. A. (1962). [Effect of 2,3,5-tri-iodobenzoic acid on free auxin content in beans.] In Russian. *Byul. Gl. bot. Suda Akad. Nauk. SSSR.*, **45**, 81–4.
RODRIGUES PEREIRA, A. S. (1961). Flower initiation in excised stem discs of Wedgewood iris. *Science N. Y.*, **134**, 2044–5.
RODRIGUES PEREIRA, A. S. (1964). Endogenous growth factors and flower formation in Wedgewood iris bulbs. *Acta bot. neerl.*, **13**, 302–21.
ROELOFSEN, P. A. (1959). *The Plant Cell Wall*, Borntraeger, Berlin.
ROELOFSEN, P. A. & HOUWINK, A. L. (1953). Architecture and growth of the primary wall in some plant hairs and the *Phycomyces* sporangiophore. *Acta bot. neerl.*, **2**, 218–25.
ROGERS, H. T., PEARSON, R. W. & PIERRE, W. H. (1942). The source and phosphatase activity of exoenzyme systems of corn and tomato roots. *Soil Sci.*, **54**, 353.
ROGGEN, H. P. J. R. & STANLEY, R. G. (1969). Cell-wall-hydrolysing enzymes in wall formation as measured by pollen-tube extension. *Planta*, **84**, 295–303.
ROGOZINSKA, J. H., HELGESON, J. P. & SKOOG, F. (1964). Tests for kinetin-like growth-promoting activities in triacanthine and its isomer, 6-(γ, γ-dimethylallylamino)-purine. *Physiologia Pl.*, **17**, 165–76.
ROGOZINSKA, J. H., HELGESON, J. P. & SKOOG, F. (1965). Partial purification

of a cell division factor from peas. *Pl. Physiol. Lancaster*, **40**, 469–76.
ROTHWELL, K. & WAIN, R. L. (1963). Studies on plant growth-regulating substances. XVII S-esters of dithiocarbamates derived from amino acids. *Ann. appl. Biol.*, **51**, 161–7.
ROTHWELL, K. & WAIN, R. L. (1964). Studies on a growth inhibitor in yellow lupin (*Lupinus luteus* L.) In *Régulateurs naturels de la croissance végétale, Colloq. int. Centre natl. Rech. Sci.*, No. 123, 363–75.
ROTHWELL, K. & WRIGHT, S. T. C. (1967). Phytokinin activity in some new 6-substituted purines. *Proc. R. Soc. B.*, **167**, 202–23.
ROUBAIX, J. DE. & LAZAR, G. (1957). Métabolisme respiratoire de la Betterave Sucrière. VII Les substances inhibitrices contenues dans les glomerules de Betterave Sucrière. *La Sucrerie Belge.*, **5**, 285.
ROVIRA, A. D. (1956). Plant root excretions in relation to the rhizosphere effect. I. The nature of root exudate from oats and peas. *Pl. Soil.*, **7**, 178–93.
ROVIRA, A. D. & McDOUGALL, B. M. (1967). Microbiological and biochemical aspects of the rhizosphere. In McLaren, A. D. & Peterson, G. H. (eds.), *Soil Biochemistry*. Dekker, New York, pp. 417–63.
ROYCHOUDHURY, R., DATTA, A. & SEN, S. P. (1965). The mechanism of action of plant growth substances: the role of nuclear RNA in growth substance action. *Biochim. biophys. Acta*, **107**, 346–51.
ROYCHOUDHURY, R. & SEN, S, P. (1964). Studies in the mechanism of auxin action: Auxin regulation of nucleic acid metabolism in pea internodes and coconut milk nuclei. *Physiologia Pl.*, **17**, 352–62.
RUBINSTEIN, B. & ABELES, F. B. (1965). Relationship between ethylene evolution and leaf abscission. *Bot. Gaz.*, **126**, 255–9.
RUDDAT, M., HEFTMANN, E. & LANG, A. (1965*a*). Conversion of steviol to a gibberellin-like compound by *Fusarium moniliforme*. *Archs Biochem. Biophys.*, **111**, 187–90.
RUDDAT, M., HEETMANN, E. & LANG, A. (1965*b*). Chemical evidence for the mode of action of AMO-1618, a plant growth retardant. *Naturwissenschaften*, **52**, 267.
RUDDAT, M., LANG, A. & MOSETTIG, E. (1963). Gibberellin activity of steviol, a plant terpenoid. *Naturwissenschaften*, **50**, 23.
RUDICH, J., HALEVY, A. H. & KEDAR, N. (1969). Increase in femaleness in three cucurbits by treatment with ethrel, an ethylene-releasing compound. *Planta*, **86**, 69–76.
RUESINK, A. W. (1969). Polysaccharides and the control of cell wall elongation. *Planta*, **89**, 95–107.
RUGE, U. (1937*a*). Untersuchungen über den Einfluss des Hetero-auxins auf das Streckungswachstums des Hypocotyls von *Helianthus annuus*. *Z. Bot.*, **31**, 1.
RUGE, U. (1937*b*). Untersuchungen über die Änderung des osmotischen Zustandsgrössen und des Membraneigenschaften des Hypokotyls von *Helianthus annuus* beim normale Streckungswachstums. *Planta*, **27**, 352–66.
RUGE, U. (1937*c*). Zur Charakteristik einer für die Physiologie der Zellstreck-

ung wichtigen Intermicellarsubstanzen pflanzlicher Membranen. *Biochem. Z.*, **295**, 29.
RÜNGER, W. (1962). Über den Einfluss der Tageslänge auf Wachstum, Blütenbildung und -entwicklung von *Echeveria harmsii*. *Gartenbauwissenschaften*, **9**, 279–94.
RUSSELL, D. W. & GALSTON, A. W. (1967). Flavonoid complexes in *Pisum sativum*. IV The effect of red light on synthesis of kaempferol complexes and on growth in sub-apical internode tissue. *Phytochemistry*, **6**, 791–7.
RUSSELL, D. W. & GALSTON, A. W. (1969). Blockage by gibberellic acid of phytochrome effects on growth, auxin responses and flavonoid synthesis in etiolated pea internodes. *Pl. Physiol. Lancaster*, **44**, 1211–16.
RUTHERFORD, P. P., WESTON, E. W. & FLOOD, A. E. (1969). Effects of 2,4- and 3,5-dichlorophenoxyacetic acids on Jerusalem artichoke tuber tissue disks. *Phytochemistry*, **8**, 1859–66.
SABNIS, D. D. & AUDUS, L. J. (1967). Growth substance interactions during uptake by mesocotyl segments of *Zea mays* L. *Ann. Bot.*, **31**, 263–81.
SACHER, J. A. (1967*a*). Senescence: Action of auxin and kinetin in control of RNA and protein synthesis in subcellular fractions of bean endocarp. *Pl. Physiol. Lancaster*, **42**, 1334–42.
SACHER, J. A. (1967*b*). Control of synthesis of RNA and protein in subcellular fractions of *Rhoeo discolor* leaf sections. *Exp. Gerontol.*, **2**, 261–78.
SACHER, J. A. & GLASZIOU, K. T. (1959). Effects of auxins on membrane permeability and pectic substances in bean endocarp. *Nature, Lond.*, **183**, 757–8.
SACHS, J. (1865). Wirkung des Lichts auf die Blüthenbildung unter Vermittlung der Laubblätter. *Bot. Ztg.*, **23**, 117–21, 125–31 & 133–9.
SACHS, R. M. & KOFRANEK, A. M. (1963). Comparative cytohistological studies on inhibition and promotion of stem growth in *Chrysanthemum morifolium*. *Am. J. Bot.*, **50**, 772–9.
SACHS, R. M. & LANG, A. (1961). Shoot histogenesis and the sub-apical meristem: the action of gibberellic acid, AMO-1618 and maleic hydrazide. In *Plant Growth Regulation*. Iowa State University Press, pp. 567–78.
SACHS, T. & THIMANN, K. V. (1964). Release of lateral buds from apical dominance. *Nature, Lond.*, **201**, 939–40.
SAEBØ, S. (1960). The action of gibberellic acid in the *Avena* coleoptile curvature test. *Physiologia Pl.*, **13**, 839–45.
SAKURAI, A. & TAMURA, S. (1966). Synthesis of dihydrohelminthosporic and allodihydrohelminthosporic acids and their plant growth-regulating activity. *Agric. biol. Chem.*, **30**, 793–9.
SALISBURY, F. B. (1955). *Kinetic studies on the physiology of flowering*, Doctoral thesis, California Institute of Technology.
SALISBURY, F. B. (1963). *The Flowering Process*. Pergamon Press, London.
SAMISH, R. M. (1954). Dormancy in woody plants. *A. Rev. Pl. Physiol.*, **5**, 183–204.
SARKAR, S. (1958). Versuche zur Physiologie der Vernalisation. *Biol. Zbl.*, **77**, 1–49.

SARKISSIAN, I. V. (1966). Enhancement of enzymatic synthesis of citrate *in vitro* by indole-3-acetic acid. *Physiologia Pl.*, **19**, 328–34.
SASTRY, K. K. S. & MUIR, R. M. (1963). Gibberellin effect on diffusible auxin in fruit development. *Science, N. Y.*, **140**, 494–5.
SAWADA, K. (1912). Diseases of agricultural products in Japan, *Formosan agric. Rev.*, No. 63, 10, 16.
SCHERFF, R. A. (1952). *The effect of an ammonium phenylcarbamate on growth and development of bean plants*, M. A. thesis, George Washington University, Washington, D. C.
SCHEUERMANN, R. (1952). Der Einfluss wasserlöslichen Vitamine auf die Wirksamkeit von Heteroauxin im Wachstumsprozess der höheren Pflanzen. *Planta*, **40**, 265–300.
SCHIEWER, U. (1967). Auxinvorkommen und Auxinstoffwechsel bei mehrzelligen Ostseealgen. II Zur Entstehung von Indol-3-essigsäure aus Tryptophan und Berücksichtigung des Einfluss der marinen Bakterienflora. *Planta*, **75**, 152–60.
SCHINDLER, W. (1958). Indol-2-essigsäure. *Helv. chim. Acta.*, **41**, 1441–3.
SCHLENDER, K. K., BUKOVAC, M. J. & SELL, H. M. (1966). The synthesis and physiological activity of several α-alkylindole-3-acetic acids. *Phytochemistry*, **5**, 133–9.
SCHMIDT, E. L. (1951), Soil microorganisms and plant growth substances. I Historical. *Soil Sci.*, **71**, 129–40.
SCHMITZ, R. Y., SKOOG, F., HECHT, S. M. & LEONARD, N. J. (1971). Cytokinins: Synthesis and biological activity of zeatin esters and related compounds. *Phytochemistry*, **10**, 275–80.
SCHOENE, D. L. & HOFFMAN, O. L. (1949). Maleic hydrazide, a unique growth regulant. *Science, N. Y.*, **109**, 588–90.
SCHOPFER, W. H. (1943). *Plants and vitamins*, Chronica Botanica Publ., Waltham Massachussetts.
SCHRAUDOLF, H. (1962). Die Wirkung von Phytohormonen auf Keimung und Entwicklung von Farnprothallien. I Auslösung der Antheridienbildung und Dunkelkeimung bei Schizaeaceae durch Gibberellinsaure. *Biol. Zbl.*, **81**, 731–40.
SCHRAUDOLF, H. (1965). Zur Verbreitung von Glucobrassicin und Neoglucobrassicin in höheren Pflanzen. *Experientia*, **21**, 520–5.
SCHRAUDOLF, H. (1966). Die Wirkung von Phytohormonen auf Keimung und Entwicklung von Farnprothallien. IV Die Wirkung von unterschiedlichen Gibberelline und von Allo-gibberellinsäure auf die Auslösung der Antheridienbildung bei *Anemia phyllitidis* und einigen Polypodiaceae. *Pl. Cell Physiol.*, **7**, 277–89.
SCHRAUDOLF, H. (1967). Wirkung von Terpenderivaten (Helminthosporol, Helminthosporsäure, Dihydrohelminthosporsäure und Steviol) auf die Antheridienbildung in *Anemia phyllitidis. Planta*, **74**, 188–93.
SCHRAUDOLF, H. & WEBER, H. (1969). IAN-Bildung aus Glucobrassicin: pH-Abhängigkeit und wachstumsphysiologische Bedeutung. *Planta*, **88**, 136–43.

SCHREINER, O. & LATHROP, E. C. (1911). Dihydroxystearic acid in good and poor soils. *J. Am. chem. Soc.*, **33**, 1412.

SCHREINER, O. & REED, H. S. (1907). Some factors influencing soil fertility. *Bull. U. S. Div. Soils*, No. 40.

SCHREINER, O. & REED, H. S. (1908). The toxic action of certain organic plant constituents. *Bot. Gaz.*, **45**, 73–102.

SCHUMACHER, W. & MATTHAEI, H. (1955). Über den Zusammenhang zwischen Streckungswachstum und Eiweisssynthese. *Planta*, **45**, 213–6.

SCHUPHAN, W. (1948). Ein Beitrag zur physiologischen Wirkung einer Pflanze auf die andere. Gemuseversuche im Allein- und Mischanbau. *Botanica econ.*, **1**, 1–15.

SCOTT, T. K. & BRIGGS, W. R. (1960). Auxin relationships in the Alaska pea. *Am. J. Bot.*, **47**, 492–9.

SCOTT, T. K. & JACOBS, W. P. (1964). Critical assessment of techniques for identifying the physiologically significant auxins in plants. In *Régulateurs naturels de la croissance végétale, Colloq. int. Centre natl. Rech. Sci.*, No. 123, 457–74.

SCOTT, T. K. & WILKINS, M. B. (1968). Auxin transport in roots. II Polar flux of IAA in *Zea* roots. *Planta*, **83**, 323–34.

SEBANEK, J. and HINK, J. (1966). Über den Einfluss des Chlorocholinchlorids in einer nicht-inhibierenden Dosis auf das Niveau endogener Gibberelline in der Erbsenepikotylen. *Acta Univ. agron. Brno*, **3**, 393–9.

SECHET, J. (1953). Contributions a l'ètude de la printanisation. *Botaniste*, **37**, 1–289.

SEELEY, R. C., FAWCETT, C. H., WAIN, R. L. & WIGHTMAN, F. (1956). Chromatographic investigations on the metabolism of certain indole derivatives in plant tissues. In Wain. R. L. & Wightman, F. (eds.), *The Chemistry and Mode of Action of Plant Growth Substances*, Butterworth, London, pp. 234–47.

SEELEY, R. C. & WAIN, R. L. (1950). A note on the growth-regulating activity of 2,6-dichlorophenoxyacetic acid. *J. hort. Sci.*, **25**, 264–5.

SELL, H. M., WITTWER, S. H., REBSTOCK, T. K. & REDEMANN, C. T. (1953). Comparative stimulation of parthenocarpy in the tomato by various indole compounds. *Pl. Physiol. Lancaster*, **28**, 481–7.

SEMBDNER, G., SCHNEIDER, G. and SCHREIBER, K. (1965). Über die biologische Wirksamkeit einiger Gibberellinsäure-Abbauprodukte. VIII. *Planta*, **66**, 65–74.

SEN, S. P. & LEOPOLD, A. C. (1954). Paper chromatography of plant growth regulators and allied compounds. *Physiologia Pl.*, **7**, 98–108.

SEN, S. P. & SEN GUPTA, A. (1961). Studies on auxin-induced metabolic changes in plant tissues: Effect of indoleacetic acid on the metabolism of pyruvate-3-C^{14} and acetate-2-C^{14} in *Avena* coleoptile. *Trans. Bose Res. Inst.*, **24**, 95–104.

SENDEN, H. VAN (1951). Untersuchungen über den Einfluss von Heteroauxin und anderen Faktoren auf die Blütenbildung bei der Kurtztagspflanze *Ka-*

lanchoë blossfeldiana. Biol. Zbl., **70**, 537–65.
SEN GUPTA, A. & SEN, S. P. (1961). Effect of indole-3-acetic acid (IAA) on phosphorus metabolism in the *Avena* coleoptile. *Nature, Lond.*, **192**, 1290.
SEQUEIRA, L. & MINEO, L. (1966). Partial purification and kinetics of indoleacetic acid oxidase from tobacco roots. *Pl. Physiol. Lancaster*, **41**, 1200–1208.
SETH, A. & WAREING, P. F. (1965). Isolation of a kinetin-like root-factor in *Phaseolus vulgaris. Life Sci.*, **4**, 2275–80.
SHANTZ, E. M. (1966). Chemistry of naturally-occurring growth-regulating substances. *A. Rev. Pl. Physiol.*, **17**, 409–38.
SHANTZ, E. M. & STEWARD, F. C. (1952). Coconut milk factor: the growth-promoting substances in coconut milk. *J. Am. chem. Soc.*, **74**, 6133–5.
SHANTZ, E. M. & STEWARD, F. C. (1955). The identification of compound A from coconut milk as 1,3-diphenylurea. *J. Am. chem. Soc.*, **77**, 6351–3.
SHECHTER, L. & WEST, C. A. (1969). Biosynthesis of gibberellin. IV Biosynthesis of cyclic diterpenoids from *trans*-geranylgeranyl pyrophosphate. *J. biol. Chem.*, **244**, 3200–9.
SHERWIN, J. E. & PURVES, W. K. (1969). Tryptophan as an auxin precursor in cucumber seedlings. *Pl. Physiol. Lancaster*, **44**, 1303–9.
SHIBAOKA, H. & THIMANN, K. V. (1970), Antagonism between kinetin and amino acids. *Pl. Physiol. Lancaster*, **46**, 212–20.
SHIBAOKA, H. & YAMAKI, T. (1959). A sensitized *Avena* curvature test and identification of the diffusible auxin in *Avena* coleoptiles. *Bot. Mag., Tokyo*, **72**, 152–8.
SHIH, C. Y. & RAPPAPORT, L. (1969). Hormonal regulation, nucleic acid synthesis and bud rest in potato tubers. *Abstr. 11th. int. bot. Congr. Seattle.*
SHIMODA, C., MASUDA, Y. & YANAGISHIMA, N. (1967). Nucleic acid metabolism involved in auxin-induced elongation in yeast cells. *Physiologia Pl.*, **20**, 299–305.
SHIMODA, C. & YANAGISHIMA, N. (1968). Strain dependence of the cell-expanding effect of β-1,3-glucanase in yeast. *Physiologia Pl.*, **21**, 1163–9.
SIEGEL, B. Z. & GALSTON, A. W. (1967*a*). The isoperoxidases of *Pisum sativum. Pl. Physiol. Lancaster*, **42**, 221–6.
SIEGEL, B. Z. & GALSTON, A. W. (1967*b*). Idolecetic acid oxidase activity of apoperoxidase. *Science N. Y.*, **157**, 1557–9.
SIEGEL, S. M. (1950). Germination and growth inhibitors from red kidney bean seed. *Bot. Gaz.*, **111**, 353–6.
SIEGEL, S. M. & GALSTON, A. W. (1953). Experimental coupling of indoleacetic acid to pea root protein *in vivo* and *in vitro. Proc. natl. Acad. Sci. U.S.A.*, **39**, 1111–1118.
SIEGELMAN, H. W. & BUTLER, W. L. (1965). Properties of phytochrome. *A. Rev. Pl. Physiol.*, **16**, 383–92.
SIEGELMAN, H. W. & FIRER, E. M. (1964). Purification of phytochrome from oat seedlings. *Biochemistry*, N.Y., **3**, 418–23.

SIETSMA, J. H. and HASKINS, R. H. (1967). Further studies on sterol stimulation of sexual reproduction in *Pythium*. *Can. J. Microbiol.*, **13**, 361–7.

SIEVERS, A. F. & TRUE, R. H. (1912). A preliminary study of the forced curing of lemons as practiced in California. *Bull. U. S. Dept. Bureau Pl. Indust.*, No. 232.

SIMON, E. W. (1967). Types of leaf senescence. *Symp. Soc. exp. Biol.*, **21**, 215–30.

SIRONVAL, C. (1957). La photopériode et la sexualisation du fraisier de quatre-saisons a fruits rouges (métabolisme chlorophyllien et hormone florigène). *C. r. Rech. Trav. Centre Rech. Hormones vég.*, (1952–56), No. 18, February 1957.

SITTON, D., RICHMOND, D. A. & VAADIA, Y. (1967). On the synthesis of gibberellins in roots. *Phytochemistry*, **6**, 1101–5.

SKINNER, C. G., CLAYBROOK, J. R., TALBERT, F. & SHIVE, W. (1957). Effect of 6-(substituted)-thio- and amino-purines on germination of lettuce seed. *Pl. Physiol. Lancaster*, **32**, 117–20.

SKOOG, F. (1935). The effect of X-irradiation on auxin and plant growth. *J. cell. comp. Physiol.*, **7**, 227–70.

SKOOG, F. (1937). A de-seeded *Avena* test for small amounts of auxin and auxin precursors. *J. gen. Physiol.*, **20**, 311–34.

SKOOG, F. (1940). Relationship between zinc and auxin in the growth of higher plants. *Am. J. Bot.*, **27**, 939–51.

SKOOG, F. (1954). Substances involved in normal growth and differentiation of plants. *Brookhaven Symp. Biol.*, **6**, 1–21.

SKOOG, F. & ARMSTRONG, D. J. (1970). Cytokinins. *A. Rev. Pl. Physiol.*, **21**, 359–84.

SKOOG, F., HAMZI, H. Q., SZWEYKOWSKA, A. M., LEONARD, M. J., CARRAWAY, K. L., FUJII, T., HELGESON, J. P. & LOEPPKY, R. N. (1967). Cytokinins: Structure/activity relationships. *Phytochemistry*, **6**, 1169–92.

SKOOG, F. and LEONARD, N. J. (1968). Sources and structure: activity relationships of cytokinins. In Wightman, F. & Setterfield, G. (eds.), *Biochemistry and Physiology of Plant Growth Substances*. Runge Press, Ottawa, pp. 1–18.

SKOOG, F. & MILLER, C. O. (1957). Chemical regulation of growth and organ function in plant tissues cultured *in vitro*. In Porter, H. K. (ed.), *The Biological Action of Growth Substances*. *Symp. Soc. exp. Biol.*, **11**, 118–31.

SKOOG, F. & MONTALDI, E. (1961). Auxin-kinetin interaction regulating the scopoletin and scopolin levels in tobacco tissue culture. *Proc. natl. Acad. Sci. U.S.A.*, **47**, 36–49.

SKOOG, F. & ROBINSON, B. J. (1950). A direct relationship between indoleacetic acid effects on growth and reducing sugar in tobacco tissue. *Proc. Soc. exp. Biol. Med.*, **74**, 565–8.

SKOOG, F., STRONG, F. M. & MILLER, C. O. (1965). Cytokinins. *Science, N. Y.*, **148**, 532–3.

SKYTT-ANDERSEN, A. & MUIR, R. M. (1969). Gibberellin-induced changes in diffusible auxin from savoy cabbage. *Physiologia Pl.*, **22**, 354–63.

SMITH, G. M. (1946). The nature of sexuality in *Chlamydomonas*. *Am. J. Bot.*, **33**, 625.

SMITH, G. M. (1951). The sexual substances of algae. In Skoog, F. (ed.), *Plant Growth Substances*. University Wisconsin Press, pp. 315–28.

SMITH, M. S. & WAIN, R. L. (1952). The plant growth-regulating activity of *dextro* and *laevo* α-(2-naphthoxy)-propionic acid. *Proc. R. Soc. B.*, **139**, 118–27.

SMITH, M. S., WAIN, R. L. & WIGHTMAN, F. (1952). Antagonistic action of certain stereo-isomers on the plant growth-regulating activity of their enantiomorphs. *Nature, Lond.*, **169**, 883.

SMITH, P. F. (1941). Studies on the influence of colchicine and 3-indoleacetic acid on some enzymic reactions. *Proc. Okla. Acad. Sci.*, **21**, 105.

SÖDING, H. (1937). Über Wuchsstoffteste. *Jb. wiss. Bot.*, **85**, 770–87.

SÖDING, H. (1952). *Die Wuchsstofflehre*. G. Thieme, Stuttgart.

SOMOGYI, G. (1952). Notes on sugar determination. *J. biol. Chem.*, **195**, 19–23.

SONDHEIMER, E. & WALTON, D. C. (1970). Structure-activity correlations with compounds related to abscisic acid. *Pl. Physiol. Lancaster*, **45**, 244–8.

SPÄTH, E. (1937). Die natürlichen Cumarine. *Ber. dt. bot. Ges.*, **A**, **70**, 83–117.

SPECTOR, C. and PHINNEY, B. O. (1968). Gibberellin biosynthesis: Genetic studies in *Gibberella fujikuroi*. *Physiologia Pl.*, **21**, 127–36.

SPIEGEL, P. (1954). *Factors affecting the rooting of grape vine cuttings*. Doctoral thesis, Hebrew University of Rehovot, Israel. (Quoted from Samish, 1954.)

SPLITTSTOESSER, W. E. (1970). Effects of 2-chloroethylphosphonic acid and gibberellic acid on sex expression and growth of pumpkin. *Physiologia Pl.*, **23**, 762–8.

SRB, A. M., OWEN, R. D. & EDGAR, R. S. (1965). *General Genetics*. Freeman, San Francisco & London.

SRIVASTAVA, B. I. S. (1968). Increase in chromatin-associated nuclease activity of excised barley leaves during senescence and its suppression by kinetin. *Biochem. biophys. Res. Commun.*, **32**, 533–8.

SRIVASTAVA, B. I. S. & WARE, G. (1965). The effect of kinetin on nucleic acids and nucleases of excised barley leaves. *Pl. Physiol. Lancaster*, **40**, 62–4.

STAHL E. & KALDEWEY, H. (1961). Spurenanalyse physiologisch aktiver, einfacher Indolderivate. *Hoppe-Seyl. Z. physiol. Chem.*, **323**, 182–91.

STEEVES, T. A. & BRIGGS, W. R. (1960). Morphogenic studies on *Osmunda cinnamomea*: the auxin relationships of expanding fronds. *J. exp. Bot.*, **11**, 45–67.

STENLID, G. (1963). The effect of flavonoid compounds on oxidative phosphorylation and on the enzymatic destruction of indoleacetic acid. *Physiologia Pl.*, **16**, 110–20.

STEVENSON, F. J. (1967). Organic acids in soils. In McLaren, A. D. & Peterson, G. H. (eds.), *Soil Biochemistry*. Dekker, New York, pp. 119–46.

STEVENSON, I. L. (1954). Antibiotic production by Actinomyces in soil demonstrated by morphological changes induced in *Helminthosporium sativum*. *Nature, Lond.*, **174**, 598.

STEVENSON, I. L. (1956). Antibiotic activity of Actinomyces in soil as demonstrated by direct observation techniques. *J. gen. Microbiol.*, **15**, 372–80.

STEWARD, F. C. (1963), (ed.) *Plant physiology*, vol. 3. Academic Press, New York and London.

STEWARD, F. C. & CAPLIN, S. M. (1951). A tissue culture from potato tuber; the synergistic action of 2,4-D and coconut milk. *Science, N. Y.*, **113**, 518–20.

STEWARD, F. C. & CAPLIN, S. M. (1954*a*). The growth of carrot tissue explants and its relation to the growth factors present in coconut milk. I(A) The development of the quantitative method and the factors affecting the growth of carrot tissue explants. *Annls Biol.*, **30**, 385–94.

STEWARD, F. C. & CAPLIN, S. M. (1954*b*). The growth of carrot tissue explants and its relation to the growth factors present in coconut milk. I(B) The role of the coconut milk growth factor in development and its relation to proliferated and tumorous growth. *Annls Biol.*, **30**, 395–8.

STEWARD, F. C. & SHANTZ, E. M. (1959). The chemical regulation of growth (Some substances and extracts which induce growth and morphogenesis.). *A. Rev. Pl. Physiol.*, **10**, 379–404.

STIVEN, G. (1952). Production of antibiotic substances by the roots of a grass (*Trachypogon plumosus*) (H.B.K. Nees) and of *Pentanisia variabilis* (E. Mey) Harv. (Rubiaceae). *Nature, Lond.*, **170**, 712.

STODDART, J. L. & LANG, A. (1967*a*). Gibberellin synthesis in red clover. *A. Rep. M.S.U./A.E.C. Pl. Res. Lab. for 1967*, p. 44.

STODDART, J. L. & LANG, A. (1967*b*). Effects of daylength on gibberellin synthesis in red clover. *Abstr. 6th. int. Conf. Pl. Growth Substances*, p. 88.

STODDART, J. L. & LANG, A. (1969). Effects of daylength on gibberellin synthesis in leaves of red clover (*Trifolium pratense* L.). In Wightman, F. & Setterfield, G. (eds.), *Biochemistry and Physiology of Plant Growth Substances*. Runge Press, Ottawa, pp. 1371–84.

STODOLA, F. H. (1958). *Source book on Gibberellin. 1828–1957*. U.S.D.A. agric. Res. Service Publ. Peoria. Ill.

STODOLA, F. H., RAPER, K. B., FENNELL, D. I., CONWAY, H. F., SOHNS, V. E., LANGFORD, C. T. & JACKSON, R. W. (1955). The microbiological production of gibberellins A and X. *Archs Biochem. Biophys.*, **54**, 240–5.

STOLAREK, J. (1968). Ionic relations and electrophysiology of single cells of Characeae. II The effect of IAA on sodium and potassium influx in cells of *Nitella translucens*. *Acta Soc. Bot. Pol.*, **37**, 337–45.

STOWE, B. B. (1955). The production of indoleacetic acid by bacteria. *Biochem. J.*, **61**, ix–x.

STOWE, B. B., RAY, P. M. & THIMANN, K. V. (1954). The enzymatic oxidation of indoleacetic acid. *C. r. Séanc., Sect. 11, 12, 8me. int. Congr., Bot.*, pp. 135–141.

STOWE, B. B. & SCHILKE, J. F. (1964). Submicrogram identification and analysis of indole auxins by gas chromatography and spectrofluorimetry. In *Régulateurs naturels de la croissance végétale, Colloq, int. Centre natl. Rech. Sci.*, No. 123, 409–19.

STOWE, B. B. & THIMANN, K. V. (1954). The paper chromatography of indole compounds and some indole-containing auxins of plant tissues. *Archs Biochem. Biophys.*, **51**, 499–516.

STREET, H. E. (1950). Studies in plant nutrition. II Further studies of the effect of some organic supplements on the growth of plants in sand culture. *Ann. appl. Biol.*, **37**, 149–58.

STUART, N. W. (1961). Initiation of flower buds in *Rhododendron* after application of growth retardants. *Science, N. Y.*, **134**, 50–2.

STUTZ, R. F. (1957). The indole-3-acetic acid oxidase of *Lupinus albus* L. *Pl. Physiol. Lancaster*, **32**, 31–9.

SUDA, S. (1960). Mode of action of maleic hydrazide on the growth of *Escherichia coli* and *Avena* coleoptile segments. *Pl. Cell. Physiol.* **1**, 247–53.

SÜDI, J. (1966). Increase in the capacity of pea tissue to form acyl-aspartic acids specifically induced by auxins. *New Phytol.*, **65**, 9–21.

SUGE, H. & RAPPAPORT, L. (1968). Role of gibberellins in stem elongation and flowering in radish. *Pl. Physiol. Lancaster*, **43**, 1208–14.

SUGE, H. & RAPPAPORT, L. (1968). Promotion of flowering by helminthosporol. *Pl. Cell Physiol.*, **9**, 825–30.

SUZUKI, J. (1966). Maleic hydrazide and isonicotinyl hydrazide as carbonyl reagents. *Physiologia Pl.*, **19**, 257–63.

SWABY, R. J. (1942). Stimulation of plant growth by organic matter. *J. Aust. Inst. agric. Sci.*, **8**, 156–63.

SWEENEY, B. M. (1941). Conditions affecting the acceleration of protoplasmic streaming by auxin. *Am. J. Bot.*, **28**, 700–2.

SYNERHOLM, M. & ZIMMERMAN, P. W. (1947). Preparation of a series of ω-(2,4-dichlorophenoxy)-aliphatic acids and some related compounds with a consideration of their biochemical role as plant growth regulators. *Contrib. Boyce Thompson Inst.*, **14**, 369–82.

SZWEYKOWSKA, A. (1962). The effect of kinetin and IAA on shoot development in *Funaria hygrometrica* and *Ceratodon purpureum*. *Acta Soc. Bot. Pol.*, **31**, 553–7.

TAGAWA, T. & BONNER. J. (1957). Mechanical properties of the *Avena* coleoptile as related to auxin and to ionic interactions. *Pl. Physiol. Lancaster*, **32**, 207–15.

TAKAHASHI, N., KITAMURA, H., KAWARADA, A., SETA, Y., TAKAI, M., TAMURA, S. & SUMIKI, Y. (1955). Biochemical studies on 'bakanae' fungus. XXXIV Isolation of gibberellins and their properties. *Bull. agric. Chem. Soc. Japan*, **19**, 267.

TAKAHASHI, N., MUROFUSHI, N., YOKOTA, T. & TAMURA, S. (1967). Gibberellins in immature seeds of *Pharbitis nil*. *Tetrahedron Lett.*, No. 12, 1065–8.

TAKAHASHI, N., MUROFUSHI, N., YOKOTA, T., TAMURA, S., KATO, J. & SHIOTANI, Y. (1967). Structure of new gibberellins in immature seeds of *Canavalia gladiata*. *Tetrahedron Lett.*, No. 48, 4861–5.

TAKEDA, A. (1959). α-Substituted aryl-acetic acids as plant growth regulators. *Contrib. Boyce Thompson Inst.*, **20**, 197–204.

TAMURA, S. and NAGAO, M. (1969). 5-(1,2-Epoxy-2,6,6,-trimethyl-1-cyclohexyl)-3-methyl-*cis,trans*-2,4-pentadienoic acid and its esters. New plant growth inhibitors structurally related to abscisic acid. *Planta*, **85**, 209–12.

TAMURA, S., SAKURAI, A., KAINUMA, K. & TAKAI, M. (1963). Isolation of helminthosporol as a natural plant growth regulator and its chemical structure. *Agric. biol. Chem., Tokyo*, **27**, 738–9.

TAMURA, S., TAKAHASHI, N., MUROFUSHI, N., IRIUCHIJIMA, S., KATO, J., WADA, Y., WANATABE, E. & AOYAMA, T. (1966). Isolation and structure of a novel gibberellin in bamboo shoots (*Phyllostachys edulis*). *Tetrahedron Lett.*, No. 22, 2465–72.

TANAKA, K. & TOLBERT, N. E. (1966). Effect of cycocel derivatives and gibberellin on choline kinase and choline metabolism. *Pl. Physiol. Lancaster*, **41**, 313–18.

TANG, P. S. & LOO, S. W. (1940). Tests of after-effects of auxin seed treatment. *Am. J. Bot.*, **27**, 385–6.

TANG, Y. M. & BONNER, J. (1947). The enzymatic inactivation of indoleacetic acid. I Some characteristics of the enzyme contained in pea seedlings. *Archs Biochem.*, **13**, 11–25.

TANIMOTO, E. & MASUDA, Y. (1968). Effect of auxin on cell wall degrading enzymes. *Physiologia Pl.*, **21**, 820–6.

TAVARES, J. & KENDE, H. (1970). The effect of 6-benzylaminopurine on protein metabolism in senescing corn leaves. *Phytochemistry*, **9**, 1763–70.

TAYLOR, H. F. & BURDEN, R. S. (1970*a*). Xanthoxin, a new naturally occurring growth inhibitor. *Nature, Lond.*, **227**., 302–4.

TAYLOR, H. F. and BURDEN, R. S. (1970*b*). Identification of plant growth inhibitors produced by photolysis of violaxanthin. *Phytochemistry*, **9**, 2217–23.

TAYLOR, H. F. & SMITH. T. A. (1967). Production of plant growth inhibitors from xanthophylls: a possible source of dormin. *Nature, Lond.*, **215**, 1513–4.

TAYLOR, H. F. & WAIN, R. L. (1966). Studies on plant growth-regulating substances. XXI The release of pectic substances from wheat coleoptile tissue incubated with solutions of ethylenediaminetetra-acetic acid. *Ann. appl. Biol.*, **57**, 301–9.

TEPFER, S. S. (1965). The growth and development of flower buds in culture. In White, P. R. & Grove, A. R. (eds.), *Proc. int. Conf. Pl. Tissue Culture.* McCutchan, Berkeley, California.

TEPFER, S. S., GREYSON, R. T., CRAIG, W. R. & HINDMAN, J. L. (1963). *In vitro* culture of floral buds of *Aquilegia. Am. J. Bot.*, **50**, 1035–45.

TEPFER, S. S., KARPOFF, A. J. & GREYSON, R. I. (1966). Effects of growth substances on excised floral buds of *Aquilegia. Am. J. Bot.*, **53**, 148–57.

TERPSTRA, W. (1953). *Extraction and identification of growth substances.* Doctoral thesis, University of Utrecht.

TESTER, C. F. & DURE, L. S. (1967). Nucleic acid synthesis during the hormone-stimulated growth of excised oat coleoptiles. *Biochemistry. N. Y.*, **6**, 2532–8.

THIMANN, K. V. (1934). Studies on the growth hormone of plants. VI The distribution of the growth substances in plant tissue. *J. gen. Physiol.*, **18**, 23–34.

THIMANN, K. V. (1935*a*). On the plant growth hormone produced by *Rhizopus suinus*. *J. biol. Chem.*, **109**, 279–91.

THIMANN, K. V. (1935*b*). On an analysis of the activity of two growth promoting substances on plant tissues. *Proc. K. ned. Acad. Wet.*, **38**, 896–912.

THIMANN, K. V. (1936). Auxins and the growth of roots. *Am. J. Bot.*, **23**, 561–9.

THIMANN, K. V. (1951*a*). Studies on the physiology of cell enlargement. *Growth Symposium*, vol. 10, 5–22.

THIMANN, K. V. (1951*b*). The synthetic auxins: Relation between structure and activity. In Skoog, F. (ed.), *Plant Growth Substances*. University of Wisconsin Press, pp. 21–36.

THIMANN, K. V. (1952). The role of ortho-substitution in the synthetic auxins. *Pl. Physiol. Lancaster*, **27**, 392–404.

THIMANN, K. V. (1953). Hydrolysis of indoleacetonitrile in plants. *Archs Biochem. Biophys.*, **44**, 242–3.

THIMANN, K. V. (1956). Studies on the growth and inhibition of isolated plant parts. V The effects of cobalt and other metals. *Am. J. Bot.*, **43**, 241–50.

THIMANN, K. V. (1958). Auxin activity of some indole derivatives. *Pl. Physiol. Lancaster*, **33**, 311–21.

THIMANN, K. V. (1963). Plant growth substances: Past, present and future. *A. Rev. Pl. Physiol.*, **14**, 1–18.

THIMANN, K. V. & BONNER, J. (1938). Plant growth hormones. *Physiol. Rev.*, **18**, 524–53.

THIMANN, K. V. & BONNER, W. D., Jr. (1948). The action of tri-iodobenzoic acid on growth. *Pl. Physiol. Lancaster*, **23**, 158–61.

THIMANN, K. V. & GROCHOWSKA, M. (1968). The role of tryptophan and tryptamine as IAA precursors. In Wightman, F. and Setterfield, G. (eds.), *Biochemistry and Physiology of Plant Growth Substances*. Runge Press, Ottawa, pp. 231–42.

THIMANN, K. V., GROCHOWSKA, M. & AVADHANI, P. N. (1967). The role of tryptophan as an IAA precursor. *Abstr. 6th. int. Conf. Pl. Growth Substances*.

THIMANN, K. V. & LEOPOLD, A. C. (1955). Plant growth hormones. In *The Hormones*. Academic Press, New York, vol. 3.

THIMANN, K. V. & LOOS, G. M. (1957). Protein synthesis during water uptake by tuber tissue. *Pl. Physiol. Lancaster*, **32**, 274–9.

THIMANN, K. V. & MAHADEVAN, S. (1964). Nitrilase I, Occurrence, preparation and general properties of the enzyme. *Archs Biochem. Biophys.*, **105**, 133–41.

THIMANN, K. V. & SAMUEL, E. W. (1955). The permeability of potato tissue to water. *Proc. natl. Acad. Sci. U.S.A.*, **41**, 1029–33.

THIMANN, K. V. & SCHNEIDER, C. L. (1938*a*). The role of salts, hydrogen-ion concentration and agar in the response of the *Avena* coleoptile to auxins. *Am. J. Bot.*, **25**, 270–80.

THIMANN, K. V. & SCHNEIDER, C. L. (1938*b*). Differential growth in plant tissues. *Am. J. Bot.*, **25**, 627–41.

THIMANN, K. V. & SKOOG, F. (1934). On the inhibition of bud development and other functions of growth substances in *Vicia faba*. *Proc. R. Soc. B*, **114**, 317–39.

THIMANN, K. V. & TAKAHASHI, N. (1961). Interrelations between metallic ions and auxin action, and the growth promoting action of chelating agents. In *Plant Growth Regulation*. Iowa State University Press, pp. 363–80.

THIMANN, K. V. & WARDLAW, I. E. (1963). The effect of light on the uptake and transport of indoleacetic acid in the green stem of pea. *Physiologia Pl.*, **16**, 368–77.

THORNTON, R. M. & THIMANN, K. V. (1967). Transient effects of light on auxin transport in the *Avena* coleoptile. *Pl. Physiol. Lancaster*, **42**, 247–57.

THURMAN, D. A. & STREET, H. E. (1960). The auxin activity extractable from excised tomato roots by cold 80 per cent methanol. *J. exp. Bot.*, **11**, 188–97.

TIMONIN, M. I. (1941). The interactions of higher plants. III Effect of by-products of plant growth on activity of fungi and actinomycetes. *Soil Sci.*, **52**, 395

TIZIO. R. 1965. The mechanism of action of ethylenediaminetetra-acetic acid on cell elongation. *Phyton, B. Aires*, **22**, 159–64.

TOLBERT, N. E. (1960). (2-chloroethyl)-trimethylammonium chloride and related compounds as plant growth substances. I Chemical structure and bioassay. *J. biol. Chem.*, **235**, 475–9.

TOMASZEWSKI, M. (1964). The mechanism of synergism effects between auxin and some natural phenolic substances. In *Régulateurs naturels de la croissance végétale, Colloq. int. Centre natl. Rech. Sci.*, No. 123, 335–51.

TOMASZEWSKI, M. & THIMANN, K. V. (1966). Interaction of phenolic acids, metallic ions and chelating agents on auxin-induced growth. *Pl. Physiol. Lancaster*, **41**, 1443–54.

TOMITA, T. (1962). Studies on vernalisation and flowering substances. I A detection of flowering substances obtained in rye diffusates. *Tohoku J. agric. Res.*, **13**, 329–39.

TOMITA, T. (1963). Studies on vernalisation and flowering substance. II Vernalisation-like phenomena in winter wheat and radish treated with flower-promoting substances. *Tohoku J. agric. Res.*, **14**, 13–24.

TOMITA, T. (1964). Electrophoretic isolation of flowering substance from vernalised plants and some vernalisation-like phenomena caused by the application of extracts and of a pure substance. In *Régulateurs naturels de la croissance végétale, Colloq. int. Centre. natl. Rech. Sci.*, No. 123, 635–48.

TONZIG, S. (1950). The significance of ascorbic acid in plants. *Proc. 7th. int. bot. Congr.*, pp. 755–6.

TONZIG, S. & MARRÉ, E. (1955). The auxin-ascorbic acid oxidase interaction as related to the physiological activity of auxin. *Rend. Ist. Lombardo Sci. Lett. Cl. sci.*, **89**, 243–68.

TONZIG, S., & MARRÉ, E. (1961). Ascorbic acid as a growth hormone. In *Plant Growth Regulation*, Iowa State Univ. Press, pp. 725–34.

TONZIG, S. & TREZZI, F. (1950). [Researches on the physiology of ascorbic acid. II Ascorbic acid and the structural viscosity of the protoplasm. III Ascorbic acid and cell extention.] In Italian. *Nuovo. G. bot. ital.*, **57**, 515–34 & 535–48.

TONZIG, S., & MARRÉ, E. (1961). Ascorbic acid as a growth hormone. In acid.] In Italian. *Rend. Accad. naz. Lincei Cl. sci. fis. mat. nat. Ser. VIII*, **16**, 695–702.

TONZIG, S. & TREZZI, F. (1954*b*). [Researches on the action of indole-3-acetic acid.] In Italian. *Rend. Accad. naz. Lincei Cl. sci. fis. mat. nat. Ser. VIII*, **16**, 603–10.

TONZIG, S., TREZZI, E. & NAVA, E. (1952). [The influence of ascorbic acid and of indoleacetic acid on the water exchange of the plant.] In Italian. *Nuovo G. hot. ital.*, **59**, 171–3.

TOOTHILL, J. R., WAIN, R. L. & WIGHTMAN, F. (1956). Studies on plant growth-regulating substances. X The activity of some 2,6- and 3,5-substituted phenoxyalkylcarboxylic acids. *Ann. appl. Biol.*, **44**, 547–60.

TORSSELL, K. (1956). Chemistry of arylboric acids. VI Effects of arylboric acids on wheat roots and the role of boron in plants. *Physiologia Pl.*, **9**, 652–64.

TOURNOIS, J. (1911). Anomalies florales du Houblon japonais du Chanvre determinées par les semis nâtifs. *C. r. hebd. Séanc. Acad. Sci. Paris*, **153**, 1017.

TOURNOIS, J. (1912). Influence de la lumière sur la floraison du Houblon japonais et du Chanvre. *C. r. hebd. Séanc. Acad. Sci. Paris*, **155**, 297.

TOURNOIS, J. (1914). Etudes sur la sexualité du Houblon. *Annls Sci. nat.* (*Bot.*), **19**, 49.

TREWAVAS, A. (1968). The effect of 3-indoleacetic acid on the level of polysomes in etiolated pea tissue. *Phytochemistry*, **7**, 673–81.

TRUELSEN, T. A. & GALSTON, A. W. (1966). Changes in growth, auxin- and ribonucleic acid metabolism in wheat and coleoptile sections following pulse treatment with indole-3-acetic acid. *Physiologia Pl.*, **19**, 167–76.

TSUI, C. (1948). The role of zinc in auxin synthesis in the tomato plant. *Am. J. Bot.*, **35**, 172–9.

TUKEY, H. B., WENT, F. W., MUIR, R. M. & OVERBEEK, J. VAN (1954). Nomenclature of chemical plant regulators. *Pl. Physiol. Lancaster*, **29**, 307–8.

TULECKE, W., WEINSTEIN, L. H., RUTNER, A. & LAURENCOT, H. J. (1961). The biochemical composition of coconut water (coconut milk) as related to its use in plant tissue culture. *Contrib. Boyce Thompson Inst.*, **21**, 115–28.

TULI, V. & MOYED, H. S. (1966). Desensitization of regulatory enzymes by a metabolite of plant auxin. *J. biol. Chem.*, **241**, 1564–6.

TULI, V. & MOYED, H. S. (1967). Inhibitory oxidation products of indole-3-acetic acid; 3-hydroxymethyloxindole and 3-methyleneoxindole as plant metabolites. *Pl. Physiol. Lancaster*, **42**, 425–30.

ULRICH, H. (1939). Photoperiodismus und Blühhormone. *Ber. dt. bot. Ges.*, **57**, 40–52.

UPPER, C. D. & WEST, C. A. (1967). Biosynthesis of gibberellin. II. Enzyme cyclization of geranylgeranyl pyrophosphate to kaurene. *J. biol. Chem.*, **242**, 3285–92.

URSPRUNG, A. & BLUM, G. (1924). Ein Methode zur Messung des Wand- und Turgordruckes der Zelle, nebst Anwendungen. *Jb. wiss. Bot.*, **63**, 1.

VACHA, G. A. & HARVEY, R. B. (1927). The use of ethylene, propylene and similar compounds in breaking the rest period of tubers, bulbs, cuttings and seeds. *Pl. Physiol. Lancaster*, **2**, 187–94.

VALDOVINOS, J. G. & ERNEST, L. C. (1967). Effect of gibberellic acid and cycocel on tryptophan metabolism and auxin destruction in the sunflower seedling. *Physiologia Pl.*, **20**, 682–7.

VALDOVINOS, J. G., ERNEST, L. C. & HENRY, E. W. (1967). Effects of ethylene and gibberelic acid on auxin synthesis in plant tissue. *Pl. Physiol. Lancaster*, **42**, 1803–6.

VALDOVINOS, J. G., ERNEST, L. C. & PERLEY, J. E. (1967). Gibberellin effects on tryptophan metabolism, auxin destruction and abscission in *Coleus*. *Physiologia Pl.*, **20**, 600–7.

VALDOVINOS, J. G. & SASTRY, K. S. S. (1968). The effect of gibberellin on tryptophan conversion and elongation of the *Avena* coleoptile. *Physiologia Pl.*, **21**, 1280–6.

VALIO, I. F. M. & SCHWABE, W. W. (1970). Growth and dormancy in *Lunularia cruciata* (L) Dum. VII The isolation and bioassay of lunularic acid. *J. exp. Bot.*, **21**, 138–50.

VANCURA, V. & HOVÁDIK, A. (1965). Root exudates of plants. II Composition of root exudates of some vegetables. *Pl. Soil*, **22**, 21–32.

VANDENBELT, J. M. (1945). Nutritive value of coconut milk. *Nature, Lond.*, **156**, 174–5.

VAN DEN ENDE, H. — *see* ENDE, H. VAN DEN

VAN DER KERK, G. J. M. — *see* KERK, G. J. M. VAN DER

VAN DER LAAN, P. A. — *see* LAAN, P. A. VAN DER

VAN DER VEEN, R.—*see* VEEN, R. VAN DER

VAN DER WEIJ, H. G. — *see* WEIJ, H. G. VAN DER

VAN DER WESTERINGH, C. — *see* WESTERINGH, C. VAN DER

VAN HOVE, C. — *see* HOVE, C. VAN

VAN OVERBEEK, J. — *see* OVERBEEK, J. VAN

VAN RAALTE, M. H.—*see* RAALTE, M. H. VAN

VAN SENDEN, H.—*see* SENDEN, H. VAN

VARDAR, Y. (1959). Some confirmatory experiments performed with IAA-C^{14} concerning the effect of TIBA upon auxin transport. *Istanb. Univ. Fen. Fak. Mecm., Ser. B.*, **24**, 133–45.

VARDAR, Y. & DENIZCI, R. (1962). Some experiments performed with

IAA-C^{14} on the auxin transport in relation to the role of sodium glycocholate (Na-G). *Ber. schweiz. bot. Ges.*, **72**, 132.
VARGA, M. (1957). Examination of growth inhibiting substances separated by paper chromatography in fleshy fruits. II Identification of the substances of growth-inhibiting zones on the chromatograms. *Acta biol. hung.*, **3**, 213–23.
VARGA, M. (1966). The specificities of apple cultivars and of gibberellins in the induction of parthenocarpic fruit. *Proc. K. ned. Akad. Wet., Ser. C*, **69**, 641–4.
VARGA, M. & BÁLINT, I. (1966). The effect of gibberellin on the growth, indoleacetic acid content and on activity of indoleacetic acid oxidase in rice seedlings. *Acta. biol. hung.*, **16**, 243–53.
VEEN, H. & JACOBS. W. P. (1969). Movement and metabolism of kinetin-^{14}C and of adenine-^{14}C in *Coleus* petiole segments of increasing age. *Pl. Physiol. Lancaster*, **44**, 1277–84.
VEEN, R. VAN DER (1935). [Root competition in coffee and rubber plantations.] In Dutch. *Arch. Koffiecult. Ned.-Ind.*, **3**, 65–104.
VEGIS, A. (1961). Samenkeimung und vegetative Entwicklung der Knospen. *Encyclopaedia of Plant Physiology*. Springer, Berlin, Göttingen, Heidelberg, vol. 16, pp. 168–298.
VEGIS, A. (1964). Dormancy in higher plants. *A. Rev. Pl. Physiol.,* **15**, 185–224.
VELDSTRA, H. (1944). Researches on plant growth substances. IV and V. Relation between chemical structure and physiological activity. *Enzymologia*, **11**, 97–136 & 137–63.
VELDSTRA, H. (1946). [Some concepts from the chemistry and physiology of growth substances and inhibitors.] In Dutch. *LandbKund. Tijdschr.* No. 701/702, 483.
VELDSTRA, H. (1947). Considerations on the interactions of ergons and their 'substrates'. *Biochim. biophys. Acta*, **1**, 364–78.
VELDSTRA, H. (1949). On the relation structure/activity with plant growth regulators. *Proc. 2nd. int. Congr. Crop Protection.*
VELDSTRA, H. (1952). Halogenated benzoic acids and related compounds. *Recl. Trav. chim. Pays-Bas. Belg.*, **71**, 15–32.
VELDSTRA, H. (1953). The relation of chemical structure to biological activity in growth substances. *A. Rev. Pl. Physiol.*, **4**, 151–98.
VELDSTRA, H. (1955). [Stereochemistry in the living cell.] In Dutch. *Chem. Weekblad.*, **51**, 158–67.
VELDSTRA, H. (1956). On form and function of plant growth substances. In Wain, R. L. & Wightman, F. (eds.), *The Chemistry and Mode of Action of Plant Growth Substances.* Butterworths, London, pp. 117–33.
VELDSTRA, H. & ÅBERG, B. (1953). On auxin antagonists. Aryl- and aryloxyacetic acids with a 'bulky' α-substituent. *Biochim. biophys. Acta*, **12**, 593.
VELDSTRA, H. & BOOIJ, H. L. (1949). Researches on plant growth regulators. XVII On the mechanism of action III. *Biochim. biophys. Acta*, **3**, 278–312.
VELDSTRA, H., KRUYT, W., STEEN, E. J. VAN DER & ÅBERG, B. (1954). Re-

searches on plant growth regulators. XXII Structure activity VII. Sulphonic acids and related compounds. *Recl. Trav. chim. Pays-Bas. Belg.*, **74**, 23.

VELDSTRA, H. & WESTERINGH, C. VAN DER (1951). Partially hydrogenated naphthoic acid and alkyl-phenylacetic acids. *Recl. Trav. chim. Pays-Bas. Belg.*, **70**, 1113–26.

VELDSTRA, H. & WESTERINGH, C. VAN DER (1952). On the growth substance activity of substituted benzoic acids. *Recl. Trav. chim. Pays-Bas. Belg.*, **71**, 318–20.

VENDRIG, J. C. (1961). Caffeic acid, a substance with auxin activity found in extracts from *Coleus rhenaltianus. Acta bot. neerl.*, **10**, 190–8.

VENDRIG, J. C. (1964). Growth-regulating activity of some saponins. *Nature, Lond.*, **203**, 1301–2.

VENDRIG, J. C. (1967*a*). Sterol-like compounds with auxin activity in human urine. *Z. PflPhysiol.*, **57**, 113–7.

VENDRIG, J. C. (1967*b*). Steroid derivatives as native auxins in *Coleus. Ann. N. Y. Acad. Sci.*, **144**, (1), 81–93.

VENIS, M. A. (1968). Auxin-histone interactions. In Wightman, F. & Setterfield, G. (eds.), *Biochemistry and Physiology of Plant Growth Substances.* Runge Press, Ottawa, pp. 761–75.

VENIS, M. A. (1969). Auxin-induced conjugation systems in peas. *Abstr. 11th. int. bot. Congr. Seattle.*

VIEITEZ, E., SEOANE, E., GESTO, M. D. V., MATO, M. C., VASQUEZ, A. & CARNICAR, A. (1966). [Substances isolated from *Ribes rubrum* cuttings and their action on cell growth.] In Spanish. *Ann. Edafol. Agrobiol.*, **25**, 69–90.

VLIEGENTHART, J. A. & VLIEGENTHART. J. F. G. (1966). Reinvestigation of authentic samples of auxins A and B and related products by mass spectrometry. *Recl. Trav. chim. Pays-Bas. Belg.*, **85**, 1266–72.

VLITOS, A. J. & CROSBY, D. G. (1959). Isolation of fatty alcohols with plant-growth promoting activity from Maryland Mammoth tobacco. *Nature, Lond.*, **184**, 462–3.

VLITOS, A. J. & CUTLER, H. G. (1960). Plant growth-regulating activity of cuticular waxes of sugar cane. *Pl. Physiol. Lancaster*, **35**, (Suppl.), vi.

VLITOS, A. J. & MEUDT, W. (1954). The role of auxin in plant flowering. III Free indole acids in short-day plants grown under photoinductive and non-photoinductive daylengths. *Contrib. Boyce Thompson Inst.*, **17**, 413–7.

VOELLER, B. R. (1964). Antheridogens in ferns, In *Régulateurs naturels de la croissance végétale*, *Colloq. int. Centre Natl. Rech. Sci.*, No. 123, 665–84.

VON ABRAMS, G. J. — *see* ABRAMS, G. J. VON

VON DENFFER, D. — *see* DENFFER, D. VON

VON EULER, H. — *see* EULER, H. VON

VON GUTTENBERG, H. — *see* GUTTENBERG, H. VON

VON WITSCH, H. — *see* WITSCH, H. VON

WAARDE, J. DE & ROODENBURG, J. W. M. (1948). Premature flowerbud initiation in tomato seedlings caused by 2,3,5-tri-iodobenzoic acid. *Proc. K. ned. Akad. Wet.*, **51**, (2).

WADA, S., TANIMOTO, E. & MASUDA, Y. (1968). Cell elongation and metabolic

turnover of the cell wall as affected by auxin and cell wall degrading enzymes. *Pl. Cell Physiol.*, **9**, 369–76.
WAGENKNECHT, A. C. & BURRIS, R. H. (1950). Indoleacetic acid-inactivating enzymes from bean roots and pea seedlings. *Archs Biochem.*, **25**, 30–53.
WAIN, R. L. (1949). Chemical aspects of plant growth-regulating activity. *Ann. appl. Biol.*, **36**, 559–62.
WAIN, R. L. (1951). Plant growth-regulating and systemic fungicidal activity. The aryloxyalkylcarboxylic acids. *J. Sci. Fd. Agric.*, **3**, 101–6.
WAIN, R. L. (1953). Plant growth substances. *R. Inst. Chem. Lectures Monogr. Rep.*, No. 2.
WAIN, R. L. (1955*a*). A new approach to selective weed control. *Ann. appl. Biol.*, **42**, 151–7.
WAIN, R. L. (1955*b*). Herbicidal activity through specific action of plants on compounds applied. *Agric. Fd. Chem.*, **3**, 128–30.
WAIN, R. L. (1967). Some developments in research on auxins and kinins. *Ann. N. Y. Acad. Sci..*, **144**(1), 223–34.
WAIN, R. L. (1968). Chemical control of plant growth. In *Plant Growth Regulators. Soc. chem. Indust. Monogr.*, No. 31, 3–20.
WAIN, R. L., RUTHERFORD, P. P., WESTON, E. W. & GRIFFITHS, C. M. (1964). Effects of growth regulating substances on inulin-storing plant tissues. *Nature, Lond.*, **203**, 504.
WAIN, R. L. & TAYLOR, H. F. (1965). Phenols as plant growth regulators. *Nature, Lond.*, **207**, 167–9.
WAIN, R. L. & WIGHTMAN, F. (1953). Studies on plant growth-regulating substances. VII. Growth-promoting activity in the chlorophenoxyacetic acids. *Ann. appl. Biol.*, **40**, 244–9.
WAIN, R. L. & WIGHTMAN, F. (1954). The growth-regulating activity of certain ω-substituted alkyl carboxylic acids in relation to their β-oxidation within the plant. *Proc. R. Soc. B*, **142**, 525.
WAKSMAN, S. A. (1948). Antibiotics. *Biol. Rev.* **23**, 452–87.
WALLACE, R. H. (1926). The production of intumescences upon apple twigs by ethylene gas. *Bull. Torrey bot. Club*, **53**, 385–401.
WALTON, D. C., SOOF, G. S. & SONDHEIMER, E. (1970). The effect of abscisic acid on growth and nucleic acid synthesis in excised embryonic bean axes. *Pl. Physiol. Lancaster*, **45**, 37–40.
WANG, T. S. C. & CHUANG, T.-T. (1966). Soil alcohols, their dynamics and their effects upon plant growth. *Soil. Sci.*, **104**, 40–5.
WANG, T. S. C., YANG, T.-K. & CHUANG, T.-T. (1967). Some phenolic acids as plant growth inhibitors. *Soil Sci.*, **103**, 239–46.
WARDROP, A. B. (1955). The mechanism of surface growth in parenchyma of *Avena* coleoptiles. *Aust. J. Bot.*, **3**, 137–48.
WARDROP, A. B. (1956). The nature of the surface growth in plant cells. *Aust. J. Bot.*, **4**, 193–9.
WAREING, P. F. (1956). Photoperiodism in woody plants. *A. Rev. Pl. Physiol.*, **7**, 191–214.
WAREING, P. F. (1959). Photoperiodism in seeds and buds. In *Photoperiodism*

and Related Phenomena in Plants and Animals. Am. Ass. Adv. Sci., Publ. 55, Washington, pp. 73–87.

WAREING, P. F. (1969). The control of bud dormancy in seed plants. *Symp. Soc. exp. Biol.*, **23**, 241–62.

WAREING, P. F. & FODA, H. A. (1956). The possible role of growth inhibitors in the dormancy of seeds of *Xanthium* and lettuce. *Nature, Lond.*, **178**, 908.

WAREING, P. F. & FODA, H. A. (1957). Growth inhibitors and dormancy in *Xanthium* seed. *Physiologia Pl.*, **10**, 266–80.

WAREING, P. F., GOOD, J. & MANUEL, J. (1968). Some possible physiological roles of abscisic acid. In Wightman, F. & Setterfield, G. (eds.), *Biochemistry and Physiology of Plant Growth Substances.* Runge Press, Ottawa, pp. 1561–79.

WAREING, P. F. & VILLIERS, T. A. (1961). Growth substances and inhibitor changes in buds and seeds in response to chilling. In *Plant Growth Regulation.* Iowa State University Press, p. 101.

WARNER, H. L. & LEOPOLD, A. C. (1967). Plant growth regulation by stimulation of ethylene production. *Bio. Science*, **17**, 722.

WAYGOOD, E. R., OAKS, A. & MACLACHLAN, G. (1956). On the mechanism of indoleacetic acid oxidation by wheat leaf enzymes. *Can. J. Bot.*, **34**, 54.

WEDDING, R. T. & ERICKSON, L. C. (1957). The role of pH in the permeability of *Chlorella* to 2,4-D. *Pl. Physiol. Lancaster.*, **32**, 503–12.

WEIJ, H. G. VAN DER (1932). Der Mechanismus des Wuchsstofftransportes. *Recl. Trav. bot. néerl.* **29**, 379–496.

WEIJ, H. G. VAN DER (1934). Der Mechanismus des Wuchsstofftransportes II. *Recl. Trav. bot.néerl.* **31**, 810–57.

WEINSTEIN, L. H., MEISS, A. M., UHLER, R. L. & PURVIS, E. R. (1956). Growth-promoting effects of ethylenediamine tetra-acetic acid. *Nature, Lond.*, **178**, 1188.

WEINTRAUB, R. L. (1948). Influence of light on chemical inhibition of lettuce seed germination. *Smithson. misc. Collns.*, **107**(20).

WEINTRAUB, R. L. & PRICE, L. (1948). Inhibition of plant growth by emanations from oils, varnishes and woods. *Smithson. misc. Collns.*, **107**(17).

WELBANK, P. J. (1963). Toxin production during decay of *Agropyron repens* (Couch grass) and other species. *Weed Res.*, **3**, 205–14.

WELLENSIEK, S. J. (1961). Leaf vernalisation. *Nature, Lond.*, **192**, 1097–8.

WENT, F. W. (1926). On growth-accelerating substances in the coleoptile of *Avena sativa. Proc. K. ned. Akad. Wet.*, **30**, 10–19.

WENT, F. W. (1928). Wuchsstoff und Wachstum. *Recl. Trav. bot. néerl.*, **25**, 1–116.

WENT. F. W. (1934). On the pea test method for auxin, the plant growth hormone. *Proc. K. ned. Akad. Wet.*, **37**, 547–55.

WENT, F. W. (1935). Coleoptile growth as affected by auxin, ageing and food. *Proc. K. ned. Akad. Wet.*, **38**, 752–67.

WENT, F. W. (1954). Thermoperiodicity and photoperiodism. *Rep. Commun. 8th. int. bot. Congr., Sect.* **11**, 335–40.

WENT, F. W. & THIMANN, K. V. (1937). *Phytohormones.* MacMillan, London.

WENT, F. W. & WHITE R. (1939). Experiments on the transport of auxin. *Bot. Gaz.*, **100**, 465–84.

WEST, C. A., OSTER, M., ROBINSON, D., LEW, F. & MURPHY, P. (1968). Biosynthesis of gibberellin precursors and related diterpenes. In Wightman, F. & Setterfield, G. (eds.), *Biochemistry and Physiology of Plant Growth Substances*. Runge Press, Ottawa, pp. 313–33.

WEST, C. A. & PHINNEY, B. O. (1956). Properties of gibberellin-like factors from extracts of higher plants. *Pl. Physiol. Lancaster*, **31**, (Suppl.) xx.

WEST, C. A. & PHINNEY, B. O. (1959). Gibberellins from flowering plants. I Isolation and properties of a gibberellin from *Phaseolus vulgaris*. *J. Am. chem. Soc.*, **81**, 2424–7.

WEST, P. M. (1939). Excretion of thiamin and biotin by the roots of higher plants. *Nature, Lond.*, **144**, 1050.

WESTERINGH, C. VAN DER (1957) [*Researches in plant growth-substances. Variations in the polar groups and their influence on growth-substance transport.*] In Dutch. Doctoral thesis, University of Leiden.

WHEELER, A. W. (1961). Effect of light quality on the growth and growth-substance-content of plants. *J. exp. Bot.*, **12**, 217–25.

WHITE, E. P. (1944). Alkaloids of the Leguminosae. VIII-XIII. *N. Z. J. Sci. Technol.*, 25 B, 137–62.

WHITE, P. R. (1934). Potentially unlimited growth of excised tomato root tips in a liquid medium. *Pl. Physiol. Lancaster*, **9**, 585–600.

WHITE, P. R. (1939). Potentially unlimited growth of excised plant callus in an artificial nutrient. *Am. J. Bot.*, **26**, 59–64.

WHITEHEAD, D. C. (1963). Some aspects of the influence of organic matter on soil fertility. *Soils Fertil.*, **26**, 217–23.

WHYTE, P. & LUCKWILL, L. C. (1966). A sensitive bioassay for gibberellins based on retardation of leaf senescence in *Rumex obtusifolius* L. *Nature, Lond.*, **210**, 1360.

WICHNER, S. & LIBBERT, E. (1968). Interaction between plants and epiphytic bacteria regarding their auxin metabolism. I Detection of IAA-producing epiphytic bacteria and their role in long duration experiments on tryptophan metabolism in plant homogenates. *Physiologia Pl.*, **21**, 227–41.

WIELAND, O. P., ROPP, R. S. DE & AVENER, J. (1954). The identity of auxin in normal urine. *Nature, Lond.*, **173**, 776.

WIESE, L. & JONES, R. F. (1963). Studies on gamete copulation in heterothallic chlamydomonads. *J. cell. comp. Physiol.*, **61**, 265–74.

WIGHTMAN, F. (1964). Pathways of tryptophan metabolism in tomato plants. In *Régulateurs naturels de la croissance végétale, Colloq, int. Centre natl. Rech. Sci.*, No. 123, 191–212.

WIGHTMAN, F. & COHEN, D. (1967). Intermediary steps in the enzymatic conversion of tryptophan to IAA in cell-free systems from higher plants. *Abstr. 6th. int. Conf. Pl. Growth Substances, Carlton.*

WIGHTMAN, F. & COHEN, D. (1968). Intermediary steps in the enzymatic conversion of tryptophan to IAA in cell-free systems from mung bean seedlings.

In Wightman, F. & Setterfield, G. (eds.), *Biochemistry and Physiology of Plant Growth Substances*. Runge Press, Ottawa. pp. 273–88.

WILCZYNSKA, K. (1959). [The new easy method F. W. Went's '*Avena* curvature test'; 'Coleoptile-hypocotyl test'.] In Polish. *Acta Soc. Bot. Pol.*, **28**, 111–28.

WILDMAN, S. G. & BONNER, J. (1948). Observations on the chemical nature and formation of auxin in the *Avena* coleoptile. *Am. J. Bot.*, **35**, 740–6.

WILDMAN, S. G., FERRI, M. G. & BONNER, J. (1947). Enzymatic conversion of tryptophan to auxin by spinach leaves. *Archs Biochem.*, **13**, 131–44.

WILKINS, M. B. (1968). Biological clocks. *Adv, Sci.*, **24**, 273–83.

WILKINS, M. B. & MARTIN, M. (1967). Dependance of basipetal polar transport of auxin upon aerobic metabolism. *Pl. Physiol. Lancaster*, **42**, 831–9.

WILKINS, M. B. & SCOTT, T. K. (1968). Auxin transport in roots. III Dependence of the polar flux of IAA in *Zea* roots upon metabolism. *Planta*, **83**, 335–46.

WILSKE, C. & BURSTRÖM, H. (1950). The growth-inhibiting action of thiophenoxyacetic acids. *Physiologia Pl.*, **3**, 58–67.

WILSON, G. B. & BOWEN, C. C. (1951). Cytological effects of some antibiotics. *J. Hered.*, **42**, 251.

WILSON, K. (1957). Extension growth in primary cell walls with special reference to *Elodea canadensis*. *Ann. Bot.*, **21**, 1–11.

WINTER, A. (1966). A hypothetical route for the biogenesis of IAA. *Planta*, **71**, 229–39.

WINTER, A. (1968). 2,3,5-Tri-iodobenzoic acid and the transport of 3-indoleacetic acid. In Wightman, F. & Setterfield, G. (eds.), *Biochemistry and Physiology of Plant Growth Substances*. Runge Press, Ottawa, pp. 1063–76.

WINTER, A. G. (1952). Untersuchungen über die Aufnahme von Penicillin und Streptomycin durch die Wurzeln von *Lepidium sativum* L. und ihre Beständigkeit in natürlichen Böden. *Z. Bot.*, **40**, 153.

WINTER, A. G. & BUBITZ, W. (1953). Über die Keim- und Entwicklungshemmende Wirkung der Buchenstreu. *Naturwissenschaften*, **40**, 416.

WINTER, A. G. & SCHONBECK, F. (1953). Untersuchungen über die Beeinflussung der Keimung und Entwicklung von Getreidsamen durch Kaltwasserauszüge aus Getreidstroh. *Naturwissenschaften*, **40**, 168.

WIRWILLE, J. W. & MITCHELL, J. W. (1950). Six new plant-growth-inhibiting compounds. *Bot. Gaz.*, **111**, 491–4.

WITHROW, A. P. & WITHROW, R. B. (1943). Translocation of the floral stimulus in *Xanthium*. *Bot. Gaz.*, **104**, 409–16.

WITSCH, H. VON & RINTELEN, J. (1962). Die Entwicklung von Farnprothallien unter Gibberellineinfluss. *Planta*, **59**, 115–8.

WITTWER, S. H. & BUKOVAC, M. J. (1957*a*). Gibberellin effects on temperature and photoperiodic requirements for flowering of some plants. *Science N. Y.*, **126**, 30–1.

WITTWER, S. H. & BUKOVAC, M. J. (1957*b*). Gibberellin and higher plants. II

Induction of flowering in biennials. *Q. Bull. Mich. St. Univ. agric. Exp. Stn.*, **39**, 650–60.
WITTWER, S. H. & BUKOVAC. M. J. (1962). Quantitative and qualitative differences in plant responses to the gibberellins. *Am. J. Bot.*, **49**, 524–9.
WITTWER, S. H. & TOLBERT, N. E. (1960). 2-chloroethyl trimethylammonium chloride and related compounds as plant growth substances. V Growth, flowering and fruiting responses as related to those induced by auxin and gibberellin. *Pl. Physiol. Lancaster.*, **35**, 871–7.
WOLLGIEHN, R. (1967). Nucleic acid and protein metabolism of excised leaves. *Symp. Soc. exp. Biol.*, **21**, 231–46.
WOOD, H. N., BRAUN, A. C., BRANDES, H. & KENDE, H. (1969). Studies on the distribution and properties of a new class of cell division promoting substances from higher plant species. *Proc. natl. Acad. Sci. U.S.A.* **62**, 349–56.
WOOD, R. K. S. & TVEIT, M. (1955). Control of plant diseases by use of antagonistic organisms. *Bot. Rev.*, **21**, 441–92.
WOODRUFFE, P., ANTHONY, A. & STREET, H. E. (1970). Studies of the growth in culture of excised wheat roots. VIII. An Ehrlich-reacting compound released into the culture medium and inhibitory to root growth. *Physiologia Pl.*, **23**, 488–97.
WOODS, F. W. (1960). Biological antagonisms due to phytotoxic root exudates. *Bot. Rev.*, **26**, 546–69.
WOOLHOUSE, H. W. (1967). The nature of senescence in plants. *Symp. Soc. exp. Biol.*, **21**, 179–213.
WOOLEY, D. W. (1952). *A Study of Antimetabolites*, Wiley and Sons.
WORSHAM, A. D., KLINGMAN, G. C. & MORELAND, D. E. (1962). Promotion of germination of *Striga asiatica* seed by coumarin derivatives and effects on seedling development. *Nature, Lond.*, **195**, 199–201.
WORT, D. J. (1964). Effect of herbicides on plant composition and metabolism. In Audus, L. J., (ed.) *The Physiology and Biochemistry of Herbicides.* Academic Press, London and New York, pp. 291–334.
WRIGHT, J. M. (1954*a*). The production of antibiotics in soil. I. Production of gliotoxin by *Trichoderma viride. Ann. appl. Biol.*, **41**, 280.
WRIGHT, J. M. (1954*b*). (Quoted by Jefferys & Brian, 1954.)
WRIGHT, J. M. (1956). The production of antibiotics in soil. IV. Production of antibiotics in coats of seeds sown in soil. *Ann, appl. Biol.*, **44**, 561–6.
WRIGHT, S. T. C. (1961). A sequential growth response to gibberellic acid, kinetin and indolyl-3-acetic acid in the wheat coleoptile (*Triticum vulgare*). *Nature, Lond.*, **190**, 699–700
WRIGHT, S. T. C. (1963). Cellular differentiation at the molecular level with special reference to proteins. *Symp. Soc. exp. Biol.*, **17**, 18–39.
YABUTA, T. (1935). [Biochemistry of the 'bakanae' fungus of rice.] In Japanese. *Agric. Hortic.*, **10**, 17–22.
YABUTA, T. & HAYASHI, T. (1939). [Biochemical studies on the 'bakanae' fungus of the rice. Part III. Studies of physiological action in the plant.] In Japanese. *Bull. agric. chem. Soc. Japan*, **15**, 82–3.

YABUTA, T. & SUMIKI, Y. (1938). [Communications to the editor.] In Japanese. *J. agric. chem. Soc. Japan*, **14**, 1526.

YABUTA, T., SUMIKI, Y., ASO, K., TAMURA, T., IGARASHI, H. & TAMARI, K. (1941). [Biochemistry of the 'bakanae' fungus of rice. Part 12, The chemical constitution of gibberellin.] In Japanese. *J. agric. chem. Soc. Japan*, **17**, 975–84.

YAMADA, N., SUGE, H., NAKAMURA, H. & TAZIMA, K. (1964). Chemical control of plant growth and development (5). Effect of kinetin and other chemicals on the degradation of chlorophyll in rice plant. *Proc. Crop Sci. Soc. Japan*, **32**, 254–8.

YEOMANS, L. M. & AUDUS, L. J. (1964). Auxin transport in roots, *Vicia faba*. *Nature, Lond.*, **204**, 559–62.

YIN, H. C. (1941). Studies on the nyctinastic movement of leaves of *Carica papaya*. *Am. J. Bot.*, **28**, 250–61.

YOMO, H. (1960). *Hakko Kyokai Shi.*, **18**, 494–99, 500–2, 600–2, 603–6.

ZACHAU, H. G., DÜTTING, D. & FELDMAN, H. (1966). Nucleosequenzen zweier Serin-spezifischer Transferribonucleinsäuren. *Angew. Chem.*, **78**, 392.

ZEEUW, D. DE & LEOPOLD, A. C. (1955). Altering juvenility with auxin. *Science N. Y.*, **122**, 925–6.

ZEEVAART, J. A. D. (1957). Studies on flowering by means of grafting. I Photoperiodic induction as an irreversible phenomenon in *Perilla*. *Proc. K. ned. Akad. Wet., Ser. C*, **60**, 324–31.

ZEEVAART, J. A. D. (1958). Flower formation as studied by grafting. *Meded. LandbHoogesch. Wageningen*, **58**, No. 3.

ZEEVAART, J. A. D. (1966). Reduction of the gibberellin content of *Pharbitis* seeds by CCC and after-effects in the progeny. *Pl. Physiol. Lancaster*, **41**, 856–62.

ZEEVAART, J. A. D. (1967). Vernalisation and gibberellins in *Lunaria annuua*. *Abstr. 6th. int. Congr. Pl. Growth Substances, Carleton*, p. 110.

ZEEVAART, J. A. D. (1969*a*). The leaf as the site of gibberellin action in flower formation in *Bryophyllum daigremontianum*. *Planta*, **84**, 339–47.

ZEEVAART, J. A. D. (1969*b*). Gibberellin-like substances in *Bryophyllum daigremontianum*, and the distribution and persistance of applied GA_3. *Planta*, **86**, 124–33.

ZEEVAART, J. A. D. & LANG, A. (1962). The relationship between gibberellin and floral stimulus in *Bryophyllum daigremontianum*. *Planta*, **55**, 531–42.

ZEEVAART, J. A. D. & LANG, A. (1963). Suppression of floral induction in *Bryophyllum daigremontianum* by a growth retardant. *Planta*, **59**, 509–17.

ZENK, M. H. (1960). Enzymatische Aktivierung von Auxinen und ihre Konjugierung mit Glycin. *Z. Naturf.*, **15**, 436–41.

ZENK, M. H. (1961). 1-(Indole-3-acetyl)-β-D-glucose, a new compound in the metabolism of indole-3-acetic acid in plants. *Nature, Lond.*, **191**, 493–4.

ZENK, M. H. (1962). Aufnahme und Stoffwechsel von α-Naphthyl-essigsäure durch Erbsenepicotyls. *Planta*, **58**, 75–94.
ZENK. M. H. (1964). Isolation, biosynthesis and function of indoleacetic acid conjugates. In *Régulateurs naturels de la croissance végétale*, *Colloq. int. Centre natl. Rech. Sci.*, No. 123, 241–9.
ZENK, M. H. (1967). Untersuchungen zum Phototropismus der *Avena*-Koleoptile. II Photooxydationen in *vivo*. *Z. PflPhysiol.*, **56**, 57–69.
ZENK, M. H. & MÜLLER, G. (1964). Über den Einfluss der Wundflachen auf die enzymatische Oxydation der Indol-3-essigsäure *in vivo*. *Planta*, **61**, 340–51.
ZIEGLER, H., KÖHLER, D. & STREITZ, B. (1966). Ist 2-Chlor-9-fluorenol-9-carbonsäure ein Gibberellinantagonist. *Z. PflPhysiol.*, **54**, 118–24.
ZIMMERMAN, P. W. & HITCHCOCK, A. E. (1933). Initiation and stimulation of adventitous roots caused by unsaturated carbon gases. *Contrib. Boyce Thompson Inst.*, **5**, 351–69.
ZIMMERMAN, P. W. & HITCHCOCK, A. E. (1937). Comparative effectiveness of acids, esters and salts as growth substances and methods of evaluating them. *Contrib. Boyce Thompson Inst.*, **8**, 337–50.
ZIMMERMAN, P. W. & HITCHCOCK, A. E. (1942). Substituted phenoxy and benzoic acid growth substances and the relation of structure to physiological activity. *Contrib. Boyce Thompson Inst.*, **12**, 321–43.
ZIMMERMAN, P. W. & HITCHCOCK, A. E. (1944). Substances effective for increasing fruit set and inducing seedless tomatoes. *Proc. Am. Soc. hort. Sci.*, **45**, 353–61.
ZIMMERMAN, P. W., HITCHCOCK, A. E. & WILCOXON, F. (1936). Several esters as plant hormones. *Contrib. Boyce Thompson Inst.*, **8**, 105–12.
ZUCKER, M., NITSCH, C. & NITSCH, J. C. (1965). The induction of flowering in *Nicotiana*. II Photoperiodic alteration of the chlorogenic acid concentration. *Am. J. Bot.*, **52**, 271–7.
ZWAR, J. A. & BRUCE, M. I. (1970). Cytokinins from apple extracts and coconut milk. *Aust. J. biol. Sci.*, **23**, 289–97.
ZWAR, J. A. & RIJVEN, A. H. G. C. (1956). Inhibition of transport of indole-3-acetic acid in the etiolated hypocotyl of *Phaseolus vulgaris*. *Aust. J. biol. Sci.*, **9**, 528–38.
ZWAR, J. A. & SKOOG, F. (1963). Promotion of cell division by extracts of pea seedlings. *Aust. J. biol. Sci.*, **16**, 129–39.

Glossary

Abscission The process by which an organ of the plant is shed by the formation of a special separation layer of loose cells across its base.
Absorption spectroscopy A physical technique for the characterization and quantitative estimation of pure substances based on their absorption of electromagnetic radiation. There are three broad types depending on the range of wavelengths involved, visible (360–770 nm), ultra-violet (200–360 nm) and infra-red (770 nm–30 μm).
(1 nm = 10^{-9} metre; 1 μm = 10^{-6} metre). The absorption spectra (relationship of absorption to wavelength) can be an aid to chemical characterization, and the degree of absorption at a particular wavelength can be used for quantitative determinations.
Acropetal Moving in a direction from the base to the apex, or from the proximal to the distal end of an organ, i.e. from a region nearest the main plant body towards a region farthest away.
Agar (= agar-agar) A carbohydrate prepared from seaweeds. It dissolves in water to give, even in low concentrations, an inert jelly which is in general use as a medium on which to grow bacteria and fungi. For this purpose various nutrients have to be supplied in the agar, which itself is not attacked in any way by the organisms.
Allosteric inhibition A biochemical 'negative feedback' process whereby an enzyme is specifically inhibited by the end products of its own reaction. The products do not inhibit by competing with the substrate for the same enzyme reaction site but presumably combine in some other way to reduce the catalytic properties of the enzyme.
Amino acid The 'building bricks' of proteins. They contain, in addition to the elements carbon, hydrogen and oxygen (which make up the carbohydrates), the element nitrogen and, in some cases, sulphur.
Antagonism This is the phenomenon in which a given chemical substance, that may itself be inactive in promoting a particular activity in an organism, prevents or reduces the activity-promoting action of an effective substance.
Apical dominance The phenomenon in which the growth of side buds on a shoot is inhibited by the presence of a terminal (apical) bud.
Artefact Something abnormal produced during the manipulation of organisms. Used to describe a structure or a substance that is not normally present in the organism, but is caused to be formed by the observational or experimental techniques employed.

Basipetal The reverse of acropetal, i.e. moving in a direction from apex to base, *towards* the main body of the organism.
Calibration The process whereby the arbitrary readings obtained from a measuring instrument or material are related to the absolute measure of the system under observation. In hormone assay, calibration would consist of constructing a graph relating the response of the organ (curvature, increase in length, etc.) to the concentration or amount of the known hormone applied. The graph then forms a scale from which any hormone amount can be read from a knowledge of the relevant response to the material assayed.
Callus Soft undifferentiated tissue which is formed at the edge of a wound in living plant tissue and which tends, under favourable conditions, to seal off the wound and to prevent infection by disease-producing organisms.
Cambium A sheath of cells which possesses the power of active growth and division, and which is responsible for the production of new tissues in mature organs. The two most important cambia are (a) the cylinder which is situated between wood and bark and which causes growth in girth in stem and root, and (b) the sheath in the outer regions of the bark in mature stems and roots which gives rise to the protective cork tissue.
Carbohydrates Organic substances composed of the elements carbon, hydrogen and oxygen. They form the dead structural elements of the plant (cellulose in the cell-wall) and the storage units for chemical energy and raw material for growth synthesis (starch, sugars).
Cell-sap The watery solution of substances filling the vesicles found in the protoplasm of almost all mature living plant cells. In the majority of such cells, there is only one large vesicle which is called the vacuole.
Cellulose The complex carbohydrate which constitutes the main structural material of most plants' cell walls. Cotton is almost pure cellulose.
Chelate A co-ordination compound formed between a bi- or trivalent metal and an organic compound possessing a number of acid groups, whereby the metal is attached to two or more of these groups. (From the Greek chĕlĕ = a claw.) The chelates of some metals (e.g. Ca) are extremely insoluble. Chelating agents—compounds which can be used to form chelates with metals.
Chlorophyll The green colouring-matter of leaves, etc. Only in the presence of chlorophyll can the radiant energy of the sunlight be converted to molecular or chemical energy, which is necessary for the growth and vital activities of all living organisms.
Chloroplasts Microscopic protoplasmic structures which contain the green chlorophyll (*q.v*). and in which the energy conversion mentioned above is brought about. In green leaf cells the chloroplasts are discoid and about 10μm ($\frac{1}{2500}$ inch) in diameter, a large number being present in most cells.
Chromosomes Short but often thread-like protoplasmic bodies that are characteristic of the nuclei of cells. Their number, form, and constitution are constant for the cells of any one variety or species, and they are the bearers of the hereditary characters of the organism.
Coleoptile The tubular first leaf produced by a cereal or a grass seedling. The second leaf, with a flattened blade, grows up inside the coleoptile and eventually bursts through it at the tip.
Control experiment In all research in which the effect, on a system, of condition or a particular treatment is being investigated, it is vital that a check be kept on the behaviour of the system in the absence of such treatment; with this the behaviour under the treatment itself can be accurately compared. The experiment designed to follow

this normal behaviour is called a control experiment.

Cortex In stems and roots the tissue lying between the outer skin (epidermis) and the conducting (vascular) tissue.

Cotyledon The first and only leaves of the embryonic flowering plant developed in the seed. There are either two (dicotyledonous plants) or only one (monocotyledonous plants). They may store food (e.g. in the garden pea) and assume the role of true leaves after germination (e.g. castor oil plant).

Crown gall A disease of plants caused by the phytopathogenic bacterium, *Agrobacterium tumefaciens*. It gets its name from the large 'gall' of tumourous tissue formed at the 'crown' of the root where it meets the stem just below the soil surface.

Cyclosis The name given to the circulatory motion of the fluid protoplasm in the cells of organisms.

Cytochromes Iron-containing proteins (haem proteins like the blood pigments) that are vitally important oxidizing enzymes in the biochemical systems controlling respiration in organisms.

Cytoplasm That part of the living cell protoplasm which lies outside the nucleus. The protoplasm of the nucleus is generally termed nucleoplasm.

2, 4-D 2, 4-Dichlorophenoxyacetic acid.

Decapitation A term used in plant physiology to denote the removal of the apex of a plant organ, e.g. the coleoptile tip in the *Avena* assay for auxins (see Chapter II).

Decarboxylation An enzymatic process occurring in living cells whereby the elements of carbon dioxide are removed from the metabolite concerned.

Dehydrogenase An enzyme responsible for catalysing the removal of the element hydrogen from one metabolite and its transfer to another. The first metabolite is thereby oxidized and the second reduced. Dehydrogenases are essential constituents of the respiratory apparatus of living cells.

Deplasmolysis (see Plasmolysis).

Detoxication A biochemical mechanism whereby living cells can counter the harmful effects of toxic substances either externally applied or endogenously produced. The usual methods are molecular fragmentation into harmless decomposition products and combination with certain normal metabolites to yield similarly harmless 'complexes'.

Dicotyledon A plant belonging to that great division of flowering plants which have two seed-leaves. The other division consists of the remaining flowering plants; these have one seed-leaf and are called monocotyledons.

Differentiation The process whereby the cell, after reaching its full size, becomes structurally modified to play its specific role in the working of the organ or tissue of which it forms a part.

Distal An adjective describing that part of an organ which is farthest from the main body of the organism. The apex of the celeoptile or root is the distal part of the organ.

Ecology The scientific study of the interrelationships of the organism and its natural environment, and of their mutual interactions.

Electrophoresis A process whereby charged particles of molecular size can be moved in solution under an electrostatic voltage gradient. The phenomenon can be used to separate ions (e.g. proteins) on moistened paper strips placed between the poles of a suitable steady voltage source, when they will move to one or other pole depending primarily on their charge and particle size. These strips with their separated compounds are called electrophoretograms.

Endogenous An adjective indicating that the substance qualified has originated in the organism's own metabolic system (cf. exogenous = arising, usually by application, outside the system).

Endosperm The main reserve food tissue of many seeds (e.g. cereal, castor oil, ash, etc.); it forms a layer between the young embryo and the seed-coat. Seeds that have no endosperm (e.g. pea, bean, etc.) have their reserve food in their cotyledons (seed-leaves).

Enzyme An organic catalyst of the living cell. A catalyst is a substance which, when present in a chemical system, even in very small amounts, will greatly speed up the chemical reaction taking place there, without itself being changed in the process.

Epicotyl The name given to that part of the young stem of a seedling which arises above the cotyledons (reserve-containing seed leaves).

Epinasty The phenomenon of plant growth movements in relation to an *internal* stimulus, e.g. unequal distribution of a growth substance arising from a local external application. In the epinastic response of leaves, the upper side of the stalk grows faster than the lower side, so that the leaf appears to sink under its own weight.

Epiphytic Describes organisms which live on the surface of plants but which obtain no nutrient from them.

Etiolation A condition produced in plants by continued growth in very weak light or in darkness. The plant grows long and spindly, has very small undeveloped leaves, and forms no green chlorophyll.

Explant A portion of a living organism removed from the main body and made to grow in culture under artificial conditions.

Freeze-drying A method of preparation, for storage or extraction, of material, from living organisms or organs, in a state as far as possible unchanged in biochemical constitution from that in the living state. It consists essentially of freezing the living tissue instantaneously by plunging into a freezing mixture or liquid nitrogen and then subliming the ice from the tissue while it is still frozen in a vacuum.

Gene The name given to an individual unit of structure, making up the chromosome, which controls the expression of a hereditary character.

Genetic (adjective) Having to do with genes and their action in heredity.

Genetics The study of the genes and of heredity.

Geotropism The response of plant organs to gravity. A growth response whereby the organ curves towards (positive geotropism) or away from (negative geotropism) the earth. The response is probably caused by the induction of a differential distribution of a growth hormone (auxin) whereby the growth of the lowermost half of a horizontal organ is either retarded (positive geotropism) or accelerated (negative geotropism).

Haulm The name given to the stem of a grass or cereal.

Hemiparasite A plant parasite which grows on a host plant and abstracts therefrom *some* of its nutrients (e.g. mineral elements) but which contains chlorophyll and synthesizes its own carbohydrates by photosynthesis.

Humus The non-living organic material of the soil, produced by the decay of dead plants and animals. A very important constituent for the maintenance of desirable physical properties such as lightness of texture, etc.

Hypha The term for the microscopic threads which are the basic structural units of almost all fungi. In the moulds, etc. these threads are separate and clearly visible under the high-powered lens. In the large fruiting-bodies, as found in the mushroom, these threads are twisted together almost indistinguishably into a compact 'tissue'.

Hypocotyl The name given to the young stem of those seedlings in which, on germination, the cotyledons (reserve-containing seed-leaves) are carried into the air, expand and become green. It is the portion of the stem between the cotyledons and the root.

IAA Indol-3yl-acetic acid (Fig. 5 I)

Inflorescence A collection of flowers borne on one stem.

Internode That part of the stem of a plant between the points where the leaves are borne.

Ion-exchange A reversible interchange between ions in solution and ions of like charge on the surface of a solid in contact with the solution. Ion-exchange resins are special synthetic solids which are either acidic (i.e. with exchangeable H^+ ions) or basic (i.e. with exchangeable OH^- ions) which can be used to take up cations or anions from solution thus releasing H^+ or OH^- ions respectively. The process can be reversed by the use of strong acids or strong bases respectively. The resins are used for the separation of components of mixtures on a basis of the ionizing properties of their components.

Kinetics That branch of science which is concerned with the rates of processes. Extended (as in the context of this book) to mean those properties of a biochemical or physiological system which relate to the changes of that system in time and the rates of those changes.

Lignin A chemical substance of as yet uncertain constitution which, in combination with cellulose, is responsible for the remarkable mechanical strength of wood.

Meristem A plant tissue in which the cells are all capable of, and, under favourable growth conditions usually in process of, active division to produce new daughter cells.

Mesocotyl The first internode of a seedling of the cereal family (Graminae). It is that first portion of the solid stem axis which extends between the base of the coleoptile and the grain. Subsequent internodes are produced above the point of insertion of the coleoptile.

Mesophyll The loose spongy tissue in the interior of the leaf blade. Its cells contain the chloroplasts. In it by far the greater part of the net photosynthesis of the plant goes on.

Metabolism The sum-total of the chemical processes taking place in all their complexity in the normal life and growth of the organism.

Metabolite A term applied to any substance which is produced or used up in the metabolism of the organism.

Middle lamella The thin layer of cementing material (mainly calcium pectates) which occurs between contiguous cell-walls.

Mitochondrion(-ia) An intercellular particle (organelle) of considerable structural and biochemical complexity which is concerned with the energy transformations associated with the oxidation of carbon compounds (respiration) in the cell.

Monocotyledon (see Dicotyledon).

Morphology The study of the form or shape of living organisms. Often used (rather loosely) to mean the shape or the form itself.

Multicellular Composed of many cells – as contrasted with unicellular, which describes those organisms whose complete body is made up of only one cell.

Mutant An individual organism differing in one or more characteristics from closely related individuals and arising from a spontaneous or induced change in its genetic material.

Mycelium The interwoven threads (hyphae) that make up the vegetative body of a fungus.

NAA Naphth-1yl-acetic acid (Fig. 8 XXIII).

NAD Nicotinamide adenine dinucleotide.
NADP Nicotinamide adenine dinucleotide phosphate.
Nitrifying organisms Bacteria which are responsible for the oxidation of ammonia to nitrite and nitrate in the soil. *Nitrosomonas* oxidizes ammonia to nitrite and *Nitrobacter* oxidizes nitrite to nitrate.
Nitrogen-fixing organisms Bacteria or algae which convert elemental nitrogen (nitrogen of the air) into organic nitrogen. The bacteria include free-living soil organisms (*Azotobacter* and *Clostridium*) and the nodule bacteria (*Rhizobium*) which invade the roots of leguminous plants and induce the formation of root nodules in which they live. These organisms are extremely important components of the nitrogen cycle in nature.
NMSP α-(Naphth-1yl-methylthio)-propionic acid (Fig. 44 CLVI)
NPA *N*-Naphth-1yl-phthalamic acid (Fig. 43 CLI)
Nucleus The spherical or ovoid protoplasmic body which is present in all normal cells and which controls their vital activities. It contains the chromosomes, the bearers of the hereditary characters.
Parenchyma Tissue composed of thin-walled isodiametric cells, i.e. cells of approximately the same diameter in all directions.
Parthenocarpy The development of a fruit from an unfertilized ovary. Such fruits are necessarily seedless.
Petiole The stalk of a leaf.
Petri Dish A circular glass dish with a loosely fitting glass lid, the whole having the shape of a flat pill-box commonly about $3\frac{1}{2}$ inches (90 mm) in diameter. It is the standard vessel used for the cultivation of bacteria and fungi on a thin layer of nutrient agar with which the experimenter lines the bottom.
Phloem That tissue which is concerned with the transport of organic food materials from one plant organ to another. In woody dicotyledonous plants it forms the innermost layer of the bark.
Photoperiodism The growth reaction of plants to day-length.
Photosynthesis Synthesis of organic compounds from carbon dioxide and water, using the radiant energy of sunlight. This radiant energy is absorbed by the green colouring-matter of plants (chlorophyll) in chloroplasts and is there converted into the chemical energy essential for the synthesis.
Phyto- Prefix meaning plant, as phytotoxic = toxic to plants
Phytohormone A plant hormone.
Plasmolysis The shrinkage of the protoplast of a plant cell away from contact with the cellulose cell wall when water is withdrawn from the cell vacuole by an external solution of high osmotic concentration (i.e. hypertonic to the solution in the cell vacuole). Deplasmolysis is the act of recovery when the hypertonic external solution is replaced by a hypotonic solution or water.
Precursor A forerunner. The chemical compound from which the substance under consideration (e.g. hormone) is formed in the organism.
Primordium The name given to an organ or group of organs (e.g. a flower) in the earliest stages of development.
Protein monolayers Many biochemical and physical properties of proteins may be studied by spreading purified samples in very thin films on the surface of very clean water. These films may be only one molecule thick and are then called monolayers.
Protoplasm The living matter of the organism.
Proximal That part of an organ which is nearest to the main body of the organism. The base of a coleoptile is the proximal part of the organ.

Respiration The chemical breakdown processes taking place in the living organism and resulting in the release of large quantities of chemical energy necessary for the maintenance of life and the continuance of growth. In the majority of organisms under normal conditions these processes also involve the release of the gas carbon dioxide.

Rf value A measure of the degree to which a substance moves on a chromatogram. The ratio of the distance moved from the origin (point of application) to the distance of the solvent front from that origin.

Rhizome An underground horizontally creeping stem, by which a plant may spread vegetatively. It is also an organ of perennation.

Sequestration In this context, the removal of ions from a biochemical system (e.g. by the use of chelating agents).

Staling products Products of the metabolism of an organ or organism which accumulate in the medium in which growth is taking place and which depresses the activity of the growing cells.

Synergism This is the phenomenon in which a given chemical substance, inactive itself in promoting a particular activity in an organism, brings about an increase in the activity-promoting action of an effective substance. Often extended to include the interaction of two active substances where the response to a mixture of those two substances is greater than the sum of the responses to the two substances acting alone.

Terminal oxidation The last stages of the process of respiration in which electrons and protons are transferred from intermediate metabolites to molecular oxygen thereby forming water.

TIBA 2, 3, 5-Tri-iodobenzoic acid.

Transpiration The loss of water by evaporation from the aerial parts of plants.

Travelling microscope A microscope mounted on a special stage enabling it to be moved in a direction perpendicular to its optical axis by means of a graduated micrometer screw. Its use is to measure with great accuracy the distance between objects or points on an object observed through the microscope.

Tropism The growth curvature of plant organs induced by unilateral stimuli such as gravity (geotropism) and unilateral light (phototropism).

Unicellular Describes an organism composed of one cell only. A bacterium is a unicellular organism.

Vacuole The central vesicle of a mature plant cell containing a watery solution of various substances – the cell-sap.

Vernalization Temperature-induced changes in the plant causing modifications in reproductive behaviour.

Xylem The tissues of the wood.

Conversion Factors

	Measures of Weight	
Avoirdupois to Metric		
1 dram (dr)	27·344 grains	1·772 grams
1 ounce (oz)	16 drams	28·3 grams
1 pound (lb)	16 ounces	0·454 kilogram
1 stone (st)	14 pounds	6·350 kilograms
1 quarter (qr)	2 stones	12·701 kilograms
1 hundredweight (cwt)	4 quarters	50·802 kilograms
1 (long) ton	20 hundredweight	1·016 tonnes
Metric to Avoirdupois		
1 milligram (mg)		0·015 grain
1 gram (gm)	1000 milligrams	0·564 dram
1 kilogram (kg)	1000 grams	2·205 pounds
1 quintal (q)	100 kilograms	220·5 pounds
1 tonne	1000 kilograms	0·984 ton
U.S. Weights to Metric		
1 pound	16 ounces	453·592 grams
1 cental	100 pounds	45·359 kilograms
1 (short) ton	20 centals	0·907 tonne
Metric to U.S. Weights		
1 quintal (q)	100 kilograms	2·205 centals
1 tonne	1000 kilograms	1·102 (short) tons
British to Metric	*Measures of Length*	
1 inch (in)		25.400 millimetres
1 foot (ft)	12 inches	30.480 centimetres
1 yard (yd)	3 feet	0·914 metre
1 mile	1760 yards	1·609 kilometres
Metric to British		
1 micron (μ)	1/1000 mm (1/1 000 000 m)	1/25 400 inch
1 millimetre (mm)		0·039 inch
1 centimetre (cm)	10 mm	0·394 inch
1 decimetre (dm)	10 cm	3·937 inches
1 metre (m)	10 dm	1·094 yards 3·281 feet 39·370 inches
1 kilometre (km)	1000 m	0·621 mile

Measures of Area (Based on 1 *metre* = 39·370 *inches)*

British to Metric		
1 square inch (sq in)		6·452 sq centimetres
1 square foot (sq ft)	144 sq in	0·093 sq metre
1 square yard (sq yd)	9 sq ft	0·836 sq metre
1 acre	4840 sq yd	0·405 hectare
1 square mile	640 acres	2·590 sq kilometres
		258·998 hectares
Metric to British		
1 square millimetre (mm^2)		0·00155 sq inch
1 square centimetre (cm^2)	100 mm^2	0·155 sq inch
1 square decimetre (dm^2)	100 cm^2	0·108 sq foot
1 square metre (m^2)	100 dm^2	1·196 sq yards
1 hectare (ha)	10 000 m^2	2·471 acres
1 square kilometre (km^2)	100 ha	0·386 sq mile

Measures of Volume

British to Metric		
1 cubic inch (cu in)		16·387 cu centimetres
1 cubic foot (cu ft)	1728 cu in	28·317 cu decimetres
1 cubic yard (cu yd)	27 cu ft	0·765 cu metre
1 bushel (bu)	2219·3 cu in	0·364 cu metre
Metric to British		
1 cubic centimetre (cm^3 = ml)		0·061 cu inch
1 cubic decimetre (cu dm)	1000 cm^3	0·035 cu foot
1 cubic metre (cu m)	1000 dm^3	1·308 cu yards
		2·750 bushels

Measures of Capacity I. *Based on* 1 *Imperial gallon (British)* = 4·546 *litres (used for both liquid and dry measure)*

British to Metric		
1 pint (pt)		0·568 litre
1 quart (qt)	2 pints	1·136 litres
1 gallon (gal)	4 quarts	4·546 litres
1 peck (pk)	2 gallons	9·092 litres
1 bushel (bu)	4 pecks	36·368 litres
Metric to British		
1 millilitre (ml = cm^3)		0·0610 cu inch
1 centilitre (cl)	10 cm^3	0·0176 pint
1 decilitre (dl)	10 cl	0·176 pint
1 litre (l)	10 dl	1·760 pints

II. *Based on* 1 *U.S. gallon (liquid measure)* = 3·785 *litres*

U.S. to Metric		
1 pint (pt)		0·473 litre
1 quart (qt)	2 pints	0·946 litre
1 gallon (gal)	4 quarts	3·785 litres
Metric to U.S.		
1 millilitre (ml = cm^3)		0·0610 cu inch
1 centilitre (cl)	10 cm^3	0·021 pint
1 decilitre (dl)	10 cl	0·211 pint
1 litre (l)	10 dl	1·057 quart

Note: 1 British pint, quart, or gallon = 1·201 U.S. (liquid) pints, quarts, or gallons respectively.

1 U.S. (liquid) pint, quart, or gallon = 0·833 British pint, quart, or gallon, respectively.

III. *Based on* 1 *U.S. quart (dry measure)* = 1·1012 *litres*

U.S. (dry measure) to Metric		
1 pint (pt)	33·600 cu in	0·5506 litre
1 quart (qt)	2 pints	1·101 litres
1 peck (pk)	8 quarts	8·810 litres
1 bushel (bu)	4 pecks	35·238 litres
Metric to U.S. (dry measure)		
1 litre (l)		0·908 quart
1 dekalitre (dkl)	10 l	0·284 bushel
1 hectolitre (hl)	10 dkl	2·838 bushels

Temperature

0° Centigrade (Celsius) = 32° Fahrenheit

The following formulae connect the two major thermometric scales:

Fahrenheit to Centigrade: $°C = 5/9\,(°F - 32)$

Centigrade (Celsius) to Fahrenheit: $°F = (9/5\,°C) + 32$

Subject Index

Page numbers in *italics* refer to Figures.